BOTTOM-INTERACTING
OCEAN ACOUSTICS

NATO CONFERENCE SERIES

I   Ecology
II   Systems Science
III Human Factors
IV Marine Sciences
V   Air–Sea Interactions
VI Materials Science

IV MARINE SCIENCES

BOTTOM-INTERACTING OCEAN ACOUSTICS

Edited by
William A. Kuperman

and
Finn B. Jensen

SACLANT ASW Research Centre
La Spezia, Italy

Springer Science+Business Media, LLC

Library of Congress Cataloging in Publication Data

Nato Conference on Bottom-Interacting Ocean Acoustics, La Spezia, 1980.
  Bottom-interacting ocean acoustics.

  (NATO conference series: IV, Marine sciences; v. 5)
  "Proceedings of the NATO Conference. . . held at the NATO SACLANT ASW Research Centre, La Spezia, Italy, June 9-12, 1980."
  Includes indexes.
  1. Underwater acoustics—Congresses. 2. Ocean bottom—Congresses. I. Kuperman, William A. II. Jensen, Finn Brunn. III. Title. IV. Series.
QC242.N37 1980   551.46'01   80-24616
ISBN 978-1-4684-9053-4   ISBN 978-1-4684-9051-0 (eBook)
DOI 10.1007/978-1-4684-9051-0

Proceedings of a conference on Bottom-Interacting Ocean
Acoustics held June 9-12, 1980, at the NATO SACLANT ASW
Research Centre, La Spezia, Italy.

© 1980 Springer Science+Business Media New York
Originally published by Plenum Pres, New York in 1980
Softcover reprint of the hardcover 1st edition 1980

P R E F A C E

This book contains the complete proceedings of a conference
held at the NATO SACLANT ASW Research Centre, La Spezia, Italy on
9-12 June, 1980. The Centre has traditionally organized many
conferences in the field of marine physics and has published the
proceedings in its own series of Conference Proceedings. For the
present conference the NATO Scientific Affairs Division has kindly
agreed to allow the proceedings to be published in their NATO
Conference Series, even though the conference was sponsored by
SACLANTCEN and not by the Scientific Affairs Division.

In recent years bottom-interacting ocean acoustics has become
a research area of intensely increasing interest. This research
area can be broken down into two basic fields: research using
acoustics to probe the ocean bottom and research aimed at
determining how the ocean bottom affects sound propagation in the
ocean. Obviously, much of the information obtained in either of
the two fields is of interest to researchers in both fields. It
was therefore the intent of this conference to assemble as much
state-of-the-art information as possible from both fields so as to
stimulate further research in this area. In order to accomplish
this, we originally called for papers in the following traditional
fields:

- Coastal-water acoustics
- Low-frequency deep-water acoustics involving bottom
        interaction
- Acoustic properties of marine sediments
- Ocean seismic studies as related to ocean acoustics
- Signal-processing techniques that include the effects
        of bottom interaction

For the conference and the present proceedings we decided to
re-categorize the papers as follows:

- Geoacoustic properties of marine sediments
- Bottom loss: reflection and refraction
- Bottom-interface and seismic-wave propagation
- Acoustic modelling
- Sound propagation: techniques and experiment vs theory
- Fluctuations, coherence and signal processing

These categories seem to represent the basic breakdown by field of present-day research in this area. Though each paper has been classified into one of these categories (for conference organization purpose), many papers overlapped two or three areas. It is also interesting to note that not only are scientific results being communicated, but the latest techniques and the state-of-the-art tools of the trade (existing and in development) are also being presented.

The forty-six papers presented at this conference represent the work of seventy scientists working at universities, government laboratories, and industrial laboratories in seven different countries. We would like to thank the contributors for their efforts and especially for their promptness in providing the editors with their final manuscripts.

William A. Kuperman
Finn B. Jensen

La Spezia, Italy
July 1980

CONTENTS

GEOACOUSTIC PROPERTIES OF MARINE SEDIMENTS

BOTTOM LOSS: REFLECTION AND REFRACTION

BOTTOM-INTERFACE AND SEISMIC-WAVE PROPAGATION

ACOUSTIC MODELLING

SOUND PROPAGATION: TECHNIQUES AND EXPERIMENT VS THEORY

FLUCTUATIONS, COHERENCE AND SIGNAL PROCESSING

ATTENUATION OF SOUND IN MARINE SEDIMENTS

Jens M. Hovem

Electronics Research Laboratory
The University of Trondheim
TRONDHEIM, NORWAY

ABSTRACT

Attenuation of sound in water-saturated sediments may be caused by frictional losses in the grain-to-grain contacts or to viscous loss due to the movement of the fluid relative to the solid frame. Both losses can be included in the Biot theory for sound propagation in porous media. This theory gives an attenuation coefficient for the frictional loss that is proportional to the frequency, f, and a viscous attenuation increasing as f^2 at low frequencies and as $f^{\frac{1}{2}}$ in the high-frequency region.

The Biot theory depends on a number of parameters which are difficult to estimate, in particular for high-porosity silt and clay. In these cases one may instead use a model for the sound velocity and attenuation based on multiple scattering theory. This model gives the same frequency behaviour as the Biot model, but fails when the concentration of suspended particles exceeds a few percent. The reason for this is discussed, a modification to the suspension model is proposed, and the result compared with the Biot model. It is shown that, depending on the grain-size distribution viscous attenuation may also increase linearly with frequency over a wide frequency band.

INTRODUCTION

The interaction with the ocean bottom makes the properties of sound propagation in marine sediments important for underwater acoustics. In predicting sonar performance one may need to know the sound velocity and attenuation with good accuracy as function of depth and location. In general, such information can only be

1

obtained from measurements and from a knowledge of the physical properties. The complications with obtaining information about the velocity and the attenuation increase significantly if one has to take into account an unknown frequency dependence.

Experimental results and theory show that the sound velocity, for all practical purposes, is independent of frequency. Attenuation will however vary with frequency, most measurements show an almost linear increase with frequency which makes the log decrement and the imaginary part of the wavenumber constants. The problem is that accurate measurements of the attenuation can only be done at relatively high frequencies and low frequency values must be obtained by extrapolation, sometimes far out of the frequency range of observation. This requires an accurate theoretical model for the attenuation of sound in sediments.

In a series of papers Biot[1,2] has given a general theory for the propagation of sound in fluid saturated porous media. A significant feature of this theory is that it allows for the movement of the fluid relative to the solid frame and thus for the possibility of a viscous loss. This loss will increase with frequency (f) proportional to f^2 for lower frequencies and poportional to $f^{\frac{1}{2}}$ at higher frequencies. The solid friction loss can be accounted for by the inclusion of imaginary components to the elastic parameters of the solid frame giving a contribution to the attenuation which increases linearly with frequency. By combining the two losses Stoll and Bryan[3] and Stoll[4,5] have shown that the attenuation given by the theory can be made to fit most experimental data. A comparison made in reference 6 between theory and experiments also indicates that viscous attenuation may be predominant in sands of high permeability with uniform and regular grains.

With grains of different sizes and shapes one should expect that the transition between low and high frequencies to be gradual resulting in a viscous attenuation increasing almost linearly with frequency over a wide band of frequencies. The effect of distributed grain sizes cannot be accounted for in the Biot model but in these cases one may instead use other models which are derived from multiple scattering theory.

In the following sections the theoretical models will be briefly reviewed and discussed with particular emphasis on the dynamic parameters which most directly have influence on the attenuation coefficient. The paper will primarily discuss the theory and reference in made to Hamilton[7,8] for a summary and discussion of experimental results.

THEORETICAL MODELS

The Biot Model

A porous medium can be considered as mineral grains in a skeleton frame which is saturated with a fluid. The bulk modulus K of the composite medium will have three components: the bulk modulus K_f of the fluid, the bulk modulus K_s of the solid grains, and the bulk modulus K_b of the frame. In addition, there is also a shear modulus μ of the frame.

The stress-strain relations can be expressed as

$$\sigma = Ke - C\xi \tag{1}$$

$$p = Ce - M\xi \tag{2}$$

where σ is the hydrostatic pressure on the bulk material and p is a pore pressure that is effective on the fluid and the grains. The dilation of the bulk material is e = dV/V and the dilation of the fluid is $\xi = dV_f/V$ with the relative volume of the fluid given by the porosity $\phi = V_f/V$. The elastic coefficients K, C and M are given by

$$K = K_s(K_b + Q)/(K_s + Q) \tag{3}$$

$$Q = (K_f/\phi)(K_s - K_b)/(K_s - K_f) \tag{4}$$

$$C = QK_s/(K_s + Q) \tag{5}$$

$$M = CK_s/(K_s - K_b) \tag{6}$$

with

$$H = K + 4\mu/3 \tag{7}$$

the equations which govern the propagation of compressional waves can be written as

$$\nabla^2(He - C\xi) = \frac{\partial^2}{\partial t^2}(\rho e - \rho_f \xi) \tag{8}$$

$$\nabla^2(Ce - M\xi) = \frac{\partial^2}{\partial t^2}(\rho_f e - \rho_c \xi) - \frac{\eta}{B} F(\kappa) \frac{\partial \xi}{\partial t} \tag{9}$$

The last term of Eq 9 represents the viscous loss associated with the flow of fluid relative to the solid frame where η is the viscosity and B is the absolute permeability of the medium, η/B is thus the flow resistance. The function $F(\kappa)$ accounts for the deviation from Poiseuille flow at higher frequencies where the argument κ is

$$\kappa = a_p(\omega\rho_f/\eta)^{\frac{1}{2}} \tag{10}$$

The average radius of the pores is a_p, ρ_f is the density of the fluid and ρ_s is the density of the grains, the composite density being $\rho = \phi\rho_f + (1 - \phi)\rho_s$. The density ρ_c in Eq 9 is introduced instead of ρ_f/ϕ to account for the possibility that not all the fluid is attached to the frame, this effect is also included by the last term of Eq 9 where the imaginary part gives an effective addition to the mass which is attached to the solid[6].

The solution of Eqs 8 and 9 gives two compressional waves, the first wave is the "normal" compressional wave, the second wave is a consequence of the movement of the fluid relative to the frame. The attenuation of this wave is very high and the second wave has therefore limited practical importance except that it is directly connected to the viscous loss of the first wave and that its experimental determination may be importent in the verification of the theory. In addition to the two compressional waves there is also a shear wave which will not be discussed here.

The frictional loss can be included in the model by adding imaginary components to the bulk modulus K_b and the shear modulus μ of the frame. In order to see the effect of this, Fig 1 shows the sound velocity and attenuation as fuctions of frequency for the two waves with different amounts of frictional loss. The other parameters used in the calculation are believed to be typical for well sorted sand of medium grain size. The results show a small dispersion of the first wave and a strong dispersion of the

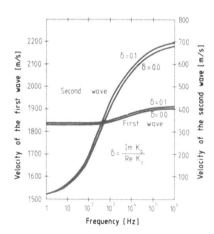 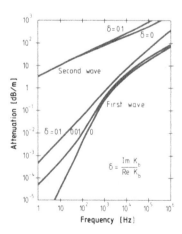

Figure 1

Sound velocities (Fig 1a) and attenuations (Fig 1b) for the first and second wave as functions of frequency.

second wave. The dispersion is mainly caused by a decrease in the effective density at higher frequencies. It should be noted that even with a dominating contribution from the frictional loss the effects on the second wave and the velocity of the first wave are very small.

The viscous loss of the Biot model is mainly governed by the pore size parameter a_p and the permeability B and a major difficulty is to find reasonable values for these parameters. It can be shown[9] that in the limiting cases of low and high frequencies the viscous attenuations α_0 and α_∞ respectively can be expressed by

$$\alpha_0 \approx \beta_0 \ B\omega^2 / (\eta v) \tag{11}$$

$$\alpha_\infty = \beta_\infty \ (\omega\eta/\rho_f)^{\frac{1}{2}} / (a_p v) \tag{12}$$

where β_0 and β_∞ are dimensionless coefficients given by the porosity and the desities and v is the velocity of the first wave.

For well sorted sand the permeability appears to agree well with the Kozeny-Carman equation[10] which gives B as

$$B = \phi m^2 / k \tag{13}$$

where k is a coefficient which takes into account the pore shape and the tortuosity of the pores and m is the hydraulic radius. For a medium composed of uniform spherical grains of diameter d, $k \approx 5$, and $m = (d/6)\phi/(1 - \phi)$. When Eq 13 is valid the logical value for the pore size parameter will be[6]

$$a_p = 2m = (d/3)\phi/(1 - \phi) \tag{14}$$

The effect of using different pore size parameters, when calculating the viscous attenuation is shown in Fig 2. With a_p and B as given above there is gradual transition from the asymtotic behaviour of Eq 11 to the behaviour of Eq 12, with a smaller value for a_p the transition is more irregular. It is important to note that in a rather broad intermediate frequency interval the increase with frequency may be quite different from both the f^2 and the $f^{\frac{1}{2}}$ law.

Fig 3 shows how the theoretical viscous attenuation will depend on the grain size (in phi-units) using Eq 13 and Eq 14 and Hamilton's[7] empirical relationship between porosity and the mean grain size. This calculation brings forward an other problem with the application of the Biot theory, Eq 13 and 14 are only valid for sands of rather uniform spherical grains in which case the porosity will be independent of the size of the grains. This is not in agreement with experience, in real sediments both the distribution and the shapes of the grains will vary with the mean grain size thereby making Eq 13 and Eq 14 in valid.

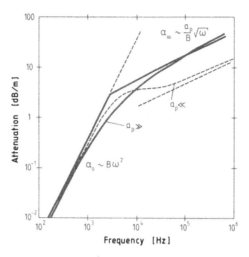

Figure 2.

Viscous attenuation as a function of frequency with
asymptotic approximations calculated with two different
pore size parameters.

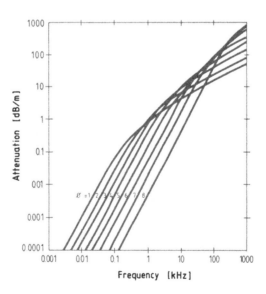

Figure 3.

Viscous attenuation as frequency and grain size, using
Hamiltons relation between porosity and grain size.

The Suspension Models

Theoretical models for the propagation of sound in suspensions can be derived from multiple scattering theory. When a plane wave of pressure amplitude p_o is incident on a solid spherical particle with diameter d the scattered wave at a location specified by the coordinates r and θ is given by

$$P_s (\theta,r) = p_o \frac{e^{i\ell_f r}}{r} \Phi(\theta) \qquad (15)$$

where

$$\Phi(\theta) = \frac{1}{6} \ell_f^2 d \ (\gamma_o + \gamma_1 \cos\theta) \qquad (16)$$

where γ_o represents the monopole scattering given by the difference in compressibility between the solid and the fluid and γ_1 is the dipole scattering governed by the density difference but with a frequency dependent correction which takes into account the characteristic thickness of the shear boundary lager in the fluid around the particle. The wavenumber in the fluid is ℓ_f. The sound propagation through a composite material with a number of randomly spaced particles is charterized by a complex wavenumber ℓ given by

$$(\ell/\omega)^2 = \frac{\rho_f}{K_f} (1 + \delta\gamma_1)(1 + \delta\gamma_1) \qquad (17)$$

where $\delta = (1 - \phi)$ is the volume concentration of particles.

The dipole moment will be a complex function of frequency, $\gamma_1 = \gamma_r + i\gamma_i$, and the sound velocity v and the attenuation coefficient α will therefor be

$$v = \{(K_f/\rho)/[(1 + \delta\gamma_o)(1 + \delta\gamma_r)]\}^{\frac{1}{2}} \qquad (18)$$

$$\alpha = (\omega/2v)(\delta\gamma_i)/(1 + \delta\gamma_r) \qquad (19)$$

It should be noted that in Eq 17 multiple scattering is included except those cases where the propagation path goes through the same particle more than once[11].

The main difference between the models which have been derived is in the calculation of the dipole moment. In the work by Urick[12] and by Urick and Ament[13] γ_1 is calculated assuming that the flow of fluid around a particle is completely unaffected by the presence of the other particles. The result is that α will increase almost linearly with the concentration which can only be valid at very low concentrations.

The same result was appearently obtained by Rytov as a special case[14] but as has been pointed out in reference 15, in the original work there is a correction factor which multiplies Eq 19 with the factor $(1-\delta)^2$ causing the viscous attenuation to disappear when $\delta = 1$ as one should expect.

The Urick and the Rytov models give almost the same results as the Biot model when applied to suspensions. At low frequency the models will be identical with particular expressions for the permeability. The Urick model will be identical to the Biot model if the latter uses a permeability B_U given by

$$B_U = (d^2/18)/(1 - \phi) \tag{20}$$

To be identical to the Rytov model requires

$$B_R = (d^2/18)\phi^2/(1 - \phi) \tag{21}$$

These implied permeabilities should be compared with Eq 13 which can be written

$$B = d^2/(36k)\phi^3/(1 - \phi)^2 \tag{22}$$

The observation that the main difference between the models are the different expressions used for the permeability suggests a modification to the suspension theory[9] such that the implied permeability will be that of Eq 22. This modified model gives almost the same results as the Biot model but has the advantage that it can be used when the medium has distributed grain sizes.

In Figs 4 and 5 viscous attenuation has been calculated as a function of volume concentration using the different models and is compared with the measurements by Hampton[16] and Urick[12]. Except for the Urick model the agreement with the measurement is fair which indicates that the suspension models may be used at such high concentrations as found in silt and clay.

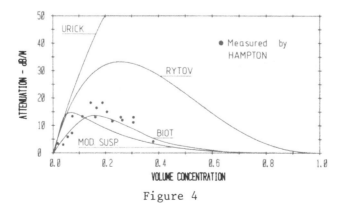

Figure 4

Attenuation in a suspension as a function of volume concentration.

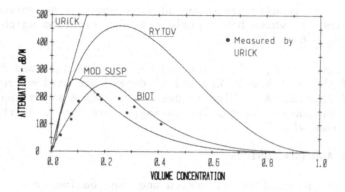

Figure 5

Attenuation in a suspension as a function of volume concentration.

    The suspension models give the same frequency dependence as
the Biot model when all the particles have the same size. The
suspension models can however easily be applied to media with a
distribution of sizes. The effect of this is shown in Fig 6 where
the modified suspension model has been used to calculate viscous
attenuation as a function of frequency for a Gaussian distribution
of sizes (in phi units) with a mean value of $m_\phi = 6$ and different
standard deviations σ_ϕ.

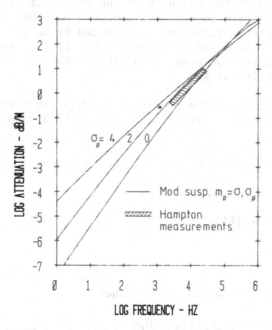

Figure 6   Viscous attenuation as a function of frequency
and grain size distribution.

The theoretical results are compared with Hamptons measurements for "black sediments" which has a grain size distribution which fits well with $m_\phi = 6$ and $\sigma_\phi = 2$.

The curve for $\sigma_\phi = 0$, which can not be distinguished from the results of the Biot model, has the f^2 dependence at low frequencies and the $f^{\frac{1}{2}}$ dependence at high frequencies. Fig 6 shows that the frequency dependence becomes increasingly more linear as the distribution becomes wider.

DISCUSSION AND CONCLUSIONS

From the preceeding discussion one can conclude that the viscous attenuation given by the Biot theory can increase almost linearly with the frequency in a limited frequency range. When applied to real sediments the Biot theory is too simple since it cannot be applied to media with distributed grain sizes. When this is taken into account by using the suspensions models derived from scattering theory one obtains, as expected, an almost linear dependence over a wide frequency range which increases as the distribution becomes wider. It is important to note this as a function of concentration, the suspension models give values in fair agreement with measurements.

These models can only be used for suspensions but McCann[17] has derived a modification which shows the effect of interparticle forces and which is related to the frame bulk modulus, K_b. The strength of the interparticle force estimated by McCann is probably too high[15] since he used the Urick model, but it is interesting to note that his modification predicts a minimum for the viscous loss for a particular value of K_b.

The same minimum is also predicted by the Biot theory as can be seen from Eqs 8 and 9. The viscous loss will disappear when $\xi = 0$, and then

$$H/C = \rho/\rho_f \tag{23}$$

When $K_s \gg K_f$ and $K_s \gg K_b$

$$H \approx K_b + K_f/\phi + (4/3)\mu \tag{24}$$

and

$$C \approx K_f/\phi \tag{25}$$

and the minimum in attenuation occurs when

$$K_b + 4/3\ \mu \approx K_f\left(\frac{1-\phi}{\phi}\right)\left(\frac{\rho_s - \rho_f}{\rho_f}\right) \tag{26}$$

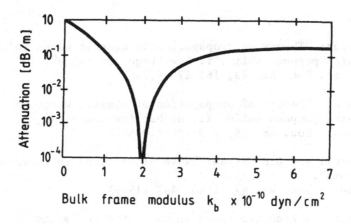

Figure 7

Viscous attenuation as a function of bulk frame modulus.

Fig 7 shows an excact calculation of the viscous attenuation as a function of K_b assumed to be equal to μ.

The preceeding discussion has concentrated on the viscous loss but in addition there is also a frictional loss which can be included by assuming complex values for K_b and μ giving a contribution to the attenuation coefficient which increases linearly with frequency. Since the two loss mechanisms may have very similar frequency dependence it is difficult to assess their relative importance by measuring attenuation as a function of frequency except when the measurements cover a wide range of frequencies and the particles are spherical and of the same size[6]. In order to verify the theory other methods should be sought, for instance direct detection of the second wave by reflection and transmission experiments[18] or by the use of Eq 26 measuring attenuation as a function of K_b and μ which depend on the static pressure on the frame.

ACKNOWLEDGEMENT

This work was partly carried out under contract with the Office of Naval Research while the author was visiting the Applied Research Laboratories, The University of Texas at Austin. The author is indepthed to Dr. L. Hampton of ARL : UT for valuable discussions and encouragement.

REFERENCES

1. M.A. Biot: "Theory of propagation of elastic waves in a fluid-
    saturated porpous solid, I. Low-frequency range".
    J. Acoust. Soc. Am. 28, 168-178 (1956).

2. M.A. Biot: "Theory of propagation of elastic waves in a fluid-
    saturated porpous solid, II. Higher frequency range".
    J. Acoust. Soc. Am. 28, 179-191 (1956).

3. R.D. Stoll and G.M. Bryan: "Wave attenuation in saturated
    sediments".
    J. Acoust. Soc. Am. 47, 1440-1447 (1969).

4. R.D. Stoll: "Acoustic waves in saturated sediments" in
    Physics of Sound in Marine Sediments, edited by L.D. Hampton
    (Plenum Press, New York, 1974).

5. R.D. Stoll: Acoustic waves in ocean sediments",
    Geophysics 42, 715-725 (1977).

6. J.M. Hovem and G.D. Ingram: "Viscous attenuation of sound in
    saturated sand".
    J. Acoust. Soc. Am. 66, 1807-1812 (1979).

7. E.L. Hamilton: "Compressional-wave attenuation in marine
    sediments".
    Geophysics 37, 620-645 (1972).

8. E.L. Hamilton: "Sound attenuation as a function of depth in
    the sea floor".
    J. Acoust. Soc. Am. 59, 528-536 (1976).

9. J.M. Hovem: "Viscous attenuation of sound in suspensions and
    high porosity marine sediments".
    To be published in the J. Acoust. Soc. Am.

10. P.C. Carman: "Flow of gases through porous media".
    Academic Press, Inc., New York, (1956).

11. P.C. Waterman and R. Truell: "Multiple scattering of waves".
    J. of Mathematical Physics 2, 512-537 (1961).

12. R.J. Urick: "The absorption of sound in suspensions of irregular
    particles".
    J. Acoust. Soc. Am. 20, 283-289 (1948).

13. R.J. Urick and W.S. Ament: "The propagation of sound in
    composite media".
    J. Acoust. Soc. Am. 21, 115-119 (1949).

14. M.E. Gültepe: "Comments on "Lamb-Urick" equation for sound attenuation in aqueous suspensions".
    Acoustica, 29, 357-358 (1973).

15. M.A. Barrett Gültepe, D.H. Everett, and M.E. Gültepe: "The influence of interparticle forces on acoustic attenuation in sediments".
    Proceedings of the Institute of Acoustics (1976).

16. L.D. Hampton: "Acoustic properties of sediments".
    J. Acoust. Soc. Am. 42, 882-890 (1967).

17. C. McCann: "Compressional wave attenuation in concentrated clay suspensions".
    Acustica 22, 352-356 (1969-70).

18. T.J. Plona: "Observation of a second bulk compressional wave in a porous medium at ultrasonic frequencies".
    Appl. Phys. Lett 36 (4), 259-261 (1980).

DIRECTIVITY AND RADIATION IMPEDANCE OF A TRANSDUCER

EMBEDDED IN A LOSSY MEDIUM

G. H. Ziehm

Forschungsanstalt der Bundeswehr für Wasserschall-
und Geophysik
Kiel, Germany

ABSTRACT

For many reasons in underwater acoustics there is a need for
in-situ measurement methods that describe the local acoustical
properties of the sea floor. Knowledge of these properties enables
one to derive the complex reflection factor under arbitrary angles
of incidence, which is an important quantity in modelling sound
propagation. Though preference has been given to all methods that
operate remotely, no successful breakthrough has yet been achieved
in providing a shipborne method that is reliable and easy to handle.
In 1977 D. J. Shirley demonstrated that when an acoustical transducer is driven while embedded in the sediment, its radiation
impedance depends on the specific acoustical properties of the
surrounding environment. Though he gave only laboratory results
and did not take into account the attenuation of the sediment,
this proposal seems to offer a basis for the development of an
appropriate shipborne in-situ measurement method. A prerequisite
for such a development is the detailed knowledge of the relationship between the specific complex acoustic data of the sediment
and the radiation impedance of the transducer. An analytical
treatment of this problem is rather complicated, even under
the assumption of compressional waves only. A mixed method of
analysis and numerical evaluation is presented.

1. INTRODUCTION

In 1974 E. Schunk[1] has described a measurement method to obtain the
acoustical reflection factor of the sea floor. This method looked
quite promising in the first instant but later on a number of
difficulties appeared which could not be overcome till now. Therefore further activities were stopped in 1978.

The desire to obtain an in-situ method for the local acoustic-
specific data of the bottom - preferably by remote-sensing - still
continues. Presumable this is a common requirement of all those
which are members of the community of shallow water acousticans.

In 1977 D. J. Shirley[2] reported a method in which the electrical
driving impedance of a transducer when in contact with the sediment
is a measure for the specific acoustical properties of the medium.
The present report refers to the work of Shirley[2] but emphasis is
given to a thorough survey of interrelations between the acoustic-
specific sediment data on one side and the directional pattern and
radiation impedance on the other. The result obtained by mathematical
analysis show the possibility in principle to develop a practical
measurement method for the above mentioned purpose.

2. ASSUMPTIONS AND RESTRICTIONS

●.) The sediment has a homogeneous and isotropic structure.
●.) Compressional waves exist only.
    The influence of shear waves is negligible.
●.) The Huygens-Rayleigh-Integral, which forms the fundamental
    relation between the acoustical field and the velocity of the
    radiating elements, is valid also when the transducer is
    embedded in a lossy medium.

The mathematical treatment presented in this paper - is restricted to the circular piston in which the whole surface is vibrating with a constant harmonic amplitude. It is reminded that no exact analytical solutions exist for the radiation impedance of other membrans of simple shapes like squares or rectangles.

3. WAVE NUMBER, SOUND VELOCITY AND CHARACTERISTICAL IMPEDANCE OF A LOSSY MEDIUM

In connection with the present task, the same electro-acoustical transducer is driven either in lossless sea-water or in sediment with sound absorbing features. Therefore most of the quantities introduced in the remaining text must be indicated by indices: 1 for water and 2 for sediment.

3.1 The wave number

In water the wave number is a real quantity

$$k_1 = \frac{\omega}{c_1} = \frac{2\pi}{\lambda_1} \quad , \tag{1}$$

ω = angular frequency; c_1 = sound velocity and λ_1 = wave length.

In sediment the wave number is a complex quantity

$$\underline{k}_2 = k_{20} - j\,\alpha_2 = k_{20}\left(1 - j\,\frac{\alpha_2}{k_{20}}\right) = k_{20}\left(1 - j\,\overline{a}\right) , \tag{2 a}$$

where

$$k_{20} = \frac{\omega}{c_{20}} = k_1 \frac{c_1}{c_{20}} \tag{2 b}$$

k_{20} and α_2 = real and imaginary component of the complex wave number $\underline{k}_2$. c_{20} = phase velocity of the sound wave in the sediment. $\overline{a} \approx$ loss angele of the sediment. Exact solution yields a loss angle $= - \tan^{-1} \bar{a}$. The selection of the sign of the imaginary component of the wave number is due to the arbitrarily choosen argument of the propagation function $\omega t - \underline{k}_2 r$, in which with increasing distance r, the sound wave's intensity decreases exponentially.

To visualize the influence of the loss angle $\bar{a}$ on the transmission loss, TL, for a distance Δr, when geometrical spread is neglected, one can easily derive

$$\frac{TL}{[dB]} = 2 \pi \quad 8.686 \; \bar{a} \frac{c_1}{c_{20}} \frac{\Delta r}{\lambda_1} \quad . \tag{3}$$

Fig. 1 shows TL for a distance Δr equal to the wave length λ_1 in water.

3.2 The characteristic impedance

The characteristic impedance is defined as the ratio of the sound pressure to the particle velocity. For a plane wave propagating through sea water it is independent of frequency and a real quantity

$$Z_1 = \varrho_1 c_1 \quad , \tag{4}$$

where ϱ_1 is the density.

For plane waves in a lossy medium the characteristic impedance becomes complex

$$\underline{Z}_2 = \omega \frac{\varrho_2}{\underline{k}_2} = \omega \frac{\varrho_2}{k_{20}} \frac{1}{1 - j \bar{a}} = \frac{\varrho_2 c_{20}}{1 - j \bar{a}} = \frac{Z_{20}}{1 - j \bar{a}} \quad . \tag{5}$$

Z_{20} is also a characteristic impedance, taking into account the phase velocity of the wave, or the real component of the wave number only.

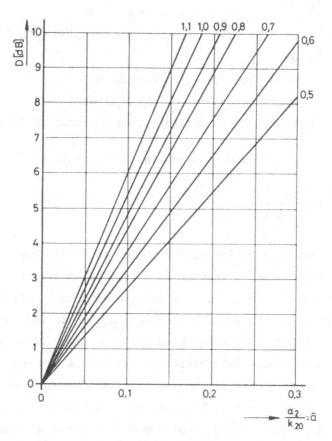

Fig. 1. Attenuation in dB for a distance in the sediment which e-
quals the acoustical wavelength in water

Parameter $\dfrac{c_1}{c_{20}}$.

4. THE STANDARDIZED DIRECTIONAL PATTERN

By the Huygens-Rayleigh-Integral the standardized directional
pattern for the circular piston is obtained as

$$\underline{N} = \frac{1}{A_M} \int_{r_M=0}^{R_M} \int_{\varphi_M=0}^{2\pi} r_M \, e^{jk_1 r_M \sin\theta \cos(\varphi_M - \emptyset)} \, dr_M \, d\varphi_M \,. \qquad (6)$$

$R_M \; ; \; A_M$ = radius and area of the membran ;

$r_M \; ; \; \varphi_M$ = polar coordinates on the membran ;

$\theta \;\; ; \; \emptyset$ = spherical coordinate system as shown in fig. 2

The solution of (6) can be found in the literature[3..6]

$$N = 2 \, \frac{J_1 (k_1 R_M \sin \emptyset)}{k_1 R_M \sin \theta} \qquad (7)$$

J_1 is the Bessel function of first kind and order one. In the case of
the circular piston, vibrating in a lossfree medium the shape of
the directional pattern is of axial symmetry and real. Solving (6)
usually at first the integration over φ_M is carried out, whereby
the Bessel function of first kind and zero order is obtained,
followed by a second integration over r_M. This sequence of integration
is not applicable when in the integrand of (6) the wave number becomes a complex quantity, because the Bessel function of the first
kind is only tabulated for real arguments. Therefore one changes
the sequence of integration. At first the integration over r_M is
fulfilled in an elementary manner. Using symmetries one obtaines,
when $\beta = \varphi_M - \emptyset$

$$\underline{N} = \frac{2}{\pi} \, \frac{1}{j\underline{k}_2 R_M \sin \theta} \int_0^{\pi} \cos \beta \, e^{j\underline{k}_2 R_M \sin \theta \cos \beta} \, d\beta \quad . \qquad (8)$$

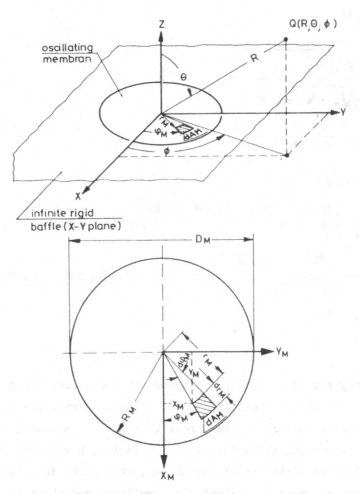

Fig. 2.   Coordinates of the oscillating element and its positioning
in a spherical coordinate system.

The exponential function in the integrand is split into a product
of a real and an imaginary exponential function. The latter is then
written as sum of a cos- and sin-component. Introducing some further
notations the standardized directional pattern becomes complex

$$\underline{N} = N\, e^{jv}$$

where

$$N = \frac{2\sqrt{R_N^{\,2} + I_N^{\,2}}}{\pi q\sqrt{1 + \bar{a}^2}\,\sin\theta} \qquad ; \qquad\qquad (9\ a)$$

and

$$v = \tan^{-1}\bar{a} + \tan^{-1}\frac{I_N}{R_N} \qquad\qquad . \qquad\qquad (9\ b)$$

The notations are

$$R_N = \frac{1}{\pi}\int_0^\pi \cos\beta\; e^{\pi q\,\bar{a}\,\sin\theta\,\cos\beta}\;\sin(\pi q\,\sin\theta\,\cos\beta)\,d\beta \quad ,$$

$$I_N = -\frac{1}{\pi}\int_0^\pi \cos\beta\; e^{\pi q\,\bar{a}\,\sin\theta\,\cos\beta}\;\cos(\pi q\,\sin\theta\,\cos\beta)\,d\beta \quad ;$$

$$q = \frac{D_M}{\lambda_1}\frac{c_1}{c_{20}} \qquad ; \qquad D_M = 2\,R_M \quad .$$

The integrals of R_N and I_N including the two parameters q and $\bar{a}$
contain the specific acoustic data of the sediment. These integrals
can be solved easily and very fast by numerical methods, much
faster if expanding (7) into a Taylor series, putting $\underline{k}_2$ instead
of k_1 into each element of the series, separating into real and
imaginary elements and calculating amplitude and phase of the
whole series. Fig. 3 presents the amplitude of $\underline{N}$ obtained by (9 a)
and depicted in polar coordinates with a linear scale. It is
obvious that the main lobe is only weakly influenced by the
loss angle $\bar{a}$ of the sediment, while the behaviour of the side

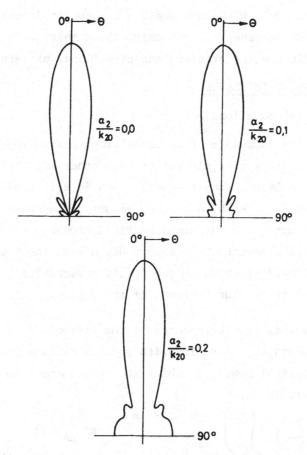

Fig. 3. The directional pattern of a circular piston oscillating in a lossy medium q = 3; Parameter $\bar{a}$.

lobes is completely changed. The side lobes are melting together and the level increases repidly for $\theta = 90^\circ$. This is not surprising. In Fig. 4 the amplitude N and phase v of the directional pattern are shown separately using cartesian coordinates. In the lossless case the amplitude has zeros and the phase can take only the values 0° and 180°. With loss angle $\bar{a}$ the amplitude only has minima and the phase increases monotonously. The density of the sediment does not influence the standardized directional pattern at all.

5. THE RADIATION IMPEDANCE

5.1 Fundamental relations

To evaluate the impedance of an oscillating piston the pressure effect of each radiating element to each other must be considered. Otherwise formulated, each element, which is driven with a velocity $\dot{z}$ must be fed by a certain amount of mechanical energy to maintain its motion in the pressure field produced by all other moving surface elements. This energy depends on the features of the transducer itself and also on the specific mechanical (acoustical) properties of the medium in front of the piston.

With a sinusoidal time dependence of the piston's motion and if all surface elements are vibrating with equal amplitude and phase, the complex mechanical power, by which the membrane must be driven, when in water, is

$$\underline{P}_{mech} = \dot{z}^2 \frac{Z_1}{\lambda_1} j \int\limits_{A_M^{(n)}} \int\limits_{A_M^{(m)}} \frac{e^{-jk_1 \vartheta_{mn}}}{\vartheta_{mn}} \, dA_M^{(m)} \, dA_M^{(n)} \quad , \qquad (10)$$

where

$dA_M^{(m)}$; $dA_M^{(n)}$ = m-th resp. n-th radiating element of the piston,

ϑ_{mn} = geometrical distance between the elements $dA_M^{(m)}$ and $dA_M^{(n)}$.

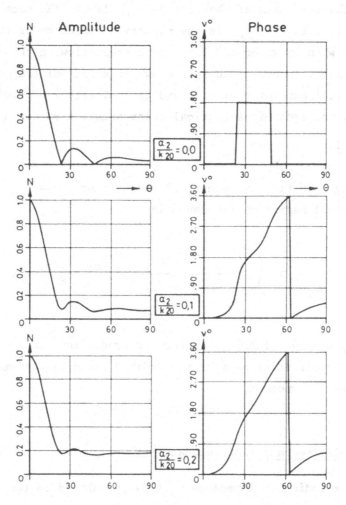

Fig. 4. Amplitude and phase of the direction pattern of a circular
         piston oscillating in a lossy medium q=3; Parameter $\bar{a}$.

Neglecting friction losses, the complex electrical power

$$\underline{P}_{el} = I^2 \, \underline{Z}_{R1} \tag{11}$$

fed into the terminals of the transducer, balances the mechanical power of (10). In (11) $\underline{Z}_{R1}$ is the radiation impedance of the transducer when in water and I is the current flowing through this impedance. According to well known relations of transducers[8;9] the electrical current I of (11) and the velocity $\dot{z}$ of (10) are related by the electric-acoustical transformation ration a.

$$\dot{z} = \frac{I}{a} \quad . \tag{12}$$

Eliminating $\dot{z}$ in (10) by using (12) and setting $\underline{P}_{el} = \underline{P}_{mech}$ the electrical radiation impedance is

$$\underline{Z}_{R1} = \frac{Z_1}{a^2} \, \frac{1}{\lambda_1} \, j \, \int\!\!\int \frac{e^{-jk_1 \mathcal{G}_{mn}}}{\mathcal{G}_{mn}} \, dA_M^{(m)} \, dA_M^{(n)} \quad . \tag{13}$$

The real component of $\underline{Z}_{R1}$ is called the "radiation resistance" while the "oscillating mass of the medium" could be derived from the imaginary component.

5.2 The relative radiation impedance

Usually the radiation impedance of (13) is related to the "acoustical surface impedance Z_{N1}". Translating the latter to the electrical side of the transducer, it becomes

$$Z_{N1} = \frac{Z_1}{a^2} \, A_M \qquad . \tag{14}$$

It can be shown that $\underline{Z}_{R1}$ converges against Z_{N1} when the dimensions
of the piston are very large compared with the acoustical
wavelength λ_1 (high frequency limiting case). For the
signification of relative radiation impedances small letters with
two indices will be used e.g.

$$\underline{z}_{11} = \frac{\underline{Z}_{R1}}{Z_{N1}} = r_{11} + j\, x_{11} \quad . \tag{15}$$

The first index refers to the medium of the radiation impedance,
the second to that of the surface impedance. In the high frequency
case $r_{11} \longrightarrow 1$ and $x_{11} \longrightarrow 0$. This convergence does not
hold if the indices are different.

5.3 The relative radiation impedance of the circular piston in water

An exact solution of (13) exists only for transducers with a circular
piston in an infinite rigid baffle, operating in a lossless medium.
This solution was first developed about 100 years ago by
Lord Rayleigh.[7] A number of well known authors[10;11] have described
Rayleigh's method in detail. In general a pure numerical treatment
of (13) for arbitrary piston shapes is possible, but in view of
the 2-fold surface integral, behind which hides a 4-fold integral,
one can imagine the consumption of computer time. The whole
situation becomes more complicated, if the treatment should be
widened for lossy media.

The solution for circular pistons to be described differs from
Rayleigh's method presented in[10] only in the sequence of the
integrations. This alteration is necessary if solutions under
complex wave numbers should be obtained.

Fig. 5 explains the geometrical conditions on the circular membran.
If tentatively the double integral of (13) is assumed to be a
double sum, one could recognize, that because of the

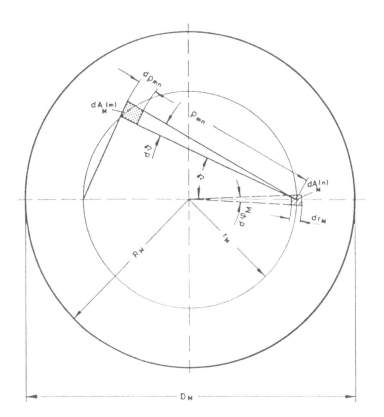

Fig. 5. Geometrical relations on the circular piston used in evaluating the radiation impedance.

interchangeability of m and n the two summations become equal.
With respect to the double integral this means: under an arbitrary
position of the surface element $dA_M^{(n)}$ the integration with respect
to $dA_M^{(m)}$ is only to be taken for the interior circle of radius r_M
on which $dA_M^{(n)}$ is positioned. From $dA_M^{(n)}$ the element $dA_M^{(m)}$
is located in a distance ϱ_{mn} at an angle ϑ . Its area is
$dA_M^{(m)} = \varrho_{mn} \, d\varrho_{mn} \, d\vartheta$, with $0 \leq \varrho_{mn} \leq 2 \, r_M \cos \vartheta$ and
$-\frac{\pi}{2} \leq \vartheta \leq \frac{\pi}{2}$.

When inserting $dA_M^{(m)}$ in (13) ϱ_{mn} cancels and the integration
over ϱ_{mn} could be simply completed. If standardization corresponding
to (15) is introduced the result is

$$\underline{z}_{11} = \frac{1}{\pi} \, \frac{1}{A_M} \int_{A_M^{(n)}} \int_{-\frac{\pi}{2}}^{\frac{\pi}{2}} (1 - e^{jk_1 r_M \cos \vartheta}) \, d\vartheta \, dA_M^{(n)} \quad ,$$

where $dA_M^{(n)} = r_M \, dr_M \, d\varphi_M$, with $0 \leq r_M \leq R_M$ and $0 \leq \varphi_M \leq 2\pi$.
The integration with respect to φ_M results in a factor 2π . With
numerous interchanges, including two integrations in parts also
the solution for r_M could be obtained easily.

$$\underline{z}_{11} = 1 - \frac{2}{\pi} \, \frac{1}{jk_1 R_M} + \frac{2}{\pi} \, \frac{1}{jk_1 R_M} \int_0^{\frac{\pi}{2}} \cos \vartheta \, e^{-j2k_1 R_M \cos \vartheta} \, d\vartheta .$$

(16)

With a real wave number k_1 there exists a well know solution[10] for
this integral

$$\underline{z}_{11} = 1 - \frac{J_1 (2k_1 R_M)}{k_1 R_M} + j \, \frac{\mathbf{H}_1 (2k_1 R_M)}{k_1 R_M} \quad . \qquad (17)$$

The real part is a Bessel function of first kind and order one, while the imaginary component contains the Struve function of the first kind which is tabulated e.g.[12;13]. Fig. 6 shows real and imaginary component of the relative radiation impedance with respect to the ratio of diameter to wavelength. The above mentioned convergence is obvious.

5.4 The relative radiation impedance of the circular piston in a lossy medium

To derive a corresponding expression to (17) if the wave number becomes complex, it is possible to expand the cylindrical functions J_1 and $\mathbf{H}_1$ in power series. Then in each term a complex wave number $\underline{k}_2$ could be introduced. It has been proven that this method, because of the special form of each term, is very complicated, even with long computer word lengths, inaccurat and time consuming. Therefore another method has been selected. Following (16) the absolute radiation impedance is

$$
\underline{Z}_{R2} = \frac{\underline{Z}_2 \, A_M}{a^2} \left[1 - \frac{2}{\pi} \frac{1}{j\underline{k}_2 R_M} + \frac{2}{\pi} \frac{1}{j\underline{k}_2 R_M} \int_0^{\frac{\pi}{2}} \cos\vartheta \, e^{-j2\underline{k}_2 R_M \cos\vartheta} \, d\vartheta \right].
$$

In both $\underline{Z}_2$ and $\underline{k}_2$, corresponding to (5) and (2 a) the notation $\underline{Q} = (1 - j \, \bar{a})^{-1}$ could be used. The integrand will be separated into real- and imaginary parts. Then the last two summands in the brackets, without considering $\underline{Q}$ could be split into real (R_R) and imaginary (I_R) component

$$
R_R = - \frac{2}{\pi^2} \frac{1}{q} \int_0^{\frac{\pi}{2}} \cos\vartheta \, e^{-2\pi \, q \, \bar{a} \, \cos\vartheta} \sin(2\pi \, q \cos\vartheta) \, d\vartheta \quad ,
$$

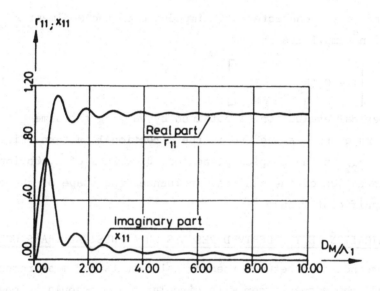

Fig. 6.   Real and imaginary component of the relative radiation
          impedance of the circular piston membrane. $\bar{a} = 0$.

$$I_R = \frac{2}{\pi^2} \; \frac{1}{q} \left[1 - \int_0^{\frac{\pi}{2}} \cos\vartheta \; e^{-2\pi \, q \, \bar{a} \, \cos\vartheta} \; \cos(2\pi \, q \, \cos\vartheta) \, d\vartheta \right].$$

With the notations $\underline{Q}$; R_R and I_R and referring to Z_{N1}, the following result for the relative radiation impedance is obtained

$$\underline{z}_{21} = \frac{Z_{20}}{Z_1} \; \underline{Q} \left[1 + \underline{Q} \, (R_R + j \, I_R) \right].$$
                                                                            (18 a)

Referring to the characteristic impedance of the sediment Z_{20} the expression simplifies to

$$\underline{z}_{22} = \underline{Q} \left[1 + \underline{Q} \, (R_R + j \, I_R) \right].$$
                                                                            (18 b)

The numerical evaluation is obtained by varying the same parameters q and $\bar{a}$ as before in the directional pattern. Fig. 7 presents $\underline{z}_{22}$ in the complex plane for 3 values of $\bar{a}$. Obviously the attenuation of the sediment influences the shape of the curve in a significant manner.

6. APPLICATION OF THE DERIVED RESULTS ON A PHYSICAL TRANSDUCER

The correspondence between theoretical results and measurement data obtained from a transducer with circular piston should be checked. In fig. 8 the relevant transducer data for this comparison are compiled.

6.1 The derivation of the electrical-acoustical transformation ratio with measured and calculated radiation resistance

Iserting the piston's diameter and the resonance frequency into (17) the relative radiation impedance becomes: $\underline{z}_{11}$ = 1,0470 + j 0,5185. Taking into account the reference value (14) for standardization, the radiaiton impedance itself is:

$$a^2 \, \underline{Z}_{R1} = (1,4183 + j \, 0,7023) \, 10^6 \, \mu bar^2 \, m^2 \, W^{-1}.$$

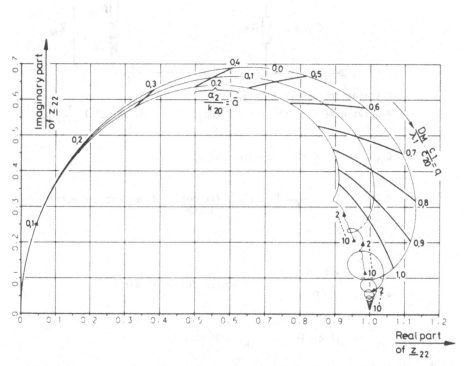

Fig. 7. The relative impedance of the circular piston in the complex plane. Parameter $\bar{a}$.

Manufacturer: HONEYWELL-ELAC-NAUTIK GmbH-Kiel
Type: TSE 4 (Ceramic)

Effective area A_M = 88,25 cm^2 ; Piston diameter D_M = 10,6 cm

Equivalent circuit diagram

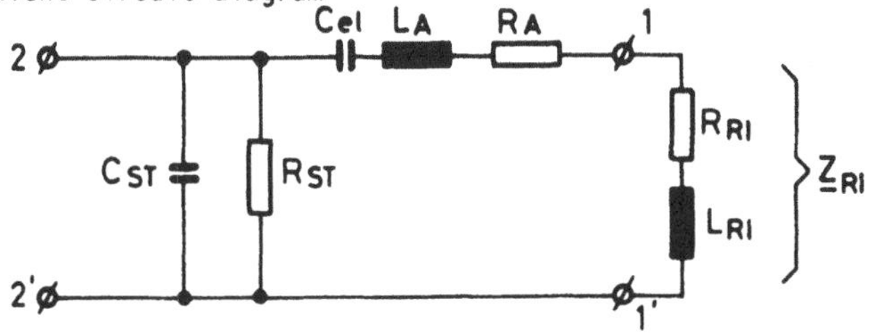

Resonant frequency: f_0 = 9,2 kHz

Receiving sensitivity (at 9,2 kHz): GE = - 69 $\left[\text{dB rel 1 } \frac{V}{\mu bar}\right]$

C_{ST} = 7 400 pF (inclusive cable)

R_{ST} = 140 $\left(\dfrac{f_0}{f}\right)$ kΩ (Frequency dependent dielectric losses)

C_{el} = 600 pF

L_A = 405 mH   )
                              Transducer driven in air
R_A = 300 Ω)

R_{R1} = 11,2 kΩ ; L_{R1} = 95 mH

Fig. 8. Compilation of the relevant transducer data for comparison
         of measured and calculated results.

Comparing the real component of which with the measured value
of 11,2 kΩ , the electrical-acoustical transformation ratio is
obtained as: a = 11,25 μbar m^2 V^{-1} . Using the value of the ratio,
a,to evaluate the oscillating medium mass - or the inductivity -
one gets L_{R1} = 96 mH. This correspondes surprisingly good to the
measured value.

6.2 The derivation of the electrical-acoustical transformation ratio with the measured receiving sensitivity of the transducer

For comparison the electrical-acoustical transformation ratio, a ,
could also be derived from the receiving sensitivity GE and further
transducer data. Following[8] or[9] in addition to the relation (12)
between electrical current and mechanical velocity of the piston
there exist another proportionality which is often used in the
receiving case. The force acting on the piston and the internal
open circuit-voltage of an ideal transducer (neglecting all
equivalent-circuit-elements on the left of terminals 1 1' in fig. 8
are connected by: a $U_{1\ 1}$, = p A_M . In reality only the
terminals 2 2' of the transducer are available. The voltages $\underline{U}_{2\ 2}$,
and $\underline{U}_{1\ 1}$, are connected by a complex factor $\underline{b}$ of which only the
amplitude is of interest. $U_{2\ 2}$, = b $U_{1\ 1}$, . The data sheet of
any transducer gives the receiving sensitivity in a logarithmic
scale GE = 20 lg g_E , where   p g_E = $U_{2\ 2}$, .

Using all the above mentioned interrelations the electrical-
acoustical transformation ratio, a , is obtained as: a = A_M (b $g_E)^{-1}$
Given the equivalent circuit elements in fig. 8, this ratio is
easily derived: b = 5 and from the mentioned GE-value:
g_E = 3,5681 10^{-4} V (μbar)$^{-1}$.
With the well known piston's area A_M the electrical-acoustical
transformation ratio results in: a $\approx$ 5 μbar m^2 V^{-1} .
This value is approximately a factor 2 smaller than that derived
via the radiation resistance. There exist a lot of reasons for

this discrepancy. Inaccuracies in the calibration method by
which GE is gained, are the most probable.

6.3 The admittance diagram of the transducer

The admittance measured between the transducer's terminals 2 2'
depicted in the complex plain offers an informative impression of
the frequency response of the tansducer. Due to the capacitor C_{ST}
and the resistance R_{ST} the almost circle-shaped admittance diagram
of C_{el}; L_A ; R_A and $\underline{Z}_{R1}$ is displaced and deformed. This effect
disturbes the recognition of admittance changes when the transducer
is driven once in water and on the other in sediment. To avoid this,
at each frequency C_{ST} will be tuned by an inductance of high
quality. The effect of R_{ST} alone results only in a displacement
of the admittance circle on the real axis which is unimportant.
Introducing numbers and using the transformation ratio derived
from the radiation resistance the admittance results in

$$\underline{G} = \left[0,3 + j\, (23,32\ x - \frac{28,87}{x}) + 10,7\ \underline{z}_{21} \right]^{-1} \quad mS \quad , \quad (19)$$

where $x = \dfrac{f}{f_o}$.

It is rather cumbersome to show the influence of all possible
combinations of the acoustic specific sediment data on the
admittance. Therefore three diagrams depicted in fig. 9 have
been plotted. In each diagram one parameter is only varied.
In fig. 9 a the loss angle, 9 b the density-ratio and 9 c the
velocity-ratio. The admittance variations caused by the
acoustical sediment properties are of an order, which might form
the foundations for an in-situ measurement method.

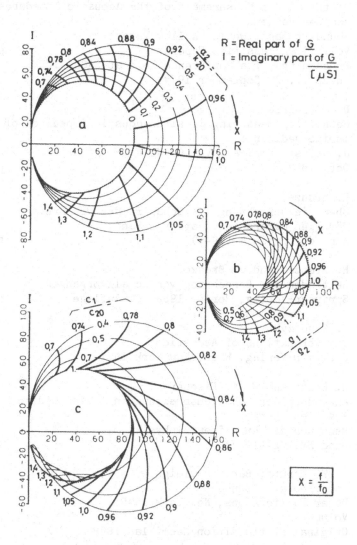

Fig. 9. Calculated admittance diagrams of the transducer mentioned in Fig. 8 operating in a lossy medium.

Literature

1.   E. Schunk
      On the Direct Measurement of the Acoustic Impedance of
      the Sea Bottom.
      Sound Propagation in Shallow Water
      Vol. II Unclassified Papers   p.   14
      Proceedings of a Conference held at SACLANTCEN
      on 23 ... 27 September 1974

2.   D. J. Shirley
      Method for measuring in situ acoustic impedance of
      marine sediments
      J. Acoust. Soc. Am. Vol 62 No 4
      Oct. 1977   p. 1028

3.   H. Stenzel
      Über die Berechnung des Schallfeldes von kreis-
      förmigen Membranen in starrer Wand
      Ann. d. Physik 4 (1949)   p.   303

4.   H. Stenzel und O. Brosze
      Leitfaden zur Berechnung von Schallvorgängen
      Springer-Verlag, Berlin 1958, 2. Auflage   p.   11

5.   E. Skudrzyk
      The Foundations of Acoustics
      Springer-Verlag, Wien, New York 1971   p.   603

6.   H. Hecht und F. A. Fischer
      Anwendung der Schallausbreitung in freien Medien
      Aus Wien, Harms:
      Handbuch der Experimentalphysik
      Band XVII/2 (1934)   p.   391

7.   J. W. Strutt; Baron Rayleigh
      The Theory of Sound
      Dover Publications, New York 1945
      Volume II   p.   163
      Original First Edition Macmillan 1878

8.   D. G. Tucker and B. K. Gazey
      Applied Underwater Acoustics
      Pergamon Press, Oxford 1966   p.   134

9.   F. A. Fischer
Grundzüge der Elektroakustik
Fachverlag Schiele & Schöne GmbH
Berlin 1959   p.   88

10.   as in 5   p.   663

11.   as in 9   p.   155

12.   M. Abramowitz and I. A. Stegun
Handbook of Mathematical Functions
National Bureau of Standards
Applied Mathematics Series 55
Fifth Printing Aug. 1966   p.   495

13.   G. N. Watson
A Treatise on the Theory of Bessel Functions
Cambridge University Press
Cambridge, England 1958   p.   666

ELASTIC PROPERTIES RELATED TO DEPTH OF BURIAL, STRONTIUM CONTENT

AND AGE, AND DIAGENETIC STAGE IN PELAGIC CARBONATE SEDIMENTS

Murli H. Manghnani, Seymour O. Schlanger and
Phillip D. Milholland

Hawaii Institute of Geophysics
University of Hawaii
Honolulu, Hawaii 96822

ABSTRACT

Laboratory measurements of density (ρ), compressional (V_p) and shear (V_s) velocities, and velocity anisotropies (A_p) and (A_s) in a pelagic ooze-chalk-limestone sequence from DSDP Site 289 are viewed in light of its present depth of burial, sediment age and diagenetic stage.

Changes in the elastic properties with increasing depth are more or less systematic; in general, ρ, V_p, V_s, A_p and A_s increase and σ decreases. In certain cases, the depths at which large variations are noted correlate well with the interpreted depths of acoustic reflectors.

The changes in elastic properties with depth are gradational from ooze to chalk, and chalk to limestone. Chert and siliceous limestones do not exhibit uniform properties as a function of depth.

Sr content of pelagic carbonates varies inversely with sediment age. It is shown that in addition to the present depth of burial the sediment age is a useful parameter for proposing geoacoustic models of the pelagic carbonate sequences.

INTRODUCTION

One of the thickest and most complete pelagic carbonate sediment sequences displaying an ooze-chalk-limestone transition with depth and sediment age is found on the Ontong Java Plateau in the western Pacific (Fig. 1). This sequence, ranging in age

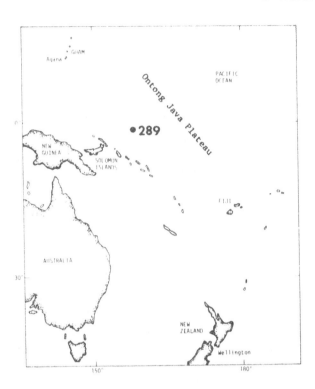

Fig. 1.   Location of DSDP Site 289 on the Ontong Java
          Plateau, and Site 210 in the Coral Sea Basin (after
          Andrews and Packham et al., 1975[1]).

from Recent to Early Cretaceous (∿110 m.y.) is a classic example
of rapid carbonate sedimentation with low terrigeneous input.[1,2]

    Recent laboratory study of physical and acoustic properties of
130 core samples obtained at DSDP (Deep Sea Drilling Project) Site
289 on the Ontong Java Plateau has provided useful geophysical
parameters (ρ, V_p and V_s) as a function of depth and enabled construction of a geoacoustic model of deep-sea carbonate sediments.[3]
This study determined that velocity anisotropy (horizontal velocity
usually faster than vertical) for both compressional (A_p) and
shear (A_s) waves generally increases with depth and diagenesis.

    In this paper, the measured geophysical parameters of the
carbonate sedimentary sequence at Site 289 are related to depth
of burial, age and diagenetic stage. Consideration of sediment
age as deduced from regional geologic settings in addition to
depth of burial, will be valuable in constructing more accurate
and realistic geoacoustic models of deep-sea sediments.

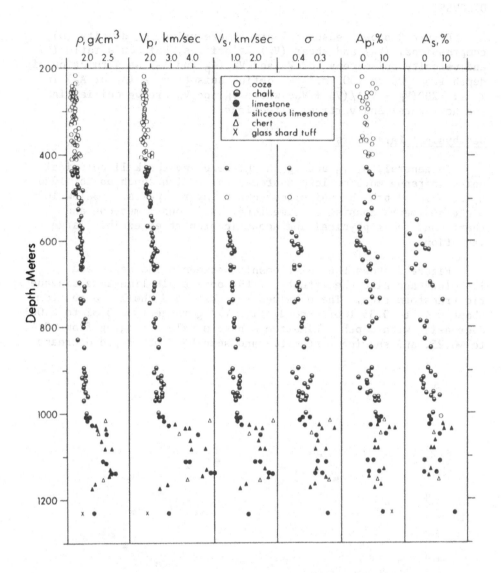

Fig. 2.   Measured values of density (ρ), compressional (V_p) and
          shear velocity (V_s), Poisson's ratio (σ), compressional
          (A_p) and shear velocity anisotropy (A_s) as a function of
          depth for the ooze-chalk-limestone sequence from
          Site 289.

DISCUSSION

Figure 2 shows measured 1-atmosphere values of density (ρ), compressional (V_p) and shear (V_s) velocities, Poisson's ratio (σ) and velocity anisotropy for V_p and V_s (A_p and A_s) as a function of depth from 200 to 1300 m.[3] We define anisotropy A (A_p or A_s) in % as: $200(\bar{V}_h - \bar{V}_v)/(\bar{V}_h + \bar{V}_v)$ where $\bar{V}_h$ and $\bar{V}_v$ are velocities in the horizontal and vertical directions.

Geophysical Parameters

In general, ρ, V_p and V_s in the ooze and chalk lithological units increase more or less systematically with depth up to $\sim$1020 m, at which depth the ubiquitous chert layer appears.[1] Below 1020 m the sediments consist of limestone, siliceous limestone and chert, and their physical and acoustic properties exhibit large variations.

Figure 3 shows the relationship between σ and depth and lithology and also the effects of the ooze-chalk-limestone diagenetic transformation. The σ values for oozes and chalks range from about 0.45 to 0.35 (correspondingly, V_p/V_s range from 3.32 to 2.08) decreasing with depth. Limestones have σ values ranging from $\sim$0.40 to $\sim$0.25, and showing systematic and somewhat more rapid decrease

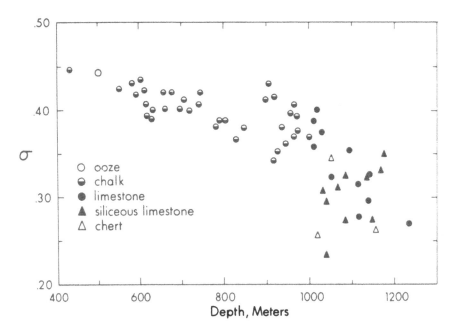

Fig. 3. Poisson's ratio (σ) versus depth for the ooze-chalk-limestone sequence from DSDP Site 289.

with increasing depth and diagenesis. The relationship between depth and σ is somewhat obscured at depths >1000 m possibly because of large variations in silica content. The amount of chert varies significantly between 1000 and 1200 m.[1]

There are noticeable variations in velocity and/or density at several depths (e.g., at ∿650 and ∿820 m) that have been related to paleo-oceanographic events such as oceanic cooling and sea level changes.[4,5] These depths also appear to correlate well with the interpreted depths of acoustic reflectors at Site 289.[1]

Velocity Anisotropy

Velocity anisotropy in marine sediments has been reported previously by several workers.[1,6-9] Most of these studies have been confined to investigation of A_p only except those of Milholland[7] and Milholland et al.[3] who have also investigated A_s.

For both V_p and V_s, the horizontal velocities are generally higher than vertical (see Figs. 4 and 5). But at several depths (e.g., at ∿600 and ∿950 m) negative anisotropy values have also been observed.[3] As can be seen in Figures 4 and 5, A_p and A_s are variable for the ooze and chalk samples (-4 to 8% for A_p, -8 to 6% for A_s); but indicate a trend of increasing anisotropy with increasing depth and diagenesis. Carbonate limestone, siliceous limestone and chert below 1020 m show appreciable anisotropy $A_p \sim 8 \pm 4\%$ and $A_s \sim 6.0 \pm 4\%$. Figure 6 shows the positive correlation between A_p and A_s. Note that the A_p values (in %) are generally larger than the A_s values.

Several possible causes of velocity anisotropy in carbonate sediments have been discussed.[7-9] One of the most probable causes of anisotropy in carbonate sediments is the preferred orientation of calcite crystals that have high elastic anisotropy. Single-crystal velocities along c-axis are slowest (V_p = 5.62 km/sec and V_s = 3.54 km/sec) and fastest along a-axis (V_p = 6.34 km/sec and V_s = 4.01 km/sec).[10] Preferred orientation of the c-axis of calcite in the vertical direction (perpendicular to bedding plane) would account for the measured anisotropy.[7,8] Our preliminary x-ray petrofabric data for a few carbonate sediment core samples from DSDP Site 289 indeed show that the c-axis is preferentially oriented. More study is needed, however, to make a quantative analysis of any correlation between A_p, A_s and preferred orientation of calcite grains in such sediments.

Age of Pelagic Carbonate Sediments as a Factor in Constructing Predictive Geoacoustic Models

In general, the depth of burial is one of the geologic factors commonly considered in the construction of geoacoustic models that

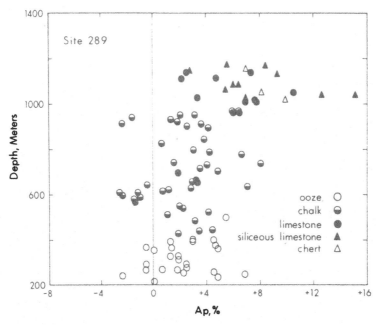

Fig. 4. Compressional velocity anisotropy (A_p) as a function of
depth for the ooze-chalk-limestone sequence (DSDP Site 289).

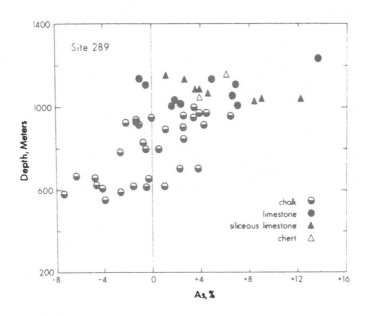

Fig. 5. Shear velocity anisotropy (A_s) as a function of depth for
the ooze-chalk-limestone sequence (DSDP Site 289).

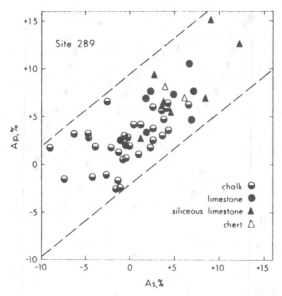

Fig. 6. Relationship between compressional (A_p) and shear
         velocity (A_s) anisotropies for the ooze-chalk-limestone
         sequence (DSDP Site 289).

predict geophysical parameters such as ρ, V_p, and V_s for marine
sedimentary sequences. We propose that the geologic age of a
pelagic carbonate layer is an equally important factor for developing predictive geoacoustic models.

     As discussed in detail by Schlanger and Douglas[4] the major
diagenetic process in the transformation of carbonate ooze to
chalk to limestone is the solution of original biogenic calcite
produced by planktonic organisms and the reprecipitation of this
carbonate as cement. The calcite precipitated by organisms such
as coccoliths, discoasters and foraminifers contains approximately
1400 ppm of Sr which substitutes for Ca in the calcite crystal
structure. Upon solution of this biogenic calcite, Sr is released
into the pore water and is largely excluded from the precipitated
cement. Therefore, the Sr content decreases with increasing
diagenesis and thus serves as an indicator of the degree to which
the original biogenic calcite has been dissolved and calcite
cement added to the sediment. Because the amount of calcite
cement influences the geophysical parameters it is important in
predicting the degree of cementation that occurred during diagenesis. Figures 7, 8, and 9 illustrate depth versus age, Sr content
versus depth, and Sr content versus age at several DSDP sites in
the Pacific. A plot of depth versus age (Fig. 7) shows a separate
trend of age with depth for each DSDP site. This is due to the
variation of sedimentation rates both at individual sites

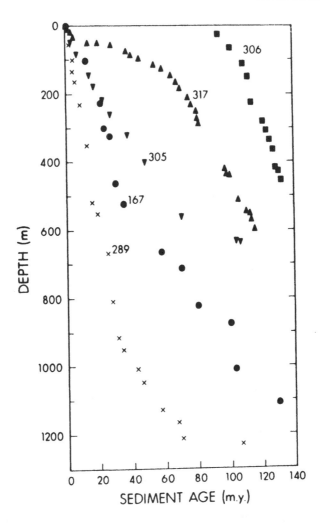

Fig. 7. Sediment age plotted against depth of burial at DSDP Sites
         289, 167, 305, 317 and 306 in the Pacific. These sediment
         sequences are largely carbonate. Each site displays a
         unique age-depth relationship in the sequence drilled.
         Data from Initial Reports of the Deep Sea Drilling Project,
         Vols. 17, 30, 32 and 33.

and between different sites. In addition, erosional episodes indicated by sedimentary hiatuses removed pre-existing layers within
individual sequences. For example, sediments as old as 90 m.y.
can be found at shallow depths, e.g., at Sites 306 and 317. Figure
8 shows that there is a poor correlation between Sr content and
depth, the coefficient of correlation r = -0.419. However, the
plot of Sr content versus age (Fig. 9) shows a strong negative
correlation (r = -0.875). This is consistent with the Schlanger

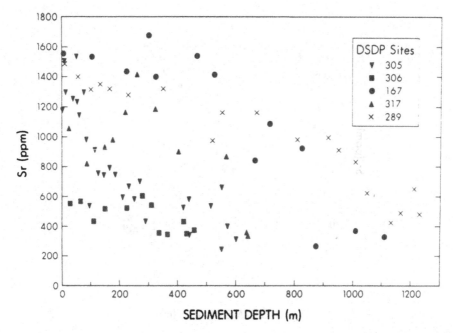

Fig. 8. Sr content plotted against depth of burial at DSDP Sites
289, 167, 305, 317 and 306. All of the samples analyzed
by atomic absorption spectrophotometry are largely carbonate. The low Sr values of 500 to 600 ppm at shallow
depths at Site 306 reflects the 90- to 100-m.y. age of
these sediments and their advanced state of diagenesis.

and Douglas[4] model for pelagic carbonate diagenesis in which they
regarded degree of cementation as time dependent because the
driving force behind the cementation process is the tendency of
the crystal system to reorganize toward a state of lowest free
energy. Elimination of the crystal defects (e.g., removal of Sr),
the dissolution of small crystals and the growth of larger ones
lower the free energy of the system. The diagenetic changes,
which in turn determine the changes in geophysical parameters, are
therefore both time and depth dependent. Riech and von Rad[12] have
also applied the "diagenetic potential" concept of Schlanger and
Douglas[4] to silica diagenesis and have shown that the development
of chert is also time as well as depth dependent.

The geological age and lithology of a pelagic sediment
sequence at any point on a lithospheric plate can be deduced by
using the "backtracking" method of Berger and Winterer.[11] With
this deduced geological history and the general Sr/age relationship discussed above it is possible to predict the diagenetic
stage and geoacoustic properties within the carbonate sequence at
any depth. In our research program we are attempting to determine

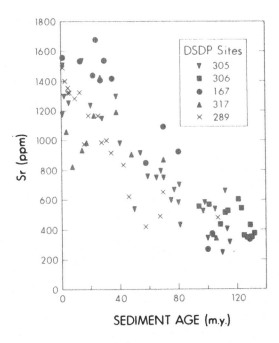

Fig. 9. Sr content plotted against age of sediment at DSDP Sites
         289, 167, 305, 317 and 306. A least-squares fit to the
         data shows the Sr/age relationship to be Sr(ppm) =
         1356 – 8.288 (age m.y.); r = – .875).

the relative importance of the two factors. Our working hypothesis
is that burial effects such as gravitational compaction are relatively more important than age effects at depths of less than
∿250 m, whereas age effects such as cementation are relatively
more important than depth of burial below 250 m.[4]

ACKNOWLEDGMENTS

     This research was supported by the Office of Naval Research.
The authors thank Susan Gaffey for Sr and petrofabric analyses,
and the Deep Sea Drilling Project which is funded by the U.S.
National Science Foundation for supplying the sediment samples.
Hawaii Institute of Geophysics Contribution No. 1065.

REFERENCES

1. J. E. Andrews, G. Packham, et al., "Initial Reports of the
    Deep Sea Drilling Project," vol. 30 (U.S. Government Print-
    ing Office), Washington, D.C. (1975).
2. W. H. Berger, T. C. Johnson, and E. L. Hamilton, Sedimentation
    in Ontong Java Plateau: Observations on a classic 'carbo-
    nate monitor,' in: "The Fate of Fossil Fuel CO_2 in the
    Oceans," N. Anderson and A. Malahoff, eds., Plenum, New
    York (1977).
3. P. Milholland, M. H. Manghnani, S. O. Schlanger, and G. H.
    Sutton, Geoacoustic modeling of deep-sea carbonate sedi-
    ments, J. Am. Acoust. Soc. (in press).
4. S. O. Schlanger and R. G. Douglas, The pelagic ooze-chalk-
    limestone transition and its implications for marine strati-
    graphy, in: "Pelagic Sediments: On Land and Under the
    Sea," Spec. Pub. No. 1, K. J. Hsü and H. C. Jenkyns, eds.,
    Int. Assoc. Sed. (1974).
5. S. O. Schlanger and I. Premoli-Silva, Tectonic, volcanic and
    paleogeographic implications of redeposited reef faunas of
    Late Cretaceous and Tertiary age from the Nauru Basin and
    the Line Islands, in: "Initial Reports of the Deep Sea
    Drilling Project," vol. 61, R. Larsen, S. O. Schlanger, et
    al., eds. (U.S. Government Printing Office), Washington,
    D.C. (in press).
6. B. E. Tucholke, N. T. Edgar, and R. E. Boyce, Physical Proper-
    ties of sediments and correlations with acoustic strati-
    graphy, Leg 35, Deep Sea Drilling Project, in: "Initial
    Reports of the Deep Sea Drilling Project," vol. 35, C. D.
    Hollister et al., eds. (U.S. Government Printing Office),
    Washington, D.C. (1976).
7. P. Milholland, Geoacoustic Model of Deep Sea Carbonate Sedi-
    ments, M.S. Thesis, Univ. of Hawaii, Honolulu (1978).
8. R. L. Carlson, N. I. Christensen, Velocity anisotropy in semi-
    indurated calcareous deep sea sediment, J. Geophys. Res.,
    84:205 (1979).
9. R. T. Bachman, Acoustic anisotropy in marine sediments and
    sedimentary rocks, J. Geophys. Res., 84(B13):7661 (1979).
10. L. Peselnick and R. A. Robie, The elastic constants of calcite,
    J. Appl. Phys., 34:2494 (1963).
11. W. H. Berger and E. L. Winterer, Plate stratigraphy and the
    fluctuating carbonate line, in: "Pelagic Sediments: On
    Land and Under the Sea," Spec. Pub. No. 1, K. J. Hsü and
    H. C. Jenkyns, eds., Int. Assoc. Sed. (1974).
12. V. Riech and V. von Rad, Silica diagenesis in the Atlantic
    Ocean: Diagenetic potential and transformations in: "Deep
    Drilling Results in the Atlantic Ocean: Continental Margins
    and Paleoenvironments," Maurice Ewing Ser. No. 3, M. Talwani,
    W. Hay, and W. Ryan, eds., Amer. Geophys. Union (1979).

APPLICATION OF GEOPHYSICAL METHODS AND EQUIPMENT TO

EXPLORE THE SEA BOTTOM

Helmut F. Weichart

PRAKLA-SEISMOS GMBH
Haarstr. 5
D 3000 Hannover 1

ABSTRACT

In marine seismic surveys the acoustic signal is
starting and spreading in the water, reflected or
refracted at discontinuities of the bottom and subbottom,
and then returning into the water to the detecting
sensors. The exploration for hydrocarbons usually requires
deep penetration by application of low frequencies
(between 5 and 150 Hz) and does not need detailed information about the sea bottom. Increased frequencies lead to
higher resolution of structures but to decreased penetration. In that direction the advanced seismic methods and
equipment can be developed to determine the interesting
acoustic parameters of the sea bottom by continuous
profiling.

This paper will give a summary of the seismic
technique and references for practicable developments in
respect to research of the sea bottom.

INTRODUCTION

Sound or acoustical events are the basis of geophysical exploration with the aim to recognize the structures
of the earth. Seismic methods are used to discover and to
explore oil and gas fields or mineral resources. The
methods have developed rapidly in the last years to increase the efficiency and reliability of field data
acquisition and to use the enormous efficiency of data
processing by computers.

Especially the marine data acquisition advanced to a high technical level of instrumentation in the geophysical equipment as well as in navigation. Marine seismic work is carried out offshore, in shallow waters, rivers, and lakes.

The normal seismic problems require deep penetration to several 1 000 m, and suitable frequencies are between 5 and 200 Hz. But some recent aspects are pointing in the direction of higher resolution and three dimensional research.

Further increase of frequencies leads to a special profiling technique which would be able to measure some interesting parameters of the sea bottom.

SEISMIC METHODS[1-3]

The basic method is the reflection method (Fig. 1A) because of its high efficiency in recognizing the structures of layers.

The other method, the refraction method (Fig. 1B), is used to explore discontinuities of the acoustic velocity when material with higher velocity is superimposed by material with lower velocity. Both velocities can be determined easily. The refraction method is used to follow strong deep discontinuities as well as to determine structures and velocities of shallow layers which are necessary for corrections of reflection measurements.

A special case of refracted waves are diving waves (Fig. 1C) if the velocity in the lower medium is increasing continuously with depth.

The basic equipment in the mentioned methods is an acoustic source and one or more acoustic sensors. The appearence of refracted waves is a question of the distance between source and sensor.

In marine seismic work source and sensors are towed behind the ship along a profile line (Fig. 2). In areas of sufficient water depth the ship steams with constant speed while sending and recording. In areas of shallow water the sensor cable (streamer) can be towed over the sea bottom in continuous profiling or - to reduce noise - in start-stop operation.

The normal way is to collect data along a given seismic profile line. But advanced migration processing

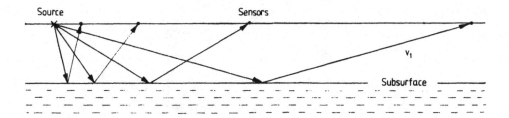

Fig. 1A   Reflected waves

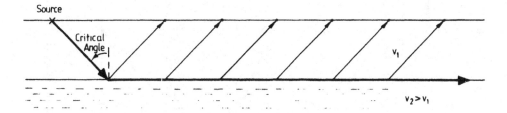

Fig. 1B   Refracted waves

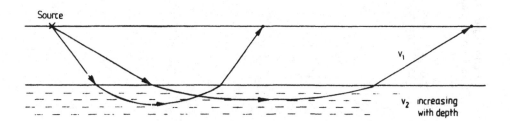

Fig. 1C   Diving waves

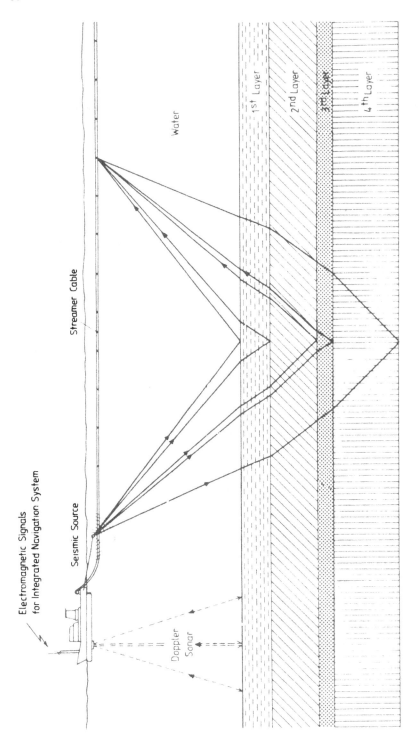

Fig. 2   Marine seismic profiling

requires the dips of structures across the profiles which can be measured by 3D-techniques.

The most important point of all seismic methods is the high multiplicity of sensor channels. It guarantees that the different wavetypes can be distinguished and that reflectors lost due to poor reflectivity or high attenuation of signals can be picked up again. Furthermore, the redundancy of data allows a lot of computer processing with the result to improve the signal/noise ratio and to extract some parameters of the media passed by acoustic rays.

Another method to increase the multiplicity of data is the so called CDP-(Common Depth Point)-method (Fig. 3), the multiple use of the same reflection point by different ray paths.

Preferably compressional waves are applied in the seismics but in the last years we observe an increasing interest in shear waves.

SEISMIC INSTRUMENTS

Sources[4-7, 32]

Sources of highest efficiency are still explosives regarding the strength as well as the broad frequency band. Developments aimed at increased safety, at increased effectiveness by more and smaller charges and at special applications in shallow water by strings of dynamite.

But the problems of supply and handling started the search for more advanced sources. The top aim was a source fed by materials always in or around a ship, and with high effectiveness in the seismic frequencies. At least these materials were found in diesel fuel, air and water. Two principles are applied to generate seismic pulses, either by compression or by cavitation of the water.

The list of Fig. 4 should be sufficient to give a short presentation of the main sources and their data.

At the moment the compression and cavitation type (Fig. 5) are commonly used. In the case of air-guns tuned arrays are able to generate highly efficient pulses at seismic frequencies. For instance an array of 27 guns will give 40 bar-meters peak to peak signals at a total

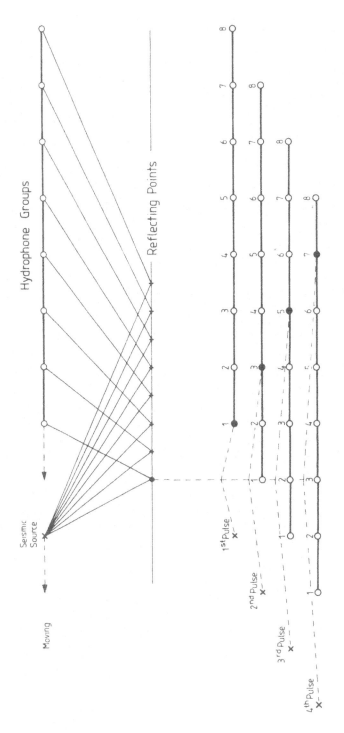

Fig. 3 CDP – (Common depth point)–Method (Principle with 4 fold coverage)

Type of Source	Type of first pulse	Storage of Energy by	Range of stored Energy	Range of Frequencies	No. in "References"
Explosives	Compression	Dynamite	4 kJ/gramm	< 1... >1000 Hz	4
Air Gun	Compression	Compressed Air	4...1000 kJ	5... 300 Hz	5
Water Gun	Cavitation	Compressed Air	≈ 20 kJ	10... 300 Hz	7
Flexichoc	Cavitation	Compressed Oil	≈ 20 kJ	10... 300 Hz	6
Gas Exploder	Compression	Gas (Propane)	40... 300 kJ	20... 300 Hz	4
Sparker	Compression	Electrical Charge	1... 100 kJ	20... 5000 Hz	4 + 32
Boomer	Compression	Electrical Charge	≈ 1 kJ	500... 10000 Hz	4 + 32

Fig. 4 Seismic sources

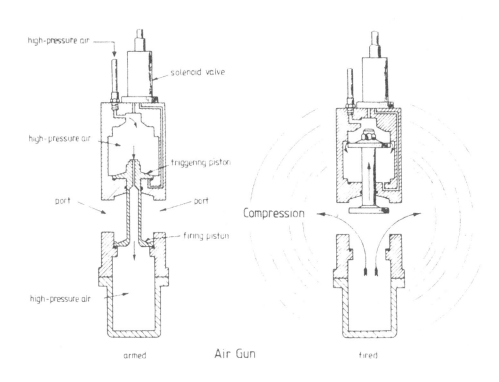

high-pressure air

solenoid valve

high-pressure air

triggering piston

port   port

firing piston

high-pressure air

Compression

armed   Air Gun   fired

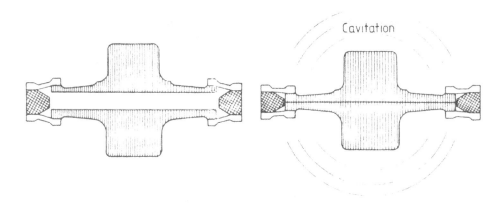

Cavitation

Fig. 5 Compression and cavitation type sources

gun volume of 30 liters at 150 bar feeding pressure in
a sequence of about 10 seconds.

For lower penetration and higher resolution sparker
arrays are operated with capacitor banks up to 100 kJoule
with few seconds' pulse intervals.

Sensors[8]

In deep and shallow waters pressure sensitive
sensors, so called hydrophones, are used. The hydrophones
are assembled in a neutrally buoyant, fluid filled
streamer cable. Common in use are piezoelectric, accelera-
tion cancelling hydrophones. Some 10 hydrophones are each
connected to arrays with acoustic directivity in seismic
frequencies. A streamer cable towed behind the ship
contains 48, 96, 200 and more seismic arrays. The electrical matching is performed either by transformers in
the streamer or by charge amplifiers on board of the ship.
Typical streamers are put together from sections of 50 m
or 100 m each to a total length of about 2 500 m or more.
The operating depth is between 5 m and 30 m, depending on
the acceptable high cut frequency of the distance to the
water/air surface. Automatic or remote controllable buoys
("birds") stabilize the depth over the whole length.

In very shallow waters and for connection with land
seismics, bay cables are used with attached electro-
dynamic sensors, so called geophones. Gimbal mounting
always aligns the geophones' axis of sensitivity parallel
to the direction of gravity.

For special shallow water areas as well as for deep
water refraction - operated parallel with reflection
equipment - hydrophones carried by buoys are used with
radio transmission of data.

Recording Instruments

For at least 10 years the digital type recording
instruments have been used, the so called DFS (Digital
Field System), with high performance and high accuracy.
Latest developments tend to more and more channels and
higher recording density on the magnetic tape. The number
of channels recorded in real time depends on the sampling
interval. At present 120 channels at 2 ms sampling are
typical. This allows to record the time functions in the
seismic band from about 5 Hz to 120 Hz. Floating point
amplifiers with 0 to 84 dB range and an A/D converter with 14
bit guarantee optimal recording of input data even at very
high dynamic voltage ranges.

Navigation and Locating Systems[9]

The high costs and accuracy of the seismic data require, of course, high standards in the navigation. Latest equipment of satellite-, radio- and acoustic navigation systems are in use. The selection of methods depends on the area. Generally integrated systems (Fig. 6) are able to increase the accuracy of navigation at any point of the earth and at any time.

Latest 3D-data acquisition and processing require to know the exact position of sources relative to the positions of all hydrophone arrays. Some waterbreak detectors and heading sensors distributed along the streamer cable and a direction finder system for the end buoy are provided for this purpose.

SEISMIC DATA PROCESSING[10,11]

Results of the data processing should be clear presentations of the structures of the explored area, mostly as cross-sections (Fig. 7).

In the cross-sections all reflecting elements are plotted in that depth as if source and sensor would have been commonly located over the associated element.

It would take a separate, very extensive paper to present the processing techniques applicable to seismic data. The following will mention the captions only:
- Static and dynamic corrections
- Stacking
- Attenuation of Multiples
- Deconvolution
- Filtering in Frequencies and Wavenumbers
- Ray Tracing
- Migration
- 3D-Seismics
- Real Amplitudes
- Extraction of Velocities

POSSIBILITIES TO INVESTIGATE THE SEA BOTTOM

If we think about the acoustic water-bottom inter-action we have to ask for the morphology as well as for the stratigraphy of the sea floor, and we should know the distribution of some acoustic parameters therein.

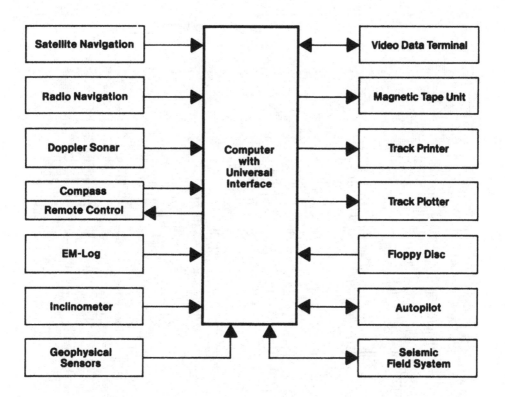

Fig. 6   Integrated satellite navigation system INDAS V

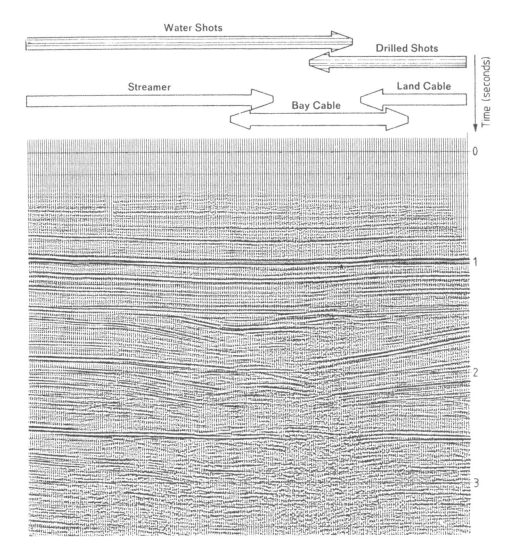

Fig. 7 Seismic section gained by different survey techniques
         (Land connections)

The depth of investigations depends on the wavelength of the interesting interaction, on the distance between source and receiver, and on the structures and properties of the bottom. The area of exploration depends on the problem which has to be solved, but generally the knowledge of larger areas will be necessary.

The morphology can easily be profiled by the well known echosounder and sidescan-sonar.

The stratigraphy can be explored by a narrow beam sounder[13] (ELAC) or better by multichannel seismic profiling. The frequencies have to be adapted to the wanted resolution, and the methods of seismic processing and interpretation can be applied.

The measurements of acoustic parameters of the strata of the bottom or subbottom can require some modifications of the known seismic methods and techniques.

Interesting parameters are:

- Coefficients of reflection and refraction depending on angle and frequency.

And more detailed:

- Velocity and attenuation in the media,
- for compressional and shear waves,
- distribution in vertical and horizontal direction.

It is not possible to give here only one method and equipment to measure all mentioned data because of the dependency on the specific problems. But it is possible to give here some general views.

A general rule in the physics of waves is that the wavelength is the most important measure. Considerations to solve a given problem have to start with the choice of suitable wavelengths.

The resolution depends on the wavelength. The wavelength should be smaller than the distance between the boundaries of the structures to be found, but larger than of the unwanted ones.

The penetration is depending on the wavelength, because the attenuation in a medium is nearly constant if related to the wavelength. We learnt from some papers[27-31] that in marine sediments the attenuation of compressional waves in different media varies between about 0.1 and

1.0 dB/wavelength. The attenuation of shear waves in
some media was measured higher by different factors up
to 10. If we pick up a signal by different sensors so
that the difference of the passed ways in the medium is
in the range of 10 wavelengths we have the chance for
good measurements of attenuation.

But it should be mentioned again that the best
chance for proper in situ measurements seems to be
obtainable by continuous profiling with high multiplicity
of receiving channels. The distance between two adjacent
receivers (separately recorded) should be smaller than a
quarter of the shortest received apparent wavelength if
we want the best correlation between them. Many other
principles of exploration seismics should be applied
too.

Finally we regard the Fig. 8 to learn that a multichannel arrangement will be able to recognize and
distinguish different kinds of waves as applied in
seismics:

If source and sensors are towed in water, whose
sound velocity is designated by v_1, and if the top layer
of the bottom has the velocity v_2 we observe in direction
of increasing distance:

- The direct wave can be detected as long as the
  attenuation is not too high.

- In the case $v_2 < v_1$ no refracted waves appear and we
  recognize reflected waves with travel times always
  longer than of the direct wave.

- In the case $v_2 > v_1$ refracted waves appear additionally,
  beginning at a distance depending on the critical
  angle and the distance H to the bottom. If v_2 varies
  close to v_1 the distance of appearence increases
  rapidly. The refracted waves always arrive earlier
  than the reflected waves, and from a certain distance
  on they overtake the direct wave.

These facts are well known and applied in exploration
seismics as demonstrated by Fig. 9.

From Fig. 8 we see that the optimal geometry of the
source/sensor arrangement depends on the distance H to
the sea floor. We know from literature that the velocity
in some bottom media can be expected close to the velocity
in water. Then we should tow the equipment in a closer
distance to the bottom or directly along the bottom. This

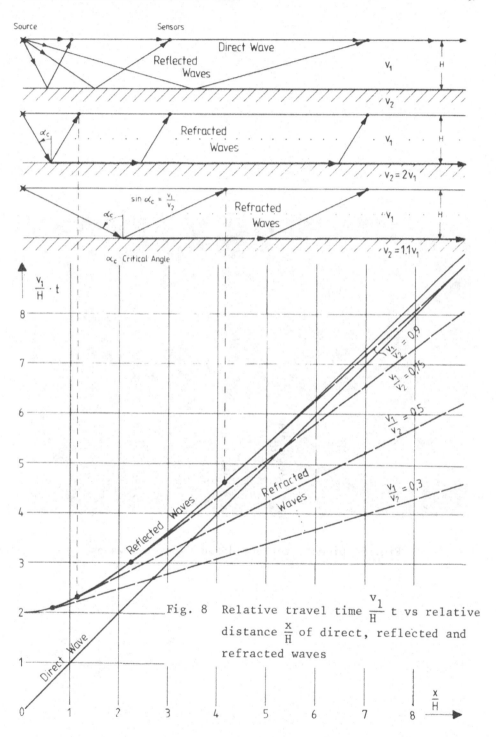

Fig. 8 Relative travel time $\frac{v_1}{H} t$ vs relative distance $\frac{x}{H}$ of direct, reflected and refracted waves

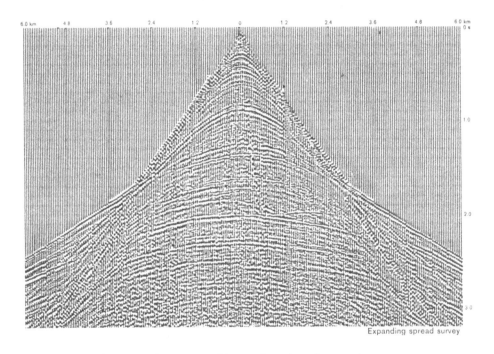

Fig. 9 Direct, reflected and refracted waves

reminds of another seismic technique, the sonic log for measurements in boreholes[12,14].

CONCLUSION

It was not possible to present practical results regarding water-bottom interaction. But an impulse should be given to think about if the advanced methods and equipment of exploration seismics could help to solve some of the interesting problems.

REFERENCES

1. R. Meissner and L. Stegena, "Praxis der seismischen Feldmessung und Auswertung", Borntraeger, Berlin + Stuttgart (1977).
2. B. S. Evenden, D. R. Stone, and N. A. Anstey, "Seismic Prospecting Instruments", Vol. 1 + 2, Borntraeger, Berlin + Stuttgart (1970/1971).
3. R. E. Sheriff, "Encyclopedic Dictionary of Exploration Geophysics", Society of Exploration Geophysicists, Tulsa/Oklahoma (1973).
4. F. S. Kramer, R. A. Peterson, and W. C. Walter, "Seismic Energy Sources 1968 Handbook", Bendix/United Geophysical Corporation.
5. H. A. K. Edelmann, "Applications of Air Gun Energy Source for Offshore Seismic Work", OTC 2513:937 (1976).
6. N. N., "Flexichoc, A High Resolution Implosion Seismic Source", Brochures from Geomecanique, Rueil, France (1976).
7. W. S. French and C. G. Henson, "Signature Measurements on the Water Gun Marine Seismic Source", OTC 3124:631 (1978).
8. H. F. Weichart, "Acoustic Waves along oilfilled Streamer Cables", Geophysical Prospecting, 21(2):281 (1973).
9. H. Rehmert, "Better Quality in Integrated Navigation by Optimization of Hard- and Software", Preprint, 49th SEG Meeting (1979).
10. T. Krey, "Modern Aspects in Exploration Seismics", PRAKLA-SEISMOS Report 3:3 (1974).
11. R. Bading, "3D-Seismics", PRAKLA-SEISMOS Report 3+4:11 (1979).
12. H. Beckmann, "Erkundung mariner Lagerstätten mit Bodenschleppsonden", Glückauf-Forschungshefte 38(1):34 (1977).
13. H. Drenkelfort, "Sediment Echolote", Information 1/80, Honeywell-Elac-Nautic, Kiel.

14. J. I. Myung and W. M. Sturdevant, "Introduction to the Three Dimensional Velocity Log", Birdwell Div./SSC, Tulsa/Oklahoma (1970).

15. K. V. Mackenzie, "Reflection of Sound from Coastal Bottoms", J. Acoust. Soc. Am. 32(2):221 (1960).

16. R. D. Tooley, T. W. Spencer, and H. F. Sagoci, "Reflection and Transmission of Plane Compressional Waves", Geophysics 30(4):552 (1965).

17. R. W. Faas, "Analysis of the Relationship between Acoustic Reflectivity and Sediment Porosity", Geophysics 34(4):546 (1969).

18. H. E. Morris, "Bottom-Reflection-Loss Model with a Velocity Gradient", J. Acoust. Soc. Am. 48(5):1198 (1970).

19. G. Mott, "Reflection and Refraction Coefficients at a Fluid-Solid Interface", J. Acoust. Soc. Am. 50(3):819 (1970).

20. S. N. Domenico, "Effect of Water Saturation on Seismic Reflectivity of Sand Reservoirs encased in Shale", Geophysics 39(6):759 (1974).

21. S. R. Rutherford and K. E. Hawker, "Effects of Density Gradients on Bottom Reflection Loss for a Class of Marine Sediments", J. Acoust. Soc. Am. 63(3):750 (1978).

22. K. E. Hawker, "Influence of Stoneley Waves on Plane-wave Reflection Coefficients: Characteristics of Bottom Reflection Loss", J. Acoust. Soc. Am. 64(2):548 (1978).

23. M. Schoenberg, "Nonparametric Estimation of the Ocean Bottom Reflection Coefficient", J. Acoust. Soc. Am. 64(4):1165 (1978).

24. F. Schirmer, "Eine Untersuchung akustischer Eigenschaften von Sedimenten der Nord- und Ostsee", Dissertation, University Hamburg (1971).

25. N. A. Morgan, "Physical Properties of Marine Sediments as related to Seismic Velocities", Geophysics 34(4):529 (1969).

26. E. L. Hamilton, "Sound Velocity-Density Relations in Sea-Floor Sediments and Rocks", J. Acoust. Soc. Am. 63(2):366 (1978).

27. F. N. Tullos and A. C. Reid, "Seismic-Attenuation of Gulf Coast Sediments", Geophysics 34(4):516 (1969).

28. C. McCann and D. M. McCann, "The Attenuation of Compressional Waves in Marine Sediments", Geophysics 34(6):882 (1969).

29. E. L. Hamilton, "Compressional-wave Attenuation in Marine Sediments", Geophysics 37(4):620 (1972).

30. E. L. Hamilton, "Sound Attenuation as a Function of Depth in the Sea Floor", J. Acoust. Soc. Am. 59(3):528 (1976).

31. E. L. Hamilton, "Attenuation of Shear Waves in
     Marine Sediments", J. Acoust. Soc. Am. 60(2):334
     (1976).
32. N. N., "Fundamentals of High Resolution Seismic
     Profiling", The Environmental Equipment Div.
     EG & G, Waltham (Mass.) 1977.

THE ACOUSTIC RESPONSE OF SOME GAS-CHARGED SEDIMENTS IN THE NORTHERN ADRIATIC SEA

Antonio STEFANON

Istituto di Biologia del Mare
C.N.R.
Riva dei 7 Martiri, Venezia, Italy.

Abstract.

High-resolution profiles obtained in the northern Adriatic Sea with a Uniboom (EG&G) sub-bottom profiling system
revealed peculiar acoustic features down to a depth of
about 70 m below the sediment surface. Several different
structures due to high gas contents are described and explained for the first time in the northern Adriatic. They
range from the so-called "basin effect" and cone-like structures to acoustic voids. Gas seeping from the bottom
has also been observed. The abrupt disappearance of acoustically significant horizons is discussed, and a tentative
explanation is given. Possible migration of gases and
trapping effect are also discussed. The importance of high-
-quality, high-resolution recording is stressed in order
to detect the anomalies - due to gases - in the sea floor,
and to evaluate their potential as geologic hazards for
offshore installations.

Introduction.

High-resolution, shallow seismic profiles run concurrently with Side Scan Sonar and Echosounder revealed
in the northern Adriatic Sea a number of acoustic structures due to gas-charged sediments.

The presence of gas has already been detected (Colantoni
ni et al. 1978), but no specific structures have so far
been reported and described in the area. The present paper deals with the gas structures we observed, and an attempt is made to explain the most interesting ones.
The results were obtained as part of a research program
of the Progetto Finalizzato Oceanografia e Fondi Marini
of the Italian National Research Council.

Methods and instrumentations.

High-resolution profiles have been made with a Side
Scan Sonar (EG&G, mod. 259-4), a high-resolution sound
source (Uniboom, EG&G, mod. 230-1 and mod. 234) coupled
with an eight element hydrophone (EG&G, mod. 265) run
concurrently with an Elac echosounder equipped with a
"sediment amplifier". EPC 3200 and Ocean-Sonics GDR-T
graphic recorders, with standard Krohn-Hite filters, H.P.
amplifiers, EPC programmer, T.V.G. and delay unit have
been used to produce the Uniboom records. In parallel the
unprocessed signals have been recorded on a Nakamichi cas-
sette recorder for subsequent playback and elaboration.
Special care has been taken in the towing techniques,
especially in regards to the relative position of the
hydrophone to the sound source and the sea surface. Thus,
single line echoes have been obtained in most cases, with
little or no ringing.
Uniboom records only are hereby presented; corrected bottom and penetration depths have been calculated on a poc-
ket-size T.I. 59 calculator using a special program (Boldrin et al. in preparation).

Results.

Gas-charged sediments have been observed in most
areas of the northern Adriatic, reducing drastically in
some cases the sound penetration.
Several different acoustic features have been observed,
some of which are difficult to interpret, especially
within the acoustic void type. Pock mark like features
have also probably been encountered. Further research is
needed to clarify the most peculiar images, and to map
the areal distribution of the gas-induced features.

Fig. 1 shows the investigated area and the locations where the features hereby presented have been recorded.

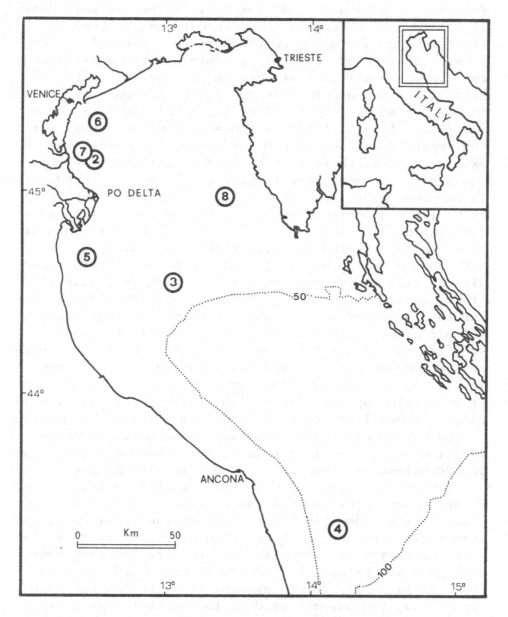

Fig. 1. Index map of the northern Adriatic Sea. The cir-
         cled numbers refer to the position of the illus-
         trated features and their text reference.

Definite, strong "basin effect", similar to the classic
example of the Baltic Sea has been encountered in a few
locations only, inside the Lagoon of Venice and on the
flanks of the Po delta. As the basin effect is well known
it will not be described here. In our area it seem to
start very close to the bottom surface, where at the most
a meter of layered sediment is visible above gas-charged
ones in which all real structures are masked. In the
Baltic, the basin effect is due to the presence of methane,
in bubbles mostly, up to 5% of the total sediment volume
(Whiticar, 1978). It has also been calculated that, at
the upper interface between gas-charged sediments and the
oxidation zone, there is a seepage of methane of about
$0.8 \ g/m^2/y$. No measurements have so far been carried out
in our area, but from the acoustic imagery we doubt that
such high gas content can be reached in the northen Adriatic. According to our records, the gas appears in most
cases to be diffused in well defined bodies, which in
some instances can be similar to the basin effect, where
all the structures are masked by the high reflectivity
of the whole mass of the gassy sediments (fig. 2). Single
strata charged with gas - where a phase reversal seems
to take place - are usually associated with the sides of
such gas bodies. Another type of gas accumulation, typical
of the deepest parts of the investigated area, is shown
in fig. 3. In such case the gas seems to be concentrated
at a certain horizon, from which it locally escapes, building up cloud-like features. Their boundaries are usually
marked by strong echoes, sometimes hyperbolic in shape,
reminiscent in some instances of the appearance of so-called "cobblestone topography". Acoustic windows are usually
not present in these bodies, and bedding is difficult to
detect in or below them. A phase reversal seems to take
place at the boundaries of the gas-charged sediments, and
some reflectors appear to be enhanced in the first multiple. Other similar acoustic features are scattere over
a large area, between 50 and 100 meter depth. In these
cases (fig. 4) it is more evident that the gas has come
from below, and has migrated up to somewhat less permeable
horizon, where accumulation occurred. The ringings (see
arrow on fig. 4) confirm the presence of a gas-charged
zone.

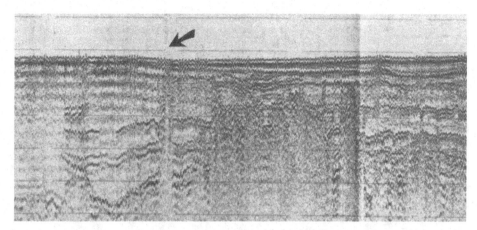

Fig. 2. Typical gas concentration off Chioggia. Note the
        high reflectivity of the gas-charged sediments up
        to the centre of the structure, and how some str-
        ong acoustic reflectors are truncated abruptly on
        the left. A phase reversal seems to take place at
        these reflectors. The arrow indicates a probable
        gas vent. Scale lines are interrupted every third
        of a mile, and spaced two fathoms apart.

Pocket-like, gas-charged features of different form can
be found at various locations and depths within the investigated area. The acoustic images suggest that they are
of two main types. The first seems to be related to an
accumulation of locally generated gas, where there is
little to suggest migration from below. In several cases
there is a resemblance to a buried channel, but closer
examination reveals (as in fig. 5) that the feature is due
to the presence of gas, due also to the pull-down effect
visible at the sides of the gas pocket and on the reflectors just below, as the sound velocity decrease within
the gassy sediments. Since buried channels are very common all over the northern Adriatic, great care must be
taken in evaluating such features.
The second type is related mostly to the cone-like structures described by Sieck (1973). In our examples some of
the cones have a high internal reflection coefficient,
while others show instead the classic acoustic void.

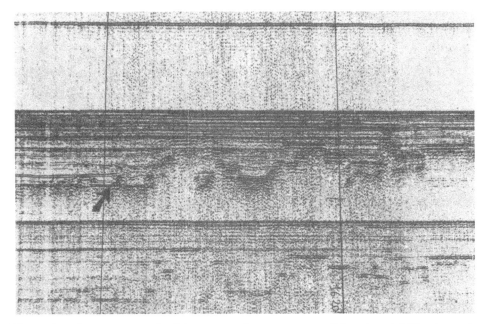

Fig. 3. Gas accumulations under very well stratified sed-
        iments. Note how the boundary of the gas-charged
        sediments are marked by very strong reflectors,
        which are somewhat locally enhanced in the first
        multiple. The rounded surfaces seem to be formed
        by single, small but strong, irregular reflectors
        as a sequence of "bright spots". Half a mile bet-
        ween the two markers, water depth 40.8 meters,
        depth of the reflector at the arrow 28.3 meters
        at a sound velocity of 1500 m/s.

The Uniboom record of fig. 6, and some Side Scan Sonar
images not presented here, clearly show that most of the
cones vent if they reach the surface. In one location bubbling through the sediment surface has been observed
during diving. Another vented gas-charged sediment cone
is shown on fig. 7. In this case some strata seem to be
displaced, and this can be explained as a feature due to
the release of pressure of gas moving upward through the
sediment column to vent.

A large area off Venice is characterized by the presence of deep, significant and very strong acoustic

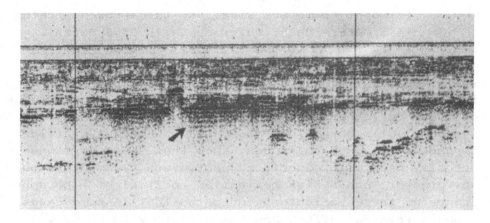

Fig. 4. Typical image of gas migration and accumulation
in the southern part of the investigated area.
Surface unstratified sediments belonging to present Po river sedimentation are 4.3 meters thick,
and overlie an erosional surface. Gas migration
and accumulation below a certain horizon is evident, and the arrowed ringings confirm that such
echoes are due to gas. 0.8 miles between markers.

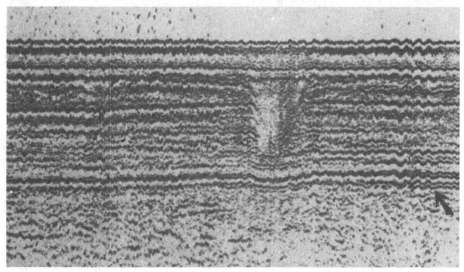

Fig. 5. Gas pocket within well stratified sediments. Note
within and below the gas pocket the "pull-down"
effect on the strata, due to the lower sound velocity in the gas-charged sediments. The gas pocket

is about 65 meters wide, and the arrowed stratum
14.5 meters deep. Water depth is 25.8 meters.

reflectors, which on the Uniboom records appear to be
frequently interrupted. No pull down effect is detectable
on other reflectors crossing below such voids. Some stratigraphic wells bored in Venice, and connecting profiles
through the lagoon entrances, show that these horizons
are clay beds with associated peat. Such sediments should
not present such a high reflection coefficient. The appearance of these reflectors indicates that in some cases
a phase reversal effect has taken place within them.
Typical Uniboom images of these phenomena are shown on
fig. 7 and fig. 8. As a working hypothesis the writer

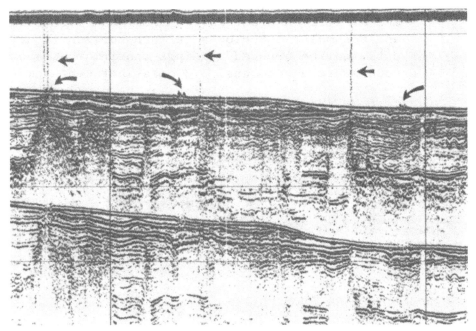

Fig. 6. Venting of gas-charged sediment cones through
        the sea bottom at a depth of about 16 meters.
        Cone structures are clearly visible and straight
        arrows indicate their bubbling. The bottom feat-
        ures indicated by the bent arrows are mounds of
        muds and clays dredged in the Venice lagoon and
        dumped offshore. A mile between the markers.

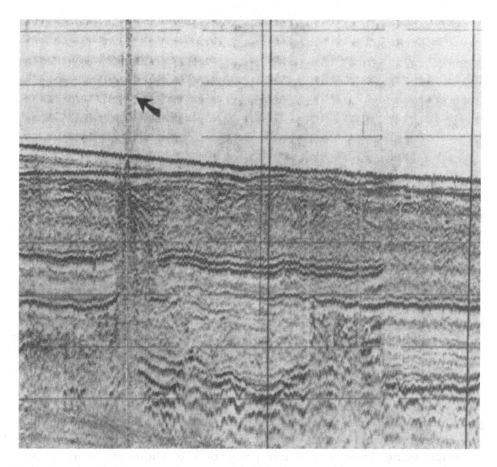

Fig. 7. Vented structure located on the flanks of the Po
delta. The strata seem to have been pushed up by
the releasing gas pressure. The arrow indicates
the bubbling. Scale lines are interrupted every
third of a mile, and spaced every two fathoms.
Note the "bright spot" effect and the phase reversal on some reflectors.

thinks that peat layers, even when very thin, produce some
gas, which becomes confined within the peat by the impermeable clays. The peat horizons therefore have an impedance
distinctly lower than the surrounding sediments, and the
double phase reversal which takes place as the sound imp-
pulse passes through, enhance and reveal a stratum which
otherwise could be beyond the resolution limit of the

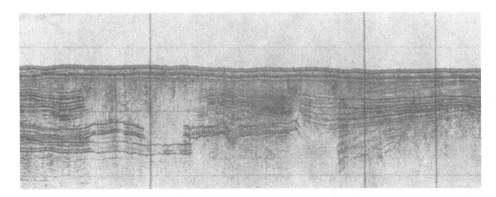

Fig. 8. A sequence of very well stratified sediments,
        showing some abrupt horizontal interruptions of
        the beds sequence. The high reflectivity - espec-
        ialy of the deep horizons - is here interpreted
        to be due to the gas content of very thin strata
        which, having a low impedance, are acoustically
        enhanced by a double phase reversal effect. The
        strata seem to disappear when this gas content is
        lower. Scale line at 1/80 of a second interval,
        0.8 miles between the two central markers.

profiling system. Such enhancement probably take place in
a variety of different stratigraphic situations, and could
be much more frequent than previously thought. The abrupt
disappearance of some strata therefore should not neces-
sarly be interpreted to be a result of an acoustic void,
but may instead indicate the lack of an enhancing effect
of gas in very thin layers, acting as a sequence of bright
spots.
The gas-peat-clay relationship described above is most
probably responsable for the interruptions of the strata
on fig. 7. Fig. 8 shows a different stratigraphic situation, in which the same effect could be produced simply by
an alternation of strata with high and lower organic (and
gas) content.

Conclusions.

        The investigated area has shown an unespected number
of different gas-induced acoustic structures. Some of them

probably represent a geological hazard to offshore drilling and construction because - as it is well known - gas-charged sediments always have peculiar geotechnical characteristics.

Apart from the case in which the gas seems to be concentrated at the site of formation -as shown in fig. 5 - evidence has been presented of gas migration through the sediments and accumulation below and within less permeable horizons relatively close to the surface. While the origin of these gases is not clear, it is the opinion of the writer that in most cases they are generated in large organic-rich layers, which have probably been deposited during and between the last couple of glaciations. The acoustic imagery of these migrating gases and the enhancement of the strata in which they are accumulating, suggest the hypothesis given on the previous pages. The interpretation of the peat-clay example off Venice is certainly valid only in such local conditions, but helps to explain a phenomenon that, if the hypothesis is correct, has been misunderstood in several cases.

Evidence has been given that only very high resolution profiling systems and properly tuned equipment can produce records suitable for the definition and interpretation of different structures, such the ones hereby presented. Long cores are needed to proceed with the research.

Acknowledgments.

The work could not have been done without the great help of colleagues of the marine geology group at the Istituto di Biologia del Mare, C.N.R., Venice. Special thanks to the Captains and Crews of the R.V. BANNOCK, D'ANCONA and MARSILI of the C.N.R.. Excellent suggestions, remarks and positive criticism have been given by several American colleagues, especially: H. Berryhillut and L. Garrison (U.S.G.S., Corpus Christi), G. Green (U.S.G.S. Menlo Park), D. Prior (L.S.U., Baton Rouge).

Selected references.

Boldrin, A., Rabitti, S, and Stefanon, A., in preparation,
     Program for the T.I. 59 calculator to obtain

      corrected water and strata depths from Uniboom and
      sparker records.
Colantoni, P., and Gallignani, P., 1978, Ricerche sulla
      piattaforma continentale dell'Alto Adriatico.
      Quaderno n° 1, Consiglic Nazionale delle Ricerche,
      Roma.
Payton, C., E., 1977,"Seismic stratigraphy - applications
      to hydrocarbon exploration", The American Associat-
      ion of Petroleum Geologists, Tulsa.
Sieck, H.,C., 1973, Gas-charged sediment cones post pos-
      sible hazards to offshore drilling, Oil and Gas J.
      July 16.
Whiticar, M. J., 1978, Relationships of interstitial gases
      and fluids during early diagenesis in some marine
      sediments, "Report S. F. B. 95" Kiel Universyty,
      Kiel.

SIMULTANEOUS APPLICATION OF REFLECTION STRENGTH RECORDER, SIDESCAN

SONAR AND SUB-BOTTOM PROFILER IN SEAFLOOR SEDIMENT MAPPING

K. Winn, F.C. Kögler and F. Werner

Geologisch-Paläontologisches Institut der Universität
Kiel, Germany

ABSTRACT

The results of mapping the seafloor sediment distribution by
simultaneous application of the following acoustical methods:
measuring reflection intensities from the echosounder returns,
sidescan sonar and sub-bottom profiling, supplemented by surface
samples and cores, are compared. The reflection intensities are
integrated from the echoes of the top 75 cm of sediment. The instrument is very sensitive to slight changes in the composition
of muds and clearly defines the boundaries between muddy sediments
and sands. However, the reflection intensities do not bring out
all the heterogenities of the sediment distribution in sandy areas.

The sidescan sonar records with their higher lateral coverage
and resolution in sandy-rocky terrain provide data on the composition, smaller morphological features and areal distribution;
clay and mud generally show no returns. Interpreted together with
the reflection intensities, the sediment distribution in the area
as well as their relations to the physical parameters of the sediments can be deduced. The sub-bottom profiler (Uniboom) has better
resolution in harder sediments than the echosounder and furnishes
the spatial distribution of the lithological units, permitting
further interpretation of provenance, tectonics and the geological
causes for this distribution.

INTRODUCTION

Previous attempts to map the surface sediment distribution in
marine areas using conventional grab, echosounder and sidescan
sonar, has indicated the need to develop a survey system that will

provide practically all the physical parameters required for a
complete geological interpretation. The system should yield data
not only on the sediment type but also on its areal extent, the
structure of the sub-bottom and its influence on the distribution
and the effects of the various hydrodynamical processes on the
sediments. The need for such a survey system was particularly felt
in the Baltic Sea where the distribution patterns are complex due
to the inhomogeniety of the sediments (Seibold et al., 1971). A
combined survey system based on acoustics was therefore tested by
the marine geology group of the Geological Institute of Kiel University. The system consists of an echosounder 18 kHz fitted with
a "sediment amplifier" and a device for measuring the reflection
intensity (Honeywell-Elac, Kiel), a sidescan sonar 1o5 $\pm$ 1o kHz
(EG & G, U.S.A.) and a sub-bottom profiler (EG & G, U.S.A.). The
sidescan sonar gives the areal coverage and physical characters
of the surface sediment features, while the echosounder reflection
intensities provide information on the physical parameters of the
sediments. The echosounder and the sub-bottom profiler furnish the
morphology of the seafloor and the sub-bottom structural configuration.

     This paper evaluates the initial testing results in two areas
in the Western Baltic Sea where the combined survey system has been
applied. The records from the various instruments have been compared and interpreted with the aim of mapping the surface sediment
distribution and sub-bottom structure for geological interpretation.
Cores and grab samples were also analysed and a few underwater
television profiles have also been run to aid in the interpretation.

SYSTEM DESCRIPTION

     The echosounder unit consists of an echosounder 18 kHz fitted
with a sediment amplifier (Hinz et al., 1969) and an additional
device for measuring the intensities of the echoes (EMG). The
transducer, which can either be mounted in the ship's belly or
towed beside the ship, can send out impulses with durations of
o.6, 3 or 1o milliseconds and pulse power of o.5 or 5.o KW. The
sound cone is transmitted with an angle of 13°. Echoes with a two
way travel time of 1ms in sediment are integrated and presented on
a scale of 0 to 1o volts. The echosounder gives the water depth,
the sea bottom morphology, and in soft sediments, reflections with
a resolution of about 5o cm of the sub-bottom with a penetration
of approximately 1o m. In coarser sediments, resolution is poor
since most of the energy is reflected at the sediment-water interface. Muds and clays have very low reflection intensities because
of their lower attenuation. A transition to sandy muds or fine
sands is indicated by a "jump" in the intensity curve. Hamilton
(1972) found that the jump in his case occurs within the mean
grain size range between 2o and 4o microns, but with further

increase in grain size, the reflection intensities again decrease. Faas (1969) reported similar relationships within the sand fraction.

The sidescan sonar 1o5 + 1o kHz consists of a fish which is towed behind the ship (2o to 6o m in shallow waters) and a dual channel recorder. The two transducers mounted on the sides of the towfish send out fan shaped sonic beams and the backscatter from the seafloor is then processed by the recorder and presented as plan view images, somewhat analogous to aerial photographs. Depending upon the scanning range selected and the ship's speed, the images are distorted and have to be corrected to their natural perspective. The resolution of the sidescan also depends upon the range, speed, object shape and its distance from the fish. Usually with 5 knots speed and 9o m lateral range per side, objects 3o cm or larger are readily identifiable. Most of the incident sonar energy is reflected away by muddy sediments or by smooth bottoms and there is little backscatter. Consequently, the record is very light. In coarser sediments, because of higher roughness, a greater percentage of the incident energy is backscattered, resulting in a darker record. Changes in the grazing angle of the sonic beam due to variations in the seafloor can secondarily effect the record density. Since the backscatter effect for any given material is constant, sediments with similar physical properties give similar images and thus the nature and extent of the sediment bodies can be readily mapped. For interpretation of smaller objects and patterns on the seafloor, a higher resolution sidescan 5oo kHz (Klein, U.S.A.) has been additionally set out.

The sub-bottom profiler (uniboom) consists of a broad spectrum sound source (o.5 to 14 kHz) mounted on a catamaran and towed just behind the ship, hydrophones and a graphic recorder. Either 2oo or 3oo Joules can be given per shot. The average main frequency is usually between 8oo – 12oo Hz giving a resolution between 5o cm and 1 meter, but depending upon the sediment type and succession, resolution up to 3o cm was sometimes attained. Because of its lower frequency compared to the echosounder, more reflections are registered from the sub-bottom in coarser and harder sediment areas.

Examples of sidescan sonograph, echogram with reflection intensity curve and uniboom record are shown in fig. 1. The sidescan sonograph displays the lateral extent, sediment patterns and types in the sandy-rocky terrain but shows no returns from the very fine to muddy sediments. It however, discloses details of the boundary between the sands and mud which is not very clear from either the echogram or uniboom records, since the trace of the glacial till – Holocene mud boundary in the last 5o cm before it reaches the surface lies beyond the resolution of these instruments.

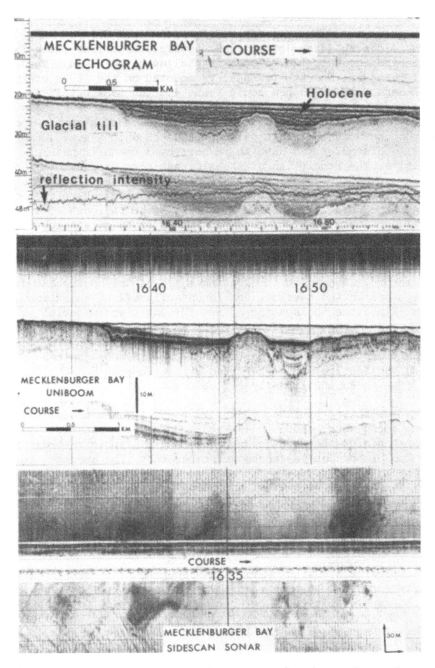

Fig. 1. Examples of echogram with reflection intensity, side-
      scan sonar and uniboom.

The echogram shows the seabottom configuration, good resolution in
the soft Holocene sediments down to the boundary with the glacial
deposits (boulder clay and outwash sands), but almost no internal
reflections from the glacial deposits themselves. The uniboom record
is similar to the echogram and provides additional data on the
structure of the glacial underground. Other high resolution seismic
devices such as tuned transducer systems with frequencies between
3.5 and 7 kHz were also tested in our area. They did not by far
yield such good results in these glacial sediments as the uniboom
system.

TESTING RESULTS

The practical applicability of the combined system was tested
in two areas of the Western Baltic. The first area lies in the
northwestern part of the Kiel Bay, south of the Danish island of
Aeroe, known as Vejsnaes Flach and Vejsnaes Channel. The second
test area is situated to the east of the Oldenburg Peninsula in
the north-western part of the Mecklenburger Bay referred here as
the Sagas Bank area. Both areas have Holocene sediments underlain
by Quaternary glacial deposits. The Vejsnaes area is swept by the
inflowing und outflowing currents of the Baltic through the Belt
seas. It is more susceptible to waves caused by east winds because
of the longer easterly fetch. In the Sagas Bank area, waves caused
by westerly winds have negligible influence while those from strong
east winds affects the sediments in the shallower parts of the
test area, but there are no strong currents in the area.

The Vejsnaes Flach rise from about 2o m to less than 8 m at
its shallowest part. It is bounded by channels both to the north
and the south. Large plateaus with a thin veneer of Holocene sediments are present in the southern part of the area (fig. 2).
Surveys were carried out in late 1977 and in 1979 on a rectangular
grid of 2.5 km spacing. Because of the complexity of the sediment
distribution on the Vejsnaes shoal, an additional oblique angle
grid with o.5 km spacing was laid in 198o.

The sediment echostrength contours (fig. 3) were drawn only
on the reflection intensities of the sediments measured with the
EMG without using sidescan controls. The extent of the muddy sediments and their boundary with the sandy deposits are well defined. The slightly higher values of the reflection intensities
in the channels are due to the presence of a small percentage of
coarser materials in the muds, and partly to the shallow position
of the glacial deposits below the Holocene sediments.

The boundary between the sands and muddy sediments is controlled by the bathymetry and the hydrodynamic conditions in the
area and lies between 2o m and 22 m. This boundary is indicated

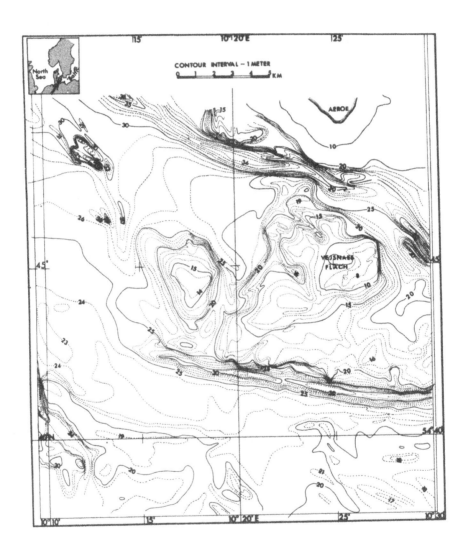

Fig. 2 Bathymetry NW Kiel Bay (Vejsnaes area)

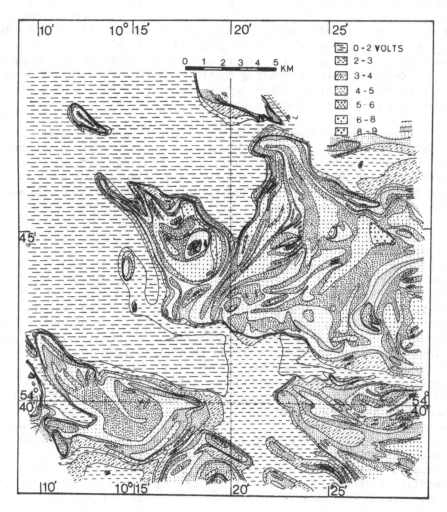

Fig. 3 Sediment echostrengths NW Kiel Bay

by a sharp rise in the reflection intensities from 3 to 5 volts.
In areas shallower than 2o meters, the influence of the underlying
glacial tills and meltwater deposits on the sediment distribution
increases and controls the type of sediment in combination with
the current regimes in the area. Two trends are shown by the echostrength contours; a northwest-southeast trend which is also displayed by the submarine topography of the areas adjacent to the
shallowest part of the Vejsnaes shoal, and a northeast-southwest
trend which reflects the morphology of the main shoal. This is due
to the strict dependance of the grain size distribution on the
bathymetry. Seibold et al. (1971) have shown that the median grain
sizes decrease with increasing depth in the Kiel Bay. Exceptions
to this rule are areas where lag sediment covers the seafloor or
where strong current regimes prevail.

     The surface sediment distribution map interpreted from the
sidescan sonographs presented in fig. 4. shows a very good correlation with the sand-mud boundary of the EMG. However, the sediment trends disclosed by the sidescan are neither expressed in
the bathymetry nor in the sediment echostrength contours. Because
the distribution patterns are very complex, a sidescan coverage
of 8o% was necessary on the Vejsnaes Flach before the sediment
distribution map could be drawn. On the adjacent plateaus, the
coverage is 3o% to 5o%. The nature, type and extent of the sand-
gravel bodies are controlled by the availability of sand and by
the composition and structure of the underlying glacial deposit.
Thus the northeast-southwest trend of the echostrength contours
could be explained by the abundance of boulders along this trend,
although the actual surface sedimentary trends are dominantly WNW-
ESE with a subsidiary north-south trend. The areas with lag sediment cover are very well defined on the sidescan sonographs.
These are related to the shallow position of the underlying
glacial tills and are generally connected with the sub-bottom
structure. Such lag sediment covers are not restricted to shallow
areas but are also present in deeper waters. On the adjacent
plateaus south and west of the Vejsnaes Flach, the lag sediments
are partially covered by a thin layer of fine to medium grained
sands. Nevertheless, the trends of the boulders are still
traceable. In the southwestern area, the NNE-SSW trends form
parallel patterns. These directions are again not revealed by the
echostrength contours which tend to follow the bathymetry.

     The second test area in the western part of the Mecklenburger
Bay east   of the Oldenburg Peninsula was surveyed in 1979 and
in 198o using a 35o meter spaced, oblique angle grid of the south
eastern part and a larger 1.5 km grid in the rest of the area. The
sidescan coverage was theoretically 1oo% in the southeast. The
bathymetric map prepared from the echosounder records shows that
the shallowest portion of the area (fig. 5) has less than 9 m and
the deeper areas in the east over 21 m water depth. Using the

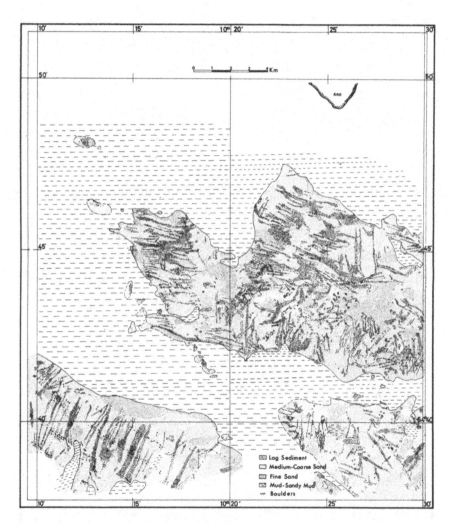

Fig. 4   Surface sediment distribution map NW Kiel Bay

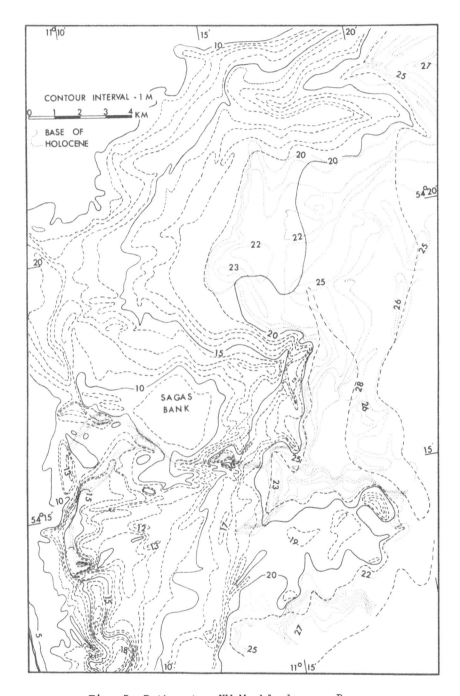

Fig. 5 Bathymetry NW Mecklenburger Bay

uniboom results additionally, the base of the Holocene could also
be contoured in fig. 5. This horizon forms a distinct reflector
in water depths over 2o meters. In the shallower portions, it is
generally not traceable because the Holocene sediments are either
very thin or the sediment cover is so coarse that most of the
energy is reflected at the surface. A few deep channels filled
with sandy sediments are observed on the uniboom records. However,
the channels are too narrow and deep to permit accurate presentation on the present scale. These channels are the seaward extensions of the Oldenburg Sound which had separated the northern
part of the peninsula from the mainland till the late Holocene.

In fig. 6, the sediment echostrengths have been contoured
using the sediment trends of the sidescan as a controlling factor
in correlating between profiles, in contrast to the Vejsnaes area
where the echostrengths are independantly contoured. The results
indicated that the sediment echostrengths are correlable with the
sediment boundaries in most cases, in spite of the fact that the
EMG integrates one millisecond two way travel time (approx. 75 cm)
of the uppermost sediment cover. Similar to the Vejsnaes Flach,
the boundary between sand-mud is very well defined. Since most
of the southern part of the test area has water depths close to
this boundary which lies between 18 and 22 m, the sediment echostrengths are particularly useful in defining areas with very
fine muddy sands and sandy muds respectively, as these sediments
have little backscattering effect and therefore do not yield characteristic patterns on the sidescan records.

The surface sediment distribution map shown in fig. 7 was
prepared using the sidescan sonar records for the sandy/rocky
areas and by comparing the sidescan with the echograms and the
uniboom for the muddy areas. There are some deviations in the
form of the sediment bodies between this map and the sediment
echostrength contours, especially close to the sand-mud boundary,
but this is expected because the EMG values are confined to the
line of profile while the sidescan delivers plan view images.
The distribution patterns are again found to be governed by the
underlying glacial deposits. With the exception of local fills
and banks, the sand bodies are thin. Areas covered by boulders
are narrow but can be traced laterally over large distances, but
the smaller number of large boulders in the area may also be due
to non-geological causes, since there is active dredging for
boulders in the area. The trends of the sediment distribution
patterns in the area are predominantly ENE-WSW except towards the
western part of the area, where some northwest-southeast
lineations interfere. The main trend reflects the structure of
the glacial moraines which are related to general glacial advances and retreats during the last glaciation (Kolp, 1965). However, the detailed interpretation shows some deviations from the
regional trend.

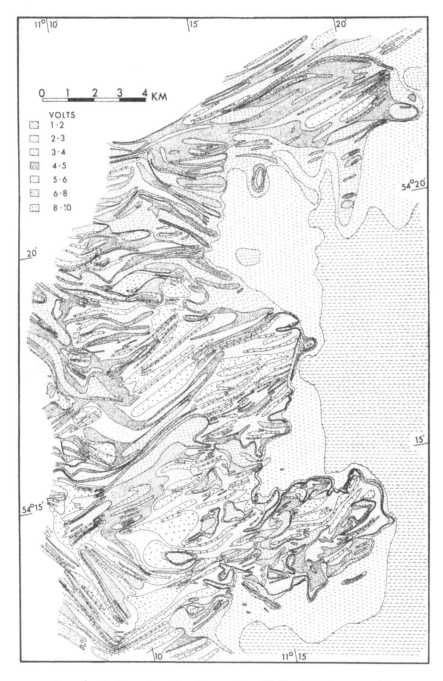

Fig. 6   Sediment echostrengths NW Mecklenburger Bay

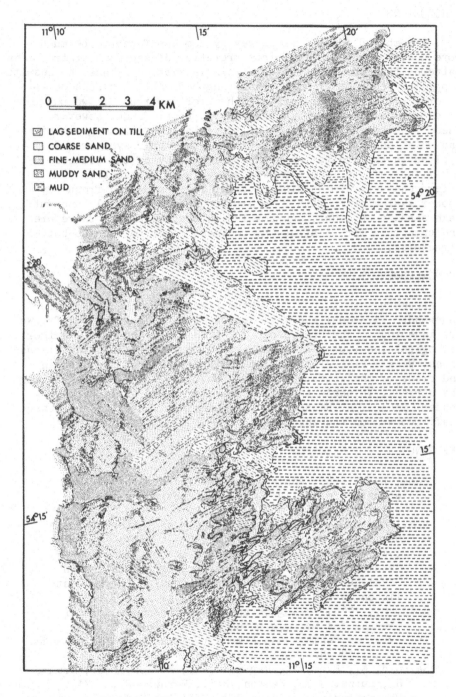

Fig. 7 Sediment distribution NW Mecklenburger Bay

CONCLUSIONS

A combined acoustical survey system was found to be much
more useful for geological interpretation of sediment distribution
patterns than the application of single systems, since the results
supplement each other. The echostrength signals are related to
several sediment parameters and therefore ambiguous. Two examples
of close-spaced surveys in the Western Baltic show, however,
these results can reflect detailed variations in the grain-size
and relationships of the sediment, if the other systems are
applied in the evaluation. The applicability of the sidescan
sonar is confined to areas with appreciable backscatter and
heterogeneous sediment distribution. The broad-band frequency
boomer system gives better resolution in glacial tills and outwash sands than other high-resolution systems and is therefore
indispensable for relating the sediment cover to the sub-bottom
structures.

ACKNOWLEDGEMENTS

Research was supported by the Fraunhofer Gesellschaft für
Angewandte Forschung and the Deutsche Forschungsgemeinschaft. We
are indebted to the Professors Dr. E. Seibold and Dr. E. Walger,
Geological Institute, Kiel University for the stimulating discussions. We are also grateful for the assistance of the captains
and crews of the research vessels "Poseidon" and "Planet" during
the surveys. Valuable technical assistance was given by Mrs. W.
Rehder, Messrs. H. Hensel, H. Langmaack, W. Jungeblut and other
staff members of the institute.

REFERENCES

Faas, R.W., 1969, Analysis of the relationship between
    acoustic reflectivity and sediment porosity, Geo-
    physics, 34: 546-553.
Hamilton, E.L., 1972, Compressional-wave attenuation in
    marine sediments, Geophysics, 37: 620-646.
Hinz, K., Kögler, F.C., Seibold, E., 1969, Reflexions-
    seismische Untersuchungen mit einer pneumatischen
    Schallquelle und einem Sedimentecholot in der West-
    lichen Ostsee, Meyniana, 19: 91-1o2.
Kolp, O., 1965, Paläogeographische Ergebnisse der Kar-
    tierung des Meeresgrundes der Westlichen Ostsee
    zwischen Fehmarn und Arkona, Beiträge zur Meereskunde,
    12-14: 19-59.
Seibold, E., Exon, N., Hartmann, M., Kögler, F.C., Krumm,
    H., Lutze, G.F., Newton, R.S., Werner, F., 1971,
    Marine geology of Kiel Bay, VIII Int.Sed.Congr.,
    Guidebook: 2o9-235.

RESULTS AND METHODS USED TO DETERMINE THE ACOUSTIC PROPERTIES OF

THE SOUTHEAST ASIAN MARGINS

Robert E. Houtz

Lamont-Doherty Geological Observatory
of Columbia University
Palisades, New York 10964

ABSTRACT

More than 850 new sonobuoy sound velocity solutions and some published solutions have been analysed statistically to develop 16 regional velocity functions from the Southeast Asian shelves and marginal basins. The velocity functions are least-squares regressions of the form $V = V_0 + Kt$, where t is one-way vertical travel-time to layer mid-points. The value of K shows surprisingly little variability on the shelves and adjoining basins from the Bay of Bengal to the Japan Sea. The average value is $1.86 \pm .30$ km/s^2. Very large values of 3.2 km/s^2 from Western Australia (not included in the average) are unique, and may result from the special combination of very low sedimentation rates on quite old oceanic crust. Paired comparisons of interval velocities and refraction velocities from the same sonobuoys show that the two measurements are interchangeable if they are compared at equal depths at layer mid-points. More than 30 high-resolution velocity inversions from the Southeast Asian shelf data have been compiled to yield 5 average seafloor velocity gradients. Gradients predicted from statistically derived functions, based on refraction solutions from relatively thick layers, are not in satisfactory agreement with these averaged values. These differences can be reconciled by picking thinner layers or by using greater (and more realistic) seafloor sound velocities than those obtained as intercepts from the statistical regressions.

Lamont-Doherty Geological Observatory of Columbia University, Palisades, New York, 10964, Contribution #2633

INTRODUCTION

The velocity characteristics of the Southeast Asian margin from
the Bay of Bengal to the Sea of Japan are rather uniform and therefore lend themselves to regional generalizations. The area is particularly favorable for measuring near-surface velocity gradients from
high-resolution velocity inversion techniques. This results from
the fact that in many of the depositional centers the accumulation
rates are quite rapid, which produces a smooth increase of velocity
with depth (Houtz, 1978). Smooth velocity increases without velocity
cusps or low velocity zones can therefore be inverted with excellent
precision. For this reason the South China shelf, the East China
basin, the Irawaddi delta south of Rangoon, and the Sarawak and
Natuna basins on the Sunda shelf are ideal locations to study and
compare velocity-depth profiles.

Sonobuoy records from the East China Basin, and to a lesser
extent from the Sunda and South China shelves, contain sub-bottom
reflection curves as well as critically refracted arrivals. This
kind of information can be used to test the interchangeability of
refraction and interval velocity solutions. Since there is no
theoretical reason why the two types of solution should not be in
agreement, a lack of agreement could indicate the widespread occurrence of undetected velocity discontinuities. At present our disk
files contain 14,000 solutions, about half of which are interval
velocities from sediment. If it can be demonstrated that refraction velocities from the sedimentary layers can be safely included,
our file of reliable sedimentary velocity solutions would increase
to about 10,000.

DATA REDUCTION METHODS

Velocity functions of the form $V = V_0 + K t$, where t (sec) is
the one-way travel time to the layer mid-points, are determined by
least-squares from sonobuoy solutions of layer velocities from a
geographical region of presumed uniformity. These regressions of
layer velocities and t provide a slope $K(km/s^2)$, an intercept V_0
(km/s), the standard error of estimate (km/s), and the correlation
coefficient. Variance ratio tests (using the standard error of
estimate) can be performed on the regressions to make decisions
about sub-dividing a given region or whether to combine adjacent
regions. Similarly, these procedures can be used to test if one
velocity function fits the data better than another. Houtz (1974)
has reported that polynomials of order higher than one rarely
decrease the standard error of estimate significantly. In the
present work, it was found repeatedly that functions of form
$V = V_0 + A h$, where h (km) is the depth to a particular velocity,
do not improve the fit to the data. In fact, the correlation
coefficient and the standard error of estimate actually degrade
somewhat when the velocities are assumed to be linear functions of

depth.

Velocity functions from three South American basins (Houtz, 1977) were obtained from critical refraction solutions. In these same basins, interval velocity solutions were also obtained, but not from the same sonobuoys. The functions based on the two methods have been compared, and it can be shown that there is no significant difference between them. However, since the basins may have been poorly sampled, there is no assurance that the agreement does not arise from fortuitous sampling. In order to avoid sampling error, it is appropriate to make paired comparisons of the two methods from the same sonobuoy and at the same depth.

The method is demonstrated in Figure 1 where the events labelled R in the photocopy are the reflection curves that were used to compute the interval velocities, shown as open circles in the line drawing. The refraction events labelled G in the photocopy have been picked as short, straight-line segments, and their solutions are shown as small circles connected by lines in the photocopy. In both cases the solutions have been plotted at layer mid-points. The mismatch between the two techniques at a particular depth is shown as ΔV. The average mis-match, taking into account the correct sign of each ΔV, is .020 $\pm$ .23 km/s from 52 measurements. By use of the t statistic it can be shown that this value is different from zero at the 50% level of significance. It is quite clear that there is no significant difference between the two techniques when the solutions are plotted at layer mid-points. The credibility of this finding is enhanced by the use of paired comparisons, which are independent of sampling errors.

The agreement between the two independent measuring techniques does not necessarily indicate a lack of low velocity zones. This is because a hidden low velocity zone within a given layer yields an interval velocity that is less than the refraction velocity, but the delay in the intercept (resulting from the low velocity zone) gives a thicker layer when the refraction method is applied. Consequently the refraction solutions (see Fig. 1) would be displaced to the right and downwards. Computations with realistic models show that the introduction of low-velocity zones would have a negligible effect on velocity functions determined with refraction solutions compared to those from interval velocity solutions.

Accurate velocity inversions can be achieved wherever the velocity depth relations are smoothly varying. These areas are principally in depositional centers where terrigenous sediments have accumulated rapidly. The Herglotz-Bateman technique was found to give quite reliable results in these basins throughout the study area. The ray parameter was determined from the least-squares fit of a parabola to 5 points, which were retrieved with a 'sliding window' technique. The sampling interval varies with ship's speed, but is

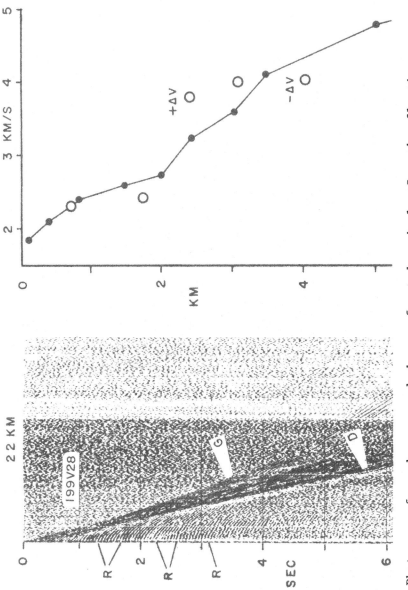

Fig. 1.   Photocopy of sonobuoy record shows refracted arrivals, G, and reflection curves, R.
Line drawing shows interval velocities, based on reflection curves, as open circles.
Refraction solutions connected with a line through velocity at layer mid-points. ΔV
represents the mis-match between refraction and interval velocity solutions.

about equal to a measurement over every 100 m of range.

REGIONAL VELOCITY FUNCTIONS

Least-squares solutions of the form $V = V_0 + Kt$ are listed in Table 1. The geographical areas designated with letters and identified in the table are located in Figure 2. The value of K is also shown for each region in the figure. The Southeast Asian data show a three-fold variation in K values from about 1 to 3 km/s^2, whereas V_0 is very stable with an average value of $1.59 \pm .07$ km/s. As a result, K is a better indicator, i.e. more sensitive, to the properties of the sediments. The standard error of estimate, the correlation coefficient, and the number of measurements are also shown in the table. All the correlations are significant at better than the 1% level. The error in K varies from .1 to .4 km/s^2, and a good average value can be taken as about .25 km/s^2.

The reliability of the regressions can be judged with the standard error of estimate, but some well-sampled sediment bodies with a uniform appearance show a lot of scatter as if it results not so much from poor sampling as an inherent property of that particular sediment body. Although it is much more difficult to assess, the sampling is probably the more crucial factor.

With the exception of Area L, which is based exclusively on data from two-ship refraction results (Curray et al., 1977), all the measurements are from airgun/sonobuoy data. A major part of the data from area B was taken from McDonald (1977). Most of the regional data sets required some editing, chiefly in areas where the consolidated and older high-velocity sediments are covered by lesser amounts of overburden; e.g., on the flanks of basins.

The principal fact that emerges from Figure 2 is the uniformity of K in a belt that extends from the Bay of Bengal to the Sea of Japan, exclusive of the region west of Australia. In this uniform belt the average value of K is $1.86 \pm .30$ km/s^2, which represents a spread that is about equal to the typical error on K (.25 km/s^2).

About 440 solutions are available from the South China shelf and rise. The region was divided into two parts on the basis of water depth: Area O represents the shelf with water depths less than 200 m and area P represents the rise between depths of 200 and 3000 m. The rather high values of K on the China shelf may extend northwards to the Yellow Sea and connect with the Korean Straits, which have similar values of K. However, in the Korean Straits the larger values may result from sampling volcanoclastic rocks with rather higher velocities than deltaic sediments at a similar depth of burial. Tertiary trachytic tuffs are exposed to the west on the coast of Korea (Um et al., 1964), and Ludwig et al. (1975) have interpreted a high-velocity layer in the Japan Sea as Miocene green

Table 1. Southeast Asian Velocity Functions
$$V = V_o + Kt$$

Area	V_o km/s	K km/s^2	σv km/s	r	n	Max Refl. Sec.	
A	1.64	1.61	.26	.91	80	3.1	
B*	1.66	1.41	.18	.88	32	1.8	McDonald (1977)
C	1.57	1.68	.20	.97	40	4.3	Houtz (1974)
D	1.58	1.87	.20	.96	56	3.0	
E	1.57	1.95	.17	.91	52	2.0	Ludwig & Houtz (1979)
F	1.63	1.45	.27	.79	18	1.8	
G	1.46	1.94	.21	.94	12	2.1	
H	1.59	1.52	.21	.92	46	3.0	Ludwig et al. (1979)
I	1.57	1.57	.22	.96	136	4.1	
J	1.61	2.19	.18	.97	36	2.2	
K	1.43	1.91	.20	.91	64	1.7	Ludwig et al. (1975)
L*	1.56	2.15	.22	.96	18	2.2	Curray et al. (1977)
M	1.50	3.21	.26	.83	21	1.0	
N	1.47	3.22	.25	.95	28	1.3	
O	1.62	2.37	.17	.98	204	3.1	
P	1.68	2.14	.21	.96	233	3.1	

*Regression based on data that includes the referenced source.
Otherwise the references indicate where the regressions were
previously published.

Area	Geographical Name
A	Southern Bengal Fan
B	Nicobar Fan
C	Bay of Bengal
D	Irawaddi Delta
E	Sunda Shelf
F	Sulu Sea
G	Celebes Sea
H	South China Basin
I	East China Shelf
J	Korean Straits
K	Japan Sea
L	Sunda Trench
M	West of Australia > 4 s
N	West of Australia < 4 s
O	South China shelf
P	South China Continental Rise

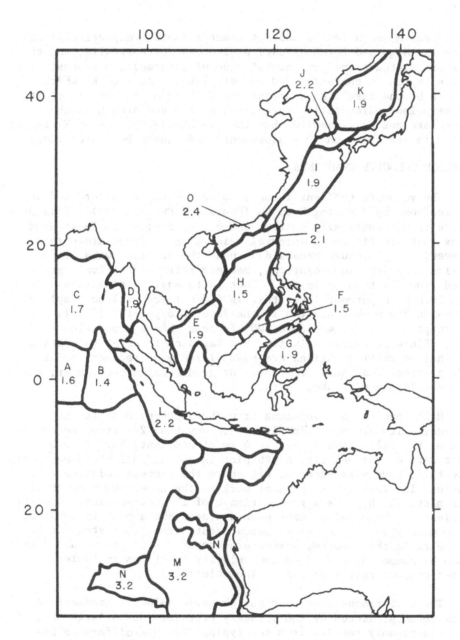

Fig. 2 Regions whose sound velocity functions are listed in Table 1.
The value of K (km/s^2) is shown for each region. Except for
unusually high values west of Australia, the value of K
varies little in the general region.

tuffs.

Areas M and N (> 4 s and < 4 s water depth, respectively) include the Argo and Perth abyssal plains west of Australia, and the
offshore plateaus and continental rise of intermediate depths. The
entire region is characterized by very large values of K, which
seem to be the result of a unique combination of rather old crust
and very low sedimentation rates. This area and area L in the Sunda
trench are not well sampled, but the two similar values of K west of
Australia are based on 49 measurements and cannot be disregarded.

SEAFLOOR VELOCITY GRADIENTS

The velocity gradient in units of sec^{-1} can be determined at
the seafloor by dividing K by V_0 (Houtz and Ewing, 1964). This procedure is mathematically correct and can be applied to least squares
lines that are fitted to interval velocities at layer mid-points.
However, the procedure breaks down near the seafloor when it is
applied to refraction solutions, because refracted arrivals are measured from the tops of layers. That is to say, a refraction solution from an uppermost layer that is 400 m thick will be displaced
downwards 200 m when plotted at the layer mid-point. If K=1, the
intercept value, V_0, would be 200 m/s less than a true value of 2
km/s. Since deepwater solutions are based on interval velocities
and shallow water solutions are based largely on refraction data,
K/V_0 provides reasonable estimates of gradients in deep water, but
is not reliable with shelf data.

High-resolution inversions from the sonobuoy data in the
Irawaddi delta are shown in Figure 3. Sonobuoy 276 from the central
part of the delta reveals nearly 6 km of sediment, but only the
upper part of the velocity depth plot is shown. All the other plots
except 273 have also been cut off above the deepest solution. Of
greatest interest in the present work is the upper 500 m or so of
each plot, which gives a good estimate of the near-surface velocity
gradients. These values have been averaged and appear in Table 2,
along with similarly obtained measurements from other study areas.
In the table the observed values are compared with the values determined by computing K/V_0 from the velocity functions in Table 1
(based preponderantly on refraction solutions).

The comparisons show that the observed velocity gradients are
below those predicted by the velocity regressions. A large part of
this discrepancy results from the typical 120 m/s difference between V_0 predicted in Table 1 (1.58 km/s) and that observed in the
velocity depth plots (about 1.7 km/s; see Fig. 3). Since this number is divided into K, it yields an erroneously large seafloor velocity gradient. It is clear from Table 2 that the intercepts are not
reliable measures of seafloor sound velocity, and that more realistic
seafloor velocities of 1.7 to 1.9 km/s would mostly eliminate the

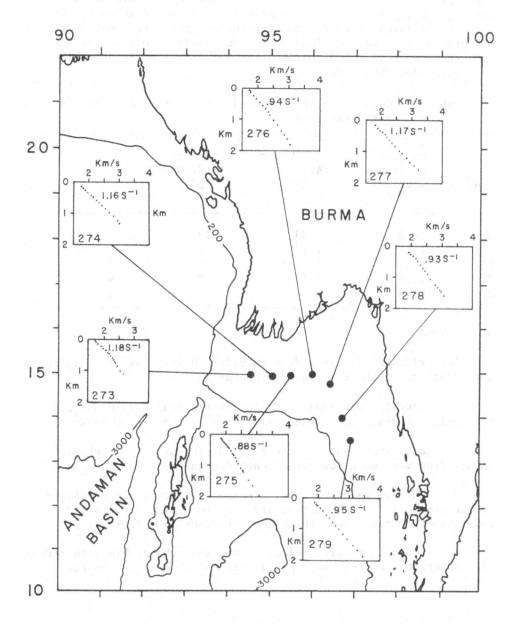

Fig.3. High-resolution velocity inversions shown in the insets
       from the Irawaddi delta. The velocity gradients are esti-
       mated from the near-surface data and are shown in the
       insets of units of s⁻¹.

discrepancies.   This source of discrepancy can be minimized by
picking the layers as thin as possible.   This was done on the South
China shelf (area O) where the near-surface refraction data were
intentionally picked to yield very thin layers.   This is the one
area (see Table 2) where the observed velocity gradients are in
good agreement with K/V_O.

Table 2.   Comparison of Predicted and Observed Seafloor Velocity Gradients

| Area | Computed Gradients K/V_O s^{-1} | Observed Gradients | | n |
		Average s^{-1}	Std. Dev. s^{-1}	
D	1.18	1.03	.13	7
E	1.24	1.15	.14	5
I	1.23	1.01	.16	8
J	1.36	1.12	.48	4
O	1.46	1.47	.27	11

BRIEF SUMMARY

The principal findings from detailed analysis of velocity depth
distributions on the Southeast Asian margin are:

1. Velocity functions along the continental margin are surprisingly uniform; the average value of K is $1.86 \pm .30$ km/s^2.

2. Much higher values of K off western Australia probably
result from very much lower sedimentation rates.

3. Seafloor velocity gradients are not accurately predicted
by the statistical methods that are used to determine velocity
functions from relatively thick layer refraction solutions.

4. Interval velocity solutions are interchangeable with refraction solutions, except for minor discrepancies near the seafloor.

ACKNOWLEDGEMENTS

This research was supported by the Earth Physics Program of the
Office of Naval Research contract no. N00014-75-C-1126.   This paper
was improved as a result of suggestions by Drs. John Ladd and
George Bryan, who kindly acted as reviewers.

REFERENCES

Curray, J., Shor, G., Raitt, R., and Henry, M., 1977, Seismic refrac-
    tion and reflection studies of crustal structure of the eastern
    Dunda and western Banda area, J. Geophys. Res., 62(17), 2479-
    2489.
Houtz, R., 1974, Preliminary study of global sound velocities from
    sonobuoy data, in: "Physics of Sound in Marine Sediments"
    (Ed. by L. Hampton), Plenum Press, 519-535.
Houtz, R., 1977, Sound-velocity characteristics of sediment from
    the eastern South American margin, Geol. Soc. Am. Bull., 88,
    720-722.
Houtz, R., 1978, Preliminary sonobuoy study of rapidly accumulating
    shelf sediments, J. Geophys. Res., 83(B11), 5397-5404.
Houtz, R., and Ewing, J., 1964, Sedimentary velocities of the west-
    ern North Atlantic margin, Bull. Seism. Soc. Am., 54(3), 867-
    895.
Ludwig, W., Kumar, N., and Houtz, R., 1979, Profiler-Sonobuoy mea-
    surements in the South China Sea basin, J. Geophys. Res., 84
    (B7), 3505-3518.
Ludwig, W., and Houtz, R., 1979, Isopach map of sediments in the
    Pacific Ocean basin, Am. Assoc. of Petroleum Geologists.
Ludwig, W., Murauchi, S., and Houtz, R., 1975, Sediments and struc-
    ture of the Japan Sea, Geol. Soc. Am. Bull., 86, 651-664.
McDonald, J., 1964, "Sediments and structure of the Nicobar fan,
    northwest Indian Ocean", Doctoral dissertation, Univ. of Calif.,
    San Diego, 148pp.
Um, S., Lee, D., and Bak, B., 1964, Geological map of Pohang, Sheet
    7022-2, Geological Survey of Korea.

CIRCULAR STRUCTURES OBSERVED IN THE DEEP SEA BY THE

SWATHMAP LONG-RANGE SIDESCAN SONAR

James E. Andrews and Peter Humphrey

Department of Oceanography
University of Hawaii at Manoa
Honolulu, Hawaii 96822

The marine geologist/geophysicist's principal tools for studying the seafloor are acoustic. The first major increase in understanding the seafloor following the original Challeger Expedition came in the 1950's with the routine application of precision echo sounding techniques from research vessels.

Since morphology and structure are closely related these applications led to early theories concerning the evolution of the seafloor. The physiographic charts of Heezen and Tharp (Heezen et al., 1959) reflect both these theories and the available data. The evolution of these charts to their present form was a function of both 1) evolving hypotheses leading to plate tectonic theories (global tectonics), and 2) the increasing density of bathymetric data to support the interpretation. Thus the charts of the North Atlantic from 1959 and 1977 (Heezen and Tharp, 1977) are distinctly different. Fracture zones, transform faults and linear abyssal hills all result from the mechanics of seafloor spreading, and they are notably absent in the morphologic view of the 1950's.

The major problem in surveying for such details of structure and morphology has always been data density. Normal echosounding systems provide only a single profile immediately beneath the ship's track. Data density is then a function of ship time available and the precision of the navigational control. At distances of over 2000 km from land this has meant celestial and satellite control. As a result track densities of more than one track per two km are difficult to obtain even when ship time is available to support the work. This inherent interpretation problem together with the navigational uncertainty has made it

difficult to correlate--or even to recognize--linearity in structures
of small to loderate relief such as abyssal hills and old fracture
zones. Magnetic anomaly data has provided most of the correlations
for the investigation of seafloor spreading, in spite of the fact
that detailed understanding of the processes, as forinstance near
triple junctions, requires morphologic and structural control.

Sonar applications have been advancing to meet these requirements. Multibeam systems (Glenn, 1970) and Seabeam (Renard et al.,
1979) can now provide detailed and rapid surveys of stripes of
seafloor (equivalent to about 75% of the water depth in width),
and because of the richness of detail which makes pattern recognition
easy and rapid it is possible to overcome navigational problems
by simply fitting map segments for a high precision chart. Ultimate
position accuracy for the location of the completed chart is, of
course, dependent on the accuracy of the primary navigational
system.

Sidescan sonar applications have been common for many years,
although principally in "shallow" depths (0-2000 m), with the
fish towed near the seafloor. Operating at relatively high
frequencies these systems give morphologic and textural detail
at ranges of a few hundred meters. For longrange surveys of the
seafloor the Institute of Oceanographic Sciences of Great Britian
has developed Gloria II (Somers et al., 1978) as a near surface
towed sidescan system operating to ranges of 30+ km to either
side of the ship's track. This is a truly rapid reconnaissance
technique which permits recognition of large and small features
and their continuity or association in structural and morphologic
trends.

Swathmap is a similar longrange sidescan sonar. It is the
application of high speed sidescan techniques from military sonar
of surface vessels. At vessel speeds of 20 kts (about 100 seconds
per km) and maximum ranges of 40 seconds (round trip travel time)
there are approximately two distinct scans per km. This results
in some loss of detail and a high degree of lateral (or cross track)
distortion. This distortion is currently removed by photographic
techniques during post-processing of the records. In the future
we hope to control this during data acquisition via computor
controll of a digital recorder. The removal of distortion is a
major factor in record interpretation for accurate determination
of azimuths.

Acoustic calibration of the records has been discussed in
previous publications (eg., Andrews et al., 1977). Horizontal
range determination is good, being affected by depth of the target
only at short ranges--although there is a weak dependence on mean
bottom depth (at ranges greater than 11 km the excess time to a
given range increases about 0.05 seconds per 100 m). Refraction

does not markedly affect ray angles between surface and the seafloor.
The 10°-12° ray normally intersects the mean seafloor at 18 km.
Refraction, while not important in locating scatterers, is important in determining the uniformity of insonification of the seafloor--
and thus directly the intensity of the returns printed on the record.
Initially we used a a standard log TVG to provide the gain compen-
sation for the higher transmission losses at longer ranges. We
are now experimenting with a more controllable TVG in order to
enhance sidelobe data beneath the ship to provide a quantitative
tie point for the observed morphology.

The records as printed are a function of transmission loss
as modified by the local velocity profile and the acoustic scattering at the seafloor. The latter is a function of composition,
orientation and roughness of the topographic elements of the seafloor.
We assume that the features appearing on our records are the result
of small and large scale relief presenting varying slopes and surface
types to the sonar beam. We have also observed gradational changes
in backscatter from areas lacking apparent relief. These are
presumed to be due factors such as sediment type, geotechnical
properties of the sediment (such as water content) and effects
of varying basement depth. Such observations have not yet been
more closely examined or quantified.

Geologically the most remarkable features identified to date
in the Swathmap work are summit depressions on seamounts and
large circular structures on the Palau-Kyushu Ridge. Three
specific types of structures have been identified. These are
1) limited summit depressions on isolated volcanic peaks, 2) elon-
gate structures with shallow central depressions and 3) very
large concentric structures of low apparent relief.

Figure 1 shows two isolated structures of type 1 observed in
the Parece Vela Basin. Cross track exaggeration is 2:1, so the
structures are essentially circular. Neither of the structures casts
an acoustic shadow, which suggests that they are small seamounts
of 500-1000 m relief. At the range at which these features occur
only structures of more than 1500 m relief will produce shadowing
by blocking part of the acoustic beam. In this case with the whole
structure insonified it is possible to observe small blank areas
at the center of each. These are summit depressions. The size
of the seamounts and of the summit depressions are similar to the
dimensions of those reported by Hollister et al; (1978), and the
depressions are most likely collapsed calderas. Feature "A" has
what appears to be a small parasitic cone on its southern flank.
Although Swathmap does not provide the detail seen in multibeam
mosaics, the impression of the morphology is that the seamounts are more
similar to the dome-shaped variety reported as typical of the Philli-
pine Sea than they are to seamounts observed in the western Pacific
which are more angular or cone-shaped (Hollister et al., 1978).

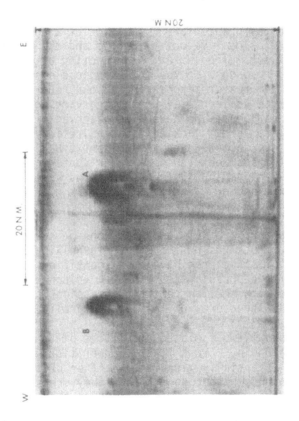

Figure 1.   Swathmap record from the Parece Vela Basin.   Course is 270° and the view is to the south (port) of the track.   The top of the record is beneath the ship, and the bottom of the record farthest from the ship.   Along track and across track ranges are indicated--cross track exaggeration is 2:1.   The two isolated seamounts exhibit summit depressions which may be collapsed calderas.   Feature "A" has a parasitic cone on its southern flank.

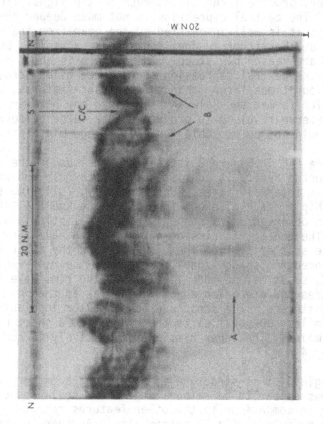

Figure 2. Swathmap record from the Palau-Kyushu Ridge. The view is always to port of the ship's track. Course from the left edge of the record to the course change (c/c) is 170°. Course from c/c to the right edge of the record is 010°. Top of the record is beneath the ship. Cross track exaggeration is 2:1. The two large elliptical features are on the ridge and sub-parallel to it. Feature "A" is the largest circular feature seen, and may represent a cone sheet or a ring dike complex.

Figure 2 presents features of type 2 and type 3 observed on the Palau-Kyushu Ridge. We know of no comparable features reported from the deep sea. Type 2 features are represented by the structure marked "B". This feature was observed from two angles thanks to a course change of 160° (from 170° to 010°) which gave first a view from the west-southwest and then a view from the east-southeast. Here, given the 2:1 cross track exaggeration the structure is eliptical in outline. The eastern rim is higher than the western rim as it is well insonified from the west, and viewed from the east it is very prominent and blocks much of the signal from the western rim. The central depression is not much deeper than the western rim as it is well insonified by the beam from the west. Total relief on the structure does not appear excessive, but the outer slopes are reasonably steep. The feature is about eight miles across, with the central depression about three miles across. This is three to four times larger than the seamount features, but the volcanism which formed the Palau-Kyushu Ridge may have provided a sufficient reservoir volume to produce such a large caldera structure where this would not be possible for the average mid-plate seamount.

Structure "A" in Figure 2 is the most curious of the three features reported here. It consists of an outer rim and at least two concentric internal ridges. Unfortunately, following the course change the ship's track diverged from the ridge and the feature was not again observed. Its shape is roughly eliptical, with a 2:1 ratio. The concentric rims do not appear to have very great relief as the scattering from them is weak. On the other hand there are three concentric shadow zones created by the rims--which suggests relief, so it may be more likely that the rims have relatively smooth surfaces and are oriented at low angles to the beam, making them poor scatterers. The structure has an outer long axis dimension of about 20 miles sub-parallel to the ridge, and a short dimension of about 12 miles. The inner rims are respectively 15 n.m. by 9 n.m. and 8 n.m. by 4 n.m.

The origin of this structure is at present a matter of speculation. It does appear to be rather large for a collapsed caldera, particularly in comparison to the other features reported, although this can not be ruled out as a possibility. Another intriguing thought is that the structure might represent a submarine cone sheet or ring dike complex. In such structures intrusion of layered low angle sills or concentric vertical dikes is followed by collapse and/or erosion to provide the concentric circular structures of low relief. Dike structures and sills have been suggested as being the more probable forms of volcanism where the crust is old and well covered by sediments (McBirney, 1967). Thus an intrusive origin for this feature might imply that it is much younger than the bulk of the Palau-Kyushu Ridge. The presence on the rims of either dominantly extrusive or dominantly intrusive rocks could make it possible to test for a caldera or cone sheet origin. In the latter case the possibility of substantial submarine erosion of the ridge surface to expose the structure may also be novel in terms of submarine processes.

Clearly much of the seafloor is less than perfectly known morphologically, and the application of rapid long range survey techniques is revealing structures previously unknown in the marine environment. The expansion of this work together with well planned local studies to clarify structures will be one of the more effective tools in marine geology/geophysics.

Acknowledgements

This work was supported by the Office of Naval Research, Code 483. We gratefully acknowledge the support and interest of ComNavSurfGruMidPac in carrying out the surveys. This paper is a contribution of the Hawaii Institute of Geophysics.

References

Andrews, J.E., Craig, J.D., and Hardy, W., 1977, Investigations of the deep sea floor by sidescan sonar techniques: central eastern Pacific, Deep Sea Res., 24:975.

Glenn, M.F., 1970, Introducing an operational multi-beam array sonar, Int. Hydrogr. Rev., 47:35.

Heezen, B.C., Tharp, M., and Ewing, M., 1959, The floors of the ocean I. The north Atlantic, Geol. Soc. Am. Spec. Paper 65.

Heezen, B.C., and Tharp, M., 1977, World ocean floors, Pub. U.S. Navy, Office of Naval Research.

Hollister, C.D., Glenn, M.F., and Lonsdale, P.F., 1978, Morphology of seamounts in the western Pacific and Philippine Basin from multi-beam sonar data, Earth and Planetary Sci. Letters, 41:405.

Renard, V., and Allenou, J-P., 1979, Le Sea-Beam, Sondeur multi-faisceaux du N/O Jean Charcot: Description, evaluation et premiers résultats, Revue Hydrographique Internationale, 56:35.

Somers, M., Carson, R., Revie, J., Edge, R., Andrews,A., 1978, Gloria II--An improved long range sidescan sonar, Ocean. International, 78:16.

A PERSPECTIVE ON BOTTOM REFLECTIVITY AND BACKSCATTERING

John J. Hanrahan

Naval Underwater Systems Center
New London Laboratory
New London, Connecticut 06320 U.S.A.

ABSTRACT

A review of available data and description of observed effects permit an assessment of progress made since the 1970 SACLANT Conference on this same topic. Physical models have been developed and refined which can both explain the observations and serve as a basis for predictions. Examples will be presented which contrast values determined from the physical models with measured data in order to identify problem areas and to recommend directions for future efforts.

INTRODUCTION

In 1970, SACLANT hosted a similar conference on bottom interactions[1] which, in my opinion and to this day, ranks as the most stimulating that I have ever attended. One reason for the success of that conference was the prevailing circumstances. Such progressive concepts as self-calibrating measurement geometries, bottom transfer functions, bottom impulse responses, and deconvolution were all topical at SACLANT and elsewhere.

The interest level was further heightened by the completion in the sixties of two large scale acoustic bottom loss surveys - the United Kingdom's NAVADO effort and the U.S.A.'s Marine Geophysical Survey. Still at issue in the underwater acoustics community was the processing and interpretation of this huge body of data as well as the preliminary generalization to a global picture. Since 1970, the effort in the United States on bottom interactions has been greatly reduced and has been primarily directed toward lower frequencies.

This paper, therefore, will focus on the available data, describe the principal observed effects and associated physical modeling, and list some of the remaining problems. The treatment will be from a viewpoint which is concerned with the needs of sonar systems.

There is a danger in doing a paper of this nature, and that is of slighting some of the major contributors to the current understanding. Be assured that the author regrets that time does not allow a more comprehensive coverage.

AVAILABLE DATA

The data base that has been compiled on bottom reflectivity and bottom scattering strength is summarized in Figure 1 as regards frequencies, grazing angle, sound sources, geographic areas sampled, signal processing applied to the data, units, and finally the defining equation itself. Most of the data were gathered prior to 1970.

The entries are self-evident with the exception of processing where three terms are listed; namely, peak, energy, and sonar simulator. A peak value is based upon measuring the highest value observed in a given analysis bandwidth. This is in contrast to an energy value where the total signal duration in the analysis band is integrated and is then used in the determination. The third means is the process whereby the characteristics of a benefitting sonar system are used to specify the signal treatment. This procedure, which is referred to as the "sonar simulator" is intended to produce values of bottom loss appropriate for a sonar system.

	BOTTOM REFLECTIVITY	BOTTOM SCATTERING STRENGTH
FREQUENCIES	50 – 4000 Hz	1000 – 4000 Hz
GRAZING ANGLE	0 – 90°	15 – 90°
SOURCES	EXPLOSIVE PULSED CW CODED	EXPLOSIVE PULSED CW CODED
GEOGRAPHIC AREAS	TENS OF THOUSANDS	FEW HUNDRED
PROCESSING	PEAK ENERGY "SONAR-SIMULATOR"	PEAK
UNITS	DECIBELS	DECIBELS
DEFINING EQUATION	$10 \text{ LOG} \left(\frac{\text{REFLECTED INTENSITY}}{\text{INCIDENT INTENSITY}} \right)$	$10 \text{ LOG} \left(\frac{\text{SCATTERED INTENSITY}}{\text{INCIDENT INTENSITY}} \right)$

Figure 1. Data Base on Bottom Reflectivity and Backscattering

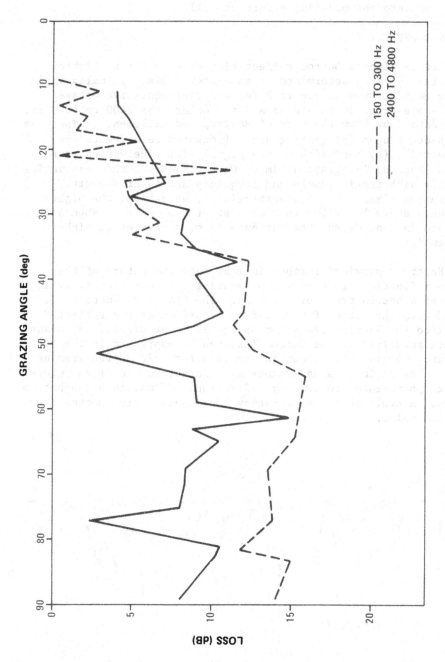

Figure 2.   Reflection Loss as a Function of Frequency and Grazing Angle

The bottom reflectivity data are much more extensive than the information on bottom scattering and this imbalance will be seen to extend into the modeling effort as well.

OBSERVED EFFECTS

Let us discuss bottom reflectivity effects first. Bottom reflectivity, when determined on an energy basis, typically varies in the manner shown on Figure 2 for two representative octave bands - one from 150 to 300 Hz and the other from 2400 to 4800 Hz. These data come from the work of Hastrup and Lallement[2]. The loss is dependent upon (a) grazing angle (measured relative to the horizontal), (b) bandwidth of the signal, (c) receiver character- istics, and (d) integration time. On the average, the loss varies directly with grazing angle and frequency and inversely with integration times. On this particular illustration, the higher frequency shows lower losses over most of the angles. When all the data is considered, the variance is much greater at higher frequencies.

Hastrup[3] provided further insight into the nature of the bottom reflection process by representing bottom reflectivity in terms of a bottom transfer function. The transfer function was formed from the ratio of the Fourier transform of the reflected signal to the Fourier transform of the incident signal. By means of this formulation, the bottom loss can be expressed in the frequency domain as is shown in Figure 3 for a 72 degree grazing angle. The rapid loss variations are characteristic of an inter- ference phenomenon produced by reflections off multiple sub-bottom layers. A small shift in frequency can change a high bottom loss to a low value.

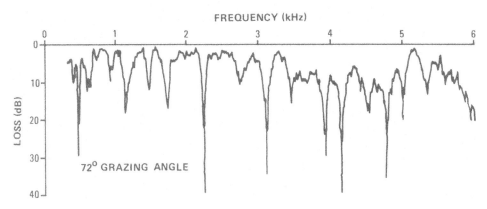

Figure 3. A Bottom Transfer Function Showing Major Changes in Reflectivity with Frequency

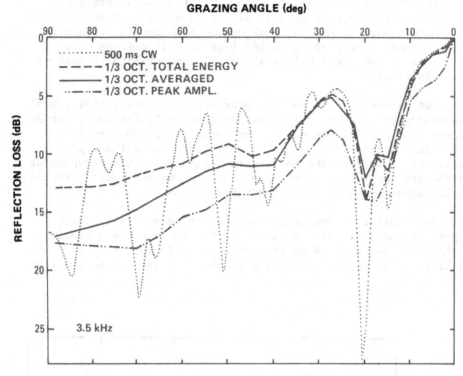

Figure 4. Bottom Loss Depends Upon Signal Treatment

The absolute value of bottom loss obtained from signals reflected off the bottom depends upon the manner in which the signal is treated. Bottom loss as a function of grazing angle is shown in Figure 4 for four types of signal treatment. Again Hastrup's work is being used to illustrate the effect.[1]

All but the dotted curve were obtained with an explosive source and 1/3 octave filtering of the bottom reflected signals.

(1) The dotted curve was obtained with a 500 ms CW pulse. With CW, it is common to observe interference type behavior and the preceding illustration on the bottom transfer function offers a satisfactory explanation.

(2) The loss curve entitled "1/3 octave peak amplitude" was obtained by reading the peak level output from a 1/3 octave filter. Over most of the angular region, this process yields the largest magnitude for bottom loss.

(3) The "1/3 octave averaged" was obtained by applying a 10 ms integration period to the 1/3 octave record. For this illustration, it can be regarded as the simulator value.

(4) The "1/3 octave total energy" curve was determined by
integrating over the total signal duration. It should be
noted that all but the peak bottom loss are in reasonable
agreement for grazing angles less than 50 degrees. For
this reason, peak values are of limited utility to systems
applications.

Let us now turn to bottom backscattering effects (Figure 5).
The values observed appear to be insensitive to frequency but do
depend upon grazing angle and geographic area. The range of
variation extends from -20 to -50 dB. In most of the reported
data, the backscattering was observed along angles corresponding
to the grazing angle. Therefore, the grazing and backscattering
angles were identical for these measurements.

As a matter of interest, backscattering and reflectivity
appear to inversely related as shown in Figure 6 for a grazing
angle of 26 degrees. When the bottom loss is low, the scattering
is high. Less bottom scattering is observed in high bottom loss
regions.

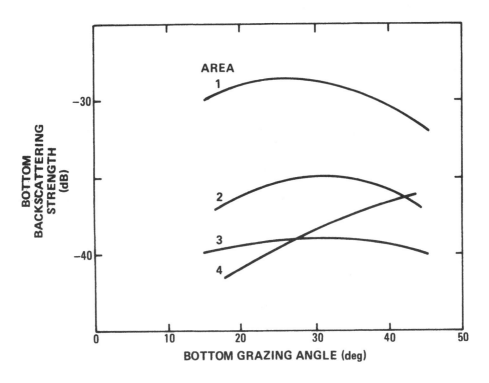

Figure 5. Bottom Backscattering Behavior

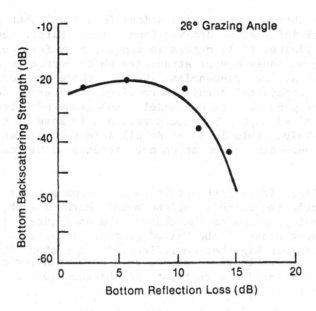

Figure 6.   Inverse Relationship Between Bottom Reflectivity and
            Backscattering

PHYSICAL MODELING

      Again, let's treat bottom loss first. The best success in
physically modeling the bottom reflectivity phenomenon has been
obtained with the Rayleigh reflection equation suitably modified
to include absorption effects. The implicit assumptions are (a)
the bottom is composed of multiple plane, absorbing layers, each
of which has its own characteristic impedance, and (b) the bottom
is capable of supporting both longitudinal and shear waves. Inputs for the model evaluation are primarily based upon information
derived from cores. Of particular interest in the core analysis
are the sound velocity and wet density within each layer. The
product of these two parameters is termed acoustic impedance.
Differences in impedance between the various bottom layers control
the magnitude of reflectivity in this model.

      There is one additional effect that must be considered. As
the frequency approaches extremely low values, experimental
evidence demands that the model be expanded to include contributions from bottom refracted paths.

There are three inherent limitations in a model that relies upon geophysical information derived from cores. First, because most cores are limited to 10 meters in length, they fail to gain any information on those deeper structures which become of paramount importance at low frequencies. Second, there are occasions when excessive geophysical detail is required in order to achieve reasonable correspondence between modeled and measured bottom loss. Upwards of 40 layers have been retained in some of these calculations. Third, this level of detail is not consistent with the knowledge concerning attenuation coefficients within each of the layers.

Nevertheless, it is instructive to assess on a large geographical scale the adequacy of the model. Horn, Delach, and Horn (reference (4)) prepared the chart shown on Figure 7 for the areas off the east coast of the United States. Those areas considered to exhibit high bottom reflectivity are shown in white. The category of favorable was assigned to a core whenever a specified acoustic impedance contrast and thickness was encountered anywhere within the core.

A representation of the results from the U.S. acoustic survey is shown in Figure 8 where the surveyed region is indicated by the dashed line. Within this boundary, the region has been classified into acoustic provinces. A province is defined as a region where the bottom reflectivity is similar. Lower province numbers are associated with higher reflectivity and, therefore, less bottom loss. For example, the province numbered one exhibited the lowest bottom loss values. The very high loss regions within the dashed boundary are unnumbered.

In order to compare core predictions with actual acoustic measurements, the two preceding illustrations were superimposed (Figure 9). In general, there is agreement on provinces 1 and 2. However, the cores generally fail to indicate the intermediate categories of provinces 3 and 4. In these intermediate provinces, we suspect that the predominant reflector is deeper than the core data and is covered by absorbing layers of sediment.

Although this comparison is admittedly disappointing, these and similar charts are still the best basis for estimating bottom reflectivity in the unsurveyed regions of the world. It should be apparent that such estimates are obviously biased to the conservative side.

While there has been a large amount of qualitative speculation on the effect of bottom roughness on bottom loss, there has been a marked lack of any quantitiative theories that would permit the computation of the actual loss figure due to the roughness whether it be on a large or a small scale. This makes

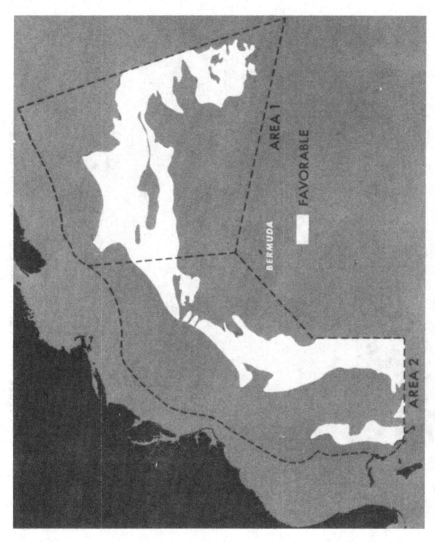

Figure 7 Bottom reflectivity based on cores

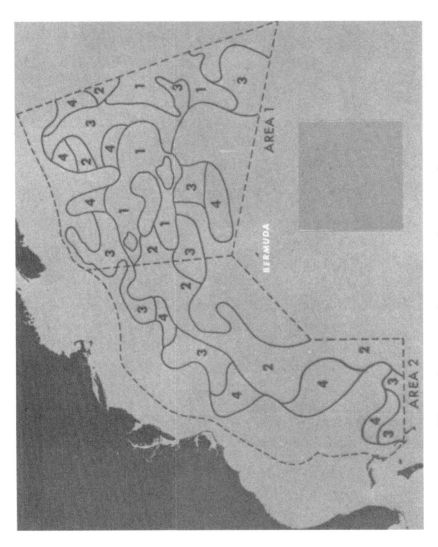

Figure 8   Acoustic bottom loss provinces

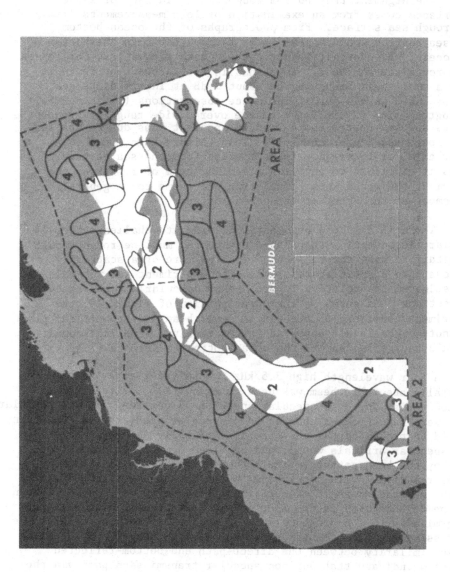

Figure 9   Acoustic provinces compared with core-based charts

it difficult to contrast the relative importance of bottom impedance
with bottom roughness. Intuitively, a reasonable hypothesis is
that both mechanisms on occasion can predominate. There are, however, several arguments that relegate roughness to a secondary
position behind impedance.

One argument that bottom roughness is in fact of lesser
importance comes from an examination of loss measurements from
the rough sea surface. From photographs of the ocean bottom,
the sea surface would appear to be far rougher than the bottom of
the ocean. However, it is not easy to make a good case for losses
from rough sea surfaces. Hayes has in fact shown that there is
even a small gain amounting to about 3 dB at intermediate sea
states based on his normal incidence observations (reference (5)).
Adlington using an explosive source over a wide range of frequencies,
angles and sea states could find no apparent loss due to sea surface roughness (reference (6)). This failure then, to show a
clear case for losses due to scattering from the sea surface,
makes it difficult to envision a similar type of loss from the
bottom of the ocean whose fine structure seems to be generally
far smoother than that of the ocean surface.

There is a second argument. In order to challenge the
popular misconception that the ocean bottom must be rough enough
in detail to make the reflection process more of a scattering
process than a specular reflection, an experiment was run which
investigated the relative importance of specular and non-specular
reflections. As shown in the upper portion of Figure 10, the
experiment consisted of steering a sonar beam in the vertical plane
and noting the levels observed at the receiver after the acoustic
signals had reflected off the ocean bottom (reference (7)).

A four wavelength high 3.5 kHz transmitting array with a
vertially steerable beam was used as a source. The receiver was
19,000 yards away on a specular angle of 30 degrees. If the specular
path were far stronger than any non-specular path, the beam pattern
should be traced out as the beam is vertically steered. Strong
non-specular arrivals would tend to widen or perhaps completely
destroy the beam pattern. The bottom half of the illustration
shows relative levels received from the hydrophone at 19,000 yards
as the depression angle of our transmission beam is changed. Shown
as the dashed curve is the short-range, direct path beam pattern
determined with a hydrophone at a 30 degree depression angle as the
depression angle of the transmission beam is changed. The evident
close similarity between the direct-path and bottom-reflected
patterns indicate that any non-specular transmission path via the
bottom was far weaker than the specular path in this location.
This particular experiment has been run in both smooth and moder-
ately rough regions with the same results.

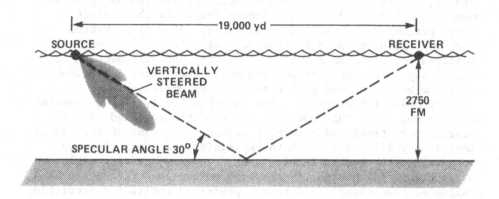

Figure 10a. Test Setup for Determining Importance of Non-Specular
Reflection in the Forward Direction

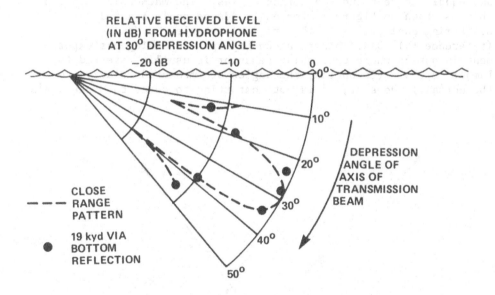

Figure 10b. Measurement of Vertical Directivity Pattern on Bottom
Reflected Sound Pulses

A third argument against bottom roughness is that the principal interest in bottom reflectivity, outside oil exploration efforts, arises from the intended naval application of receiving energy from a desired path and indeed realizing that objective a high percentage of the time. In regions of extreme bottom roughness, such as the Mid-Atlantic Ridge, it would appear fruitless to rely upon a bottom bounce type of path because of the uncertainty in controlling performance. Therefore, in regions where the bottom bounce path appears to be exploitable, an adequate model for predicting bottom reflectivity in the specular direction exists. This model consists of the modified Rayleigh equation for stratified absorbing layers. By means of this model there is a solid basis for both explaining and predicting the dependence of bottom reflectivity on frequency, grazing angle, and location. Of course, if measured bottom loss values are available, then these would be the preferred choice for prediction purposes.

The opposite situation exists with bottom backscattering. Very little is available in the literature except for information of one type - the MacKenzie type of experiment where bottom scattering strength is obtained for angles between 30° and 90°, and equal incident and scattering angles. The MacKenzie relationship is shown on Figure 11 for an incidence angle ϕ and a scattering angle θ.   In the original article, MacKenzie (reference (8)) based his relationship upon an analoguous experiment in optics where the cosine behavior is usually referred to as Lambert's Law. (In optics, the angles are measured relative to the normal.) However, it is not comforting to refer to MacKenzie's

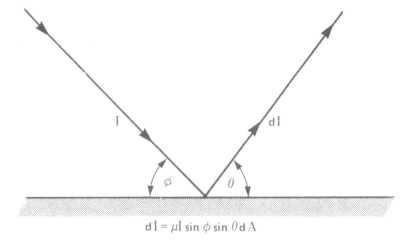

$$d\mathrm{I} = \mu\mathrm{I} \sin\phi \sin\theta\, d\mathrm{A}$$

Figure 11. MacKenzie Model for Bottom Backscattering

reference, <u>A Treatise of Light</u>, by R. A. Houstoun, a book first
printed in 1915 and an apparent bestseller as evidenced by being
reprinted at least 10 times up to 1938. My point in detailing all
this background is to quote the following statement made by
Houstoun in regard to the illumination of various surfaces with
light. "Lambert's cosine law is obeyed very well by a scraped
surface of plaster of Paris, but with varying degrees of approximation in the case of other surfaces". Surprisingly enough,
MacKenzie's relationship seems to hold for bottom backscattering
in a good many locations.

On Figure 12, the MacKenzie relation is superimposed on the
data previously presented. What is needed urgently is an improved
theoretical model and field data for low angle scattering and for
geometries where the backscattering angle does not match the grazing
angle.

SUMMARY

This review of available data and modeling efforts on bottom
reflectivity and bottom backscattering indicates the following
status:

(1) <u>Bottom Reflectivity</u>. In those areas where acoustic
bottom loss surveys have been conducted, there is an

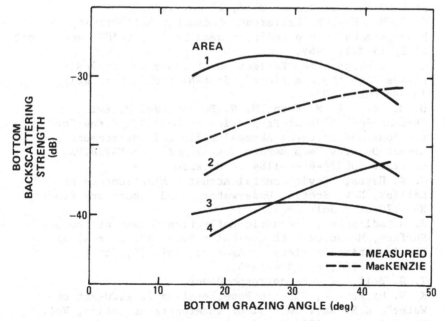

Figure 12. Measured Values Compared With MacKenzie's Equation

adequate basis for estimating sonar performance. However, in unsurveyed regions, reliance must be placed upon the modified Rayleigh reflection model and core data. This approach, in general, produces conservative estimates, especially at low frequencies. The recommended course of action is to extend the surveys into areas of potential importance.

(2) Bottom Backscattering. The need here is more modest since a complete global understanding is not required. It is high time for innovation and a release from an experimental procedure that goes back to the 1950's and a theoretical treatment that lacks rigor. Data at low grazing angles is urgently needed. Perhaps a three-dimensional accounting of the scattered field is un-attainable. But, a measurement program which examines geometries with unequal incident and scattering angles, could satisfy a long standing need.

REFERENCES

1.   R. Hall and L. Mellberg, "Proceedings of a Conference on Reflection and Scattering of Sound from the Sea Bottom Held at SACLANTCEN 21-23 April 1970", SACLANT Special Report M-65, 15 May 1970.

2.   O. F. Hastrup, B. Lallement, "Acoustic Reflectivity Measurements in the Mediterranean Sea", SACLANT Tech Report #152, 15 July 1969.

3.   O. F. Hastrup, "The Reflection of Sonar and Explosive Pulses from the Sea Floor", SACLANT Tech Report #216, 15 September 1972.

4.   D. R. Horn, B. M. Horn, M. N. Delach, and M. Ewing, "Prediction of Sonar Properties of Deep Sea Cores for the Sohm and Hatteras Abyssal Plain and Environment", Lamont-Doherty Tech Report Nos. 1 and 2 on NAVSHIPS Contract N00024-69-C-1184 of November 1969.

5.   G. B. Hayes, "Environmental Acoustic Measurements at Halifax, Nova Scotia, Underwater Sound Laboratory Report No. 670 of 8 July 1965.

6.   R. H. Adlington, "Acoustic-Reflection Losses at the Sea Surface, Measured With Explosive Sources", Journal of the Acoustical Society of America, Vol. 35, No. 11, November 1963, pp 1834-1835.

7.   T. G. Bell, private correspondence.

8.   K. V. MacKenzie, "Bottom Reverberation in 2100-Fathom Water", U.S. Navy Journal of Underwater Acoustics, Vol. 6, No. 1, January 1956, pp 24-36.

SOME BOTTOM-REFLECTION LOSS ANOMALIES NEAR GRAZING

AND THEIR EFFECT ON PROPAGATION IN SHALLOW WATER

Ole F. Hastrup

SACLANT ASW Research Centre

La Spezia, Italy

ABSTRACT

The bottom-reflection loss near grazing from a hard bottom,
or from one that has a hard sub-bottom, can in certain cases be
fairly high. This can result in higher propagation losses in
shallow water than are usually expected. Two frequently observed
cases have been studied, one where the bottom can propagate shear
waves and another where the bottom is covered by a layer of soft
unconsolidated sediments. The first case causes a low-frequency
attenuation whereby an optimum frequency is created for the propagation, whereas the second case creates selected frequencies for
which the propagation is poor.

INTRODUCTION

The reflection of acoustic signals from the sea floor and the
propagation in shallow water are two closely-related physical
phenomena. An understanding of the characteristics of the bottom-
reflection coefficient is therefore often useful in understanding
the sometimes odd behaviour of acoustic propagation in shallow
water.

With the ever-increasing interest in shallow-water acoustics,
several empirical propagation laws and classification systems have
been proposed. These have not always been totally successful
because of a lack of understanding of the acoustic effect of the
sea floor, which is the most important boundary controlling the
propagation except in the presence of internal ducts. As an
example, it is frequently believed that with a hard bottom or sub-

bottom and for small grazing angles the reflection will always
be close to perfect, with no or very little loss. Although this
can be true, there are situations where the reflection coefficient
behaves differently.

The purpose of this paper is to study two typical sea-floor
types found in shallow water. The first consists of a soft layer of
unconsolidated sediments, such as mud, on top of a hard sub-bottom
(slow bottom currents), for which it will be shown that for certain
conditions very high reflection losses occur very close to grazing.
The second bottom type consists of hard sediments such as sand and
gravel (fast bottom currents), where the influence of shear waves
creates finite losses for small grazing angles.

Since the character of the two examples is very different we
will treat them separately.

I   HIGH LOSS AT GRAZING

The sea floor under consideration is a soft, low-velocity layer
on top of a harder, high-velocity half-space as indicated in
Fig. 1. For the three layers, 1, 2 and 3, c_1, c_2 and c_3 are the
compressional wave velocities, ρ_1, ρ_2 and ρ_3 the wet densities, d
the layer thickness, and R_{12}, R_{23} the local Rayleigh reflection
coefficients from interfaces 1-2 and 2-3. $c_3 > c_1 > c_2$, which means
that we have an intromission angle case for R_{12} and a critical
angle case for R_{23}. Further, we will consider the case where θ is
small and $\theta < \arccos(c_1/c_3)$, in other words the reflection from
interface 2-3 is total.

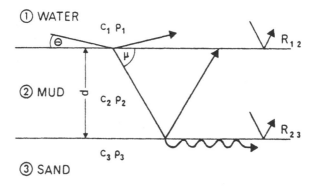

Fig. 1 Layering geometry

The reflection coefficient for the complete layering can be written as[1]:

$$R(\theta,k) = \frac{R_{12} + R_{23} \cdot e^{i\phi}}{i + R_{12} \cdot R_{23} \cdot e^{i\phi}},\qquad(1)$$

where $\phi = 2kd \sin\mu$ is the geometrical phase shift through layer 2, with k being the wavenumber. We will study this expression in more detail for $\theta \to 0$. From Snell's Law we have

$$\frac{\cos\theta}{\cos\mu} = \frac{C_1}{C_2} \quad \text{or}$$

$$\mu = \arccos(C_2/C_1 \cos\theta),\qquad(2)$$

$$\frac{d\mu}{d\theta} = \frac{C_2/C_1 \sin\theta}{\sqrt{1 - (C_2/C_1)^2 \cos^2\theta}},$$

which for $\theta \to 0$ gives $\frac{d\mu}{d\theta} \to 0$.

This means μ will vary very little with θ for θ close to zero and we will consider it constant and equal to μ_0.

For small grazing angles the local reflection coefficients can be expressed by exponential functions[1] and we have:

$$R_{12} = -e^{-Q\theta}$$
$$R_{23} = -e^{-S'\mu},\qquad(3)$$

with

$$Q = \frac{2 \rho_2/\rho_1}{\sqrt{(C_1/C_2)^2 - 1}}$$

$$S' = \frac{2 \rho_3/\rho_2}{\sqrt{(C_2/C_3)^2 - 1}}$$

For $C_3 > C_1 > C_2$, Q is real and positive and S' is imaginary and negative

$$S' = -\frac{i\, 2 \rho_3/\rho_2}{\sqrt{1 - (C_2/C_3)^2}} = -iS$$

and

$$R_{23} = -e^{iS\mu} = e^{i(S\mu-\pi)} \tag{4}$$

Inserting Eqs. 3 and 4 into Eq. 1, we get:

$$R(\theta,k) = \frac{-e^{Q\theta} + e^{i(S\mu-\pi)} \cdot e^{i\phi}}{1 - e^{Q\theta} \cdot e^{i(S\mu-\pi)} \cdot e^{i\phi}}$$

or

$$R(\theta,k) = \frac{-e^{Q\theta} + e^{i(S\mu-\pi+2dk\cdot\sin\pi)}}{1 - e^{i(S\mu-\pi+2dk\cdot\sin\mu)}} ,$$

which for $\theta \to 0$ and $\mu \to \mu_0$ gives

$$R(k)_{\theta\to0} = - \frac{1 - e^{i(S\mu_0-\pi+2dk\cdot\sin\mu_0)}}{1 - e^{i(S\mu_0-\pi+2dk\cdot\sin\mu_0)}} = -1$$

as expected: total reflection with a 180° phase shift.

But when $e^{i(S\mu_0-\pi+2dk\cdot\sin\mu_0)} = 1$ (5)

we have singularity with $R_{\theta\to0} \to \frac{0}{0}$.

From physical reasons we know that $|R| < 1$ and we can therefore expect R to have a minimum. This can also be shown by numerical calculations since the work involved in the analytical study of R(k) at this singularity is very tedious.

From Eq. 5 we get $S\mu_0-\pi+2kd\cdot\sin\mu_0=2n\pi$, n=0,1,2,.. and with $k = 2\pi/\lambda$.

$$\left(\frac{d}{\lambda}\right)_n = \frac{(2n+1)\pi-S\mu_0}{4\pi\sin\mu_0} \tag{6}$$

For these values the reflection coefficient will be small even very close to grazing. Figure 2 shows the reflection loss L = -20 log R as function of θ and d/λ for the following layering.

$C_1 = 1500$ m/s $\rho_1 = 1$ g/cm^3

$C_2 = 1455$ m/s $\rho_2 = 1.45$ g/cm^3

$C_3 = 1575$ m/s $\rho_3 = 1.85$ g/cm^3

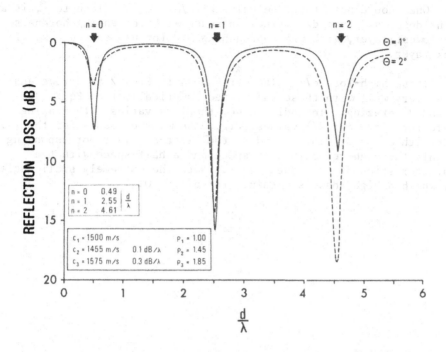

*Fig. 2 Reflection loss as function of wavelength for
          1° and 2° grazing angle*

To enhance the singularities a small amount of attenuation in both layers has been introduced.

Using the above parameters at Eq. 6 we can then calculate the d/λ values for which high losses are expected:

$$\left(\frac{d}{\lambda}\right)_n = \frac{(2n+1)\pi - 1.637}{3.055}$$

This yields

$$\frac{d}{\lambda} = \begin{cases} 0.49 & n = 0 \\ 2.55 & n = 1 \\ 4.61 & n = 2 \end{cases}$$

One should note that the value of $d/\lambda = 0.49$, close to $\frac{1}{2}$, is a coincidence and that d/λ values increase with increasing hardness of the lowest layer, with d/λ approaching 0.8 for n=0 in the case of a rock layer.

These high-loss d/λ values are shown in Fig. 2 as arrows and agree very well with those calculated numerically from Eq. 1 at 1° and 2° grazing. To indicate how the loss varies with grazing angle for different d/λ values, Fig. 3 shows the losses for the same case with $d/\lambda = 0$, 0.5, 2, and ∞, the first and last corresponding to only the high-velocity half-space and a half-space with the characteristics of the upper layer. Note the extremely small angle for which a high loss is obtained for $d/\lambda = 0.5$.

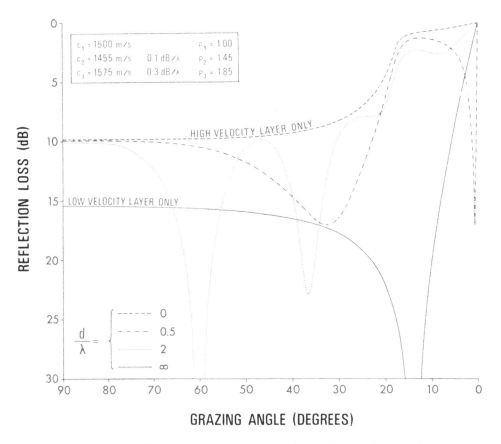

Fig. 3 *Reflection loss as function of grazing angle*

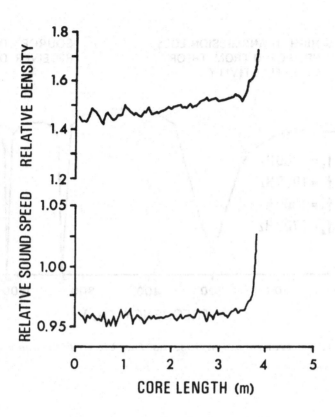

Fig. 4 Sea floor characteristics

    To show that the frequencies corresponding to such conditions
are found in the frequency range of interest for sonar detection,
Fig. 4 shows as an example the relative sound speed* and relative
wet density** measured on a core taken on the Italian continental
shelf. Using these acoustic parameters with a water sound speed of
1500 m/s and d = 3.7 m we find from Eq. 6 that high losses near

---

* The relative sound speed is defined as the ratio between the
  sound speed in the bottom and the sound speed in the water at
  the bottom.

**The relative wet density is defined as the ratio between the
  wet density in the bottom and the density of the water at the
  bottom.

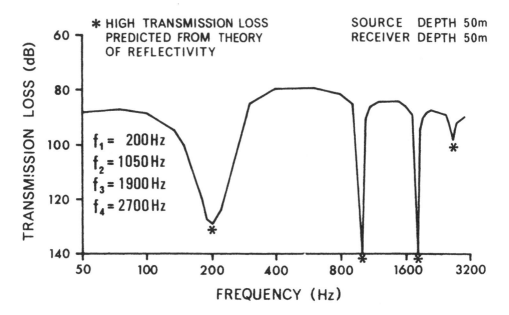

Fig. 5 *Transmission loss as function of frequency*

grazing are expected for

$$f_1 = \quad 200 \text{ Hz},$$
$$f_2 = 1050 \text{ Hz},$$
$$f_3 = 1900 \text{ Hz},$$
$$f_4 = 2700 \text{ Hz},$$

which are all within the frequency ranges of both active and passive sonar systems.

As a further illustration, the transmission losses for iso-velocity conditions at a range of 35 km and for a water depth of 115 m with bottom characteristics corresponding to the above core have been calculated[2][3]. As seen in Fig. 5, it very markedly shows the effect on shallow-water transmission.

How can we explain these reflection loss anomalies for discrete d/λ values?

Let us look at the waves being reflected inside the first layer, as seen in Fig. 6. When $\theta \to 0$ the local reflection coefficient $R_{21} \to 1$ with a zero-degree phase shift, since we have a plane wave coming from medium 2 being reflected from the higher velocity medium 1, where μ_0 is in fact the critical angle. Writing the equation for conditions under which the wave fronts interfere constructively in layer 2 we get:

$$(AB + BC) \cdot k + \psi_1 + \psi_2 = 2\pi n \ ,$$

where ψ_1 and ψ_2 are the phase shifts at the two interfaces. From the above, $\psi_1 = 0$ and $\psi_2 = S\mu_0 - \pi$, AB + BC is easily expressed by d, and μ_0 as:

$$AB + BC = 2d \ \sin\mu_0 \ \ or$$

$$S\mu_0 + 2dk \ \sin\mu_0 - \pi = 2\pi n,$$

which is exactly the same criterion for the singularities in the reflection coefficient.

This means that we are dealing with the propagation of trapped modes in the top layer and their characteristic equation is Eq. 6. With just a small amount of attenuation in the layer it absorbs most of the incident energy and thereby creates a low reflection coefficient just close to grazing.

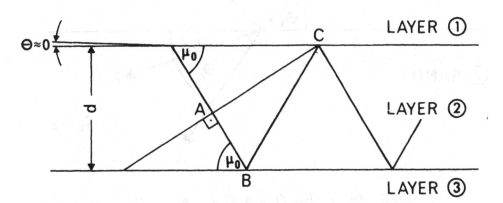

Fig. 6 Wave path in the bottom layer

II   SHEAR-WAVE EFFECTS

Where the bottom is hard it is able to transmit not only compressional waves but also shear waves, whose velocities increase with the bottom hardness. These shear waves or transverse waves affect the reflection loss near grazing and thereby also the shallow-water propagation over such a hard bottom, as is frequently observed in measurements.

The Rayleigh reflection coefficient from a solid half-space, see Fig. 7, can be expressed in several ways, of which one of the least complicated is[1]

$$R = \frac{Z_2\sin^2 2\theta_s + Z_s\cos^2 2\theta_s - Z_1}{Z_2\sin^2 2\theta_s + Z_s\cos^2 2\theta_s + Z_1} , \qquad (7)$$

Z being the impedances $Z_1 = \dfrac{\rho_1 C_1}{\cos\theta}$, $Z_2 = \dfrac{\rho_2 C_2}{\cos\theta_2}$, $Z_s = \dfrac{\rho_2 C_s}{\cos\theta_s}$,

C the velocities, ρ the densities, and θ the grazing angles related by Snell's law as

$$\frac{C_1}{\cos\theta_1} = \frac{C_2}{\cos\theta_2} = \frac{C_s}{\cos\theta_s}$$

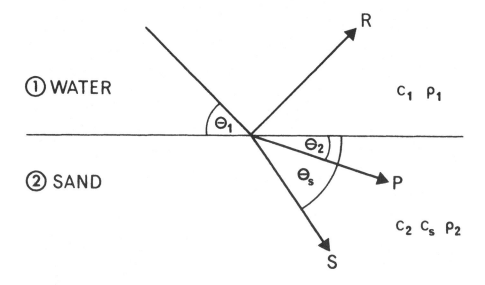

Fig. 7   Reflection from a solid sea floor

Since a mathematical analysis of Eq. 7 is rather complicated, let us look numerically at the reflection coefficient for two distinct cases: one with two critical angles when $C_2 > C_s > C_1$, and one with one critical angle when $C_2 > C_1 > C_s$.

The reflection loss, defined as $-20 \log R$, is seen in Figs. 8a and 8b, respectively, for the two cases. We notice that for $C_1 > C_s$ no total reflection is observed between the critical angle of the compressional wave and grazing, giving higher losses than when there are no shear waves. This can be explained physically by having shear waves carrying energy away from the boundary, giving the feeling that the bottom has been "softened". The influence of bottom attenuation will sometimes mask this effect to a certain degree, as we will see in the following.

$C_1 > C_s$ is the most frequently observed condition on the continental shelf and corresponds to a sedimentary bottom. We will consider three types of bottom: soft, medium, and hard, consisting of silt, sand-silt, and sand having acoustic characteristics as shown in Table 1.

TABLE 1

	Bottom type	C_2/C_1	C_s/C_2	(a) dB/λ	(b) dB/λ	ρ_2/ρ_1
A	SOFT	1.03	0.23	0.2	0.6	1.6
B	MEDIUM	1.11	0.27	1.0	2.0	1.9
C	HARD	1.24	0.35	0.7	1.5	2.1

Although only a few values of shear-wave velocity are known for shelf sediments, we know that the velocity increases with bottom hardness. From what is available[4], estimated values of C_s/C_2, together with values for a characteristic attenuation and density, are also given in Table 1, in which (a) and (b) are, respectively, the attenuation of the compressional and shear waves. Using these values, the three reflection-loss curves in the $0°$ to $35°$ grazing angle interval have been computed and displayed in Fig. 9. The result is quite astonishing, with the hardest bottom giving the highest losses and the softest bottom the lowest losses close to grazing, which are just the angles of interest for shallow-water propagation.

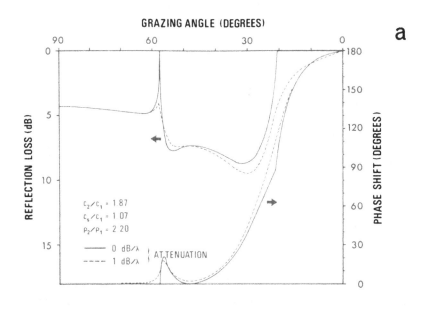

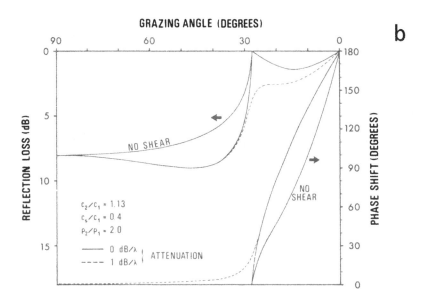

Figs. 8a), b)   Reflection loss and phase shift as function of
                grazing angle for two types of solid sea floors

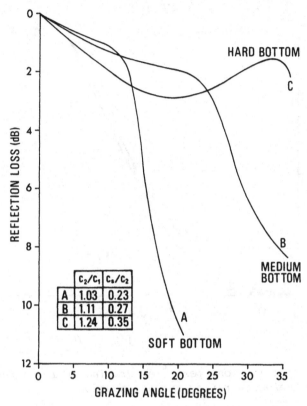

Fig. 9 *Reflection loss as function of grazing angle for three types of sedimentary sea floors*

One of the ways to characterize the importance of the reflection coefficient for shallow-water propagation is to look, for example, at the loss in decibels per degree at grazing. Such values have been calculated for the three bottom types of Table 1 with and without attenuation and as a function of C_s/C_2; these curves are shown on Fig. 10.

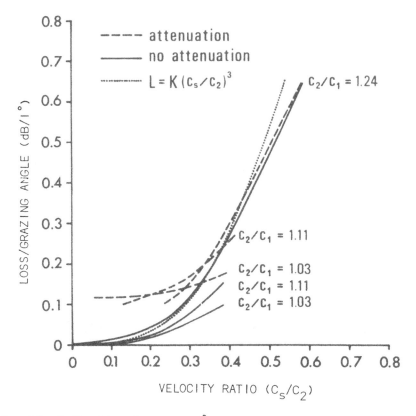

Fig. 10 *Reflection loss per 1° at grazing angle as function*
of velocity ratios and attenuation

From this, the following conclusions can be drawn. First, we
see that when bottom attenuation is ignored, the losses increase
with bottom hardness (compressional wave velocity) for a given
c_s/c_2 ratio, rather opposite to what is usually believed, and
that, second, the losses strongly increase with increasing c_s/c_2
ratio. In real conditions, where attenuation is present, the
above conclusions hold for higher c_s/c_2 values. We can say that
for low c_s/c_2 values the attenuation is the dominating mechanism,
whereas for higher c_s/c_2 values the shear-waves effect dominates
and the effect of attenuation is only small. From this we can
expect that the c_s/c_2 ratio will influence shallow-water

propagation and that the effect will be strongest at lower frequencies where the acoustic penetration in the bottom is larger.

Using the wave equation for shallow water with a bottom
sustaining shear waves and attenuation, we find that the eigenvalues for the propagation are complex[5], meaning that we have
attenuated modes and that the mode attenuation coefficient K for
a given depth can be written as

$$K \propto n^2 \lambda^2 c_s^3 \; ,$$

where n is the mode number and λ the wavelength. The assumption
made to arrive at this expression was that we were far away from
cut-off, a condition corresponding to small grazing angles. This
agrees well with the conclusions from the reflection-loss study and
on Fig. 10, for comparison, is shown a third-degree polynomium that
quite closely approximates the c_s/c_2 dependence.

Therefore to prove the effect of shear waves experimentally
one evidently has to look at the low-frequency shallow-water propagation, but at frequencies well above cut-off. To do this we
will consider the propagation along two tracks, M and N, with
almost identical bathymetry and sound-speed profiles (Fig. 11) but
with very different bottom characteristics (Fig. 12), one being
hard because of the very high sand content of the bottom and one
being much softer because its bottom contains only about 50% sand.
Comparing the transmission losses, Fig. 13, along the two tracks,
M and N, for frequencies of 50, 100, 800 Hz shows a very marked
low-frequency attenuation over the hard bottom (M), but almost the
same high-frequency attenuation over both tracks, showing the
effect of shear waves. This low-frequency attenuation combined
with the usual high-frequency attenuation (volume absorption and
scattering) creates an optimum frequency feature very often observed
in shallow-water propagation. We can say that the higher the shear-
wave velocity, the higher the low-frequency attenuation and the
higher the optimum frequency.

We have seen that shear waves can play an important role in
shallow-water propagation, so let us look at the losses predicted
by normal-mode theory in shallow water for different shear-
wave velocities[2,3] and compare them with measured data.

Figure 14 shows the transmission losses as function of
frequency at a range of 50 km, calculated for the values in
Table 1, together with an additional hard bottom of type D with
higher shear-wave velocities. Also shown are the measured losses.
A good agreeement is found with the calculated losses for a hard
bottom with a c_s/c_2 ratio of about 0.40 to 0.45, clearly showing
the low-frequency attenuation caused by shear waves.

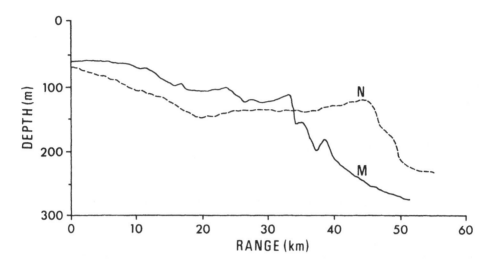

Fig. 11   Bathymetry along tracks M and N

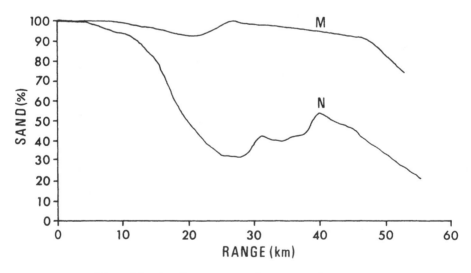

Fig. 12   Sand content along tracks M and N

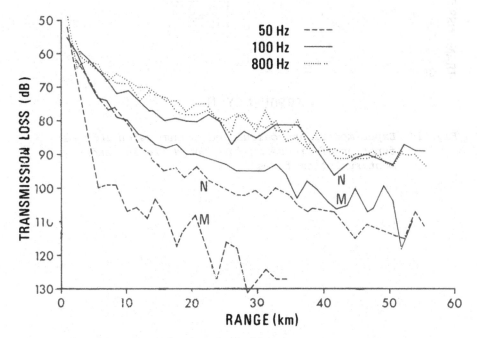

Fig. 13 Transmission loss along tracks M and N
for selected frequencies

CONCLUSION

When trying to understand acoustic reflection from a sedimentary bottom or the shallow-water propagation over such a bottom,
one has to be prepared to find effects that are often contrary to
what is generally believed. Guessing can therefore be quite mis-
leading and only more in-depth studies will provide the right
guidance for understanding and interpretating experimental results
in such an environment. The best reflecting bottom is still one
with an extremely high attenuation; the optical analogy is the
common mirror.

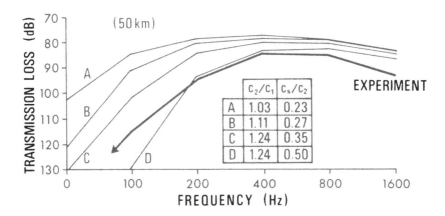

Fig. 14 *Experimental and calculated transmission loss as function of frequency for four different sedimentary bottom types*

REFERENCES

1. L. M. Brekjovskikh, "Waves in Layered Media", Academic Press, New York, N.Y. (1960).

2. F. B. Jensen and M. C. Ferla, SNAP: the SACLANTCEN normal-mode acoustic propagation model, SACLANTCEN SM-121, SACLANT ASW Research Centre, Italy (1979). [AD A 067 256]

3. F. B. Jensen and W. A. Kuperman, Environmental acoustic modelling at SACLANTCEN, SACLANTCEN SR-34, SACLANT ASW Research Centre, La Spezia, Italy (1979). [AD A 081 853]

4. E. L. Hamilton, Acoustic properties of the sea floor: A review, in: "Ocean Acoustic Modelling", W. Bachmann and R.B. Williams, eds., Proceedings of a Conference at La Spezia, Italy, 8-11 Sep 1975, Part 4: Sea bottom, SACLANTCEN CP-17, SACLANT ASW Research Centre La Spezia, Italy (1975), pp. 18-1 - 18-96. [AD A 020 936/1G1]

5. E. T. Kornhausen and W. P. Raney, Attenuation in shallow-water propagation due to an absorbing bottom, J. Acoustical Society America 27:689-692 (1955).

DETERMINATION OF SEDIMENT SOUND SPEED

PROFILES USING CAUSTIC RANGE INFORMATION

George V. Frisk

Woods Hole Oceanographic Institution
Woods Hole, Massachusetts 02543
U.S.A.

ABSTRACT

A method for determining sediment sound speed profiles is described. It employs measurements, obtained at various receiver
heights, of ranges to the caustic which is formed due to a positive
gradient in the sediment. The bottom profile can then be obtained
from equations, derived using the WKB approximation, which relate
the parameters of the water-bottom profile to the caustic range and
source/receiver heights. Using an existing result, the theory is
presented for the case of an isovelocity ocean overlying a sediment
half-space with a linear gradient and continuous sound speed at the
water-bottom interface. A new theoretical result is derived for
the linear gradient case with a discontinuity at the water-bottom
interface. The method is illustrated using data at 220 Hz.

INTRODUCTION

Positive sound speed gradients in ocean bottom sediments can
cause the formation of caustics in the acoustic field due to a point
source located in the water. The caustics can extend into the water
column and manifest themselves as high intensity regions in the
field measured near the bottom. The horizontal range to the caustic
depends upon the source/receiver heights above the bottom and the
parameters of both the water and bottom sound speed profiles. For
certain types of profiles, the analytic relationship among these
quantities can be determined using the WKB approximation. We propose a method for determining the bottom sound speed profile which
involves measuring ranges to the caustic at different source/re-
ceiver heights and using the WKB equations to calculate the parameters of the profile. We present the inversion equations for a

one- and two-parameter bottom profile with a linear gradient and
illustrate the method using data at 220 Hz.

THEORY

A characteristic water-bottom sound speed profile and the corresponding ray diagram illustrating caustic formation are shown in
Figure 1. The principal features of the profile are: a weak positive gradient ($\sim$ 0.016 s^{-1}) in the water; a drop in sound speed
($\sim$ 1-2% relative to the water sound speed) at the water-bottom
interface; a strong positive gradient ($\sim$ 1-2 s^{-1}) in the top sediment layer; and a strong subbottom reflector. The ray paths shown
are those which dominate the bottom interaction at low frequencies
and small grazing angles[1]. These are the direct path, the sub-
bottom reflected path, and the refracted path.

The refracted paths give rise to a caustic with two branches,
one of which extends into the water column. The range to the
caustic in the water has been calculated using the WKB approximation
for the one-parameter bottom profile[2,3] depicted in Figure 2a. This
model consists of an isovelocity water half-space with sound speed
c_0 overlying a sediment half-space with a linear sound speed gradient β. The result for the caustic range r_c is[2,3]:

$$r_c = \frac{2}{\beta} \left[2\beta c_0 (z + z_0) \right]^{\frac{1}{2}} \tag{1}$$

where z_0 and z are the source and receiver heights, respectively.

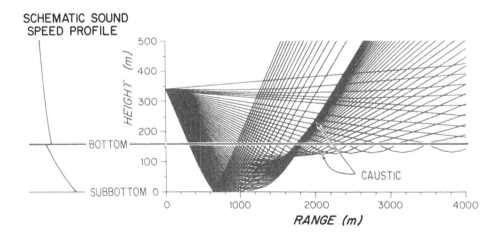

Fig. 1. Characteristic water-bottom sound speed profile and ray
        diagram.

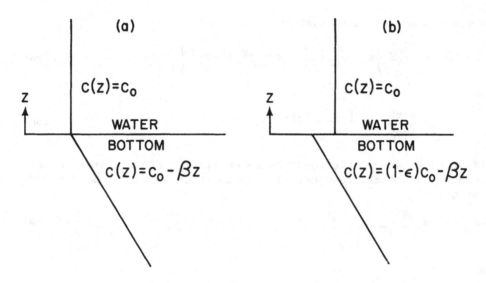

Fig. 2.   Sound speed profiles with (a) one-parameter and (b) two-
          parameter bottom profiles.

Equation (1) can be directly inverted to obtain the single parameter
β as a function of r_c, c_o, and the source/receiver heights:

$$\beta = \frac{8c_o}{r_c^2} (z + z_o).$$ (2)

Thus if r_c is measured, β can be readily calculated.

    Using the WKB approximation, we have calculated r_c for the two-
parameter bottom profile shown in Figure 2b.   This model differs
from the one in Figure 2a in that the sound speed in the bottom at
the water-bottom interface differs from the water sound speed, thus
introducing a second parameter ε.   The result for r_c is:

$$r_c = \frac{2}{\beta} \left[2\beta c_o(z + z_o) + 4\varepsilon c_o^2 \right]^{\frac{1}{2}}.$$ (3)

The calculation leading to Equation (3) involves, in addition to the
WKB approximation, the following approximation:

$$\varepsilon << \frac{\sin\theta_c}{\cos^2\theta_c},$$ (4)

where θ_c is the minimum grazing angle in the water associated with
caustic formation.   Usually $\varepsilon << 1$, so that for typical gradients
and the associated θ_c this condition is satisfied.   If we have one
measurement of r_c and know either β or ε, we can solve for the

second parameter using either

$$\beta = \frac{4c_0}{r_c^2} \left\{ z + z_0 + \left[(z + z_0)^2 + \varepsilon r_c^2 \right]^{\frac{1}{2}} \right\} \tag{5a}$$

or

$$\varepsilon = \frac{\beta^2 r_c^2}{16 c_0^2} - \frac{\beta (z + z_0)}{2 c_0} . \tag{5b}$$

If we have two caustic range measurements r_{c1} and r_{c2} at receiver heights z_1 and z_2, respectively, we can then solve for both β and ε using

$$\beta = \frac{8 c_0 (z_1 - z_2)}{r_{c1}^2 - r_{c2}^2} \tag{6a}$$

and

$$\varepsilon = \frac{1}{16 c_0^2} \left[\beta^2 r_{c1}^2 - 8 \beta c_0 (z_1 + z_0) \right] . \tag{6b}$$

EXPERIMENT

     Let us illustrate the method using 220 Hz data obtained in the Hatteras Abyssal Plain at 34°N, 67°W in 5150 m of water. The details of the experiment are described in Reference 1. The observed amplitude versus range for a source height of 180 m and receiver height of 2.7 m is shown in Figure 3. The amplitude exhibits a strong spatial interference pattern which arises from the combination of the direct and bottom-interacting steady-state signals. A peak is evident in region B of the data. This peak, in addition to having the maximum amplitude, has a shape which is clearly distinguishable from other features of the data. We therefore identify this peak as that due to the caustic with a corresponding range of 1750 m. Furthermore, we assume that ε is known to be 0.016 from other work[1]. Using Equation (5a), we then obtain a value of 0.94 s^{-1} for β. This result is in close agreement with results obtained from a considerably more detailed analysis of the data[1].

CONCLUSIONS

     The success of the method in the case described above suggests that the idea may be extended to the case of a general bottom profile:

$$c(z) = \sum_{n=0}^{N} \beta_n z^n . \tag{7}$$

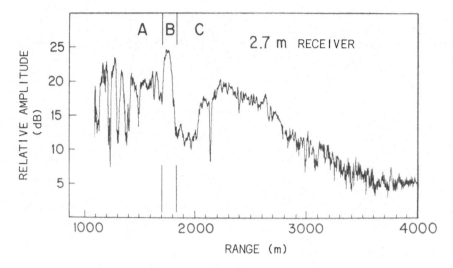

Fig. 3. Measured amplitude versus range for 180 m source height
and 2.7 m receiver height.

Then although a simple analytic, invertible relationship between r_c
and the coefficients β_n may not exist, a numerical solution may be
feasible. The number N of coefficients which can in principle be
determined depends on the number of receivers used in a vertical
array. In pursuing the technique further, we must examine the ease
of identifying the caustic peaks in the data in general and also
determine optimal receiver spacing. The proposed method may provide
a relatively simple way for determining the sediment sound speed
profile to an arbitrary degree of accuracy.

REFERENCES

1. G.V. Frisk, J.A. Doutt, and E.E. Hays, "Bottom Interaction of
    Low-Frequency Acoustic Signals at Small Grazing Angles in the
    Deep Ocean," submitted to J. Acoust. Soc. Am.
2. D.A. Sachs and A. Silbiger, "Focusing and Refraction of Harmonic
    Sound and Transient Pulses in Stratified Media," J. Acoust.
    Soc. Am. 49, 824-840 (1971).
3. D.A. Sachs, "Sound Propagation in Shadow Zones," J. Acoust. Soc.
    Am. 51, 1091-1097 (1972).

INFERENCE OF GEO—ACOUSTIC PARAMETERS

FROM BOTTOM—LOSS DATA[*]

C. W. Spofford

Ocean Acoustics Division
Science Applications, Inc.
McLean, Virginia 22102

ABSTRACT

In areas of thick sediments the received level of shallow-angle low-frequency bottom interacting signals tends to be dominated by a bottom-refracted rather than bottom-reflected path. The loss along this path, assuming a constant gradient, g, and attenuation, α, in the sediment is proportional to the ratio α/g. Assuming that α varies linearly with frequency, the shallow angle loss appears to increase linearly with both grazing angle and frequency. Hence bottom loss data appear to be incapable of separating α and g, providing only estimates of their ratio.

This paper describes a technique for estimating both α and g from classical bottom-loss versus grazing-angle data. In such data acquired in thick-sediment areas the lowest frequencies exhibit an abrupt increase in loss at an apparent grazing angle, $\hat{\theta}_c$, corresponding to the development of a minimum-range caustic in the bottom-refracted paths. The gradient may be estimated from $\hat{\theta}_c$ and the measurement geometry. The attenuation, α, is then estimated from the linearity of loss versus angle and frequency.

Examples using actual bottom-loss data and seismic time-domain data are shown. The implications for depth-dependent gradients and attenuations are also discussed.

---

[*]Work supported by U.S. Naval Electronic Systems Command Bottom Interaction Program under direction of NORDA Code 500.

INTRODUCTION

A large growing body of data on seafloor properties has substantially increased our understanding of bottom-interacting acoustics. As summarized by Hamilton,[1] these data would seem to suggest that in areas of thick (~1 km) sediment cover one ought to be able to reliably estimate bottom-interaction losses. In response to pressing operational needs a number of U.S. Navy organizations collected direct measures of "bottom loss" versus grazing angle and frequency in a wide number of ocean areas. Attempts by the collecting organizations to identify topographic or geologic correlations were largely unsuccessful. NAVOCEANO observed significant differences in reflectivity between areas of thick and thin sediments. They were unable, as was NADC with the other large data set, to find key correlations beyond sediment thickness.

The disappointment of being unable to correlate the measurements with the geo-acoustic data was further aggravated by the substantial complexity of the data when compared with results from geo-acoustic models. The implications of these early comparisons were: (1) that we did not yet know enough about the acoustics of the bottom to reliably use geo-acoustics models; and (2) that the bottom might be too complex to ever permit such modeling.

Hanna[2] was able to show that a particularly vexing complexity in the data was attributable to processing artifacts. In a few measurements for which enough accuracy was maintained to remove this artifact ARL/UT obtained much simpler bottom-loss curves. While these seemed to be more consistent with the geo-acoustic models, they retained a feature which appeared to contradict the models. For the lower frequencies (< 100 Hz) the loss increased quite markedly over only a few degrees for grazing angles near 40 degrees. The effect was apparent and quite consistent in many data sets. It was initially attributed to the onset of basement interaction; however, in many areas the sediments were much too thick to permit this at only 40 degrees grazing.

This paper addresses the resolution of this problem. The point of rapid increase in bottom-loss is shown to provide a reliable estimate of the sound-speed gradient in the sediment. Given the gradient, the sediment attenuation may be estimated from the observed loss versus grazing angle and frequency. The results are quite simple for constant gradient and attenuation. The effects of a decreasing gradient and/or an increasing attenuation are illustrated. Typically the depth-dependence of both must be estimated as shown in a comparison with data. As a result of this analysis much of the existing bottom-loss survey data may be used to provide estimates of the sound-speed and attenuation structure of thick, deep-ocean sediments.

ESTIMATING THE SOUND-SPEED GRADIENT

Figure 1a illustrates the ray geometry for a shallow source to a shallow receiver at several ranges over a simplified, constant-gradient, refracting bottom. At shallow grazing angles the refracted (solid) and reflected (dashed) paths between the source and receiver are nearly collinear. As the range decreases the grazing angle increases and the refracted path penetrates more deeply, becoming increasingly more separated. At sufficiently short ranges (comparable to a few water depths) the refracting paths reach a minimum achievable range. For steeper paths the range begins to increase, and inside this range only the reflected path is present. Beyond this range the shallow refracted path actually has a much deeper counterpart at a very high angle.

The angle-versus-range behavior of these paths is illustrated in Figure 1b. The caustic at range R_c corresponds to the minimum refracted-path range and occurs for an angle θ_c. The reflected path at this range has grazing angle $\hat{\theta}_c$. The corresponding intensity versus range for the paths is illustrated in Figure 1c for low (<100 Hz), medium (~400 Hz), and high (>1 KHz) frequencies. The reflected ray's intensity is frequency independent, controlled by spreading losses and the impedance mismatch at the water-sediment interface. The refracted paths lose energy to volume effects in the sediment which increase with frequency and accumulate along the ray paths.

At the lower frequencies the shallow refracted path dominates and the intensity will be high for ranges greater than R_c (and apparent grazing angles less than $\hat{\theta}_c$). For apparent grazing angles greater than θ_c (i.e. ranges less than R_c) the loss should appear to increase substantially (by 10-15 dB, typically) with the disappearance of the refracted path.

While the caustic in the bottom has been observed by seismologists for years, its importance at these acoustical frequencies has gone largely unnoticed. Frisk et al.[3] have observed it at ~200 Hz for a near-bottom geometry where bottom penetration was slight. Two important points are worth noting:

1. The rapid, large increase in loss does not reflect any change in bottom properties at the turning depth of the corresponding ray; and

2. The actual angle of the refracted ray path is substantially greater than the reflected grazing ray to the same range.

The caustic may be used to estimate the gradient as shown in Figure 2. Consider an idealized homogeneous water column of

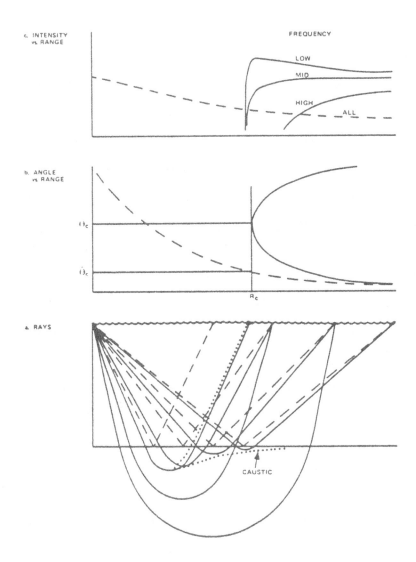

Fig. 1. Ray geometry with angle and intensity versus range for paths interacting once with a strongly refracting bottom.

depth, D, and sound-speed, c_0, over a constant-gradient, g, bottom. (At the steep angles of interest for the caustic analysis, refraction in the water column and sound-speed discontinuities at the water-sediment interface have negligible effects.)   For a source and receiver as shown, the range as a function of refracting angle, θ, is a relatively simple function.

The caustic corresponds to $R'(\theta) = 0$, leading to θ_c and R_c as in Figure 2.   When the grazing angle, $\hat{\theta}_c$, is computed corresponding to R_c, it is easily related to θ_c and the gradient.   Hence the gradient, g, may be estimated from the grazing angle of the onset of high loss, $\hat{\theta}_c$, and the average height of source and receiver off the bottom, $\bar{D}$.   For multiple bounce paths $\bar{D}$ asymptotically approaches the water depth, D.

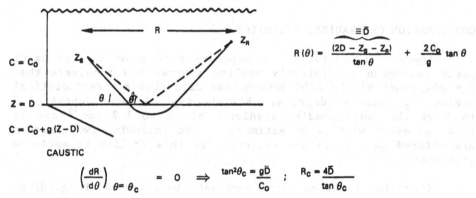

$$R(\theta) = \frac{(2D - Z_S - Z_R)}{\tan\theta} + \frac{2C_0}{g}\tan\theta$$

$$\left(\frac{dR}{d\theta}\right)_{\theta = \theta_c} = 0 \implies \tan^2\theta_c = \frac{g\bar{D}}{C_0} \quad ; \quad R_c = \frac{4\bar{D}}{\tan\theta_c}$$

CORRESPONDING GRAZING ANGLE

$$\tan\hat{\theta}_c = \frac{R_c}{2\bar{D}} = \tfrac{1}{2}\tan\theta_c \implies g = \frac{4C_0}{\bar{D}}\tan^2\hat{\theta}_c$$

Fig. 2.   Estimation of the sound-speed gradient, g, as a function of geometry, D, and apparent caustic angle, $\hat{\theta}_c$ for a homogeneous water column over a constant-gradient bottom.

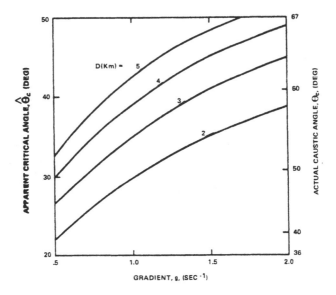

Fig. 3.   Dependence of apparent and actual caustic angles on
          gradient and water depth.

CONFIRMATION OF GRADIENT ESTIMATES

Figure 3 illustrates the dependence of g on $\hat{\theta}_c$ and water
depth (assuming sufficiently shallow sources and receivers that
$\bar{D} = D$). Much of the NADC bottom-loss data show apparent critical
angles, $\hat{\theta}_c$, near 40 degrees. For the typical water depths of 4
to 5 km the corresponding gradients of .8 to 1.2 sec^{-1} are in
good agreement with other estimates. The following two examples
are offered as more direct evidence for this ability to estimate
gradients.

Dicus[4] has reported using shot data to estimate the gradient
in the Tagus Abyssal Plain using seismic techniques. Figure 4
(borrowed from Dicus) illustrates the low-pass filtered waveforms
from a deep series of shots to a deep hydrophone. The shot
waveform has been removed by deconvolution, leaving an arrival
structure consisting of a low-amplitude reflected wave followed by
a much stronger refracted wave. As the grazing angle increases
the refracted wave is increasingly delayed until at 34.5 degrees
grazing there is no evidence of it at all. The corresponding
bottom loss (not shown here) shows a 10-dB increase in low frequency loss between 30.3 and 34.5 degrees.

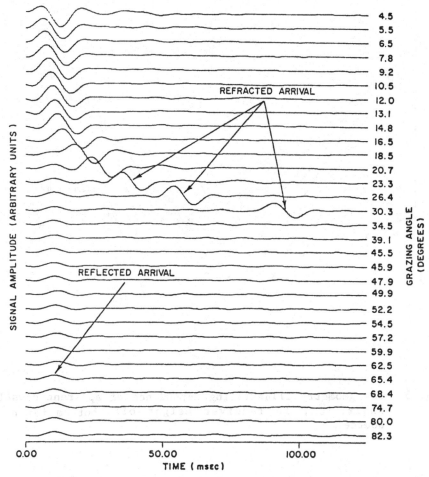

Fig.4.   Low-pass (40-90 Hz) filtered impulse response data from
the Tagus Abyssal Plain showing the delay between the
reflected and refracted arrivals and the disappearance
of the refracted arrival.[4]

Dicus used standard seismic techniques to estimate g from a
fit to the delay versus grazing angle. His value of .79 sec^{-1}
corresponds precisely to the value obtained using his geometry and
θ_c = 34.5 degrees. The actual caustic angle would be somewhere
between 30.3 and 34.5 degrees leading to a gradient estimate
between .57 and .79 sec^{-1}. Because Dicus used plane-wave
expressions in his analysis, he was unable to account for the
disappearance of the refracted path and postulated a rapid
decrease in gradient beyond the turning point depth of the 30.3
degree ray. This would not have been necessary if the point-
source nature of the problem (leading to the emergent caustic)
were considered.

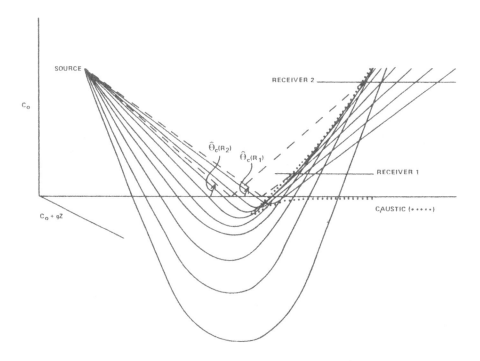

Fig. 5. Ray geometry illustrating dependence of apparent caustic angle, $\hat{\theta}_c$, on receiver height off bottom for deep source.

Another implication of this analysis is that the dependence of $\hat{\theta}_c$ on $\overline{D}$ should lead to measurably different apparent critical angles in the same area using different source/receiver depth combinations. Figure 5 illustrates the expected effect for a deep source and two receivers, one somewhat nearer to the bottom than the other. Since the reflected rays at the range of the caustic are different for different receiver depths, $\hat{\theta}_c$ should increase as the receiver moves away from the bottom.

Data collected by Herstein[5] of NUSC show just such an effect (Figure 6). The expression for $g(\hat{\theta}_c, \overline{D})$ may be used for two different values of $\overline{D}$ and $\hat{\theta}_c$ to eliminate g. For this measurement, the ratios of tan $\hat{\theta}_c$ for the two geometries should be 1.21. The observed ratio is 1.13 and in another data set is 1.20. Such differences are probably within measurement error, especially since in this geometry water-column refraction may not be negligible.

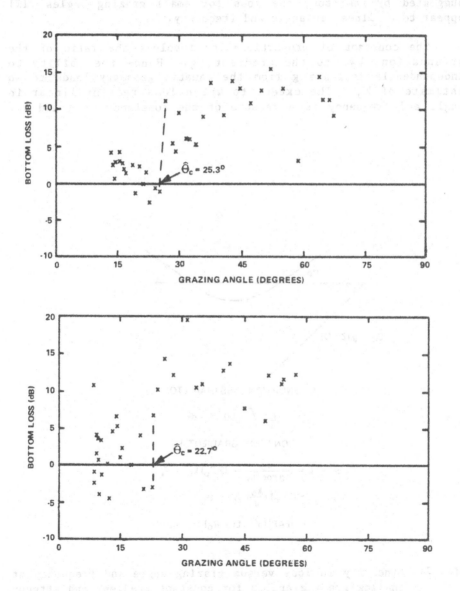

Figure 6: Data showing dependence of $\hat{\theta}_c$ on receiver depth.[5]

ESTIMATING THE ATTENUATION

When the refracting path dominates the reflecting path the difference between the observed level and the level computed assuming reflection may be used to estimate the volume attenuation along the path in the sediment. As shown in Figure 7, for a linear sound speed in the sediment (with negligible offset at the interface) and a constant, frequency-proportional attenuation as suggested by Hamilton, the loss for small grazing angles will appear to be linear in angle and frequency.

The constant of proportionality involves the ratio of the attenuation, k_0, to the gradient, g. Hence the ability to independently estimate g from the caustic geometry leads to an estimate of k_0. The extent to which loss remains linear in angle and frequency is a measure of the constancy of g and k_0.

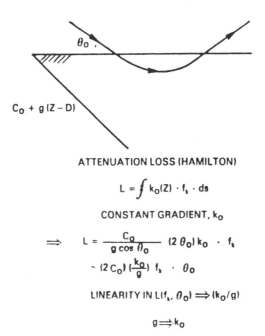

ATTENUATION LOSS (HAMILTON)

$$L = \int k_0(Z) \cdot f_k \cdot ds$$

CONSTANT GRADIENT, k_0

$$\Rightarrow \quad L = \frac{C_0}{g \cos \theta_0} \, (2\theta_0) k_0 \cdot f_k$$

$$\sim (2C_0) \left(\frac{k_0}{g}\right) f_k \cdot \theta_0$$

LINEARITY IN $L(f_k, \theta_0) \Rightarrow (k_0/g)$

$$g \Rightarrow k_0$$

Fig. 7. Linearity in loss versus grazing angle and frequency at shallow grazing angles for constant gradient and attenuation.

Figure 8 compares volume losses versus $\hat{\theta}$ on refracting paths
for different combinations of gradient and attenuation. The
relative roles of profile curvature ($g(z)$) and attenuation gradient ($k'_0(z)$) are illustrated by the solid curves at 100 Hz. The
decrease in gradient and increase in attenuation for $g(z)$ and
$k_0(z)$, respectively, are indicated by their values at 500
meters. The depth-dependent values are typical of terrigenous
sediments where the gradient decrease is strongest. The dependences at higher frequencies are indicated by the dashed curves.
Note that even for the non-constant g and k_0 the loss in the
first 20 degrees of grazing angle is nearly linear in angle and
frequency.

COMPARISON WITH DATA

A number of bottom-loss stations collected by NADC have been
analyzed using this approach to estimating g and k_0. A program
was developed to simulate the measurements including several
processing artifacts (which are not discussed here) given estimated geo-acoustic parameters. Figure 9 compares the results of

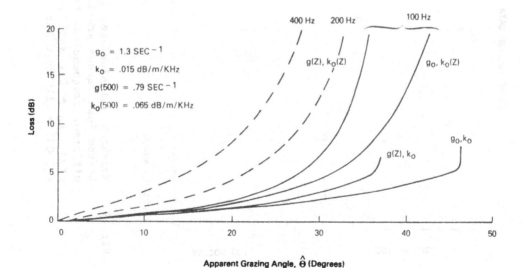

Fig. 8. Loss versus apparent grazing angle at 100 Hz (————) for
        constant and variable gradient and attenuation, and at
        200 and 400 Hz (- - -) for variable g and k_0 typical
        of terrigenous sediments.

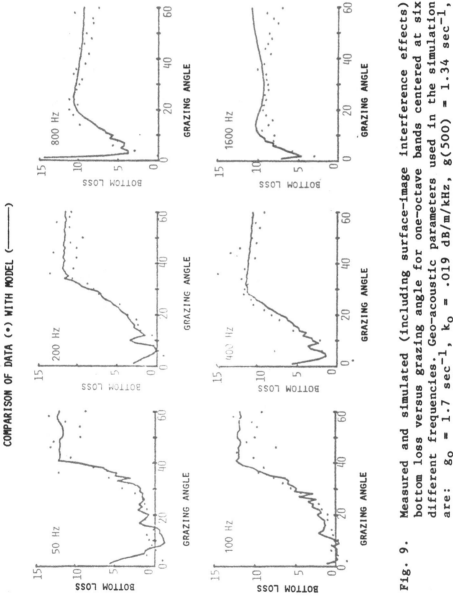

Fig. 9. Measured and simulated (including surface-image interference effects) bottom loss versus grazing angle for one-octave bands centered at six different frequencies. Geo-acoustic parameters used in the simulation are: $g_0 = 1.7$ sec^{-1}, $k_0 = .019$ dB/m/kHz, $g(500) = 1.34$ sec^{-1}, $k_0(500) = .119$ dB/m/kHz.

one such simulation for one-octave bands centered at 50, 100, 200, 400, 800, and 1600 Hz. The geo-acoustic parameters in this case are representative of very low-loss calcareous sediments where the profile curvature is less but the attenuation gradient is more than for the terrigenous sediments.

Many such analyses have been made and are being summarized in a report in preparation. Gradients and profile curvature are quite consistent with those reported by Hamilton[6]. Attenuations are found to vary linearly with frequency and be much smaller than Hamilton's[7] near the sediment surface, reaching his reported values within a few hundred meters.

SUMMARY

A technique has been developed for analyzing existing bottom-loss data in thick-sediment areas to infer geo-acoustic parameters. The sediment gradient is estimated from the rapid increase in loss associated with the bottom-refracted rays' caustic. This then permits separating the attenuation from the gradient and estimating the attenuation from the linearity of loss versus grazing angle and frequency.

ACKNOWLEDGMENTS

The author would like to thank Drs. R. R. Greene and J. B. Hersey and Messrs. W. F. Monet and J. G. Casserly of SAI for their support in this analysis.

REFERENCES

1. E. L. Hamilton, Prediction of deep-sea sediment properties: state-of-the-art, in "Deep-Sea Sediments, Physical and Mechanical Properties," A. L. Inderbitzen, ed., Plenum Press, New York (1974).
2. J. S. Hanna, Some complications in the traditional measurements of bottom loss, Science Applications, Inc. report SAI-76-644-WA (1976).
3. G. V. Frisk, J. A. Doutt, and E. E. Hays, Bottom interaction of low-frequency acoustic signals at small grazing angles in the deep ocean, submitted to J. Acoust. Soc. Am.
4. R. L. Dicus, Preliminary investigations of the ocean bottom impulse response at low frequencies, U. S. Naval Oceanographic Office, TN 6130-4-76 (1976).
5. P. D. Herstein, NUSC New London, private communication.
6. E. L. Hamilton, Sound velocity gradients in marine sediments, J. Acoust. Soc. Am. 65:909 (1979).
7. E. L. Hamilton, Sound attenuation as a function of depth in the sea floor, J. Acoust. Soc. Am. 59:528 (1976).

ATTENUATION ESTIMATES FROM HIGH RESOLUTION

SUBBOTTOM PROFILER ECHOES*

D.J. Dodds

Huntec ('70) Limited
25 Howden Road
Toronto, Ontario, Canada M1R 5A6

ABSTRACT

A sonogram expresses signal power as a function of both time and frequency. Sonograms of sea floor echoes obtained with a high resolution (0.2 ms duration) broadband (1-10 kHz) sound source show effects of surface scattering at the sea floor, frequency selectivity of some subbottom reflectors, and the frequency dependence of sound attenuation in the sediment. By assuming attenuation to be proportional to frequency, sonograms of some subbottom reflectors (targets) yield an estimate of the attenuation in the overlying sediment. The quality of these estimates depends on the frequency range over which a good signal-to-noise ratio exists, the interference of scattered energy with the target reflection, knowledge of the transmitted pulse spectrum, and the frequency characteristics of the target. The estimates of attenuation tend to be lower than those obtained by other workers using samples and in situ probes, but this may be due to the depth of burial.

INTRODUCTION

Broadband subbottom profiler echoes (which are most often used to obtain seismic cross-sections of the seabed) can be analyzed and interpreted in terms of a model of the seabed which includes the effects of sediment attenuation and scattering. Parrott et al.[1] calculated separate measures of coherent and incoherent energy in the first millisecond of a sea floor echo as an aid to geological interpretation of subbottom profiler sections. The coherent and incoherent components were ascribed to reflection

---

* Bedford Institute of Oceanography contribution No. 907.

and scattering, respectively, from the water-sediment interface.
The subsequent work which is reported here exploits the variation
of seabed acoustic response with frequency. A broadband (1-10
kHz) sound source is used. The echoes obtained are analyzed to
obtain a 'sonogram' which depicts echo power as a function of both
time and frequency, in an easily assimilated format. The attenuation and scattering effects of the seabed and the directional
response of the source are frequency-dependent, and the sonogram
can be interpreted in terms of these effects. The frequency
dependence of the energy of a reflection from a subbottom interface can be attributed to attenuation in the overlying sediment,
and the attenuation properties of the sediment estimated. Such
estimates have been made for twelve locations in an area of the
eastern Canadian shelf where sediment types are known from previous mapping, and the results compared with published sediment
attenuation measurements.

ACOUSTIC DATA

     Seismic reflection data is obtained using the Huntec DTS, a
high resolution subbottom profiler[2]. The DTS towed body ('fish')
contains an electrodynamic source ('boomer') which produces a
repeatable impulse-like output directed downward[3]. The duration
of the pressure pulse is about 0.2 ms (Fig. 1). Its spectrum has
useful components between 1 kHz and 10 kHz (Fig. 2). Typically,
the boomer is fired 1 to 4 times per second. Echoes from the sea
floor are picked up by a hydrophone mounted on the fish. The fish
is towed by a survey vessel at speeds of up to 12 km h^{-1} and
depths of up to 300 m. Placing the fish well below the surface,
and closer to the sea floor, improves resolution and signal levels
and reduces the effect of noise and unwanted echoes from the sea
surface. Changes in fish depth, especially those caused by heav-
ing of the towing vessel, can significantly advance or delay an
echo as compared with the previous echo. This can mask the actual
topography of the sea floor and destroy the registration (coherence) of one echo with the next. To correct this problem, fish
depth is continuously sensed with an accuracy on the order of
0.1 m and the boomer firing time is delayed or advanced accordingly[4]. The signal from the hydrophone is recorded on analog
magnetic tape together with necessary timing information, and
simultaneously displayed as a seismic section on a dry paper graphic recorder. An online signal processor corrects for spreading
losses.

CALCULATION OF SONOGRAMS

     Sonograms of the recorded echoes are calculated digitally.
Each echo is digitized at a 50 kHz sample rate. Then the echo
amplitude is normalized for source energy and spreading loses and
an operator is applied to correct the signal spectrum for the

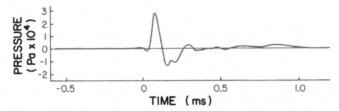

Fig. 1. On axis pressure pulse of the DTS boomer at a range of
         5 m, from an analog magnetic tape recording.

effect of the non-flat source spectrum.* The resulting normalized
echo is processed as shown in Fig. 3 to obtain the sonogram. The
calculation uses a sequence of overlapping time windows, each
window being of short duration compared to the length of the
echo. A Gaussian window function is used. A power spectrum is
calculated from the windowed echo, using each window in turn. The
spectrum values can be presented on a rectangular grid as a function of the time of the center of the window and of frequency; the
spectrum from each time window contributes a column to the grid.
The sonogram is obtained by contouring the resulting grid.

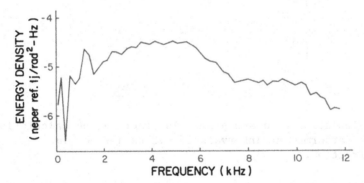

Fig. 2. Spectrum of the on axis DTS boomer output at a range of
         5 m. The quantity plotted is the energy per unit solid
         angle, per Hertz bandwidth, expressed in nepers.

---

* The operator is a Weiner optimum FIR filter designed to shape
  the on-axis pressure pulse (measured during calibration tests)
  into a Gaussian pulse with bandwidth, to the $-\frac{1}{2}$ neper (-4.3 dB)
  point, of 12 kHz, in the presence of an assumed noise spectrum.
  It therefore whitens the signal within the frequency range where
  useful signal power is available. Its length is limited to 0.4
  or 0.8 ms. Due to the length limitation, the inclusion of an
  assumed noise spectrum, and small variations in actual pulse
  shape during field operations, the operator does not precisely
  shape the source signature to the desired Gaussian pulse.

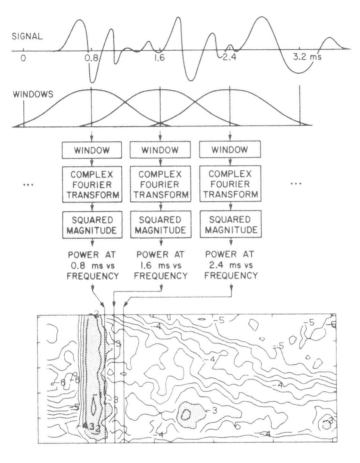

Fig. 3.   Calculation of a sonogram.   In practice the calculation
          is performed at intervals of 0.2 ms in the time
          direction.

The sonogram can also be thought of as the result of passing
the signal through a bank of narrowband filters with overlapping
pass bands. The smoothed power output of each filter contributes
a row to the grid. In fact, the two approaches to sonogram calculation are equivalent provided that the appropriate choice of
filter operator and smoothing technique is used. By considering
the relation between the time width of the window function (window
duration) and the bandwidth of the corresponding filter it can be
shown that there is a trade-off between resolution in the time
direction and resolution in the frequency direction. The filter
corresponding to the Gaussian window function has a Gaussian frequency response. Use of the Gaussian window function minimizes
the product of window duration and filter bandwidth (defined as
the standard deviation of the window function and the filter

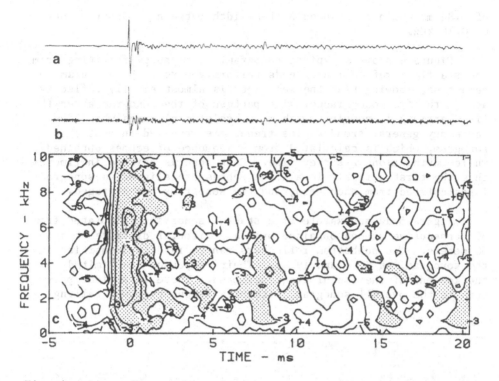

Fig. 4.   A sea floor echo and its sonogram. (a) The recorded echo
from the sea floor.   (b) The echo after application of
the whitening operator. (c) A sonogram of the echo. The
first arrival, reflected from the water-sediment interface, appears at 0 ms, and stretches across the frequency
range shown with little variation in power. A subbottom
reflection appears at 8 ms. Signal amplitude is contoured on a logarithmic scale. The major contour interval is 1 neper (8.7 dB); the minor (unlabelled) contours
are at intervals of 0.25 neper (2.2 dB) but these do not
appear in areas of high gradient. The signal amplitude
is normalized so that a perfect plane reflector would
give a peak value of 0 neper on the sonogram. Areas
inside the -3 contour have been shaded for emphasis.

response respectively)[5]. The sonograms shown here have been calculated with a window duration between $-\frac{1}{2}$ neper (-4.3 dB)* points

* The relative strength of two signals can be expressed in nepers
  or dB. The relative strength in nepers is $\log_e$ of the amplitude ratio. The relative strength in dB is $10 \log_{10}$ of the
  power ratio. The relative power expressed in dB can be found by
  multiplying the relative amplitude expressed in nepers by 8.7.

of 0.84 ms and a corresponding bandwidth between $-\frac{1}{2}$ neper points of 0.76 kHz.

Figure 4 shows a typical sonogram. The energy resulting from the sea floor reflection extends uniformly across the sonogram, at zero time, showing that the sea floor is almost equally reflective across the frequency range. The pattern of the sonogram after the first arrival is complex, with many isolated closures which obscure any general trends. The trends are revealed in a stacked sonogram, which is calculated from a sequence of echoes obtained while moving along a survey line. Each echo is shifted in time so that the first arrivals in all the echoes are aligned, a sonogram is calculated from each echo, and the sonograms are averaged.

Figure 5 is an example of a stacked sonogram, calculated from 30 consecutive echoes obtained over a horizontal distance of about 60 meters. Two subbottom reflectors can be readily identified in the sonogram, at 8 ms and 19 ms. Their responses are limited to the low end of the useful frequency band. There is a clear pattern to the overall sonogram, with high frequency power falling

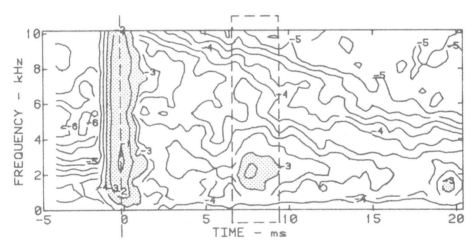

Fig. 5. A stacked sonogram. Sonograms from the echo of Fig. 4 and the 29 echoes following it have been averaged, with the first arrivals aligned. This reduces random variation in the individual echoes and emphasizes the main features. Subbottom reflectors are clearly visible at 8 ms and 19 ms. The reduction in power of the subbottom reflections at high frequencies is interpreted to be caused by attenuation in the overlying sediment. This effect has been analyzed using the spectra of the first arrival at 0 ms and of the subbottom reflector at 7 to 9 ms as indicated.

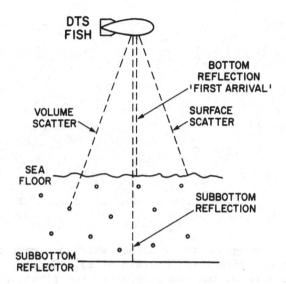

Fig. 6.   A conceptual model of the interaction of 'normal inci-
          dence' sound with the seabed. As well as normal reflec-
          tion of sound from the bottom and underlying strata
          ('subbottom reflectors') significant energy is backscat-
          tered to the DTS fish from roughness features on the
          bottom and from scatterers (e.g., boulders or local
          density changes) below the bottom. Sound is attenuated
          as it passes through the sediment.

off more quickly than low frequency power after the first arrival. In the following sections, all the sonograms which are
discussed have been calculated from 30 consecutive echoes.

QUALITATIVE INTERPRETATION OF SONOGRAMS

     The sonogram patterns can be interpreted in terms of a con-
ceptual model of the seabed (Fig. 6). This model includes the
normal incidence reflection of sound from the sea floor ('bottom')
and subbottom interfaces. In addition, it includes the scattering
of sound from the bottom, due to surface roughness ('surface scattering') and the scattering of sound from within the sediment
('volume scattering'). Scattered energy of both kinds arrives
after the reflection from the bottom (the 'first arrival'). The
model also includes the attenuation of sound during its passage
through the sediment.

     In the sonogram of Fig. 5, which was obtained from a sand
seabed, the energy reflected from the two subbottom reflectors at
8 ms and 19 ms is limited to low frequencies, mainly less than

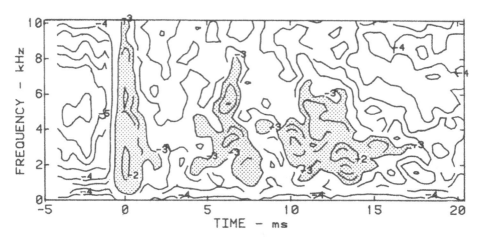

Fig. 7. Stacked sonogram from a seabed which has many coherent
subbottom reflectors (Emerald Silt). Complex patterns
can be seen on the sonogram after the first arrival.
These are interpreted to be the result of constructive
and destructive interference between the reflections from
closely spaced layers.

3 kHz. This is interpreted as an effect of attenuation, since the
rate of attenuation of sound in sediments increases with
increasing frequency[6].

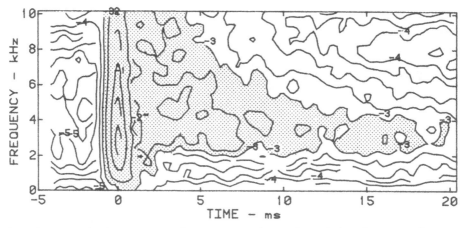

Fig. 8. Stacked sonogram from a smooth surfaced (Sambro Sand)
bottom overlying a sediment (Scotian Shelf Drift) containing many
internal scatterers. The first arrival is clearly outlined by the
−2 contour, and is of a duration comparable to the time resolution
of the sonogram. The energy which arrives after the first arrival
is interpreted to be caused by volume scattering.

Figure 7 is a sonogram from a sediment with many closely spaced coherent internal reflectors (Emerald Silt*). The sonogram shows complex patterns following the sea floor reflection, even though it is a stacked sonogram. Some of the internal reflectors show marked selectivity favouring a narrow frequency band. This type of feature can be interpreted as a result of constructive and destructive interference between closely spaced layers. Mayer[8] has observed similar effects in synthetic seismograms calculated from deep sea core density and velocity logs.

Figure 8 is a sonogram from a seabed having a smooth sand surface with many scatterers in the underlying glacial till material. (Sambro Sand surface overlying Scotian Shelf Drift.) The first arrival is clearly defined and of a duration which is comparable to the time resolution of the sonogram, as can be seen by examining the -2 neper contour. However, considerable energy arrives after the first arrival, evidently as a result of scattering. From geological knowledge of the bottom, this can be ascribed to volume scattering from boulders or perhaps local density variations in the material.

The clearly defined first arrival of Fig. 8 can be contrasted with that of Fig. 9, which is a sonogram from a rough bottom (Scotian Shelf Drift and bedrock). Here the first arrival is not so clearly defined. The -2 neper contour, for example, encloses an area with an extent in the time direction of about 8 ms. The first arrival merges with the later scattered energy. Furthermore, the high frequency content of the first arrival is noticably reduced above 6 kHz. These phenomena can be explained as the effect of strong surface scattering. Such scattering would 'stretch' the reflection from the sea floor. Also, high frequency energy which is normally confined to a narrow beam, would be scattered so as to cover a much wider beam, so reducing the high frequency energy received at the hydrophone.

---

* King[7] has mapped the surficial geology of parts of the eastern Canadian shelf, and classified the sea floor into geological units on the basis of texture, seismic character, and stratigraphic relationships. His classification is used here. It is important to note that although the geological unit called Emerald Silt consists predominantly of sediments in the silt size range, local variations in texture do occur so that no direct relation can be assumed between the unit name and grain size. This applies also to the LaHave Clay, Sambro Sand, and Sable Island Sand and Gravel units.

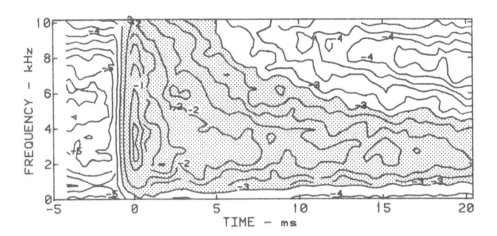

Fig. 9. Stacked sonogram from a rough (Scotian Shelf Drift and
         bedrock) bottom. There is no clear distinction between
         the first arrival and the energy which arrives after it,
         and the high frequency content of the first arrival is
         reduced. This is interpreted to be the result of strong
         surface scattering.

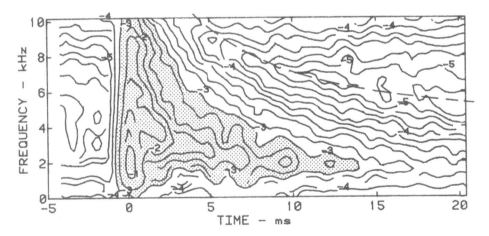

Fig. 10. Stacked sonogram from a seabed of shells overlying sand
         (Sable Island Sand and Gravel - shell facies). There is
         no clear distinction between the first arrival and the
         energy which arrives after it. This is interpreted to be
         the result of strong surface scattering from the shell
         bed. The trough extending from 4 ms, 10 kHz to 20 ms,
         5 kHz coincides with the travel time to the bottom along
         the direction of the first node in the radiation pattern
         of the source.

The effect of surface scattering is again evident in Fig. 10, which was obtained from a seabed consisting of a bed of shells overlying sand (Sable Island Sand and Gravel - shell facies)[1]. Again, the first arrival is stretched and its high frequency content reduced. An additional feature of this sonogram is the trough extending from 20 ms, 5 kHz to 4 ms, 10 kHz. The position of this trough corresponds to the travel time to the sea floor along the direction of the first node in the radiation pattern of the boomer. The radiation pattern, when measured at discrete frequencies, becomes wider with decreasing frequency. Thus the travel time to the bottom along the direction of the first node increases with decreasing frequency. The dashed line on the sonogram corresponds to the expected position of the trough*. Its close agreement with the observed position demonstrates that the scattered energy evident in this sonogram is returned from the sea floor or just below, and not from volume scatterers deep below the sea floor.

THE ATTENUATION EFFECT - THEORY

Attenuation will cause the amplitude of a plane sine wave in a linear medium to fall off exponentially with the distance travelled. Thus,

$$p = p_0 e^{-\alpha x} \tag{1}$$

where p is the sound pressure amplitude;
    p_0 is the sound pressure amplitude at x=0;
    α is the attenuation rate;
    x is the distance travelled.

This may be rewritten,

$$\log_e p = \log_e p_0 - \alpha x \tag{2}$$

In general, the attenuation rate α is a function of frequency. The nature of this function depends upon the mechanism of attenuation. At least three possible mechanisms can be identified: anelastic losses in the frame of sediment particles; viscous losses

---

* To calculate the expected position of the trough, the source is modelled as a circular piston of 0.5 m diameter[9]. The first node in the radiation pattern will then be at 47° off axis at 5 kHz, 32° at 7 kHz, and 21° at 10 kHz. The trough feature moves to later times as the height of the source above the bottom is increased. It is weak or absent in Fig. 5, 7, 8 and 9, in part because the height was greater than in Fig. 10.

due to relative movement of the sediment particles and the pore water; and losses due to volume scattering. If a simple model of anelastic losses is applicable, the attenuation rate α should be proportional to frequency[6]. A model of viscous losses[10] predicts that α is proportional to the second power of frequency in the low frequency limit (e.g. < 10 Hz in sands) and to the one-half power of frequency in the high frequency limit (e.g. > 100 kHz in sands). If volume scattering obeys the Rayleigh scattering law (which applies to objects which are small relative to the sound wavelength) scattering cross section will be proportional to the fourth power of frequency[9]. Assuming a uniform random distribution of scatterers in the medium, the amplitude of a plane sine wave will therefore fall exponentially with α proportional to the fourth power of frequency. In all cases, attenuation disappears at zero frequency. Losses in actual ocean sediments appear to be nearly proportional to frequency, at least over the frequency range being considered here[6], but there is recent evidence for significant departures from the 'first power law' in sands[11].

If we assume that the rate of attenuation is proportional to frequency (recognizing that this assumption must be treated with caution), equation (2) becomes,

$$\log_e p = \log_e p_0 - kxf \tag{3}$$

where k is the attenuation constant;
    x is the distance travelled through the sediment;
    f is the frequency.

Thus $\log_e p$ as a function of f is linear, with slope $-kx$.

ATTENUATION CONSTANT ESTIMATION

The linear dependence of $\log_e$ amplitude on frequency should be visible in a sonogram of a subbottom reflector ('target'). The spectrum of the target reflection at time 8 ms in Fig. 5 can be obtained by integrating the sonogram over a time interval which includes the reflection.* This spectrum is shown in Fig. 11(a). Its shape is the result of three effects: the residual effect of non-uniform source spectrum (which is not completely corrected by

---

* The integration is used to overcome a problem caused by the stacking of sonograms. If the target is dipping relative to the bottom, the target reflection will occupy a range of times on the stacked sonogram, and no single column of the stacked sonogram will include equal contributions from the target reflections in all the echoes.

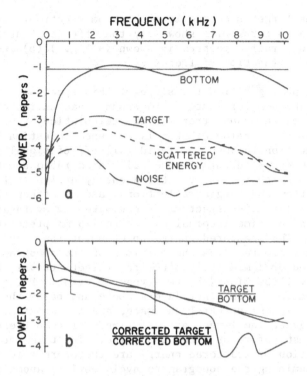

Fig. 11. An analysis of the sonogram of Fig. 5 showing an estimate of the attenuation constant. (a) Spectra of the bottom reflection, target reflection, noise before the first arrival, and 'scattered energy' around the time of the target arrival. (b) Ratio spectra of target power/bottom power. The corrected ratio spectrum is calculated from target and bottom spectra which have been corrected for scattered energy and for noise respectively. The vertical lines show the frequency range in which the uncorrected spectrum is considered valid. The slope of the straight line fitted to the uncorrected ratio spectrum is −0.22 neper/kHz (−1.9 dB/kHz). The travel time is 8.4 ms relative to the bottom arrival. Using an assumed sound velocity of 1.8 km/s this yields an attenuation constant $k = 14 \times 10^{-3}$ neper/m-kHz (0.12 dB/m-kHz).

the whitening operator); the reflection frequency response of the target interface (and the transmission frequency response of any overlying interfaces) and the effect of attenuation between the bottom and the target. Let us assume that the responses of the bottom and the target interface are 'white'; that is, their reflectivity is independent of frequency. This is the case for a plane interface between homogeneous media. With these assumptions, the effect of non-uniform source spectrum can be corrected by dividing the target spectrum by the bottom spectrum. The resulting ratio spectrum will show only the effects of attenuation. The $\frac{1}{2} \log_e$ ratio spectrum is shown in Fig. 11(b) and is close to a linear function of frequency.

Before computing k from the slope of this spectrum, it is necessary to take some precautions to ensure that extraneous effects do not cause undue errors in the calculation. The portion of the ratio spectrum below 0.7 kHz is not used in estimating sediment attenuation because the rapid fall-off of signal power at low frequencies causes errors in the calculated ratio spectrum. In addition, noise energy and scattered energy arrive at the same time as the bottom and target reflections, and so affect the measured spectra. The noise spectrum is estimated by averaging the sonogram across the time interval 4.0 ms to 1.5 ms preceding the first arrival. The scattered energy spectrum at the target arrival time can be estimated from the portions of the sonogram immediately before and/or immediately after the target arrival, although this estimate will be biased by two types of error. Energy of coherent reflections from minor layers above and below the target may be included in the spectral estimate; and the spectrum of scattered energy at the target arrival time will be different from the spectrum just before or after that time. The time interval(s) for the estimation of scattered energy are therefore selected manually by examining the sonogram to avoid obvious anomalies. The noise spectrum and the scattered energy spectrum are subtracted from the bottom spectrum and target spectrum respectively, and a ratio spectrum is computed from these 'corrected' spectra. This corrected ratio spectrum is not directly used for calculating the attenuation constant, because of the uncertainties involved in the corrections, but the uncorrected ratio spectrum is assumed to be reliable only in the frequency range where it agrees with the corrected spectrum within 0.3 neper (3 dB). Above that range, there is usually a large contribution of scattered energy to the ratio spectrum. At high frequencies, the spectrum of the target reflection may therefore be less than the ratio spectrum, although it cannot be greater.

Having observed these precautions, a straight line can be fitted to the reliable portion of the uncorrected ratio spectrum. Its measured slope is an estimate of $(-kx)$. The distance x can be estimated from the arrival time of the target reflection after the first arrival and an estimate of sound velocity in the sediment above the target reflector. Thus k can be estimated.

Since attenuation vanishes at zero frequency, the intercept of the fitted straight line is an estimate of the target reflection strength relative to the first arrival strength, with the effect of attenuation removed. If the reflectivity of the bottom

were estimated from the first arrival spectrum, the reflectivity
of the target could be estimated using the intercept value.

The example of Fig. 11 yields a spectrum which is very nearly
linear. Deviations from a linear function may indicate attenuation which is not linear with frequency. However, they may also
be due to the bottom or target reflectivity varying with frequency, or to the selective reflection of some frequencies from
layers between the bottom and the target. Therefore spectra which
do not fit a linear function are not used for attenuation estimates. No attempt has been made to fit nonlinear functions to the
ratio spectra.

ATTENUATION ESTIMATES

Using the technique described in the previous section, an
estimate of attenuation constant k was made for twelve locations
on the eastern Canadian shelf where DTS data had been recorded.
At each location, a 'target' reflector was visible on the DTS
seismic section, and interpretation indicated that the layer of
sediment between the sea floor and the target reflector was of
uniform geological classification. Locations representing a range
of sediments were selected for analysis. Between 10 and 50
stacked sonograms (calculated from data obtained over a segment of
survey line 1 to 3 km in length) were analyzed for each location.
The sea floor sediment was assigned a geological classification
and a textural classification based on previous mapping work[7].
Each line of Table 1 summarizes the results from one location,
showing the mean and standard deviation of the estimated values of
the attenuation constant k. Geological and textural classifications, and statistics on the thickness of sediment above the
target, are also shown. The table is in increasing order of estimated grain size. Mean estimated values of k vary between
0.7×10^{-3} neper/m-kHz (0.006 dB/m-kHz) and 32×10^{-3} neper/m-kHz
(0.28 dB/m-kHz). Standard deviations vary between 0.2×10^{-3} neper-
/m-kHz (0.002 dB/m-kHz) and 9×10^{-3} neper/m-kHz (0.08 dB/m-kHz).
In general, greater thicknesses of sediment give smaller standard
deviations. As a percentage of the mean value of k, the standard
deviations vary between 3% and 60%. The mean estimated attenuation constants show a positive correlation with grain size.

DISCUSSION

For comparison, Table 1 shows predicted ranges of values for
the attenuation constant k derived from measurements in the literature. For each location, a grain size range was estimated from
the geological classification of the sediment. This was combined
with Hamilton's graphical relationship[6] between grain size and
attenuation (which summarizes several workers' published results)

Table 1. Summary of Attenuation Estimates

Location number	Sediment type	Thickness – m	k Values – neper/m-kHz·10^{-3}	
			measured	predicted (Hamilton)
La Have Clay				
1	Silty Clay	23 ±2	0.7 ±0.3	3 to 17
2	Silty Clay	37 ±2	1.4 ±0.2	3 to 17
3	Silty Clay	37 ±0.5	1.6 ±0.3	3 to 17
4	Silty Clay	20 ±6	2.8 ±1.3	3 to 17
5	Clayey Silt	9 ±3	6.7 ±4	5 to 20
Emerald Silt				
6	(Silt with	35 ±2	3.1 ±0.6	5 to 20
7	(Clay, Sand,	20 ±3	2.7 ±0.8	5 to 20
8	(and Gravel	12 ±1	2.1 ±1.0	5 to 20
Sambro Sand				
9	Silty Sand	3.7 ±1	19 ±9	30 to 70
Sable Island Sand and Gravel				
10	(Sand with	11 ±2	14 ±3	30 to 70
11	(Gravel	6.1 ±0.2	31 ±5	30 to 70
12	(	15 ±0.6	32 ±1	30 to 70

to yield a predicted range for the attenuation constant. A positive correlation of attenuation constant with grain size appears
both in the present estimates and the earlier results. The estimates of k are typically half the lower limit of the range predicted from earlier results. While the correlation of the present
k values with earlier results is encouraging, the reason for the
discrepancy in the actual values is not clear. Certainly, grain
size does not by itself fully characterize a sediment. The different source and mode of deposition of the sediments of the Eastern Canadian shelf, controlled as it is by glaciation, may have
resulted in differences in sediment properties from those of the
southwestern United States coast where much of the earlier work
has been performed. Another explanation is the difference in the
conditions under which attenuation was measured. Most of the
values reported by Hamilton were obtained in situ in the first one

or two meters of the seabed or on unconfined samples in the laboratory. In both cases the static stresses would be small. On the other hand, the sonogram analyses involve sediment layers 4 to 40 m thick. Assuming that the pore water is free to move through the frame of sediment grains, the sediment frame at any depth below the sea floor supports the weight of the sediment grains above that depth (although buoyancy reduces this weight). Thus large static stresses are exerted on the sediment frame in the layers which have been investigated with the sonogram technique. Experimental work on sands has shown a reduction in attenuation as static stress increases[11]. This effect may account for the lower attenuation constants found in the present work.

To properly verify an attenuation estimate, it is necessary to make a second estimate at the same location using a different technique. A possible approach is to perform acoustic propagation modelling and sonogram analysis at the same location, matching as closely as possible the sound frequencies and sediment thicknesses involved. Preliminary steps have been taken to do this by Chapman and Ellis[12,13]. They used an attenuation constant obtained by the sonogram technique as a parameter in a propagation model and obtained fair agreement between model predictions and actual propagation loss measurements.

The reliability of attenuation estimates using sonogram analysis depends on the assumption that the bottom and target interfaces are equally reflective across the frequency range used. Hardware is currently under development to permit continuous accurate calibration of the source spectrum; this will eliminate the need to assume that the bottom is a 'white' reflector. The assumption that the rate of attenuation is proportional to frequency is a potential source of error. At present, sonogram analyses which show serious departures from the first power law are rejected, but this does not preclude the possibility that the first power law fails outside the frequency range analyzed. Recent work[11,14] may allow better definition of the importance of the viscous loss mechanism and the associated departures from the linear dependence of attenuation on frequency. A quantitative analysis of volume scattering using sonograms may permit an estimate of its importance as an attenuation mechanism which causes departures from linear dependence.

Although only the attenuation effect has been analyzed here, it is possible to describe all the phenomena of the 'conceptual model' (Fig. 6) quantitatively. Such a model would permit the calculation of a 'synthetic sonogram' from parameters of surface roughness, acoustic impedance, attenuation, and density of volume scatterers. This raises the possibility that the parameters of a synthetic sonogram could be manipulated to achieve agreement with an actual sonogram, thereby yielding estimates of the parameter

values. Such parameters would be significant to both the
geologist and the acoustician.

ACKNOWLEDGEMENTS

   This work was supported by the Canadian Government through
the Seabed Project (administered by the Atlantic Geoscience Centre
of the Geological Survey of Canada) and through the Defence
Research Establishment Atlantic. The data used was collected on
cruises of the CSS Hudson and CFAV Quest. The research was performed at the Bedford Institute of Oceanography, Dartmouth, Nova
Scotia. L.H. King provided valuable advice and assisted with geological interpretations of DTS records. The manuscript was reviewed by R.K.H. Falconer, J.M. Ross, D. Chapman and D. Ellis.
The contribution of these organizations and individuals is gratefully acknowledged.

REFERENCES

   1. D.R. Parrott, D.J. Dodds, L.H. King and P.G. Simpkin, Mea-
      surement and evaluation of the acoustic reflectivity of
      the sea floor, Canadian Journal of Earth Sciences (in
      press).
   2. R.W. Hutchins, D.L. McKeown and L.H. King, A deep tow high
      resolution system for continental shelf mapping, Geoscience
      Canada 32 (2): 95-100 (1976).
   3. D.L. McKeown, Evaluation of the Huntec ('70) Hydrosonde
      Deep Tow Seismic System, Report BI-R-75-4, Bedford Insti-
      tute of Oceanography, Dartmouth, Nova Scotia, Canada
      (1975).
   4. R.W. Hutchins, Removal of tow fish motion noise from high
      resolution seismic profiles. Presented at SEG-US Navy
      Symposium on Acoustic Imaging Technology and On Board Data
      Recording and Processing Equipment, National Space
      Technology Laboratories, Bay St. Louis, Mississippi,
      U.S.A. August 17-18, (1978).
   5. F.J. Harris, On the use of windows for harmonic analysis
      with the discrete Fourier transform, Proc. IEEE 66 (1):
      51-83 (1978).
   6. E.L. Hamilton, Compressional-wave attenuation in marine
      sediments, Geophysics 37 (4): 620-646 (1972).
   7. L.H. King, Surficial geology of the Halifax-Sable Island map
      area, Marine Sciences Branch Paper 1, Department of Energy,
      Mines and Resources, Ottawa (1970).
   8. L. Mayer, The origin of fine-scale acoustic stratigraphy in
      deep-sea carbonates, Journal of Geophysical Research 84
      (B11): 6177-6184 (1979).
   9. C.S. Clay and H. Medwin, "Acoustical Oceanography", John
      Wiley & Sons, Inc., New York (1977).

10. R.D. Stoll, Acoustic waves in ocean sediments, Geophysics 42 (4): 715-725 (1977).

11. R.D. Stoll, Experimental studies of attenuation in sediments, Journal of the Acoustical Society of America 66 (4): 1152-1160 (1979).

12. D.M.F. Chapman and D.D. Ellis, Propagation-loss modelling on the Scotian Shelf: The geo-acoustic model, Saclant ASW Research Conference on Ocean Acoustics Influenced by the Sea Floor, LaSpezia, Italy, June 9-12, (1980).

13. D.D. Ellis and D.M.F. Chapman, Propagation-loss modelling on the Scotian Shelf: Comparison of model predictions with measurements, Saclant ASW Research Conference on Ocean Acoustics Influenced by the Sea Floor, LaSpezia, Italy, June 9-12, (1980).

14. J.M. Hovem and G.D. Ingram, Viscous attenuation of sound in saturated sand, Journal of the Acoustical Society of America 66 (6): 1807-1812 (1979).

LOW FREQUENCY BOTTOM REFLECTIVITY MEASUREMENTS

IN THE TUFTS ABYSSAL PLAIN

N. R. Chapman

Defence Research Establishment Pacific
FMO Victoria, BC
VOS 1BO

ABSTRACT

Ocean bottom reflectivity has been studied at two sites over the Tufts Abyssal Plain in the northeast Pacific Ocean. Measurements of bottom loss were determined from measured propagation loss of the first, second, and third bottom bounce paths over a range of grazing angles from 5° to 75°, and for a number of 1/3 octave bands from 40 Hz to 600 Hz. At these low frequencies the bottom loss versus grazing angle indicates a critical angle behaviour, and the loss extrapolates to a value of 6 dB at normal incidence.

The theoretical bottom loss was computed using a simple two-layer model of the ocean bottom in which the sediment is modeled by a layer with a constant sound speed gradient and constant attenuation, and the underlying mantle crust is modeled by a half space with constant sound speed and attenuation. A good fit with the acoustic measurements was achieved using values of the model parameters obtained from seismic experiments carried out in the study.

INTRODUCTION

Bottom reflectivity has been studied at two sites in deep waters of the northeast Pacific. Both sites, shown in Figure 1 at 46°34'N 140°40'W (Site 1) and 46°N 143°30'W (Site 2), overlie the Tufts Abyssal Plain, a region of turbidite deposits. The experiments were carried out to provide estimates of the bottom reflection loss and to develop a geoacoustic model of the ocean bottom.

Measurements of bottom loss versus grazing angle have been determined from multipath propagation loss measurements at both sites, and are presented in third octave bands from 40 Hz to 600 Hz. The bottom loss concept provides a simple description of the interaction with the ocean bottom in terms of Rayleigh reflection; however, the nature of the acoustic interaction with the ocean bottom is more complicated than the simple picture of specular reflection from a single interface. Consequently the propagation data have been used to probe the structure of the bottom to obtain parameters of a geoacoustic model. The detailed structure of the near-surface sediments has been determined using deconvolution processing to obtain the ocean bottom impulse response. Information about the deeper sediments and substrate layers has been obtained by analyzing the travel times of the head waves from the interfaces between the layers.

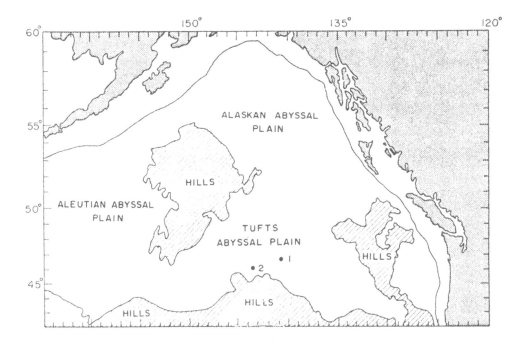

Fig. 1. Geographical location of the experiments. The sites are indicated by the closed circles.

The geophysical information from the bottom profiles and the bottom loss measurements has been used to develop a model of the bottom. In this model the ocean is described by a constant sound speed half space, and the bottom is described by a sediment layer of constant sound speed gradient overlying another half space of constant sound speed. Both regions are assumed to be fluids of constant density and attenuation. Using this model to account for the subbottom structure, the plane wave reflection coefficient and the propagation loss were calculated and compared with the experimental data.

DESCRIPTION OF THE EXPERIMENTS.

The bottom reflectivity data have been collected in experiments carried out with two ships to measure propagation loss of low frequency sound. One ship was used to monitor the receiving hydrophone which was suspended from a free floating buoy at a depth of about 400 m, and the second ship opened range on a specified course and deployed 1.8 lb SUS charges at depths of 21 m and 188 m at intervals of about 1 km. The signals from the 188 m shots were used to determine the bottom loss, and the refraction profile was taken with the 21 m shots.

The signals were digitized at a rate of 1500 Hz and bandpass filtered in 1/3 octave bands over a range of 20 Hz to 600 Hz. The energy of each multipath arrival was obtained from the filtered data using a threshold detector, and the propagation loss was determined using values of SUS source levels measured in recent experiments at DREP[1].

GEOACOUSTIC MODELING OF THE OCEAN BOTTOM

Quantitative information about the geophysical nature of the bottom at each site was obtained from the measurements of bottom loss, seismic refraction profiles, and deconvolution processing to determine the ocean bottom impulse response. The results of these measurements, described below, are used to develop the geoacoustic model.

Bottom Loss Measurements

The bottom reflection loss was determined from the measured propagation loss of the bottom bounce paths by subtracting an estimate of the water column loss along the paths. In this approach, the estimates of water column propagation loss were

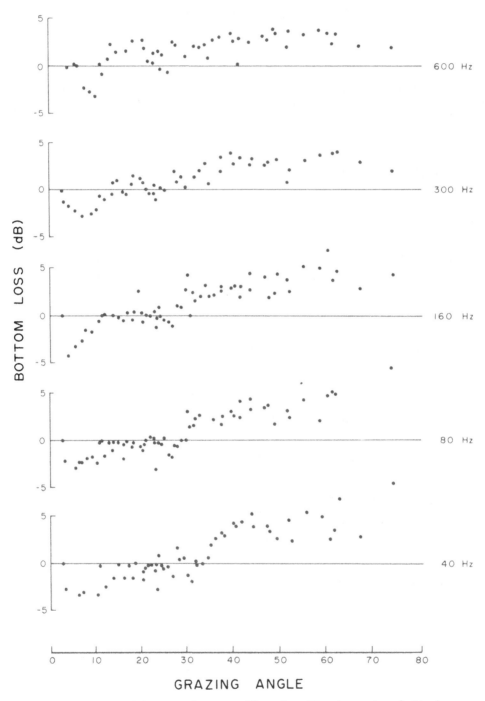

Fig. 2. Measured bottom loss at Site 1. The loss is plotted vs. grazing angle for 1/3 octave bands from 40 Hz – 600 Hz.

calculated by ray theory, assuming specular reflection with no loss from a single interface at the ocean bottom. The method is described by the expression

$$BL = (H - H_c + 6)/n \qquad\qquad (1)$$

where BL is the bottom loss (in dB), H and H_c are the measured and calculated bottom bounce propagation loss, and n is the order of the bottom bounce path. In this work, the first, second, and third bottom bounces have been used. The 6 dB correction accounts for the four bottom interacting paths associated with each order of bottom bounce, assuming a random phase contribution from signals of equal amplitude.

The 1/3 octave average bottom loss measurements are presented in Figures 2 and 3, which show bottom loss versus grazing angle at frequencies from 40 Hz to 600 Hz. At both sites the bottom loss is generally less than 6 dB at the largest grazing angles, indicating strong propagation via bottom bounce paths at these low frequencies. The frequency dependence is similar at both sites, with the loss at large grazing angles increasing from about 3 dB at 600 Hz to about 6 dB at the lowest frequency.

In the bottom loss measured at Site 1 (Figure 2), a critical angle is observed which decreases from about 38° at 40 Hz to about 12° at 600 Hz. This critical angle is probably associated with the reflection of compressional waves near the ocean-sediment interface. The frequency dependence indicates that the lower frequencies interact with a deeper sediment layer of greater sound speed than the near-surface sediments. At about 70° there is an indication of another critical angle, likely due to reflection from the substrate beneath the sediments. The negative values of bottom loss at grazing angles smaller than 10° are attributed to the use of the Rayleigh reflection model in the analysis of the data. This behaviour has been observed also by Santaniello et al. in experiments in the north Atlantic[2].

In Figure 3, the low frequency bottom loss curves for the measurements at Site 2 demonstrate a critical angle which decreases from about 38° at 40 Hz to about 30° at 160 Hz. At higher frequencies the critical angle is not well defined, although there is significant loss at some grazing angles less than 30°. This behaviour is somewhat different than that observed at Site 1 indicating that the subbottom structure is different at the two sites. The general behaviour at high frequencies can however be attributed to the interaction with near-surface reflecting layers of lesser sound speed than the deeper sediments.

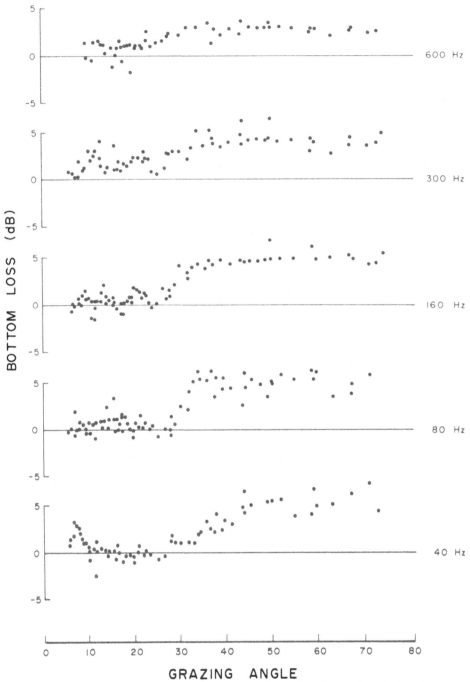

Fig. 3. Measured bottom loss at Site 2. The loss is plotted vs. grazing angle for 1/3 octave bands from 40 Hz – 600 Hz.

Seismic Refraction Profile

Seismic refraction profiles were obtained from the experiments with the 21 m shots. These experiments provided estimates of the thickness and average sound speed of the sediment column and the deeper crustal layers. The shot records were low-pass filtered at 12.5 Hz and the travel time of the first head wave arrival was plotted versus range. The curves were analyzed by assuming the bottom was a horizontally stratified medium, and the thickness and sound speed of each layer were determined by standard procedures described by Officer[3].

The bottom profiles of both sites are shown in Figure 4. The values of sound speed for the first two layers of crustal rock are in good agreement with the measurements reported by Raitt[4] for the northeast Pacific. (The first arrival head wave refracted from the deepest layer was not observed in our experiments because the shot runs extended only to about 40 km).

At Site 1, the sediment thickness and sound speed were estimated to be 0.4 km and 2.3 km/sec. These values were obtained from a slower head wave which was associated with critical refraction in the sediment column. The estimated thickness agreed closely with the results of a continuous seismic reflection profile which was carried out during our experiments with a sparker at that site. At Site 2 the sediment thickness of 0.2 km was estimated from the refraction analysis and bathymetry data; this layer is somewhat thinner than at Site 1.

Ocean Bottom Impulse Response

The detailed structure of the near-surface sediments was determined by deconvolution processing to obtain the ocean bottom impulse response. Deconvolution in the frequency domain has been commonly used to remove the interference of bubble pulses in the signals from explosive charges. The method used in this analysis is similar to the procedure described by Dicus[5], and the mathematical description of the process is given in equation 2,

$$h(t) = F^{-1} \left[\frac{R(i\omega) \, D^*(i\omega)}{|D(i\omega)|^2 + |N(i\omega)|^2/B^2} \right] \tag{2}$$

F^{-1} indicates an inverse Fourier transform of the expression within the brackets. R, D, and N are the Fourier transforms of the bottom reflected signal, the direct path signal, and a noise sample, and B^2 is an estimate of $(R(i\omega)/D(i\omega))^2$.

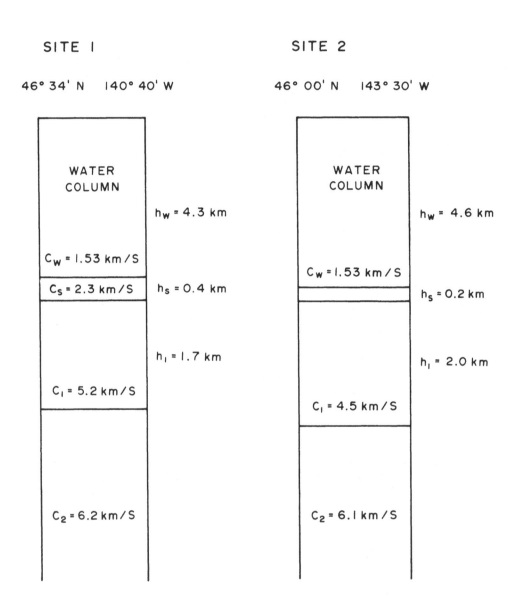

Fig. 4. Bottom profiles determined from seismic refraction
         profiles at Sites 1 and 2.

The bottom reflected signal was obtained by time windowing
the first bottom bounce arrivals from the 188 m shots to select
the first arrival which interacted only with the ocean bottom.
For this shot depth the length of the window varied from about
120 msec at the closest ranges to about 80 msec near the first
convergence zone. The waveform used for the direct path signal
was generated from earlier close range measurements of the SUS
charge waveform. This waveform was scaled using a sinx/x interpolation to compensate for the effects of small variations in
explosion depth observed between the generated direct path signal
and the received signal from each shot deployed in the propagation run. The explosion depth was measured by cepstrum techniques.

The deconvolved impulse responses of the ocean bottom at
Site 2 are shown in Figure 5 where the broadband bottom interacted signals are displayed for grazing angles from 28° to 6.5°.
The bubble pulse interference has been removed, and at least 2
prominent arrivals are observed after the reflection from the
water-sediment interface. These arrivals were interpreted as
reflections from shallow subbottom layers of constant thickness
and sound speed. Assuming that the layer thickness and sound
speed are independent of range, and that the layer is thin
compared to the water column, the arrival time difference, Δt, of
the signals from the ocean bottom and the reflecting layer is
related to the grazing angle, θ_g, at the bottom by

$$(\Delta t)^2 = \left(\frac{2h}{c_s}\right)^2 - \left(\frac{2h}{c_w}\right)^2 \cos^2\theta_g \qquad (3)$$

where h and c_s are the sediment layer thickness and sound speed,
and c_w is the sound speed at the base of the water column (c_w=
1.53 km/sec at Site 2). The values of h and c_s were determined
from a linear least squares fit of a plot of Δt^2 vs $\cos^2\theta_g$.

For the first layer, h = 7 m and c_s = 1.56 km/sec.
Using the relationship

$$\cos\theta_c = c_w/c_s, \qquad (4)$$

the critical angle, θ_c, for reflection from this layer is 12°.
Consequently, reflection from this layer could account for the
bottom loss at small grazing angles ($\theta \sim 10°$) observed at 300 Hz
and 600 Hz in Figure 3. At lower frequencies, the effect of this
shallow layer is not likely significant, and reflections will be
observed from the deeper layers.

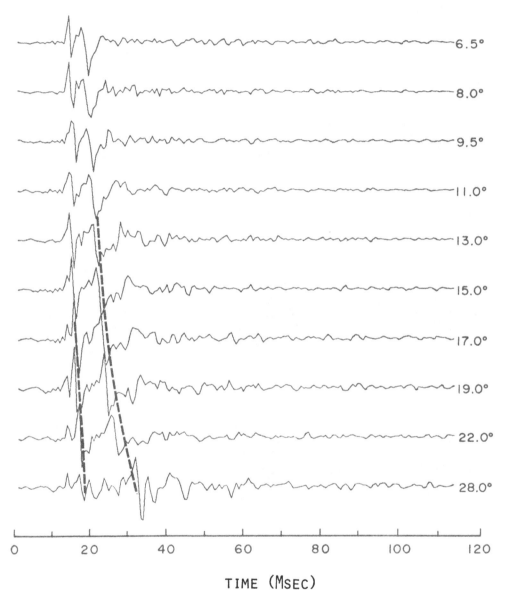

TIME (MSEC)

Fig. 5. Deconvolved pressure-time histories for grazing angles
         from 28° to 6.5°. (The signal amplitude is in arbitrary
         units). The subbottom reflections are identified by the
         broken curves.

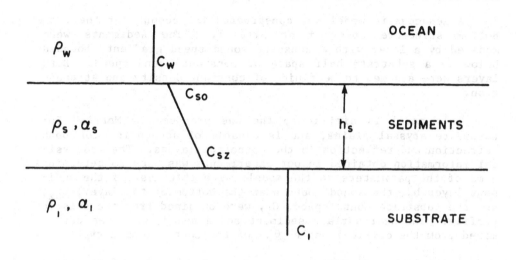

$$\rho_w = 1.03 \; gm/cm^3$$
$$C_w = 1.53 \; km/S$$

$$\rho_s = 1.6 \; gm/cm^3$$
$$C_{so} = 1.7 \; km/S$$
$$C_{sz} = 2.3 \; km/S$$
$$\alpha_s = 0.002 \; dB/m$$
$$h_s = 0.4 \; km$$

$$\rho_l = 2.2 \; gm/cm^3$$
$$C_l = 5.2 \; km/S$$
$$\alpha_l = 0.0008 \; dB/m$$

Fig. 6. Geoacoustic model of the bottom. The parameters of the ocean layer, the sediments, and the substrate are identified by the subscripts, w, s, and l, respectively.

Geoacoustic Model of the bottom

A geoacoustic model was constructed to account for the sub-bottom structure observed at Site 1. The sediments were modeled by a layer with a constant sound speed gradient, bounded below by a substrate half space of constant sound speed. Both layers were assumed to be fluids of constant density and attenuation.

This model is similar to the one proposed by Morris[6] for turbidite abyssal plains, and is capable of accounting for both refraction and reflection in the subbottom layers. The geophysical information obtained in our experiments was used to determine some of the parameters in the model. The thickness of the sediment layer h_s, the sound speed near the bottom of the layer, C_{sz}, and the substrate sound speed, C_1, were obtained from the seismic profile. The near-surface sediment sound speed, C_{so}, was determined from the critical angle, θ_c, of the bottom loss curve.

Initially, the attenuation, α, and density, ρ, of the sediment and the substrate were taken from the values reported by Hamilton for abyssal plains of turbidite composition [7,8]. The final value of the sediment attenuation was determined by fitting with the bottom loss data, as discussed below. The model is shown in Figure 6.

COMPARISON WITH EXPERIMENT

The geoacoustic model was tested by comparing calculations of the plane wave reflection coefficient and propagation loss with the measurements made at Site 1. The plane wave reflection coefficient, calculated for this model following Morris[6], is compared in Figure 7 with the measured bottom loss in the 160 Hz band. Although this calculation does not account for the interaction of spherical waves with the ocean bottom, a comparison with the plane wave reflection coefficient is meaningful for the conditions of shallow source and receiver used in the experiment[9].

The modeled bottom loss, shown in Figure 7 by the solid and broken curves, was computed from the relationship

$$BL = -20 \log |R| \tag{5}$$

where R is the plane wave reflection coefficient averaged over a 1/3 octave band. The solid curve was computed using a value of 0.002 dB/m for the sediment attenuation, whereas the broken curve was obtained using the value of 0.01 dB/m reported by Hamilton[8]. Better agreement with the data was achieved with the smaller value of sediment attenuation.

The solid curve, which accounts for refraction in the sediments and reflection from the substrate, agrees fairly well with the measurements except in the range of grazing angles from 50° to 70° where the model predicts a smaller loss. Some disagreement with the data is expected, however, since the model does not account for the effects of sediment or substrate rigidity. In addition, the sound speed variation within the sediment has been approximated by a linear gradient; consequently the effect of reflecting layers which can introduce discontinuities in the sound speed profile has been ignored.

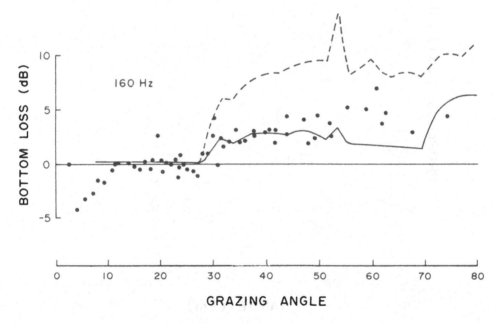

Fig. 7. Calculated and measured bottom loss for Site 1 in the 160 Hz band. The solid and broken curves are calculated using a value of sediment attenuation, α_s = 0.002 dB/m, and α_s = 0.01 dB/m, respectively. The measurements are shown by the closed circles.

The geoacoustic model was used to describe the bottom in a calculation of propagation loss, and the result was compared with the experimental measurement. The solid curve shown in Figure 8 is a normal mode calculation[10] of the propagation loss at 40 Hz based on the water column sound speed profile measured at Site 1 and the geoacoustic model of the bottom. Although the calculation does not include the continuous mode spectrum, the agreement with the measured propagation loss is good in the ranges between about 15 km and 35 km where the propagation is dominated by bottom-interacted paths. This agreement is encouraging considering the simplicity of the geoacoustic model.

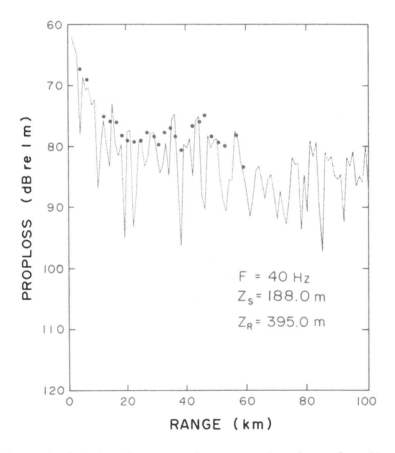

Fig. 8. Calculated and measured propagation loss for Site 1. The normal mode calculation at 40 Hz is shown by the solid curve, and the measured propagation loss in the 40 Hz – 1/3 octave band is indicated by the closed circles.

REFERENCES

1. N. R. Chapman, "Source Levels of 1.8-lb SUS Charges",
    DREP Technical Memorandum 79-10.

2. S.R. Santaniello, F.R. DiNapoli, R. W. Dullea, and P. D.
    Herstein, "Studies on the interaction of low-frequency
    acoustic signals with the ocean bottom", Geophysics, 44:
    1922 (1979).

3. C. B. Officer, "Introduction to the theory of sound
    transmission", McGraw-Hill, New York, (1958).

4. R. W. Raitt, "The Crustal Rocks" in: "The Sea, Vol. 3",
    M. N. Hill, ed., Interscience, New York (1963).

5. R. L. Dicus, "Synthetic deconvolution of explosive source in
    colored noise", Naval Oceanographic Office, Technical Note
    6130-3-76.

6. H. E. Morris, "Bottom reflection loss model with a
    velocity gradient", J. Acoust. Soc. Amer., 48: 1198 (1970)

7. E. L. Hamilton, "Sound attenuation as a function of depth in
    sea floor", J. Acoust. Soc. Amer., 59:528 (1976).

8. E. L. Hamilton, "Sound velocity-density relations in
    sea-floor sediments and rocks", J. Acoust. Soc. Amer.,
    63(2)"366 (1978).

9. D. C. Stickler, "Negative bottom loss, critical angle shift,
    and the interpretation of the bottom reflection
    coefficient", J. Acoust. Soc. Amer., 61:707 (1977).

10. D. C. Stickler, "Normal mode program with both the discrete
    and branch line contribution", J. Acoust. Soc. Amer.,
    57:856 (1975).

RESONANCES IN ACOUSTIC BOTTOM REFLECTION AND THEIR RELATION TO THE

OCEAN BOTTOM PROPERTIES

W. R. Hoover,* A. Nagl and H. Überall†

*David W. Taylor Naval Ship Research and
 Development Center, Bethesda, Maryland 20084

†Physics Department, Catholic University
 Washington, D. C. 20064

INTRODUCTION

We have initiated a program to study the resonances in the
acoustic reflection coefficient of a layered ocean bottom, patterned
after the resonances of sound reflection from a fluid or elastic
layers[1,2]. Computer programs have been written for obtaining the
reflection coefficient from multilayered fluid or elastic media,
with constant or linearly depth-dependent sound velocities in each
layer. Resonances are evident in the reflection coefficient both
as functions of frequency and of angle of incidence, and are shown
to depend on the properties of the layered ocean bottom. Results
will be presented in the form of three-dimensional graphs.

RESONANCE THEORY

Calculated reflection coefficients for the reflection of
acoustic signals from the ocean floor exhibit in general a complicated structure consisting of more or less regular sequence of
peaks and dips. A typical example[3] is provided by Fig. 1 which
shows the bottom reflection loss versus grazing angle for a 518-m-
thick turbidite layer at 20 Hz.

Features as these can usually be attributed to resonance phenomena of some sort, and it is often found that the resonance
structure, i.e. the distribution and widths of the resonance peaks,
contains all relevant information about the interacting medium
involved. Thus it appears logical to use knowledge about the reso-

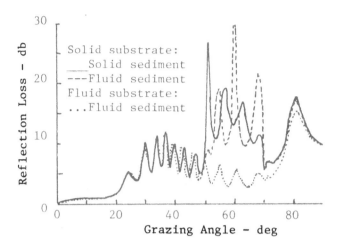

Fig. 1   Reflection loss versus grazing angle for a 518-m-thick
         hypothetical turbidite layer at 20 Hz.[3]

nance structure obtained from measurements to gain access to properties of the interacting medium, i.e. to solve the inverse scattering
problem or the inverse reflection problem, as the case may be. So
far few attempts have been made to apply this approach to sound
reflections from the ocean floor, where it might be providing a
valuable tool for acquiring information about the structure and the
properties of the reflecting medium.

We have initiated a program of systematically exploring the
feasibility of applying the resonance approach to sound reflections
from layered media in such a way that it can be directly used for
solving the problem of ocean bottom reflections. First attempts
along these lines can be found in References 2 and 3 were the cases
of a liquid and an elastic layer embedded in a liquid medium were
considered. Here, we would like to discuss another simple case,
somewhat more relevant to ocean bottom acoustics, that of a liquid
layer embedded between two different liquids.

Realistic models of the ocean floor must of course, besides
layering, also include shear wave propagation, attenuation and
density and velocity gradients. The implication of these
complications for the resonance formalism will be briefly discussed
at the end of this contribution. An example of the effects of
adding shear wave propagation e.g., is already provided in Fig. 1
where the major effect is seen to be increased absorption. Adding
more realistic features to the simple model we are presenting here
will of course considerably increase the complexity of the analysis
and the number of parameters to be dealt with. However, as will be
shown, there is sufficient redundance in the information provided

by the resonance analysis in our example that it can be expected that more realistic cases can also be treated satisfactorily.

Application to Three Fluid Case

The example chosen here for demonstrating the application of the resonance formalism is that of a liquid layer embedded between two different liquid, semi-infinite media, as indicated in Fig. 2. The reflection coefficient for this case is given by Brekhovskikh[4] and can be rewritten in the form

$$R = \frac{A \cos\delta - i(1-\tau^2)\sin\delta}{S \cos\delta - i(1+\tau^2)\sin\delta} \tag{1}$$

with

$$A = \frac{z_1 - z_3}{z_2}, \quad S = \frac{z_1 + z_3}{z_2}, \quad \tau^2 = \frac{z_1 z_3}{z_2^2}, \quad \delta = \frac{2\pi f d}{c_2} \cos\theta_2 \tag{2}$$

where

$$z_i = \frac{c_i \rho_i}{\cos\theta_i} \tag{3}$$

are the layer impedances expressed in terms of the sound speeds c_i and the densities ρ_i. It is assumed here that $z_1 > z_2 > z_3$, which corresponds to the case of a sediment layer embedded between the ocean and a high density substratum. The parameters chosen are

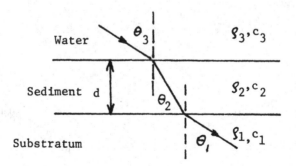

Fig. 2 Fluid layer embedded between two different fluids with incident and transmited sound wave.

c_3 = 1500 m/s, c_2 = 2544 m/s, c_1 = 5495 m/s

ρ_3 = 1 g/cm^3, ρ_2 = 2.2 g/cm^3, ρ_1 = 2.6 g/cm^3

If the square of the reflection coefficient is plotted in a three-dimensional graph simultaneously vs. the frequency-thickness product fd and the incident angle θ_3 one obtains a surface, parts of which are shown in Figs. 3 through 5. The surface is seen to consist of a series of ridges and valleys which run almost parallel to the θ_3-

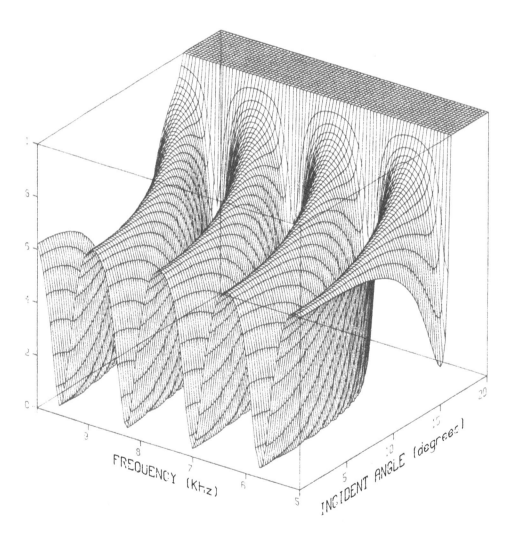

Fig. 3 Square of reflection coefficient for the 3-liquid case, plotted as function of the frequency-thickness product fd between 5 and 10 kHz-m and incident angle($0^o \leqslant \theta_3 \leqslant 20^o$)

axis for low frequency-thickness products and curve away more and more from the θ_3- axis as the value of fd is increased.

For constant θ_3 one observes a regular sequence of maxima and minima. For constant frequency-thickness products there is little structure as a function of angle for values of fd < 5KHz-m. At around 5 KHz-m the first minimum appears (Fig. 3) and from then on the increasing curvature of the ridges and valleys leads to a growing number of maxima and minima of $|R|^2$ along lines of constant fd (Fig. 4).

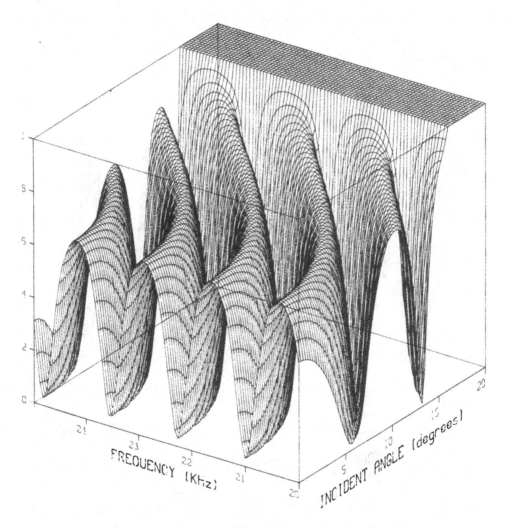

Fig. 4 Square of reflection coefficient for the 3-liquid case,
plotted as function of the frequency-thickness product fd
between 20 and 25 kHz-m and incident angle($0^o \leqslant \theta_3 \leqslant 20^o$).

Fig. 5 shows the surface of Fig. 4, covering the range of fd between
20 and 25 kHz-m, plotted upside down (thus actually showing the
square of the transition coefficient). This clearly reveals the
existence of sharp ridges reminiscent of forms describing resonance-
type processes. It is therefore suggestive to interpret the transition coefficient as a series of resonances (or the reflection
coefficient in terms of anti-resonances) and to apply the resonance
formalism developed in other areas of physics to the problem at
hand, namely the analysis of the inverse reflection problem, and
thereby simplifying its solution.

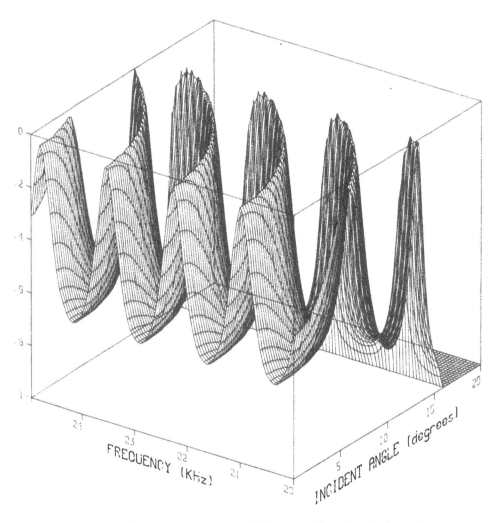

Fig. 5   Square of transmission coefficient for the 3-liquid case,
         plotted as function of the frequency-thickness product fd
         between 20 and 25 kHz-m and incident angle ($0° \leqslant \theta_3 \leqslant 20°$).

The basic assumption of the resonance formalism is that in the
vicinity of a resonance the amplitude for a process is described
essentially by a Breit-Wigner resonance form plus some slowly
varying background. Following this idea, the transmission
coefficient   can be expanded around the maxima and then written
approximately as a sum of resonance shapes, the reflection
coefficient being correspondingly represented by a sum of antireso-
nance shapes, obtained by expanding the exact form around its minima.
The sum over the resonances (or antiresonances) is to be taken only
symbolically since the expansion is assumed to be valid only in the
immediate vicinity of each resonance position, but is expected to
fail at large distances away from the resonances where the individual resonance shape is simply expected to approach the background.
For the assumed case of $Z_3 < Z_2 < Z_1$ the minima of the reflection
coefficient (1) are found to be given by the condition $\cos \delta = 0$, i.e.

$$\delta_n = (2n+1) \frac{\pi}{2} \qquad n = 0,1,\ldots \tag{4}$$

Linearizing around these positions with respect to δ leads to the
following resonance expression for the reflection coefficient:

$$R = \sum_n \frac{F(\delta-\delta_n) \pm iG \frac{1+\tau^2}{S}}{(\delta-\delta_n) \pm i \frac{1+\tau^2}{S}} \tag{5}$$

with

$$F = \frac{A}{S} = \frac{z_1-z_3}{z_1+z_3} , \qquad G = \frac{1-\tau^2}{1+\tau^2} \tag{6}$$

Interpreting $\frac{1+\tau^2}{S}$ as the resonance halfwidth, the expressions in the
denominators of (5) are recognized as the standard resonance
denominators leading to the well-known Breit-Wigner resonance shapes.
Using Snell's law and the definition for δ given in (2), the
resonance condition (4) can be written as

$$fd \sqrt{n_d^2 - \sin^2\theta_3} = (2n+1) \frac{c_3}{4} \qquad n = 0,1,\ldots \tag{7}$$

where $n_d = c_3/c_2$.

Two Types of Resonances

For the specific example chosen, the relation between the
frequency-thickness variable fd and the incident angle θ_3 as given
by (7) is shown in Fig. 6 for various values of n. As this figure
shows, one obtains regularly spaced resonances for the variable fd
at constant values for the angle θ_3, and, if the frequency-
thickness product is held constant, irregularly spaced resonances

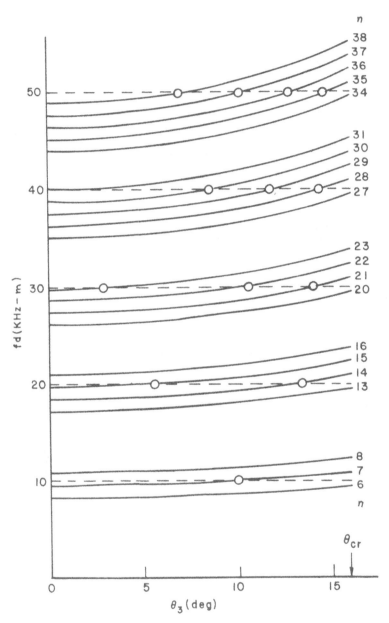

Fig. 6   Resonance positions for the three-liquid case. The circles
         indicate the position of angle resonance for various
         frequency-thickness products. The curves indicating the
         resonance positions have not been drawn for all mode numbers n.

as a function of angle (indicated in Fig. 6 by circles for a few valves of fd). For a more detailed discussion of the two types of resonances identified above, it is convenient to introduce new variables. For the case of the frequency-thickness resonances the reflection coefficient is rewritten as

$$R = \sum_n \frac{F(x-x_n) \pm iG\,\frac{\Gamma_n}{2}}{(x-x_n) \pm i\,\frac{\Gamma_n}{2}} \tag{8}$$

where the definitions

$$x_n = \frac{\delta_n}{\sqrt{n_d^2 - \sin^2\theta_3}} \tag{9}$$

and

$$\frac{\Gamma}{2} = \frac{1 + \tau^2}{s\sqrt{n_d^2 - \sin^2\theta_3}} \tag{10}$$

for the amplitudes and the half-widths were used.

One thus finds that the width of the fd- resonances is independent of the order n, but that it depends on the angle θ_3, being largest at normal incidence and decreasing with angle. The spacing of the resonances

$$\Delta x = \frac{\pi}{\sqrt{n_d^2 - \sin^2\theta_3}} \tag{11}$$

is found to increase with angle.

For the square of the reflection coefficient one obtains

$$|R|^2 = \sum_n \frac{F^2(x-x_n)^2 + G^2\Gamma^2/4}{(x-x_n)^2 + \Gamma^2/4} \;. \tag{12}$$

Fig. 7 compares the prediction of the resonance approximation (12) with the exact expression for $|R|^2$ obtained by squaring Eq. (8). It is seen that the agreement is excellent in the vicinity of the resonances and that the approximation fails as expected between the resonances.

The angular resonances are conveniently discussed in terms of the variable

$$y = \sin\theta_3 = \sqrt{n_d^2 - (\delta/x)^2}$$

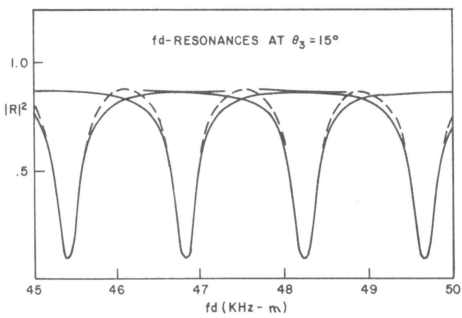

Fig. 7 The frequency-thickness resonances in the range between 45 and 50 kHz-m at θ_3=15° obtained by the resonance approximation are compared with the exact value for $|R|^2$(dashed line).

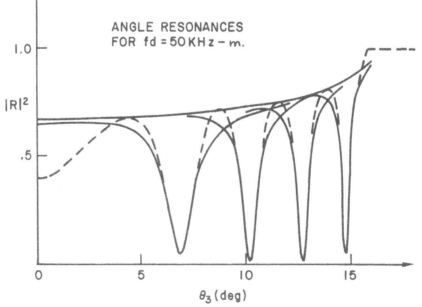

Fig. 8 The angle resonances for fd = 50 kHz-m obtained by the resonance approximation (full lines) are compared with the exact value for $|R|^2$ (dashed line).

according to which the resonance positions are obtained as

$$y_m = \sqrt{n_d^2 - (\delta_m/x)^2}$$ (13)

For $\cos\delta$ one obtains in the vicinity of y_m the expansion

$$\cos\delta = \cos\delta_m + (y-y_m) \left.\frac{d \cos\delta}{dy}\right|_{y=y_m}$$

$$\approx -(y-y_m)x^2 y_m/\delta_m$$

The reflection coefficient can thus be written as

$$R = \sum_m \frac{F(y-y_m) - iG\gamma_m/2}{(y-y_m) - i\gamma_m/2}$$ (14)

where the definition

$$\gamma_m/2 = \frac{1+\tau^2}{Sxy_m} \sqrt{n_d^2 - y_m^2}$$ (15)

for the half width was used.

The comparison of $|R|^2$ obtained from Eq. (14) with the exact result is shown in Fig. 8 for fd=const=50 kHz-m. Again the agreement in the vicinity of the resonances is excellent.

Relation between Resonance Parameters and Layer Parameters

The importance of the resonance approximation lies in the fact that it facilitates solving the inverse scattering problem. The quantities characterizing the resonances (positions and widths) can, in principle at least, be easily measured. On the other hand, the theoretical expressions relating these quantities to the parameters characterizing the reflecting layers are known (explicit expressions will be given below). Hence the experimentally measured quantities can be directly related to the properties of the reflecting medium.

By measuring the critical angle θ_{cr}, the minimum value R_m of the reflection coefficient, the spacing and width of the fd-resonances at some angle and the position and width of an angle resonance at a particular frequency, all the parameters describing the reflecting medium for the particular case considered (i.e. c_1, c_2, δ_1, δ_2, d) can be deduced.

By measuring the spacing

$$\Delta x = \frac{2\pi d}{c_3} \Delta f$$

of the fd- resonances at some particular angle θ_3 one obtains the ratio of sound velocities,

$$n_d = \frac{c_3}{c_2} = \sqrt{\left(\frac{c_3}{2\pi d \Delta f}\right)^2 + \sin^2\theta_3} \qquad (16)$$

On the other hand, measuring the position y_m of an angle resonance at a particular frequency f yields

$$n_d^2 = y_m^2 + \left(\frac{\delta_m}{x}\right)^2$$

or

$$\sqrt{n_d^2 - y_m^2} = \frac{2n_m+1}{4fd} \, c_3 \qquad (17)$$

because of

$$\delta_m = (2n_m+1) \, \frac{\pi}{2} = \frac{x}{2\pi fd} \, c_3$$

Eqs.(16) and (17) can be solved for n_d (or c_2) and d. From the critical angle θ_{cr} the sound speed c_1 of medium 1 can be found:

$$c_1 = \frac{c_3}{n_r} = \frac{c_3}{\sin\theta_{cr}}$$

Measurement of the half width $\Gamma/2$ of the fd-resonances at some fixed angle θ_3 yields

$$\frac{1+\tau^2}{S} = \frac{\Gamma}{2} \sqrt{n_d^2 - \sin^2\theta_3} \qquad (18)$$

The value of τ can e.g. be found from the minimum R_m of the reflection coefficient at θ_3:

$$\tau^2 = \frac{1-R_m}{1+R_m} \qquad (19)$$

All other quantities being known in Eq. (18), this equation can thus be solved for S which yields

$$S = \frac{z_1+z_3}{z_2} = (1+\tau^2)/(\frac{\Gamma}{2}\sqrt{n_d^2 - \sin^2\theta_3}) \qquad (20)$$

S can also be written as

$$S = \frac{z_1+z_3}{\sqrt{z_1 z_3}} \, \tau = \left(\sqrt{\frac{z_1}{z_3}} + \sqrt{\frac{z_3}{z_1}}\right)\tau \qquad (21)$$

which can be solved for $(z_1/z_3)^{\frac{1}{2}}$ since S and τ are known. From the definition (3) of the impedances one finds:

$$\sqrt{\frac{z_1}{z_3}} = \sqrt{\rho_1} \left(\frac{1 - \sin^2\theta_3}{n_r^2 - \sin^2\theta_3} \right)^{1/4} \tag{22}$$

Since n_r is already assumed to be known, Eq. (22) thus yields ρ_1. The density ρ_2 can then be found from the known value for τ^2:

$$\tau^2 = \frac{\rho_1\rho_3}{\rho_2} \frac{n_d^2 - \sin^2\theta_3}{\cos\theta_3 \sqrt{n_r^2 - \sin^2\theta_3}}$$

For the case considered here, the information available is clearly redundant since all parameters could be deduced without involving the large amount of information contained in the angular resonance except for the position of one of these resonances. Since the positions and widths of the angular resonances vary with angle it should in principle be possible to determine a much larger set of medium parameters than the five parameters involved in the present example. Thus one can hope to be able to solve the inverse reflection problem also for more realistic cases than the one considered here.

Extension of the Resonance Approach to More Complicated Cases

Adding shear waves first to the substrate then also to the sediment is seen from Fig. 1 to lead each time to significant changes in the reflection coefficient. The major differences occur in the size of the coefficients. The number of resonances remains the same, but the widths and positions of the individual resonances change somewhat.

Taking absorption into account will lead esentially only to reduced magnitudes of the reflection coefficient. Above the critical angle, however, new resonances may appear which can provide additional information on the reflecting medium.

If one or more layers are added to the basic model described in Fig. 2 the simple periodicity of the reflection coefficient with frequency is no longer maintained and also the angular dependence is expected to become more complicated. Fig. 9 shows the behavior of $|R|^2$ if the fluid layer in Fig. 2 is replaced by two fluid layers, each with thickness $\Gamma/2$ and with densities of 2 and 2.3 g/cm^3, and sound speeds of 2000 and 3000 m/s, respectively, the layers arranged such that the impedance still increases with depth. It is evident that there is a considerable amount of new structure which on the one hand complicates the analysis, but which on the other hand can be exploited to yield additional information.

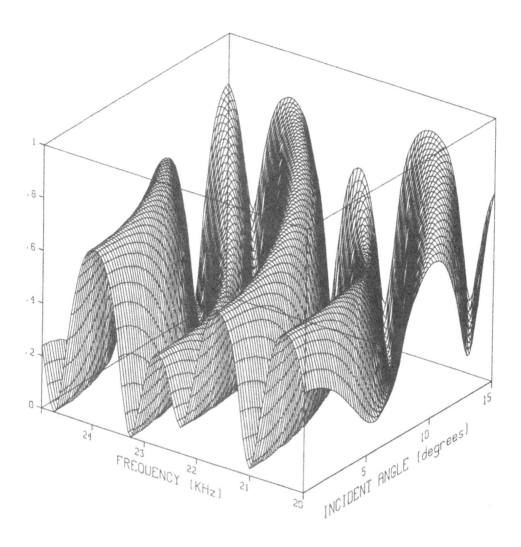

Fig. 9 Square of reflection coefficient for the 4-liquid case
        plotted as function of the frequency-thickness product   fd
        bewteen 20 and 25 kHz-m and incident angle (0º ≤ θ_3 ≤ 15º).

It is seen that the pronounced resonance structure of the reflection coefficient found in the simple model is generally
maintained even after more realistic features are added so that the
resonance approach presented here is in principle still applicable.
The question of how to relate the resonance data in the more
complicated cases to the layer parameters is presently under investigation by us. Indications are that for a large class of cases the
resonance formalism demonstrated here on a simple example is indeed
a promising tool for identifying ocean bottom properties, provided
reflectivity measurements over a wide enough angle and frequency
range are available.

References

1. R. Fiorito, and H. Überall, JASA 65, 9 (1979)
2. R. Fiorito, W. Madigosky and H. Überall, JASA 66,1857 (1979)
3. P. J. Vidmar and T. L. Foreman, JASA 66, 1830 (1979)
4. L. M. Brekhovskikh, Waves in Layered Media, Academic Press,
    New York (1960).

COMPARISON OF SYNTHETIC AND EXPERIMENTAL BOTTOM INTERACTIVE

WAVEFORMS

F. R. DiNapoli, D. Potter and P. Herstein

Naval Underwater Systems Center (NUSC)
New London, CT 06320

INTRODUCTION

Our interest in subbottom acoustics is motivated by proposed continuous wave (CW) passive systems designed to utilize more than just the energy content of signals arriving at the receiver via predominantly bottom bounce paths. To reach a point where the merits of such systems could be delineated with confidence or contributions to their conception could be made, it was first necessary to have a believable theoretical model capable of predicting the received waveform versus time produced by a transient source. The apparent contradiction that while the system concepts pertain to CW signals, the model is required to accommodate transient signals can be explained from the belief that the highest quality bottom bounce data is obtained from transient sources such as air guns or explosives thus achieving credibility for the model implies assessment against transient data.

This paper is divided into three parts: (a) a synopsis of the model, (b) brief observations regarding previous assessments against seismic data, and (c) current work on assessing the model against high quality "bottom loss" data. The term "bottom loss" with quotation marks is an abbreviation for the experimental determination of the energy lost into the bottom. This quantity differs significantly[1] from the plane wave definition of bottom loss.

THEORETICAL MODEL

The Fast Field Program (FFP)[2,3] is a direct numerical evaluation of the Fourier-Bessel integral designated as H under the heading, transfer function, in Fig. 1. The evaluation is accomplished by utilizing an FFT involving the transform variables of wavenumber and range. This technique has proven to provide accurate results for the propagation loss of a CW signal of any frequency emitted by a point source embedded in an arbitrary range independent environment.

The generalization to the broadband case is accomplished in a straightforward manner by repeatedly evaluating the transfer function at equally spaced discrete frequencies and taking the FFT of the transfer function to obtain the impulse response of the medium versus time and range for a fixed source, receiver depth configuration.

ASSESSMENT AGAINST SEISMIC DATA

The comparisons against multichannel seismic reflection data[4] were only partially satisfying for a variety of reasons.

Seismic systems are generally uncalibrated which implies that meaningful comparisons can only be made against travel time versus range results.

The peak of the seismic source frequency spectrum is low to optimize penetration depth. Thus calibrated seismic data would be of limited value to the applications of interest which extends to higher frequencies.

- **Transfer Function**
$$H(r, z, f) = \int_0^\infty G(z, z_s, \xi, f)\, H_0^{(1)}(\xi r)\, \xi\, d\xi$$

z (depth); f (frequency)

Evaluated by FFT (Wave Number (ξ), Range (r))

- **Impulse Response**
$$h(r, z, t) = \int_{-\infty}^\infty H(r, z, f)\, e^{-i2\pi f t}\, df$$

Evaluated by FFT (time (t), frequency (f))

Fig. 1 Mathematical Formulation

The most difficult problem, however, is that the environmental inputs to the model must be currently obtained from approximate inverse solutions of the experimental data. The typical result is a constant sound speed layered subbottom. Values for density and attenuation are usually not available from the approximate inverse solution and must be obtained from historical data, when available, or empirical relations.[5]

A potential dilemma is that if the prediction does not agree with the data one is not sure if the blame lies with the theoretical model or the approximate environmental input to the model. For the cases examined the agreement (travel time versus range results) was excellent, however, an assessment of the ability to predict the amplitude of subbottom arrivals was still lacking.

ASSESSMENT AGAINST "BOTTOM LOSS" DATA

To this end a set of old NUSC "bottom loss" data taken by Herstein, et al.,[6] in the Hatteras Abyssal Plain was examined. This data was appealing, in comparison to the seismic data, for the following reasons.

The experimental geometry was predetermined so that self-calibrating "bottom loss" calculations could be made. Thus it would be possible to compare energy values in addition to those for travel time versus range. This fell short of the desire for comparison with the amplitude of individual bottom and subbottom arrivals, but it provided an additional assessment measure not available from the seismic data.

The deconvolved data showed strong indications of an arrival from a subbottom gradient.

Finally the frequency spectrum of the data spanned the region of interest.

The major liability was that once again the environmental inputs to the model were obtained from an approximate inverse solution. Since the accuracy of the inferred environmental input data will dominate the remainder of the paper a more detailed discussion of the experiment seems appropriate.

The depths of the explosives and self-contained Autobuoy receiver (see Fig. 2 for a sketch of the experimental components) were predetermined so that time isolated direct and bottom interacting arrivals could be obtained over a wide range of grazing angles. With time isolated direct and bottom interacting arrivals and an accurate prediction for the propagation loss along the paths it is possible to calculate the "bottom loss" without calibrating the source[1,6]. Thus the term "self-calibrating".

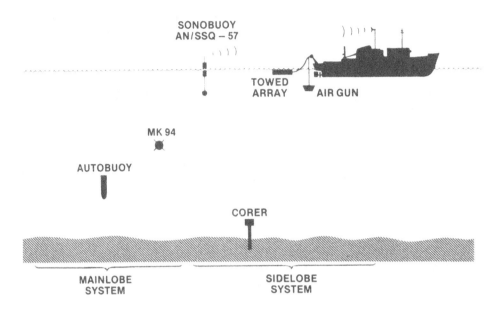

Fig. 2   Experimental Geometry

Seismic measurements were conducted in the same vicinity by
the Lamont Doherty Geological Observatory (LDGO) using an air gun,
a 122 m long line array, and a sonobuoy. In addition, core measurements[1] using a 1.22 m boomerang core and several sound speed profile
(SSP) casts were also made.

The analysis of the core data at two locations revealed the
average values for interface density and sound speeds shown in
Fig. 3 under core data. Historical data[7] from the Hatteras Abyssal
Plain shows a higher average value for density and a lower value
for the interface sound speed.

The approximate inverse solution of the LDGO seismic data
resulted in a single sound speed gradient having an interface value
of 1520 (m/s) and extending downward with a positive gradient of
1.32 (s^{-1}).

The NUSC result, inferred from higher frequency data, indicated
the presence of a thin 20.4 m layer of constant sound speed
(1580 m/s) overlaying a constant gradient (1.35 s^{-1}) layer.

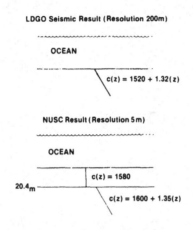

CORE DATA

Average Values	Core 1	Core 2	Historical
Density	1.13	1.30	1.77
Sound Speed	1617	1609	1572

LDGO Seismic Result (Resolution 200m)

OCEAN

$c(z) = 1520 + 1.32(z)$

NUSC Result (Resolution 5m)

OCEAN

$c(z) = 1580$

20.4$_m$

$c(z) = 1600 + 1.35(z)$

Fig. 3   Environmental Parameters Inferred from Experimental Data

The two separately inferred subbottom descriptions provide a
valuable starting point for future work aimed at determining the
viability for system analysis of seismic data which is considerably
denser than currently available "bottom loss" data.

If the bottom arrivals versus time from the NUSC data are
stacked in range three major patterns can be visually identified.
This is more clearly evident in the deconvolved results shown to
the right of Fig. 4. Proceeding from left to right, the first
pattern can be conclusively identified, from ray theory travel
times, as a reflection from the ocean bottom interface. The next
group was assumed to be a reflection from a subbottom interface.
The last pattern has the characteristics of a subbottom refraction.

Herstein arrived at the subbottom sound speed structure in
Fig. 3 by comparing the observed time differences with those
obtained from the CONGRATS[8] ray tracing program for various assumed
sound speed profiles. Comparisons between observed and predicted
arrival times versus range for the subbottom reflected and refracted
arrivals are shown in Fig. 5.

The environmental inputs to the FFP are depicted in Fig. 6.
Although the measured SSP in the water column was available and
could have been implemented an average constant sound speed of
1554 m/s was used in the water in order to reduce the required
computer time. This was felt to be justifiable for "bottom loss"

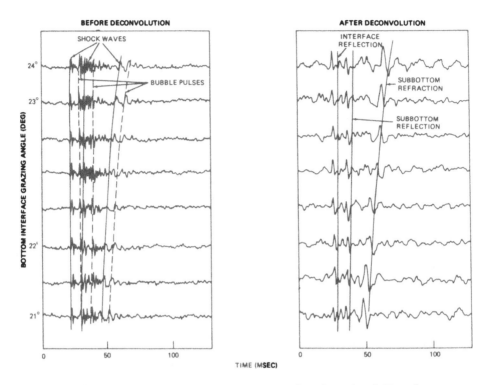

Fig. 4   Experimental Bottom Interacting Received Waveforms

comparisons since that calculation[1] removes the effect of energy
loss in the water column leaving a result representative of the
energy lost into the bottom.

The values of subbottom sound speed and interface density were
identical to those inferred by Herstein from the NUSC data.   The
value of density used for the gradient layer was obtained by a
formula[9] relating density and sound speed.

The values for attenuation in the literature show orders of
magnitude variation.   Thus it was decided to initially use zero
attenuation in order to establish a base line for subsequent
interactions.

The transfer function was calculated over the frequency band,
95.1 to 282.9 Hz, with a resolution of .367 Hz for a total of 512
frequency calculations.   At each of these frequencies the wave
number spectrum was calculated with a resolution of roughly
$.5X10^{-4}(m^{-1})$ which results in a range resolution of 243 (m) out to
approximately 124.5 (km).

The impulse response as computed by the FFP is provided in
Fig. 7. The ordinate is range increasing from the bottom and the
abscissa is time increasing from the left.

The arrival times of the direct path at each range have been
brought into alignment to form the vertical pattern at the extreme
left of Fig. 7. The remaining patterns, from left to right are
(i) the ocean bottom interface reflection, (ii) the subbottom
reflection, (iii) the subbottom refraction with one reversal in the
gradient layer, (iiii) and finally the subbottom refraction with
two reversals in the gradient layer. The straight line emerging
from the reflection pattern and proceeding upwards is a head wave.

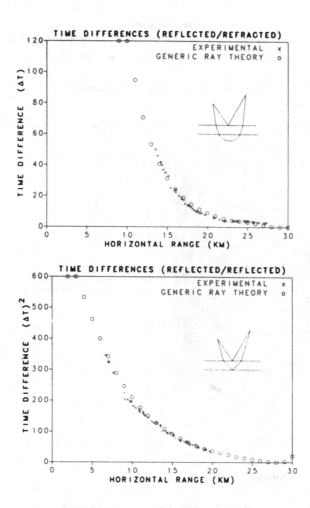

Fig. 5   Comparison Predicted and Observed Travel Time Differences

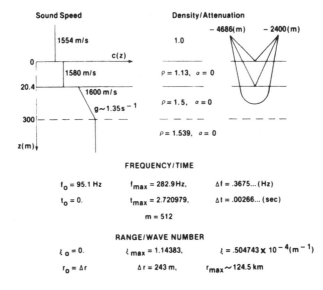

Fig. 6   Model Input Parameters

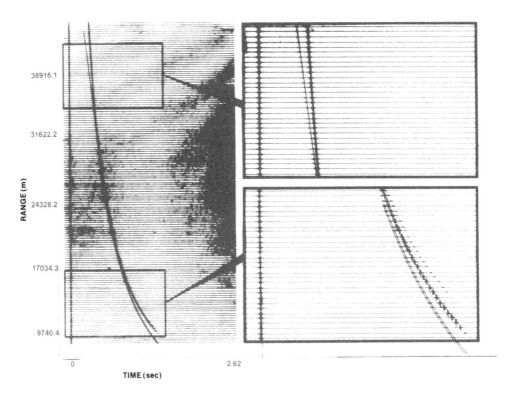

Fig. 7   FFP Impulse Response

Each trace has been scaled differently in order to show all arrivals. Thus inferences regarding the strength of the various arrivals should not be made from this plot.

Arrival times versus range for this environment have been computed using classical ray theory and are overplotted with the FFP results in Fig. 7.

The synthetic FFP impulse response data was ideally filtered and processed for "bottom loss" in a manner identical to that used to process the experimental data. The lowest band available from the experimental data for which a good signal to noise ratio (S/N~4dB) existed was from 95 to 105 Hz. The experimental values are plotted along with the FFP values in Fig. 8. The agreement is quite poor with the experimental values showing a larger positive loss at high grazing angles and considerably more scatter at the low grazing angles. In addition a major discrepancy exists between about 20 and 35 degrees grazing where the FFP is showing about 8 dB less loss on the average.

The next band for which data existed (S/N~11dB) covered the frequencies from 157 to 198 Hz. The two sets of values are plotted in Fig. 9. The agreement at low grazing angles has improved, however, the disagreement at the high angles has reversed itself in this higher frequency band. The FFP results now show more positive loss than the experimental values. The disagreement present between 20 and 35 degrees grazing is still about 8 dB as it was with the lower band.

The last band for which calculations were made covered the frequency regime from 224 to 282 Hz. The results of this comparison are shown in Fig. 10. The region from 20 to 35 degrees grazing still exhibits the same discrepancy of about 8 dB found in the other two lower bands. Aside from this good agreement is found at both low and high grazing angles.

It is of some interest to take the environmental input used in the FFP and compute the plane wave bottom loss. This has been done at the center frequency, 252 Hz, of the highest band and the results are plotted along with the point source "bottom loss" results in Fig. 10. In the plane wave definition of bottom loss one examines the ratio of the received to incident fields under the assumption that the received field is due to a single plane wave incident at a particular angle. Negative values of bottom loss violate the conservation of energy principle under this definition. The negative values of "bottom loss" are due to the fact that the received signal contains arrivals from incident fields which have left the source at different angles. This violates the plane wave definition. At small grazing angles these arrivals are time coincident at the receiver and their interference produces the negative values of

"bottom loss". At higher grazing angles the unwanted arrivals may
be separated in time but they are usually included in the integration to find energy thus violating the plane wave definition.   In
spite of these facts the plane wave bottom loss calculation can be
used as an economical gross diagnostic tool to highlight the
sensitivity of the results to changes in the input parameters, such
as density and attenuation, needed to bring the broad band predictions in closer agreement with the data. Work along these lines is
in progress. The discrepancy of 8 dB between 20 and 30 degrees
grazing will be examined first because of its consistency throughout
all three frequency bands. An examination of the arrival structure
versus grazing angle reveals that the refracted arrival and its
multiple are present only between 14 and 40 degrees grazing. In
order to bring the predictions in closer agreement with the data
these arrivals must be reduced in amplitude relative to the reflected arrivals. The addition of attenuation in the gradient
layer certainly seems called for. The assumed density value in
this layer is also highly suspect. Furthermore, changes in these
environmental parameters which result in good agreement in this
angular regime must not destroy the agreement exhibited in the
highest band for grazing angles larger than 40 degrees where only

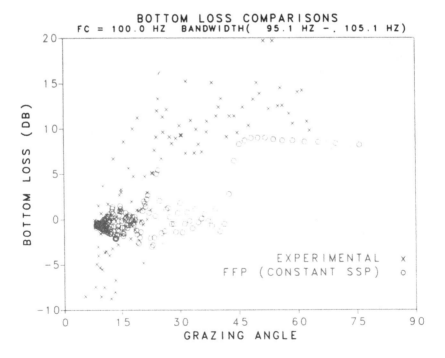

Fig. 8 "Bottom Loss" Comparison Low Band

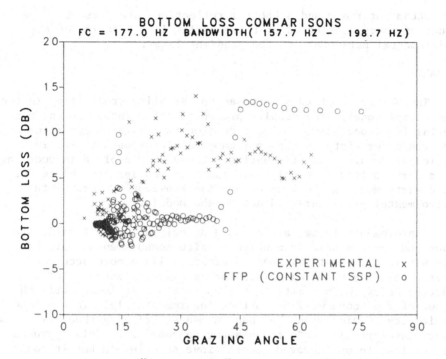

Fig. 9 "Bottom Loss" Comparison Mid Band

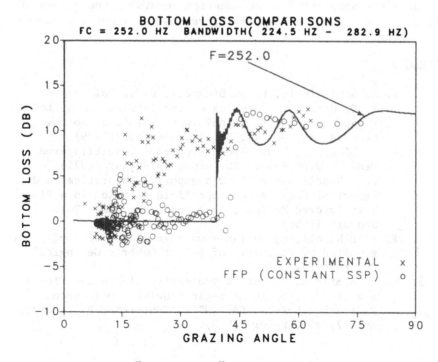

Fig. 10 "Bottom Loss" Comparison High Band

the bottom interface and subbottom reflections are present. The
amount of energy reflected at these angles is a function of the
environmental parameters in the gradient layer.

SUMMARY

The objective of this work was to establish credibility in the
theoretical model. The results presented offer encouragement but
are far from conclusive. In our opinion the disagreement stems
from the uncertainty in the environmental input parameters, in
particular density and attenuation. Those not involved in modeling
would perhaps find it predictable that the modeler lays blame for
the disagreement on the lack of precise knowledge regarding the
environmental parameters and not on the model.

Improvements in two areas would be helpful: (i) A calibrated
experiment over a hole from which in situ measurements could be
made within the first few hundred meters, (ii) a more accurate
inverse solution yielding values for density and attenuation as a
function of depth in addition to sound speed. It is unlikely that
either of the above will be forthcoming immediately. In the interim
sensitivity studies with the model aimed at bringing theoretical and
experimental results into focus seems appropriate. This approach
may not lead to an improved general inverse solution but it could
point the way to improved techniques for experimental determination
of the subbottom environmental parameters. In addition it could
illuminate the more significant question regarding the degree of
precision to which subbottom parameters must be known for system
applications.

REFERENCES

1. S. R. Santaniello, F. R. DiNapoli, R. K. Dullea and
    P. D. Herstein, "Studies on the Interaction of Low
    Frequency Acoustic Signals with the Ocean Bottom",
    Geophysics, Vol. 44, No. 12, December (1979).

2. F. R. DiNapoli, "Fast Field Program for Multilayered
    Media", NUSC Report No. 4103, 26 August (1971).

3. F. R. DiNapoli and R. L. Deavenport, "Theoretical and
    Numerical Green's Function Field Solution in a Plane
    Multilayered Medium", J. Acoust. Soc. Am., Vol. 67,
    January (1980).

4. F. R. DiNapoli, unpublished work in conjunction with data
    supplied by P. Stoffa of Lamont Doherty Geological
    Observatory.

5. E. L. Hamilton, "Acoustic Properties of the Sea Floor:
    A Review", Oceanic Acoustic Modeling Conference,
    Part 4 Sea Bottom, SACLANTCEN Conference Proceedings
    No. 17, 15 October (1975).

6. P. D. Herstein, R. K. Dullea and S. R. Santaniello, "Hatteras Abyssal Plain Low Frequency Bottom Loss Measurements", NUSC Report No. 5781, 14 April (1979).

7. D. R. Horn, M. Ewing, B. M. Horn and M. N. Delach, "A Prediction of Sonic Properties of Deep-Sea Cores, Hatteras Abyssal Plain and Environs", TR No. 1, CU-1-69 NAVSHIPS N00024-69-C-1184, November (1969).

8. J. S. Cohen and L. T. Einstein, "Continuous Gradient Ray Tracing System (CONGRATS) II: Eigenray Processing Programs", NUSC Report 1069.

9. G.H.F. Gardner, L. W. Gardner and A. R. Gregory, "Formation Velocity and Density - The Diagnostic Basics for Stratigraphic Traps", Geophysics, Vol. 39, No. 6, December (1974).

REFLECTION AND REFRACTION OF PARAMETRICALLY GENERATED SOUND

AT A WATER-SEDIMENT INTERFACE

Jacqueline Naze Tjøtta* and Sigve Tjøtta*

Applied Research Laboratories
The University of Texas at Austin
Austin, Texas 78712

ABSTRACT

The results of an experimental study of the penetration of highly
directional acoustic beams into bottom sediments were recently
reported by Muir, Horton, and Thompson [J. Sound Vib. 64, 539-551
(1979)]. Of special interest was the behavior of a narrow beam
generated by a parametric source. We have considered this problem
theoretically. Simplified equations for the reflected and refracted
beams at the water-sediment interface are derived and solved ana-
lytically subject to the nondissipative boundary conditions. The
range of validity is discussed. Results are presented that seem to
explain the experimental observations.

INTRODUCTION

The results of an experimental study of the penetration of
highly directional acoustic beams into bottom sediments were recently
reported by Muir, Horton, and Thompson.[1] Of special interest was
the behavior of a narrow beam generated by a parametric source. We
have considered this problem theoretically, including also the
reflected beams. Simplified equations are derived and solved analyt-
ically subject to the nondissipative boundary conditions. The range
of validity of these equations is discussed. Results are presented
that seem to explain the experimental observations. The basic
theory was developed in a previous paper,[2] and the main results were
presented[3] at the 99th Meeting of The Acoustical Society of America
in Atlanta, Georgia. In the present paper we derive some new results
and use these to analyze in detail the experimental observations by
Muir et al.[1] A brief outline of the theory is also given.

I.   THEORY

In this section we give a brief summary of the derivation of the equations governing the sound field produced by the impact of a highly directional sound beam on a two-fluid interface. For a detailed presentation, see Ref. 2, where analytical solutions are also obtained.

The amplitude of the incident beam at the interface is assumed weak enough that nonlinear effects are negligible. Although dissipation is accounted for in both fluids, diffraction is assumed to be the dominating effect. The shear modulus of the sediment is thus neglected and the nondissipative boundary condition is imposed at the interface, while absorption is retained in the equations of motion. The reflected and refracted sound pressures, p_r and p_t, respectively, thus satisfy four equations: two Helmholtz equations, in water and sediment, respectively, and two continuity equations at the interface for pressure and normal velocity. They must also satisfy the Sommerfeld's radiation conditions at infinity, the incident pressure p_i being given. The notations are the same as in Ref. 2, the dashed symbols referring to water and the undashed to sediment. The geometry of the problem is as shown in Fig. 1. For a highly directional beam, $p_i = \exp(i\chi'z')q_i'(x',y',z')$, where q_i' is slowly varying with z', the characteristic length for the variation with x',y' being the half beamwidth a (or 3 dB radius a_{3dB}) with $k'a$ large (y' is the coordinate transverse to the plane of incidence). In this case it can be shown that p_i is the solution of the Helmholtz equation with boundary condition $p_t = \exp(i\chi'\sin\theta_i x)T_0 q_i$ at the interface, where $q_i(x,y) = q_i'(z=0)$ and T_0 is the complex transmission coefficient for an infinite plane wave,[4,5]

$$T_o = 2\left[1 + \rho' \sqrt{\chi^2 - (\chi'\sin\theta_i)^2} \Big/ (\rho\chi'\cos\theta_i)\right]^{-1} \quad . \tag{1}$$

The solution of this boundary value problem (infinitely compliant phase-shaded piston) is in principle known, but complicated. However, at large distances from the spot, $r \gg ka^2/2\cos\theta_i$, the Born approximation can be used and leads to the simple **very farfield** expression

$$p_t(r,\theta,\phi) \simeq T_o p_o \frac{i\chi S}{2\pi r \cos\theta_i} \exp(i\chi r) F(\theta,\phi) \quad , \tag{2}$$

where $S/\cos\theta_i$ is the area of the spot, and the directivity function F depends on the model used for the incident beam. For a beam with a square cross section (2a x 2a), i.e., $q_i' = p_o$, where $|x'|, |y'| \leq a$ and zero elsewhere, we have

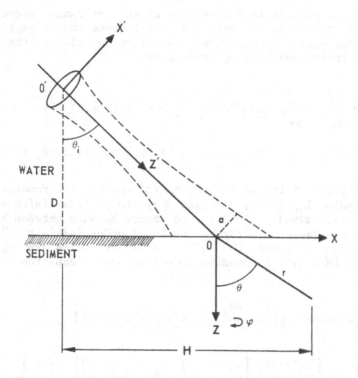

Fig. 1. Geometry of the problem.

$$F(\theta,\phi) = \cos\theta \; \frac{\sin\varepsilon_x}{\varepsilon_x} \; \frac{\sin\varepsilon_y}{\varepsilon_y} \tag{3}$$

$$\varepsilon_x = \frac{a}{\cos\theta_i} \left(\chi\sin\theta\cos\phi - \chi'\sin\theta_i\right) \; , \quad \varepsilon_y = a\chi \, \sin\theta \, \sin\phi \; . \tag{4}$$

For a beam with a Gaussian cross section,

$$q_i' = P_o \, \exp\left[-0.347(x'^2 + y'^2)/a_{3dB}^2\right] \; ,$$

$$F(\theta,\phi) = \cos\theta \, \exp\left[-\left(\varepsilon_x^2 + \varepsilon_y^2\right)/4N\right] \; . \tag{5}$$

One way to obtain simple expressions at shorter ranges where the
Born approximation is not valid is to introduce the assumption of
large ka not only in the boundary condition but also in the Helmholtz
equation itself, which is equivalent to

$$
\left[\frac{\partial^2}{\partial x^2} + \frac{\partial^2}{\partial y^2} + \frac{\partial^2}{\partial z^2} + 2i\chi'\sin\theta_i \frac{\partial}{\partial x} + \chi^2 - (\chi'\sin\theta_i)^2 \right]
$$

$$
\times \left[p_t \exp(-i\chi'\sin\theta_i x) \right] = 0 \quad . \tag{6}
$$

In the vicinity of the spot where p_t is expected to "remember" its
boundary value $T_0 p_i(z=0)$, $\partial/\partial x$ and $\partial/\partial y$ are of order $(a/\cos\theta_i)^{-1}$
and a^{-1}, respectively. In order to ensure balance between horizontal
and vertical gradients, $\partial/\partial z$ has to be of order $(a/k'\sin\theta_i)^{-1/2}$.
When $k'a\sin\theta_i$ is large, the $\partial^2/\partial x^2$ and $\partial^2/\partial y^2$ terms are negligible
and we are left with a two-dimensional parabolic equation with
solution

$$
p_{t,1}(x,y,z) = T_0 \sqrt{\frac{\chi'\sin\theta_i}{2\pi}} \; z \exp\left(i\chi'\sin\theta_i \, x - i\frac{\pi}{4} \right)
$$

$$
\times \int_{-\infty}^{x} \exp\left\{ \frac{i\chi'\sin\theta_i}{2} \left[\frac{z^2}{x-u} - \left(\frac{\beta}{\chi'\sin\theta_i} \right)^2 (x-u) \right] \right\} \frac{q_i(u,y)}{(x-u)^{3/2}} \, du \quad , \tag{7}
$$

where $\beta^2 = (\chi'\sin\theta_i)^2 - \chi^2$. In turn, at relatively large distances
from the spot, i.e., $r \gg a/\cos\theta_i$, this simplifies to

$$
p_{t,1}(r,\theta,0) \simeq T_0 p_0 \frac{2}{\cos\theta_i} \sqrt{\frac{\chi'a^2\sin\theta_i}{2\pi r}}
$$

$$
\times \exp\left\{ \frac{i\chi'\sin\theta_i}{2} r \sin\theta \left[1 + \tan^{-2}\theta + \left(\frac{\chi}{\chi'\sin\theta_i} \right)^2 \right] - \frac{i\chi}{4} \right\} F_1(\theta) \quad . \tag{8}
$$

For a beam with a square cross section, for example, the directivity
function F_1 is

$$F_1(\theta) = \frac{\cos\theta}{(\sin\theta)^{3/2}} \frac{\sin\left\{\frac{\chi'a \, \tan\theta_i}{2}\left[\tan^{-2}\theta + 1 - \left(\frac{\chi}{\chi'\sin\theta_i}\right)^2\right]\right\}}{\left\{\frac{\chi'a \, \tan\theta_i}{2}\left[\tan^{-2}\theta + 1 - \left(\frac{\chi}{\chi'\sin\theta_i}\right)^2\right]\right\}} . \quad (9)$$

We now summarize the qualitative results for the refracted sound. When $k'a \sin\theta_i$ is large, i.e., $k'a$ large and θ_i not too small, four regions are discerned:

- the nearfield, $0 \leq r \lesssim a/\cos\theta_i$,   Eq. (7)
- the farfield, $a/\cos\theta_i <<r<<(k'a^2\sin\theta_i/2)$,   Eq. (8)
- the intermediate region, $r=O(k'a^2\sin\theta_i/2)$,
- the very farfield, $k'a^2\sin\theta_i/2<<r$.   Eq. (2)

In the farfield and very farfield, directivity functions are defined. The farfield is a region of two-dimensional spreading ($r^{-1/2}$), while the spreading is spherical in the very farfield (r^{-1}). This is due to the fact that diffraction effects build up on shorter scales in the plane of incidence than transverse to it, the y dependency in the nearfield and farfield being only what is "remembered" from the incident beam. A comparison of F and F_1, however, shows that the farfield and very farfield directivities are very similar (see Fig. 2).

Thus there is only a change in amplitude and in spreading transverse to the plane of incidence when one goes from the farfield to the very farfield, the transition taking place in the intermediate region (see Ref. 2 for a detailed discussion).

The very farfield directivity, i.e., the farfield directivity of a phase-shaded (and for some cases amplitude-shaded) source, is readily discussed (see Figs. 3 and 4). Below critical incidence, the refraction angle θ_t has values close to but smaller than those predicted by Snell's law ($k'\sin\theta_i=k\sin\theta_t$ when there is no attenuation; see Ref. 4 otherwise). Snell's law is recovered in the limit ka (or $ka_{3dB}) \rightarrow \infty$. In the vicinity of and above the critical incident angle, the departure from Snell's law increases when a decreases, so that a thinner beam is refracted (and reflected) more steeply. At such incident angles, the shape of the cross section affects the farfield configuration. There are no sidelobes in the case of a Gaussian cross section; there are sidelobes in the case of a square cross section, their number and relative amplitude increasing with

244   J. N. TJØTTA AND S. TJØTTA

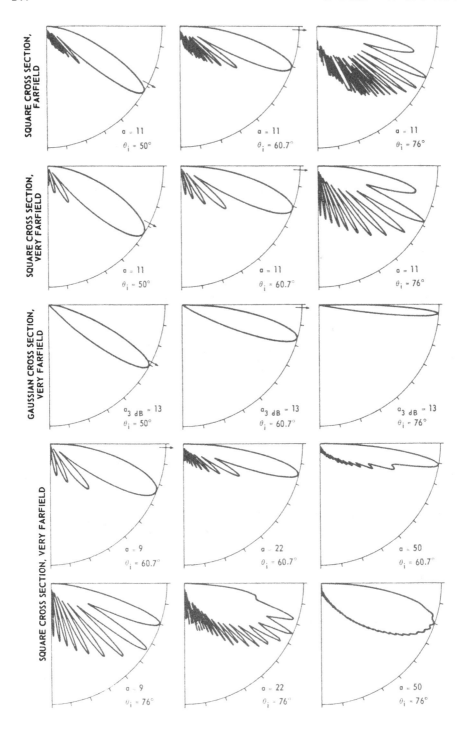

Fig. 2.   Some farfield and very farfield directivity patterns of the
refracted sound in the plane of incidence.

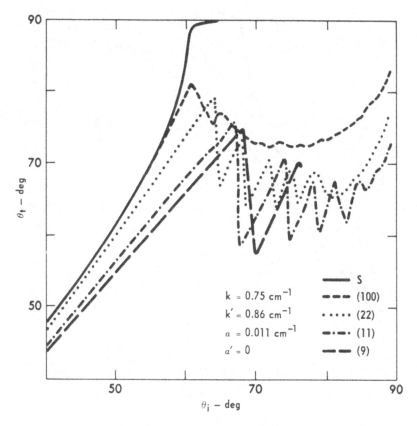

Fig. 3. Angle of refraction versus angle of incidence for incident
beams with square cross section of different width a (cm,
in parentheses). S corresponds to Snell's law (a→∞).

incident angle. This is due to the fact that interferences build
up in the regions of the spot where the amplitude gradient is sharp,
but are blurred in regions where the amplitude shading is smooth.

Figures 5(a),(b) and 6(a),(b) show constant amplitude curves
(solid lines) in the plane of incidence and in the nearfield. With
the exception of very high incident angles for the Gaussian model,
a beam structure is clearly discerned, although no analytical definition of the acoustic axis can be given at this stage. The axis is
displaced relative to the center of the spot ($\xi=\zeta=0$) by an amount
which increases with incident angle. Around critical incidence, the
displacement is about $a/\cos\theta_i$ or $a_{3dB}/\cos\theta_i$, as if the beam was
generated at the leading edge of the spot precisely where diffraction effects are most important. It is worth noting that the
Gaussian beam is refracted as a locally Gaussian beam, the 3 dB

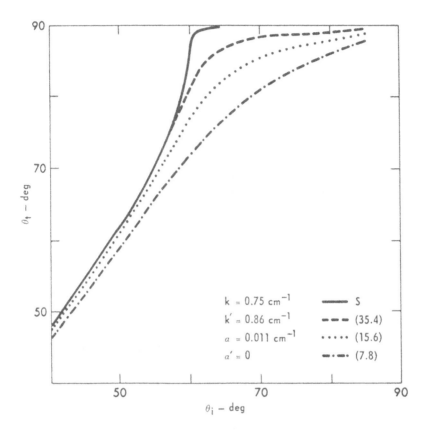

Fig. 4. Angle of refraction versus angle of incidence for incident
beams with Gaussian cross section a_{3dB}(cm, in parentheses).

width varying slowly with the distance along the axis (see also the
farfield and very farfield expressions, Eqs. (34) in Ref. 2).

Inspection of the nearfield patterns (Figs. 5(a),(b) and
6(a),(b)) also shows that along the axis (the double arrow line),
the pressure amplitude varies locally as an exponential function of
the distance, with a loss coefficient very close to the absorption
coefficient in the sediment, α (here 0.011), and practically independent of the value of θ_i. Accordingly, along the axis of the
refracted beam, the only loss mechanism predicted by our model is
absorption (with proper correction added for the geometrical
spreading in the farfield and very farfield). This is in striking

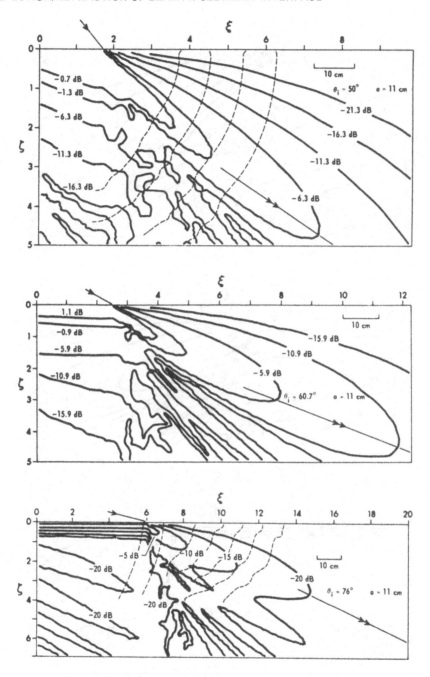

Fig. 5(a). Constant pressure amplitude (——) and constant phase
(---) curves of the refracted sound, incident beam with
square cross section. $\xi=x/h$, $\zeta=z/h$, $h=2\pi/k'\sin\theta_i$. The
arrow indicates the right edge of the incident beam
$(x=a/\cos\theta_i)$.

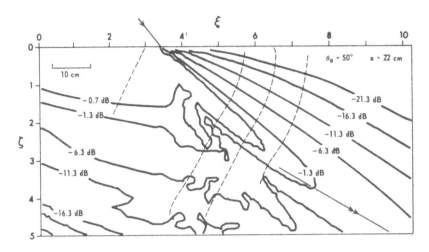

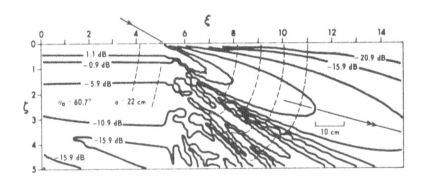

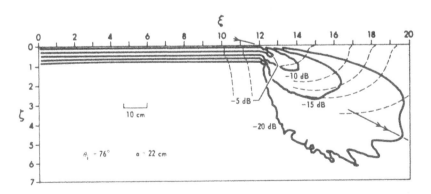

Fig. 5(b).   Constant pressure amplitude (—) and constant phase
             (- - -) curves of the refracted sound, incident beam with
             square cross section. $\xi=x/h$, $\zeta=z/h$, $h=2\pi/k'\sin\theta_i$.   The
             arrow indicates the right edge of the incident beam
             ($x=a/\cos\theta_i$).

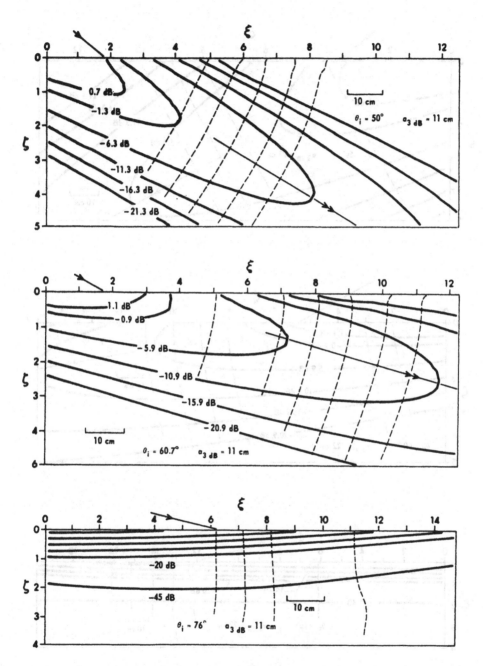

Fig. 6(a). Same as Fig. 5(a),(b) for an incident beam with
Gaussian cross section.

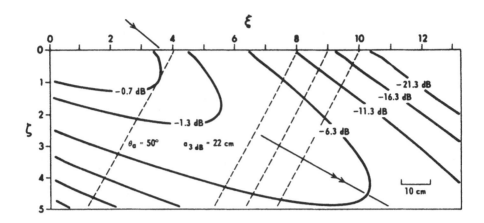

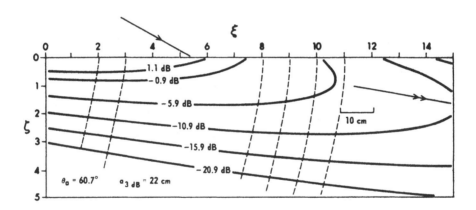

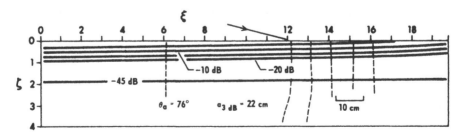

Fig. 6(b).   Same as Fig. 5(a),(b) for an incident beam with
             Gaussian cross section.

contradiction to the prediction of plane wave theory, according to
which the refracted wave should be evanescent in the z direction
when the incident angle is above the critical angle. It also
explains why Muir et al.[1] (their Fig. 10) measured amplitudes much
higher than the one predicted by the plane wave theory at 76° inci-
ence.

Finally, Fig. 7 indicates how the on-axis pressure amplitude
varies in the very farfield as a function of incident angle and beamwidth. One has to keep in mind that these curves, obtained from
Eqs. (3) and (5), are valid only at rather large distances from the
spot (they are not relevant for Muir et al.'s observations, which
are in the nearfield).

Although parts of the above results build on the assumption
of a large $k'a \sin\theta_i$, the theory is likely to apply when this condition is relaxed, the transition from the nearfield to the very
farfield being then more rapid since the different scales become
comparable.

We conclude this section by emphasizing the role of diffraction,
even when attenuation is relatively strong. Varying α affects the
shape of the sound field only slightly and quantitatively. As an
example, replacing $\alpha = 0.011$ Np/cm by $\alpha=0$ affects the directivity
curves only by making the valleys between the sidelobes go through
the origin. Due to the scaling used when deriving the governing
equations, the results are expected to be valid as long as the
attenuation length, α^{-1}, is not much shorter than a characteristic
length for diffraction effects, for example, $k'a^2\sin\theta_i/2$. For the
case of stronger attenuation, the validity of the nondissipative
boundary condition, among other factors, should be analyzed.

II. WAVEFRONTS IN THE NEARFIELD OF THE SPOT

The wavefronts of the refracted sound beam in the nearfield
have been obtained by using Eq. (7) and plotting the constant phase
curves in the plane of incidence. Some wavefronts (dashed curves)
are shown on Figs. 5(a),(b) and 6(a),(b). A look to the orientation
of the constant phase curves along the axis of the refracted sound
beam leads to the following general remarks. In the case of the
Gaussian cross section model with $a_{3dB} = 22$ cm, and at any incident
angle, the wavefronts have nearly the orientation predicted by the
infinite plane wave theory. When a_{3dB} is decreased to 11 cm, however, the wavefronts are more steeply dipping, the deviation from
the infinite plane wave theory being greater at greater incident
angles. Although the orientation may fluctuate, the wavefronts
produced by a square cross section beam with a given (a,θ_i) are
generally more horizontal than those produced by the equivalent
Gaussian beam $(a \simeq a_{3dB}$, see Ref. 2, Sec. VI). Here again the deviation increases with the incident angle. It also increases with

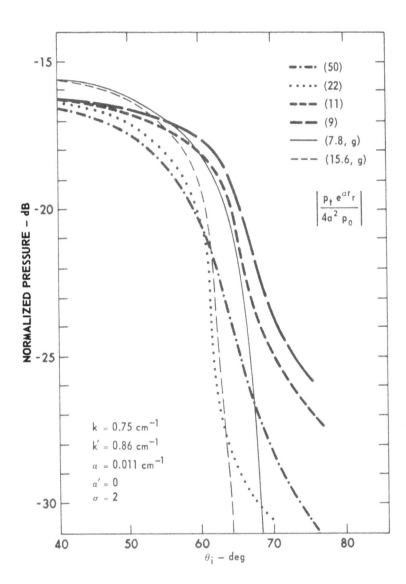

Fig. 7. Normalized on-axis refracted pressure amplitude versus
angle of incidence in the very farfield of the spot for
different beams. The numbers in parentheses are the value
(cm) of a (square cross section) or a_{3dB} (Gaussian cross
section) when followed by g.

decreasing beamwidth. We thus conclude that increasing the role of diffraction (i.e., increasing the incident angle, decreasing the beamwidth, sharpening the edges of the incident beam) always leads to more steeply dipping wavefronts. The above conclusions are qualitatively well supported by the observations.[1] We also conclude that at high incident angles the orientation of the wavefront is very sensitive to the model adopted for the amplitude profile of the incident beam (see also the discussion in Sec. III).

III. PRESSURE AMPLITUDE

The frequency used by Muir et al.[1] was 20 kHz corresponding to $k' \approx 0.86$, $k \approx 0.75$, $\alpha \approx 0.011$, and $\sigma \approx 2$ (unit: cm). The critical incident angle was thus $\theta_i^* \approx 60.7°$. The linear beamwidth (3 dB) was 10.2°, corresponding to $a_{3dB}(R) \approx R\sin(5.1°)$, where R is the distance between the source and the point of impact 0. The point of impact was in the focusing region of the parametrically generated beam. Table B1 of Ref. 1 gives five points on the curve $a_{3dB}(R)$, which is approximated by a cubic spline for the numerical computation of $|p_t(\theta_i)|$ using Eq. (7) (the hydrophone was buried at z = 17.8 cm beneath the interface, clearly in the nearfield of the spot at every incident angle).

Since the locations of the source in the water and the hydrophone in the sediment were fixed, the acoustic axis of the incident beam was tilted in a vertical plane (the plane of incidence). As seen in Fig. 1, when 0' and the observation point (x,0,z) are fixed, so are D,z, and the horizontal distance H between the source and the observation point. Then $x=H-D\tan\theta_i$, $R=D/\cos\theta_i$, and a_{3dB} (or a) also becomes a function of θ_i through R. Thus $|p_t|$ is a function of θ_i, the other parameters p_0, σ, k', k, α, D, H, and z being constant, and the function a_{3dB} (or a(R)) being given according to the source used. Under these conditions $|p_t/p_0|$ has been calculated using Eq. (7) and, assuming either a square or a Gaussian cross section model for the incident beam, the corresponding a(R) or $a_{3dB}(R)$ evaluated as explained above. The values used for D and H are those corresponding to the experiment (H=PC, see Tables 1 and 2 in Ref. 1), with the exception of the set D = 147 cm, H = 665 cm where the analogous experimental value of H is not known. The results displayed in Fig. 8 may indicate that H was larger than 665 cm.

For a given source (linear or parametric) and a given (D,H), the pressure amplitude has a maximum $|p_M|$ for some value θ_M of θ_i ($\theta_M=90°-\delta_M$ in Ref. 1). When the incident beam is assumed to have a square cross section, there is also a secondary maximum occurring when the hydrophone crosses a sidelobe (see also Fig. 5). The values of $|p_M|$ and θ_M predicted for the Gaussian and the square cross section models, respectively, are close to each other at low incident angles and become more different as θ_i increases (the

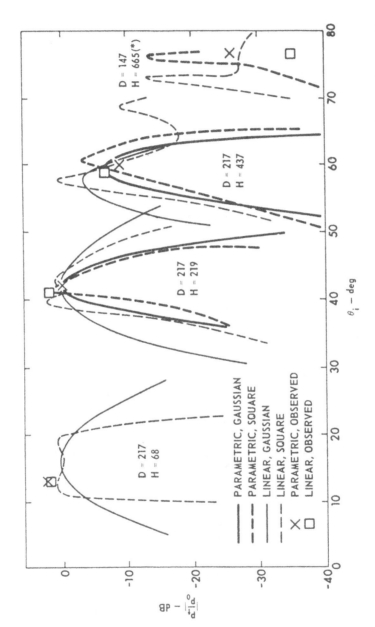

Fig. 8. Refracted pressure amplitude versus tilt angle in the plane of incidence, for different (D,H).

curves--not plotted--corresponding to the Gaussian model,
D = 147 cm and H = 665 cm, give $|p_M/p_o|$ equal to -49 dB for the
parametric source and to -52 dB for the linear one, i.e., considerably lower than with the square cross section model). This is due
to the fact that the amplitude profile of the incident beam around
its edge influences the amplitude shading of the spot in a much
stronger way at higher incidences (see also Sec. II). (It should
also be noted that Eq. (7) presupposes a spot which is symmetric
as a consequence of the assumption of q'_i varying little with z'
along the interface. This approximation is questionable when θ_i is
not far below $90°-\delta$, 2δ being the angle between the 3 dB points of
the incident beam.) Comparing the curves obtained for the parametric
and the linear sources, respectively, it follows that for a given
(D,H) θ_M is smaller for the linear than for the parametric beam.
The corresponding $|p_M/p_o|$ is greater for the linear source, with
exception of the last set, D = 147 cm, H = 665 cm, where the parametric source gives a larger amplitude. These conclusions are seen
to be in complete qualitative agreement with the observed values,
also plotted in Fig. 8. The quantitative agreement is also good,
with exception of the last set of values (H = 665 cm). In this last
case, it is clear that, even with the correct value for H, the
measured $|p_M/p_o|$ would be found between the theoretical values
corresponding to the Gaussian and the square cross section model,
respectively. On the edge of the beam, the amplitude profile may
be sharper than that of the Gaussian beam, but not quite as sharp
as the step function presumed in the square cross section model. A
better agreement should therefore be expected when using the measured
amplitude profile in the numerical evaluation of Eq. (7).

IV. REFLECTED BEAM

Results similar to those in Sec. I are obtained for the
reflected sound when replacing χ by χ' everywhere except in Eq. (1),
T_o by T_o-1, and θ being measured from the ascending vertical, in
Eqs. (2)-(7). Directivity patterns are discussed as in Sec. I, the
reflection angle θ_r having values close to but smaller than θ_i and
tending towards θ_i as ka increases. Figure 9 shows the variation of
θ_r with θ_i for different values of a in the case of the square cross
section model. The nearfield of the reflected sound has also been
computed for a Gaussian cross section model (a_{3dB} = 11 cm) using
Eq. (7) when $\theta_i=60.7°$ and $76°$, respectively. Some constant amplitude
and constant phase curves are shown in Fig. 10. The reflected
sound beam is seen to have a slightly asymmetric Gaussian profile,
the axis being practically symmetric to the axis of the incident
beam. There is no displacement, even at $76°$ incidence, due to the
fact that $\chi^2-(\chi'\sin\theta_i)^2$ is replaced by $(\chi'\cos\theta_i)^2$ and therefore
always has a positive real part.

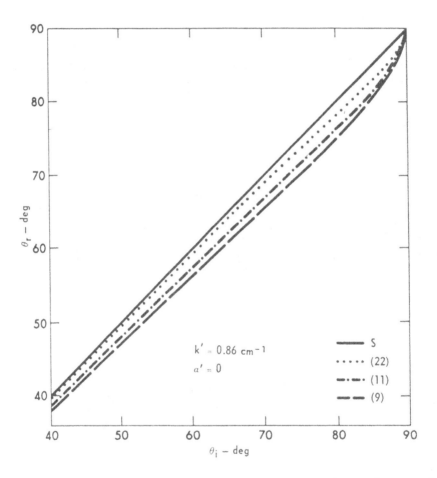

Fig. 9. Angle of reflection versus angle of incidence for incident
         beams with square cross section of different width a (cm,
         in parentheses). S corresponds to Snell's law (a→∞).

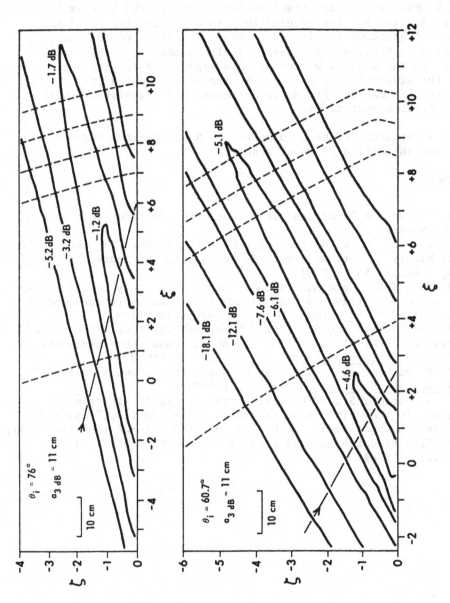

Fig. 10. Constant pressure amplitude (——) and constant phase (——) curves for the reflected sound beam. Incident beam with Gaussian cross section.

ACKNOWLEDGMENTS

    The authors wish to express their appreciation to
Dr. T. G. Muir at Applied Research Laboratories, The University of
Texas at Austin, for introducing them to a fascinating problem.
We are most grateful to him and to Prof. Claude W. Horton, Sr.,
Mr. Harlan G. Frey, and Dr. David T. Blackstock for helpful dis-
cussions on the subject matter of this paper, and for support
during the progress of the work. We are also indebted to
Mr. Keith S. Brown and Mr. Bruce A. Rousseau for assistance with
the numerical computations.

    This research has been supported by Naval Coastal Systems
Center and the Office of Naval Research.

REFERENCES

On leave from the Department of Mathematics, University of Bergen,
Bergen, Norway.
1.   T. G. Muir, C. W. Horton, Sr., L. A. Thompson, "The penetration
      of highly directional acoustic beams into sediments," J. Sound
      Vib. 64:539-551 (1979).
2.   J. Naze Tjøtta and S. Tjøtta, "Theoretical study of the pene-
      tration of highly directional acoustic beams into sediments,"
      submitted to J. Acoust. Soc. Am. for publication.
3.   J. Naze Tjøtta and S. Tjøtta, "Theoretical study of the pene-
      tration of highly directional acoustic beams into sediments,"
      J. Acoust. Soc. Am. Suppl. 1, 67:S29-S30 (1980).
4.   K. V. Mackenzie, "Reflection of sound from coastal bottoms,"
      J. Acoust. Soc. Am., 32:221-231 (1960).
5.   C. W. Horton, Sr., "The penetration of highly directional
      acoustic beams into a sedimentary bottom, Part I.," Applied
      Research Laboratories Technical Report No. 74-28 (ARL-TR-74-28),
      Applied Research Laboratories, The University of Texas at
      Austin (1974).

TRANSMISSION OF A NARROW BEAM OF SOUND

ACROSS THE BOUNDARY BETWEEN TWO FLUIDS

H. O. Berktay and A. H. A. Moustafa

University of Bath
School of Physics
Bath, BA2 7AY, UK

ABSTRACT

In studying shallow penetration of acoustic waves into saturated
marine sediments, narrow beams of sound (at kHz frequencies) have
been used under controlled conditions, (Muir, Horton, and Thompson.
J. Sound & Vibration, 64, June 1979). In order to contribute to the
understanding of the physical basis of the penetration of sound
across the boundary, a model experiment was devised at a frequency
of about 1 Mhz, using two non-mixing liquids. This paper presents
the experimental results obtained and compares these results with a
theoretical development. The theory developed also provides a means
of understanding some of the effects observed by Muir and co-workers
when a parametric source was used for penetration into a saturated
sediment.

1. INTRODUCTION

In studying the penetration of narrow beams of sound into
saturated marine sediments Muir and his co-workers obtained some
very interesting experimental results with some unexpected features[1],
particularly when a parametric source was used for the experiments.

Similar experiments[2] made by a group at the University of
Birmingham also showed some unexplained features, especially when
the acoustic axis of the beam was incident on the sediment surface
at a grazing angle at or below the critical value for total internal
reflection.

In an attempt to understand the physical basis of the phenomenon,
it was decided to undertake an investigation under laboratory

259

conditions, using two non-mixing fluids as the two media required
for transmission studies. This is a report of the theoretical and
experimental results of this investigation.

The theory presented follows along the lines developed in Refs.
3 and 4 for the penetration of a spherical wave across the boundary
between two fluids, but considers the particular constraints imposed
by the use of a narrow beam of sound obtained from a rectangular
piston. The same approach has also been developed for a truncated
end-fired line source, throwing some insight on the results obtained
using a parametric source.

2.   THEORY

As depicted in elevation in Fig. 1, a rectangular transducer
of sides 2bx2a is considered to be placed in the upper medium with
its shorter edges (2a) horizontal, and a small omnidirectional
hydrophone is used to sample the pressure field in the lower medium.

Following the development in Refs. 3 and 4, radiation from the
transducer can be represented as a spatial distribution of plane
waves, the field due to each incident plane wave calculated allowing
for the transmission coefficient at the boundary and the total
pressure at the field point obtained by intergrating over the spectrum
of plane waves.

The results are discussed below for a continuous sine-wave
transmission and for a tone burst transmission, for the particular
case of a relative refractive index less than unity.

The suffices 1 and 2 are used with ρ, c and k to represent,
respectively, the density, velocity of propagation and the wave
number for the two media.

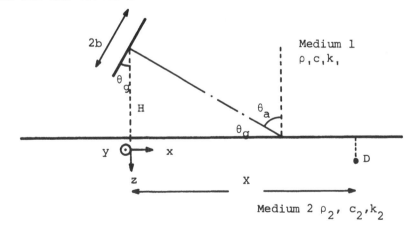

Fig. 1   The Geometry

2.1 The Field for a Continuous Sine-Wave

Assuming the rectangular transducer to be in an infinate baffle and to be driven uniformly by a sinusoidal wave at a frequency w, with surface velocity amplitude U_o, the pressure at a point $\underline{R}$ in the first medium (in the absence of the boundary with the second fluid) can be shown to be given by the following integral expression, neglecting the time-dependence, e^{-jwt}.

$$P = -P_o(jk_1/2\pi)^2 S \iint \frac{\sin(k_1 aM)}{k_1 aM} \frac{\sin(k_1 bN)}{k_1 bN} \frac{dk_x dk_y}{k_1 k_z}$$
$$\exp\left[j(\underline{k}_1 \cdot \underline{R} - k_z H)\right] \quad \ldots\ldots\ldots\ldots 1$$

Here, the geometry of Fig. 1 is used. $\underline{k}_1$ indicates the wave-vector associated with a particular plane wave, the x,y,z components of which are given by k_x, k_y, k_z, respectively. S is the surface area of the transducer. Also,

$$P_o = \rho_1 G U_o,$$

$$M = k_y/k_1,$$

$$N = (k_x \cos\theta_a + k_z \sin\theta_a)/k_1 \quad \ldots\ldots\ldots\ldots 1a$$

(The limits of integration are normally, $\pm\infty$ for both k_x and k_y. However, these limits should be modified to allow for the presence of the baffle.)

The integrand in Eq. 1 represents, essentially, a plane wave (with wave vector $\underline{k}_1$) incident on the boundary between the two fluids, which consists of the plane Z=O. The direction of $\underline{k}_1$ can be indicated by angles θ and ϕ, where θ is measured from the Z-axis (i.e., from the normal to the interface) and ϕ, in the plane of the interface, from the x-axis. In this coordinate system, the components of $\underline{k}_1$ may be written in the form

$$k_x = k_1 \sin\theta \cos\phi,$$

$$k_y = k_1 \sin\theta \sin\phi, \text{ and}$$

$$k_z = -k_1 \cos\theta.$$

Then, the field at (x,D) in the second medium, with the depth D taken as positive, can be expressed by the following integral equation:

$$P_t(x,D) = -P_o S(jk_1/2\pi)^2 \int_{-\pi/2}^{\pi/2} d\phi \int_{0}^{\pi/2-j\infty} d\theta \, \sin\theta . T(\theta) \, \Gamma(\theta,\phi)$$
$$\exp\left[jk_1 \eta(\theta,\phi)\right] \quad \ldots 2$$

Here,

$$\Gamma(\theta,\phi) \equiv \frac{\sin k_1 aM}{k_1 aM} \cdot \frac{\sin k_1 bN}{k_1 bN}, \qquad\qquad \cdots\cdots \quad 3$$

with $M = \sin\theta \sin\phi$,

$\qquad\qquad N = \sin\theta \cos\phi \cos\theta_a - \cos\theta \sin\theta_a$;

$\qquad \eta(\theta,\phi) = x \sin\theta \cos\phi + H \cos\theta + D\sqrt{(n^2 - \sin^2\theta)}, \quad 4$

with $n = k_2/k_1$, $\cdots\cdots \quad 5$

and the pressure transmission coefficient is

$$T(\theta) = 2(\rho_2/\rho_1) \cos\theta \left[(\rho_2/\rho_1)\cos\theta + \sqrt{(n^2 - \sin^2\theta)}\right]^{-1} \cdots \quad 6$$

it being understood that the value of the square-root with positive imaginary part is used over the path of integration.

Also, if either of the two media is absorptive, then the corresponding wave number, and hence n, will be complex. Here, we are only interested in values of n with the real part less than unity and the magnitude of the imaginary part much smaller than unity. (This latter condition implies that absorption per wavelength is small.)

For the present purposes, the main interest lies in geometries where the distances H and D correspond at least to many wavelengths, and the magnitude of k,x is large. In these circumstances, it is possible to evaluate the integral asymptotically.

The two-dimensional directivity function $\Gamma(\theta,\phi)$ is 'well-behaved' for real values of (θ,ϕ). However, as $|\theta|$ tends to ∞ as $\pi/2 - j\infty$, $|\Gamma| \to \infty$. Then, provided D + H is greater than b, the exponential term tends to zero at a faster rate and the integrand tends (exponentially) to zero.

In these circumstances, the most significant contributions to the double integral in Eq. 2 will come from the values of (θ,ϕ) close to the stationary-phase point (or, more generally, to the 'saddle point') which satisfies the equations.

$\qquad \eta_\theta = \eta_\phi = 0$

The corresponding values of the two variables are $\phi = 0$ and $\theta = \theta_0$, say, where

$$x \cos\theta_0 - H \sin\theta_0 - D \sin\theta_0 \cos\theta_0 / \sqrt{(n^2 - \sin^2\theta_0)} = 0 \quad ..7$$

Eq. 7 is identical to that obtained for a point source placed at the centre of the transducer. This equation has at least one solution, θ_0 the real part of which is less than that of the 'critical angle' for total internal reflection, $\theta_c = \sin^{-1} n$. When the imaginary part of n is small in magnitude, the imaginary part of θ_0 is also small, being positive or negative (with that of n) depending on whether k_1 or k_2 is complex.

The evaluation of the integral in Eq. 2 around the 'stationary' point is discussed in Appendix A. For large values of k,x, the most significant term is shown to yield (see Eq. A2) the following expression for the continuous-wave (c.w.) field in the second medium.

$$P_t/P_o \simeq -S(jk_1/2\pi)T(\theta_0)\sqrt{\sin \theta_0} \left[-x\eta_{\theta\theta}(\theta_0,0)\right]^{-\frac{1}{2}} e^{jk_1\eta(\theta_0,0)}$$

$$\frac{\sin\left[k_1 b \sin (\theta_0-\theta_a)\right]}{k_1 b \sin (\theta_0-\theta_a)} \quad \ldots\ldots 8$$

where (from Eq. 4)

$$-\eta_{\theta\theta}(\theta_0,0) = \eta(\theta_0,0) + n^2(1-n^2)D(n^2-\sin^2\theta_0)^{-3/2} \quad \ldots\ldots 9$$

The exponent in Eq. 8 is complex, the real part of which yields the absorption term. However, as discussed in Appendix B, when absorption per wavelength in the media considered is small a simpler approach is permissible.

If, initially, the media are assumed to be lossless, k_1, k_2 and n are real. Then, Eq. 7 yields a real root θ_0, which corresponds to the angle of incidence for the ray path obeying 'Snell's Law', with the corresponding angle of transmission, θ_t (also real) given by

$$\sin \theta_t = \frac{1}{n} \sin \theta_0.$$

Then, the various terms appearing in Eq. 8 are all real, and the result obtained through this equation gives the 'ray' solution for the field in the second medium in the absence of absorption. (See Fig. 2). The complete solution can be obtained by allowing for absorption along the ray path, by multiplying the 'lossless' result obtained through Eq. 8 by $e^{-\alpha_1 H/\cos\theta_0}$ or by $e^{-\alpha_2 D/\cos\theta_t}$ as appropriate.

This approach leads to a much easier interpretation of the result.

[Perhaps the first point of note is that Eq. 8 reduces to the far-field solution of Eq. 1, in the absence of the boundary, at a field-point $(\theta_0,0)$ - a result which can be obtained by substituting $n = (\rho_2/\rho_1) = 1$ in Eq. 8.]

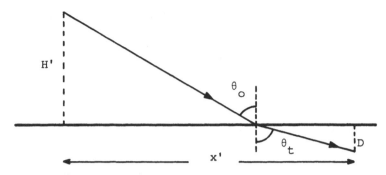

Fig. 2 The ray-path

If the distances x, H and D are fixed and the transducer is rotated around its horizontal axis (i.e., θ_a is varied) the ray angle θ_0 remains constant. Then, the absorption term and the phase of the pressure waveform remain fixed, while the pressure amplitude varies as the far-field directly function (in the first medium) in the vertical axial-plane of the transducer. Thus, even when the beam angle θ_a is greater than the critical angle for total internal reflection, the value of the ray path angle controls penetration into the second medium.

2.2 The Field for a Tone-burst Transmission

So far, the solution for a continuous sine-wave has been considered. In the experimental work to be described, a pulsed waveform was used in order to avoid multipath problems. Careful study of the waveforms observed indicated the presence of two arrivals at the hydrophone which could not be ascribed to spurious multipath effects. An understanding of this phenomenon can be obtained by considering the Fourier Transform of the result in Eq. 8, ignoring absorption.

With the exponent purely imaginary, the exponential term indicates a time retardation, τ_0, given by

$$\tilde{\tau}_0 = \left[x \sin \theta_0 + H \cos \theta_0 + D\sqrt{(n^2 - \sin^2 \theta_0)}\right]/c_1 \quad \ldots \ldots \ 10$$

The inverse-transform of the sine function, apart from an amplitude factor, is a 'window function' which is unity for time in the range $-\tau_1 < t < \tau_1$, and is zero for all other values of time, where

$$\tau_1 = (b/c_1) \sin (\theta_0 - \theta_a) \qquad \ldots \ldots \ 11$$

The factor $-j\omega$ in Eq. 8 indicates a differentiation with respect to time, and hence we obtain for the inverse-transform of the right-hand side of the equation the following expression:-

$$g(t) = (a/\pi)\ \frac{T(\theta_0)\sqrt{\sin\theta_0}}{\sin(\theta_0-\theta_d)}\left[-x\eta_{\theta\theta}(\theta_0,0)\right]^{-\frac{1}{2}}\left[\delta(t+\tau_1-\tau_0)\right.$$
$$\left.-\delta(t-\tau_1-\tau_0)\right],\quad \dots\ 12$$

where $\delta(t)$ is the Dirac impulse function.

If the surface-velocity of the transducer is of the form

$$u(t) = U_0\ f(t)$$

where $f(t)$ is a normalized function of time, then the reference pressure $p_0(t)$ can be represented in the form (see Eq. 1a)

$$p_0(t) = P_0\ f(t)$$

and the time-waveform of the field in the second medium, $p_t(t)$, can be obtained by the convolution of $f(t)$ and $g(t)$,

i.e., $p_t(t)/P_0 = f(t) * g(t)$.   13

As the expression $g(t)$ consists of two δ-functions, Eq. 13 yields two signals, each of the form $f(t)$, arriving with different delays. The delays $\tau_0 \pm \tau_1$ can be evaluated from Eqs. 10 and 11, and yield

$$c_1(\tau_0 \pm \tau_1) = (x \pm b\cos\theta_a)\sin\theta_0 + (H \pm b\sin\theta_a)\cos\theta_0 + D\sqrt{(n^2-\sin\theta_0)}$$
$$\dots\ 14$$

Hence, with reference to Fig. 1, the two δ-functions appear to be emanating respectively, from the upper and the lower edges of the transducer.

The difference between the times of arrival of the two 'edge waves' will be $2\tau_1$ with that due to the upper edge arriving first when τ_1 is positive, i.e., when θ_0 is greater than θ_a.

It should be noted that the 'edge' waveforms are, basically, the same as that of the surface velocity of the transducer, whereas the 'c.w' result is proportional to the time-derivative of the latter.

When the magnitude of the delay difference $2\tau_1$ is small compared with the period of the frequency used, the two 'edge' effects merge together, producing a waveform which is proportional to the time-derivative of $f(t-\tau_0)$, yielding the 'c.w' result.

If the transducer velocity function $f(t)$, is in the form of a tone-burst, as the value of τ_1 is altered by rotating the transducer (i.e., by varying θ_a) the two signals received at the hydrophone will 'move' relative to one another.

If the amplitude of the first arrival is monitered, it would be expected that this would correspond to that of the lower-edge wave when θ_o is less than θ_a, and to the amplitude of the upper-edge wave when θ_o is greater than θ_a, approaching the c.w. amplitude when $\theta_o \simeq \theta_a$. From the expression for the 'response function' $g(t)$ it is clear that the amplitude of either edge wave, on its own, does not have any side-lobes, but it decreases monotonically with increasing $|\theta_o - \theta_a|$.

If the amplitude of the received signal is observed at a part of the pulse where the two arrivals are present simultaniously, this should correspond to the c.w. result given in Eq. 8.

The validity of the derivation given here for the 'impulse response' may be considered to be somewhat limited in that that it assumes the ray-path angle, θ_o, to be the same for waves emanating from either edge of the transducer.

For a more rigorous approach, one may start from the integral expression for the c.w. field given in Eq. 2.

Expressing the term sin $(k_1 bN)$ in the exponential form, Eq. 2 can be represented as the sum of two integrals which can be written in the following manner:-

$$P_{te}/P_o = \pm (j/2\pi^2) \int d\phi \int d\theta \ \sin \ \theta \ T(\theta) \ \frac{\sin k_1 aM}{MN} \ e^{jk_1 \eta_e(\theta,\phi)} \quad .. \quad 15$$

where $\eta_e(\theta,\phi) = x' \sin \theta \cos \phi + H' \cos \theta + D\sqrt{(n^2-\sin^2\theta)}$

with $x' = x \pm b \cos \theta_a$,

               $H' = H \mp b \sin \theta_a$.

Clearly with the upper signs, (x', H') give the coordinates of the lower edge of the transducer, and with the lower signs one obtains the coordinates of the upper edge. Thus, each integral corresponds to one of the 'edge waves'.

Unlike the integrand in Eq. 2, each integrand in Eq. 15 tends to infinity when N is zero. The asymptotic evaluation of these integrals is considered in some detail in Appendix A1, where it is shown that (see Eq. A3) the most significant terms are of the form

$$P_{te}/P_o \simeq \mp \ \frac{aT(\theta_o)\sqrt{\sin \theta_o}}{\pi \ \sqrt{[-x'\mu_e''(\theta_o)]}} \ \frac{e^{jk_1\mu_e(\theta_o)}}{\sin \ (\theta_o-\theta_a)} \ , \qquad ..... \quad 16$$

where $\mu_e(\theta) = x' \sin \theta + H' \cos \theta + D\sqrt{(n^2-\sin^2\theta)}$,

$\mu_e''(\theta)$ indicates the second derivative with respect to θ, and θ_o is the root of the equation

$$\mu'_e(\theta) = 0$$

giving different ray-path angles θ_0 for each edge wave for a particular value of θ. This result is valid for large values of k,x', provided that θ_0 is not equal to θ_a. As discussed above, when θ_0 and θ_a are nearly equal, the two edge waves merge and cannot be identified separately. In which case, the c.w. result should be used.

As the coefficient of the exponential term in Eq. 16 is real and independent of frequency, the Fourier Transform of the field yields two impulse-functions, one for each edge-wave, as in Eq. 12 - the only difference being that the values of θ_0 are different for the two waves.

3.   EXPERIMENTAL WORK

3.1   General

The experiments were made in a tank with dimensions 305 mm x 305 mm x 610 mm.

Two non-mixing liquids (namely castor oil and ethylene glycol) were poured in and allowed to settle. A mirror-like interface (free from dirt and bubbles) was obtained between the two fluids.

The absorption coefficient for ethylene glycol was known to be small[6]. That for castor oil was determined acoustically at the frequency used (about 1MHz).

The temperatures of the two liquids were measured before each experimental run. Also, the sound velocity in each liquid was deter- mined, at each temperature by placing the projector and hydrophone in that fluid[6] (with the latter on the acoustic axis of the projector) and measuring the time-of-flight at various values of the separation between the two. The slope of the regression line was used to determine the velocity of propagation, while the off-set (on the time axis) gave a measure of the group delay in the equipment used in the experiments. The correlation coefficients obtained for the regression lines were very close to unity. The off-sets varied, the differences being less than 1μs.

Throughout the experiments, the time-of-flight measurements were made with a timer-counter. A double-pulse generator, with variable delay between the two pulses, was used to determine the period to be measured. The transmitted tone-burst was triggered from the positive-going edge of the first pulse, which also triggered the timer-counter. The second pulse was used to stop the count. Thus, by varying the delay on the double-pulse generator until the second pulse coincided with the start of the received signal, the time delay could be measured accurately, to within a fraction of 1μs.

By subtracting the off-set time mentioned above from the reading of
the counter-timer, the time-of-flight was determined.

The relevant parameters of the two fluids used are shown in
Table 1. To measure variations in the time of arrival, for example,
with changes in the attitude of the transducer, the counting technique
was used at a particular angle, and then the variations in the
arrival time were noted on the oscilloscope.

The high signal-to-noise ratio obtained during the experiments
permitted the accurate determination of the instant of arrival of
the signals.

Table 1

Parameters of the Two Fluids

	Density (gm/cm^3) [6]	Sound Velocity (m/s) (at 22.5°C)	Absorption (Np/cm) (at about 1MHz)
Castor Oil	0.95	1521	0.05
Ethylene Glycol	1.115	1667	–

3.2 Experimental Arrangement

For experiments on the transmission of acoustic waves across the
boundary of two fluids, a projecting transducer was placed in castor
oil and a hydrophone in ethylene glycol.

The transducer consisted of a rectangular (25 mm x 6 mm) single-
plate (PZT) element, mounted on a block of composite material
(pulverised fuel ash/epoxy resin) and potted in epoxy resin for
protection, the thin layer covering the radiating surface being about
0.75 mm thick. The thickness resonance frequency of the transducer
was found to be 1.065 MHz. The transducer was mounted in such a way
that the acoustic axis could be rotated in a vertical plane which
included the longer (i.e. 25 mm) axis of symmetry of the transducer.
The angle between the vertical and the surface of the transducer (θ_g)
could be read on an extended protractor fixed to the mounting assemb
assembly.

The transducer assembly was mounted in the slider of a horizontal optical bench positioned on the experimental tank. This permitted the horizontal spacing between the transducer and the hydrophone to be varied. Further, the vertical position of the hydrophone and of the transducer could be varied independently. The hydrophone consisted of a small cylindrical PZT element, about 3 mm in length, mounted on a long stalk.

Although the transducer and the hydrophone were not calibrated individually, using acoustic measurements (with both in the same medium) the results were normalized in a manner suitable for comparison with those obtained theoretically.

For this purpose, the pressure ratio P/P_0 along the axis of the transducer working in castor oil was computed as a function of range, at the frequency of interest, using Eq. 1. Absorption in castor oil was allowed for, using a nominal value.

With the transducer and the hydrophone in castor oil, the latter was moved along the acoustic axis of the former, the ratio of the voltages across the hydrophone and the transducer, V_{RX}/V_{TX}, being noted as a function of range.

The ratio $(V_{RX}/V_{TX}) \div (P/P_0)$ should be independent of range, possibly except for an exponential term due to the coefficient of absorption being different from the nominal value. After correcting for this exponential term, the constant ratio gives the overall sensitivity of the transducer/hydrophone combination.

Using this approach, the absorption coefficient in castor oil was estimated to be 0.05 Np/cm. With this value of absorption coefficient and a 'sensitivity' of 3.09 (pressure ratio divided by voltage ratio) a good fit was obtained between the measured and computed results for ranges extending from 16 cm to 42 cm.

3.3 Experiments

The pressure field in the second medium was investigated for various positions and attitudes of the transducer, at different depths from the interface. The experiments were made by fixing the hydrophone position at a particular depth and then monitoring the received signal level while either rotating or moving the transducer horizontally at a fixed attitude.

In this manner, the pressure field was monitored for the signal arriving first and also for the signal constituting the bulk of the pulse, the latter was used to indicate the 'continuous wave' (i.e., c.w.) result. All the results were converted into a normalised form 'P/P_0' using the calibration data mentioned in 3.2 above.

To monitor the pressure for the 'first arrival', the voltage difference between the first (positive) maximum and the following (negative) minimum of the waveform was measured.

For the 'c.w.' measurements, most of the results were obtained by noting the voltage differences between the fourth maximum and the following minimum of the received waveform. Also, some results were obtained by monitoring the peak-to-peak voltage of the received pulses on a compressed time scale. Good agreement was observed between the two methods of measurement.

A further problem in making the 'free field' pressure measurements at such high frequencies arises from the directivity of the hydrophone in the vertical plane. Unfortunately, the direct-ional response of the hydrophone could only be measured over a limited range of angles, indicating 3dB loss for about $\pm 15^{\circ}$ from the horizontal.

4. RESULTS

4.1 Times of flight

Possibly the most reliable measurements made were those of transit time for different geometries.

As mentioned in section 3.1, in any given run, the transit at certain reference positions were measured using a counting technique. Then, the variation in transit time was noted on oscilloscope screen (to obtain better interpolation) as the geometry was varied. Some quite typical results are shown in Figs. 3 and 4.

In Fig. 3, the measured transit times, and those estimated for waves emanating from the lower and the upper edges of the transducer, are plotted against the 'grazing angle' of the transducer axis, θ_g, for a particular transducer/hydrophone configuration. The respective values of the flight time for grazing angle of 13° were taken as references for both the measured and the estimated values for the lower edge wave. The measured and estimated values of transit time at two grazing angles are also indicated.

The transit time estimates are based on the travel time along the corresponding ray paths, computed using the changing positions of the edges of the transducer when it is rotated around its horizontal axis.

In Fig. 4, the values of transit time measured at various horizontal distances are shown for a fixed transducer angle, compared with estimated results. These results were obtained for a relatively low grazing angle θ_g, a few degrees less than the critical value for total internal reflection (which was about 24.5°). The

transit times, measured and estimated, are shown relative to those
estimated for the lower-edge wave. The two estimated curves
(included for purposes of comparison) show calculations based on
waves the phase centres of which are assumed to be at the upper-edge
and at the centre of the transducer, respectively.

These and similar results provide strong experimental evidence
to suggest there are two replicas of the transmitted signal arriving
at the field point via different paths, with different transit times.

4.2 Pressure in the lower medium - 'c.w.' results

Theoretical results presented in section 2 (Eq. 8) suggest that
when the positions of the hydrophone and of the transducer are fixed
and the latter is rotated, the field should describe the far-field
directivity function of the transducer in the first medium, the
central point being the ray-path angle, θ_0, corresponding to the
particular geometry. Experimental results obtained do verify this.

As an example, two beam patterns (obtained at different horizontal spacings, but with H and D the same) are shown in Figure 5, in a
normalized form, showing well-formed side-lobes even at 30 cm. The
results obtained in castor oil at a range of 42 cm, on the other hand,
showed a typical near-field pattern.

The two curves in Fig. 5 substantially overlap when one is
shifted by about 1.2°. It is also worth noting that the values of
θ_0 for the two cases were calculated to be 63.6° and 64.8°,
respectively, for horizontal distances of 30 cm and 42 cm.

To compare the measured and estimated values of the pressure
amplitude in the second medium, for a particular configuration, the
values of P_t/P_0 were estimated using Eq. 8. A set of results, for
the 30 cm spacing case already mentioned are shown in Fig. 6 (with
θ_0 so close to the critical value $\sin^{-1} n$, the ray-path in the
second medium is close to horizontal, causing only a small correction
for hydrophone directivity). It can be seen that the experimental
results produce a broader beam than predicted (about 3.6° instead of
3.1°) and are higher by about 2.5 dB.

Sets of pressure measurements were made keeping the transducer
height and attitude constant, and changing the horizontal spacing
between the transducer and the hydrophone, with the latter at various
depths. These experiments permitted the building up of a two-
dimensional picture of the field in the second medium. Particular
attention was paid to the possibility of a beam shift being present
in the second medium, similar to that obtained in the reflection of
a narrow beam when the angle of incidence is greater than the value
for total internal reflection. Within the capability of the
experimental arrangement (the hydrophone depth could not be reduced

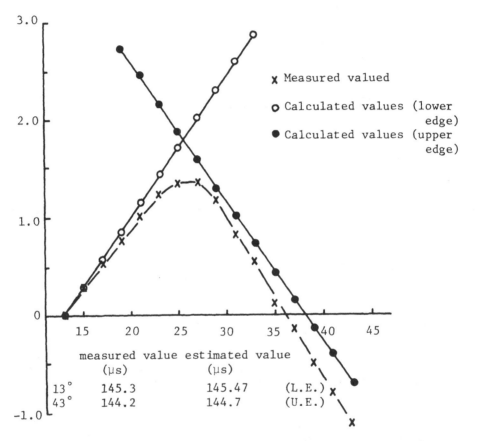

Fig. 3.   Transit time variation with transducer angle θ_g.
X = 22.0 cm, H = 5.2 cm, D = 1.4 cm.

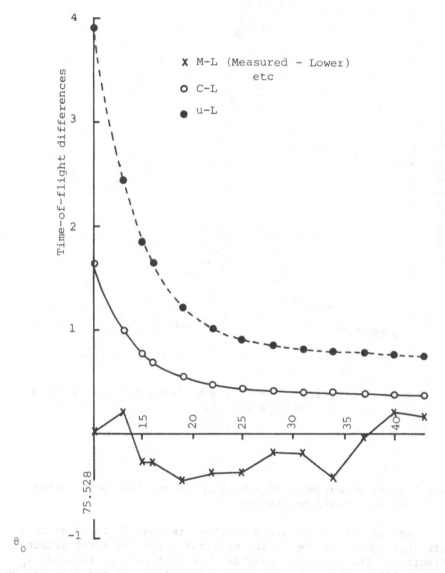

Fig. 4. Differences between measured and estimated transit times
with those calculated for the lower edge of the transducer
as reference. $H = 5.2$ cm, $D = 1.4$ cm, $\theta_g = 22°$.

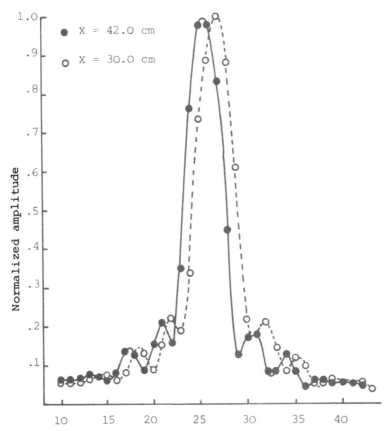

Fig. 5. Normalized directivity patterns for c.w. results
         h = 5.2 cm   D = 1.4 cm.

below 1.0 cm) no evidence of a beam shift or signs of a 'lateral
wave' component could be observed.

A set of results obtained is shown in Fig. 7, for a grazing
angle (θ_g) about 2.5° below the critical value for total internal
reflection. The predicted results show that as X is reduced, θ_o
increases, describing the successive side-lobes of the directivity
function of the transducer. The measured results confirm this trend
to a limited extent. It is thought that this is mainly due to the
rapid variation of the field with depth, the hydrophone averaging
over these variations. Table II shows the estimated values of $|P_t/P_o|$
for three values of D, at 5 mm intervals, to illustrate the extent
of the variations.

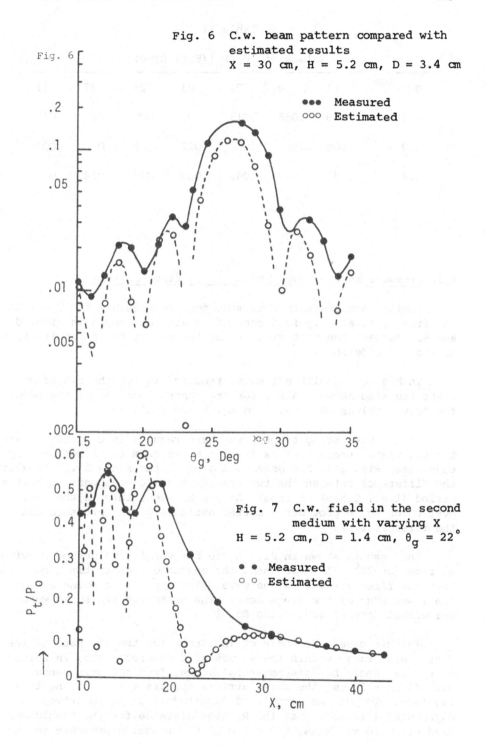

Fig. 6 C.w. beam pattern compared with estimated results
X = 30 cm, H = 5.2 cm, D = 3.4 cm

●●● Measured
○○○ Estimated

Fig. 7 C.w. field in the second medium with varying X
H = 5.2 cm, D = 1.4 cm, θ_g = 22°

●● Measured
○○ Estimated

Table II

Values of $\left| P_t/P_o \right|$ (Estimated)

Dcm \ Xcm	17	19	21	23	25	27	31
1.4	.059	.036	.013	O	.008	.011	.011
1.9	.005	.05	.038	.022	.0o8	O	.007
2.4	.032	.017	.042	.038	.026	.014	O

4.3 Pressure in the lower medium - the first arrival

Similar measurements were made for the pressure field due to the first arrival. Typical sets of results are shown in Figs. 8 and 9. Perhaps the most significant feature of these results is the absence of sidelobes.

In Fig. 8, results estimated from Eq. 16 for the two edge-waves are also shown. Also, for the central portion of the beam, the 'c.w.' values obtained from Eq. 8 are included.

It is interesting to observe these results in conjunction with the calculated transit times for the waves from the lower and upper edges (see Fig. 3). For grazing angles in the range 23.5^o to 27.5^o the difference between the two transit times corresponds to half a period (i.e., $0.5\mu s$) or less. As can be seen, this is the range over which Eq. 8 provides a better estimate of the pressure due to the first arrival.

The results shown in Fig. 9 are for varying values of X, with θ_g fixed at 22°. In this case, the measured transit times confirmed that the first arrival is the wave which appears to emanate from the lower edge of the transducer. The estimated pressures were calculated accordingly, using Eq. 16.

In this case, the angle of incidence for the ray path varies over a wide range within the values of X covered. But, in spite of the ray-angle being substantially away from the axis (about 5.1^o for x = 14 cm) the first arrival appears to come along this ray path. To put the normalized magnitude into perspective, it maybe useful to note that the Raleigh distance for the transducer used was 0.10 m. Thus, the ratio of the measured pressure to the

'source level' of the transducer (in linear units) can be found by
multiplying the ordinates in Fig. 9 by 10.

5. DISCUSSION AND CONCLUSIONS

     The results presented above must be considered in the light of
the following limitations.

     The accuracy with which the attitude of the transducer could be
determined was, probably, within 0.5°.

     The finite size of the hydrophone resulted in an average value
of the pressure being obtained. This could affect the results
significantly, especially when the pressure varied significantly
with small changes in depth.

     Theoretical work presented assumed a 'piston' in an infinate
baffle, giving a calculated 3 dB beam width of about 3.1° along its
long dimension. Experiments with the transducer used showed that
the corresponding beam width was about 3.6°, suggesting that it was
not behaving ideally. The most significant consequence of this would
be that the pressure variation along its axis would be affected,
causing uncertainties in the calibration of the transducer/hydrophone
configuration using the technique described in section 3.2.

     The flight time measurements and the trends of pressure
variation in the second medium, both for c.w. and first-arrival,
tend to confirm the theoretical predictions. The general agreement
between the estimated and measured pressure levels is acceptable,
within the constraints discussed above.

     In these circumstances, we feel justified in suggesting that
the 'ray-path' approach provides an acceptable basis for the study
of the penetration of acoustic waves across the boundary between
two fluids, even when the incident waves are confirmed within a
narrow beam. When the ray path obtained (assuming a directional
source at the phase-centre of the beam) is within the main lobe, this
approach gives acceptable results.

     However, if the ray path is far removed from the beam axis, the
constituent rays emanating from the discontinuities of the aperture
travel via significantly different paths, with different transit
times. In this situation, signals arriving via the two different
paths should be considered individually, and then the results superposed to estimate the resultant field.

     When the beam axis is incident on the boundary at an angle
greater than the critical value for total internal reflection, for
example, the ray-paths must lie outside the main lobe of a narrow
beam, which results in the two 'edge waves' arriving with

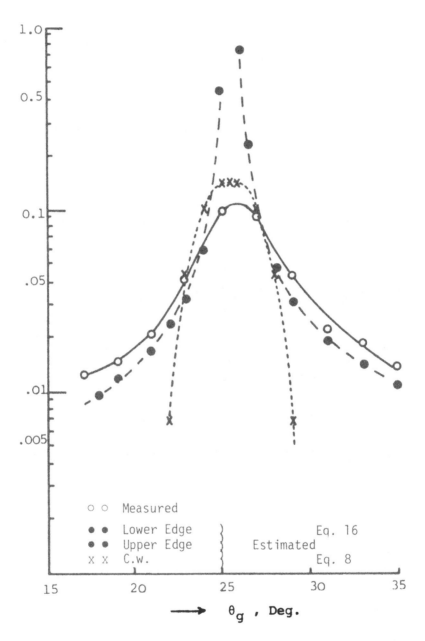

Fig.8. Beam pattern for the first arrival
        X = 22.0 cm, H = 5.2 cm, D = 1.4 cm

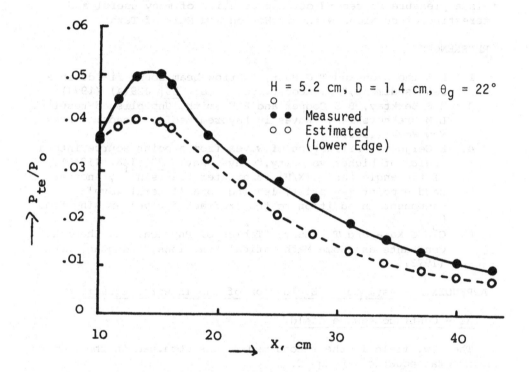

Fig. 9. Pressure due to first arrival with varying x.

significant delay difference, causing signal distortion and elongation.

ACKNOWLEDGMENTS

The experiments reported were made at the Electronic and Electrical Engineering Department of the University of Birmingham. It is a pleasure to record our appreciation of many useful and interesting discussions with Drs Horton and Muir of Texas.

REFERENCES

1. L A Thompson and T G Muir, "Narrow-beam sound fields in a sand sediment," J. Acoust. Soc. Amer. 55, 429(A) (1974).
2. H O Berktay, B S Cooper and B V Smith, Unpublished results.
3. L M Brekhovskikh, "Waves in Layered Media," Academic Press, New York (1960).
4. E Gerjouy, "Refraction of waves from a point source into a medium of higher velocity," Phys. Rev., 73, 1442 (1948).
5. If the angle $\tan^{-1}$ (X/H) is greater than $\sin^{-1}$ n, another saddle point may exist, leading to a 'lateral wave' component in addition to the 'refracted wave' arising from θ_o.
6. G W C Kaye and T H Laby, "Tables of Physical and Chemical Constants and Some Mathematical Functions," Longman, London (1973).

APPENDIX A - Asymptotic Evaluation of the Integral Equations

AI - Continuous wave field

The c.w. field in the second medium was obtained in the form of an integral equation in Eq. 2.

The most significant contribution to the integral will accrue for small values of $|\phi|$. Then, in the plane-wave spectrum $\Gamma(\theta,\phi)$, we may assume k_1aM to be small and replace the associated sinc function with unity.

Similarly for small $|\phi|$,

$$k_1bN \simeq \tfrac{1}{2} k_1b \sin\theta \cdot \cos\theta_a \frac{2\sin(\theta-\theta_a)}{\sin\theta \cos\theta_a} - \phi^2$$

and the exponent becomes

$$k_1\eta(\theta,\phi) \simeq k_1\mu(\theta) - \tfrac{1}{2} k_1x \sin\theta \cdot \phi^2$$

with

$$\mu(\theta) = x \sin\theta + H \cos\theta + D\sqrt{(n^2-\sin^2\theta)}.$$

Expanding $\sin(kbN)/(kbN)$ in ascending powers of ϕ^2 and integrating along the line $re^{-j\pi/4}$ term by term, an approximate result is obtained for the ϕ integral, as follows.

$$I_\phi = \int_{\sim -\pi/2}^{\sim \pi/2} \frac{\sin k_1 bN}{k_1 bN}\, e^{jk_1\eta(\theta,\phi)} \cdot d\phi$$

i.e., $\simeq e^{jk_1\mu(\theta)} \int_{-\infty}^{\infty} \frac{\sin k_1 bN}{k_1 b}\, e^{-jQ\phi^2} d\phi$

i.e., $\simeq e^{j(k_1\mu-\pi/4)} \sqrt{(\pi/Q)} \cdot \left\{ \frac{\sin q}{q} + j\frac{V}{2Q}\left(\frac{\cos q}{q} - \frac{\sin q}{q^2}\right)..\right\}$

with

$Q = \tfrac{1}{2}k_1 x \sin\theta,$

$V = \tfrac{1}{2}k_1 b \sin\theta\cdot\cos\theta_a,$

$q = k_1 b \sin(\theta-\theta_a).$

The neglected terms are in higher orders of V/Q, the coefficients of which are well behaved. Thus, for b/x small, only the first term needs to be used.

With this result, Eq. 2 gives

$$P_t/P_o \simeq S(k_1/2\pi)^{3/2} x^{-1/2} e^{-j\pi/4} \int_0^{\pi/2-j\infty} d\theta \cdot T(\theta)\sqrt{\sin\theta}\cdot\frac{\sin q}{q} e^{jk_1\mu(\theta)} \quad ... \tag{A1}$$

The integral can be evaluated using the steepest descent approach. The path of integration is the same as that for a point source over which the integrand is well behaved, provided $b < (H+D)$. The most significant term, which can be obtained directly from Eq. 2 through the stationary-phase evaluation of the double integral, becomes

$$P_t/P_o \simeq -jS(k_1/2\pi)T(\theta_o)\sqrt{\sin\theta_o}\cdot[-x\mu''(\theta_o)]^{-\frac{1}{2}}\cdot\frac{\sin q_o}{q_o}\cdot e^{jk_1\mu(\theta_o)} ,\,... \tag{A2}$$

where θ_o is the root of the equation

$$\mu'(\theta_o) = X\cos\theta - H\sin\theta - D\sin\theta\cdot\cos\theta\cdot(n^2-\sin^2\theta)^{-\frac{1}{2}} = 0,$$

and

$q_o = k_1 b \sin(\theta_o-\theta_a).$

AII - The Expression for the edgewaves

The ϕ integral in Eq. 15 is complicated by the presence of N in the denominator of the integrand, which may have zeros along the path of integration. With

$$N = \sin\theta \cdot \cos\phi \cdot \cos\theta_a - \cos\theta \cdot \sin\theta_a,$$

for N to vanish,

$$1 - \cos\phi = \sin(\theta-\theta_a)/\sin\theta \cdot \cos\theta_a.$$

Along the path of integration, $\sin\theta$ is positive and real. Further, the values of θ of interest will be in the vicinity of the stationary value, θ_0 say. Thus, N will have zeroes when θ_0 is greater than θ_a.

The most significant contribution to the ϕ integral will arise from small values of $|\phi|$. Accordingly, M would be small and the ϕ integral in essence would be reduced to the form

$$I \simeq \int_{-\infty}^{\infty} d\phi \cdot e^{-jB\phi^2}/(\gamma-\phi^2)$$

with

$$B = \tfrac{1}{2} k_1 x' \sin\theta,$$

$$\gamma = 2\sin(\theta-\theta_a)/\sin\theta \cdot \cos\theta_a,$$

and

$$\theta \simeq \theta_0.$$

The integrand has poles along the real axis only if $\theta_0 > \theta_a$; the integral reduces to

$$I \simeq 2j\sqrt{B} \cdot e^{-j\pi/4} \int_0^{\infty} \frac{e^{-t^2} dt}{j\gamma B - t^2}$$

i.e., $\simeq \dfrac{\pi}{\sqrt{(j\gamma)}} e^{-j\pi/4} \cdot W[\sqrt{(j\gamma B)}],$

where $W(\sigma)$ is error function with complex argument and the square-root with positive imaginary part is implied.

For $\theta > \theta_a$, an additional term is included for the integral around the real pole at $\sqrt{\gamma}$. It can be shown that these additional terms produce identical results, for the two edges, in terms of the θ integrand and thus they cancel when the two edge waves are superposed with the appropriate signs (see Eq. 15).

Hence, Eq. 15 is reduced to the form

$$\frac{P_{te}}{P_0} \simeq \mp \frac{jk_1 a}{\pi\cos\theta_a} \cdot e^{-j\pi/4} \int_0^{\pi/2-j\infty} \frac{d\theta \cdot T(\theta)}{\sqrt{(j\gamma)}} \cdot W[\sqrt{(j\gamma B)}] \cdot e^{jk_1 \mu_e(\theta)}$$

In this case, the integrand is multivalued in the complex θ plane. In fact, a new branch line is introduced, the equation for which is $\theta = \theta_a - jv$, with v real and positive. The path of steepest descent needs to be modified when $\theta_o < \theta_a$, introducing an additional term to the integral. These terms, however, are again identical for the two edge waves, thus cancelling out in the overall result.

The argument of the error function $W(\sigma)$ is

$$\sigma = [jk_1 x' \sin(\theta - \theta_a) / \cos\theta_a]^{\frac{1}{2}} .$$

With $k_1 x'$ large, provided θ_o is not equal to θ_a, $|\sigma|$ is large over the modified path of integration. Then, asymptotically,

$$W(\sigma) \rightarrow j / (\sigma \sqrt{\pi}).$$

Thus, except when $\theta_o \simeq \theta_a$ so that $|\sigma|$ is small, integration along the path of steepest descent gives

$$\frac{P_{te}}{P_o} \simeq \mp \frac{a \ T(\theta_o) \sqrt{\sin\theta_o}}{\pi \ \sqrt{[-x'\mu_e''(\theta_o)]}} \ \frac{e^{jk_1\mu_e(\theta_o)}}{\sin(\theta_o - \theta_a)} , \ldots \tag{A3}$$

where

$$-\mu_e''(\theta_o) = \mu_e(\theta_o) + n^2 D(1 - n^2) \ (n^2 - \sin^2\theta_o)^{-3/2} .$$

As discussed previously, when $\theta_o \simeq \theta_a$ the c.w. result given in Eq. A2 is more appropriate to use.

APPENDIX B - Absorption along the ray path

In the expressions developed for the field in the second medium, the exponents were of the form $jk_1\mu(\theta_o)$, with the imaginary part of $k_1\mu(\theta_o)$ indicating exponential attenuation in preparation. We assume k_2 to be real and k_1 (and hence n) to be complex to allow for absorption in the first medium.

Let $k_1 = (2\pi/\lambda_1)(1 + jA)$, with $A = \alpha_1 \lambda_1/2\pi \ll 1$. Then, $\mu'(\theta)$ is also a complex expression, resulting in a small imaginary component for θ_o as well.

With $\theta_o = u + jv$, the imaginary part of $\mu(\theta_o)$ is small, being of the order (A,v). Then,

$$k_1 \ \mu(\theta_o) \simeq (2\pi/\lambda_1) \cdot \{Re\mu(\theta_o) + j[Im\mu(\theta_o) + A \cdot Re\mu(\theta_o)]\}$$

But, to the first-order in A and in v,

$$Im\mu(\theta_o) \simeq v \cdot \mu_\theta(\theta_o) - A(\lambda_1/\lambda_2)\mu_n(\theta_o)$$

The coefficient of v is zero by the definition of the saddle point, at least to the accuracy involved. Thus,

$$\text{Im}\left[k_1 \mu(\theta_o)\right] \simeq \alpha_1 \cdot \left\{\text{Re}\mu(\theta_o) - Dn_o^2 (n_o^2 - \sin^2 u)^{-1/2}\right\}.$$

With

$$n_o = \lambda_1/\lambda_2,$$

the coefficient of α_1 becomes

$$X \sin u + H \cos u - D \sin^2 u \; (n_o^2 - \sin^2 u)^{-1/2} \; .$$

Treating u as the angle of incidence for the ray path for zero absorption, $\left[\text{i.e., from } \left(\frac{\partial \mu}{\partial \theta}\right)_{o,o} = o\right]$ we obtain

$$X \cos u - H \sin u - D \sin u \cdot \cos u \cdot (n_o^2 - \sin^2 u)^{-1/2} = 0$$

Thus, the coefficient of α_1 reduces to

$$H \cos u + \sin u \; (H \sin u/\cos u)$$

i.e., to H/cosu, which gives the path length in the absorbing medium.

Thus, for cases where absorption per wavelength is small, it is sufficient to calculate the ray path angle ignoring absorption, and then to allow for absorption over the appropriate part of the ray path.

EXPERIMENTAL DETERMINATION OF PROPERTIES OF THE SCHOLTE WAVE

IN THE BOTTOM OF THE NORTH SEA

Florian Schirmer

Universität Hamburg
Institut für Meereskunde
2ooo Hamburg 13, W.-Germany

ABSTRACT

An experiment was carried out in the North Sea with the objective of exciting the Scholte wave and investigating its propagation. Shots of 3 and 1o kg TNT at the sea bottom provided the
seismic source. 3-component seismometers were set up at distances
between 0.8 and 1.7 km from the source. The seismic signal was
transmitted by radio to the recording ship. An elastic wave with
a propagation velocity of 111 m/s was detected. The hodograph
shows a vertical, polarized, retrograde ellipse. The observed
frequencies are in the range 4.5 ± 1 Hz. The vertical component
accounts for around 64% of the elastic energy as measured at the
sediment/water interface. After subtraction of the geometric
attenuation (E_r prop. $1/R$), it was possible to determine the
order of magnitude of attenuation of the Scholte wave
(7.10^{-3} dB/m).

INTRODUCTION

Some years ago I read a theoretical paper on the Scholte
wave, which was written by Dieter Rauch[1] from the SACLANTCEN. I
found a lack of experimental data on this special type of interface wave between solid earth and water. One numerical example
from computations (not from direct measurements) is given by
Cagniard[2] in appendix II of his book. He concludes that the large
scale waves taken from ordinary seismometer records of suboceanic
earthquakes show a false Rayleigh wave which travels with a considerably higher velocity than the Scholte wave. This can be

285

tested easily, because the velocity of the Scholte wave must be
less than the velocity of the compressional wave in water or of
the shear wave in the sediment.

The best way to look for a Scholte wave is to use a set of
low frequency geophones on the sea bottom. The hodograph will be
a vertical polarised ellipse with retrograde particle motion.
Additionally the wave will be nondispersive in the case of a non-
layered bottom. Very frequently, however, the bottom is layered.
This case is discussed by H.-H. Essen in this conference.

EXPERIMENTAL SETUP

Fig. 1 shows the area of the North Sea in which the experiment
was performed. The sea bottom there is very flat and mean water
depth is 13o m. In Fig. 2 the three geophones are shown. They are
sensitive to all spatial directions and are mounted on sledges.
By towing the sledges some distance with the research ship PLANET
it was possible to adjust one of the horizontal components of each
seismic system to a straight profile from the shot point to the
sledges.

The signals of the 3 seismic components were transmitted by
special radio-buoys to the research ship. To prevent seismic transmission between the anchor weight and the geophones we used a
cable of 150 m length between each anchor and its sledge. Shooting
was done from the research ship with charges of TNT lowered to the
bottom. Charges varied between 3 and 1o kg. The exact distances
between the shot position and the geophones were computed from
the travel time of the water wave. During the experiment the
sledges remained at their positions and the ship varied its
distance to the sledges. Looking by RADAR at the three buoys the
research ship was maneuvered on a straight line between shot point
and all geophones.

Fig. 3 gives a block diagram of one of the geophone circuits,
including data transmission, and in Fig. 4 the measured sensitivity
of the geophone is plotted against excitation frequency together
with its phase behaviour. The minus 3 dB point of the geophones
is at 7 Hz.

For the experiment it would have been better to use a low
frequency geophone, say of 2 or 3 Hz, but there are two difficulties: first we did not know what the frequency of the Scholte
wave would be and, second, a low frequency system is very sensitive
to a dip in its vertical direction. A better solution to this
problem may be the digital seismometer developed at SACLANTCEN and
used by D. Rauch (see his conference paper).

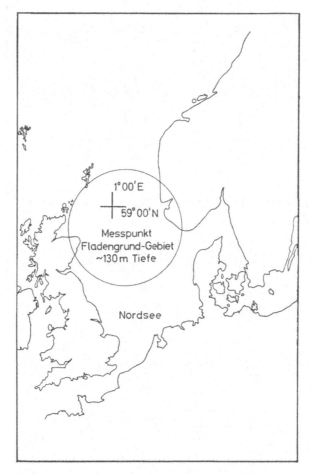

Fig. 1: Location on the European shelf where the experiment was performed

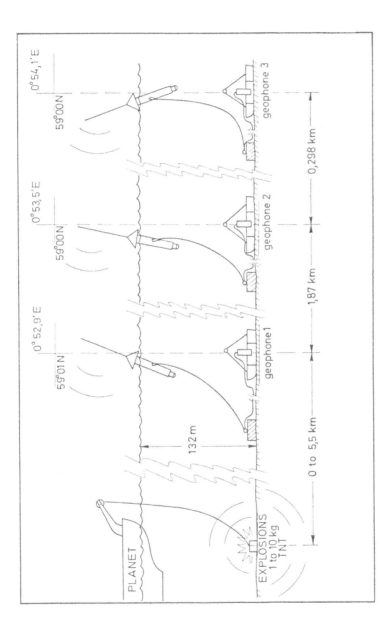

Fig. 2: Arrangement of the 3 geophone sledges at fixed positions along a profile

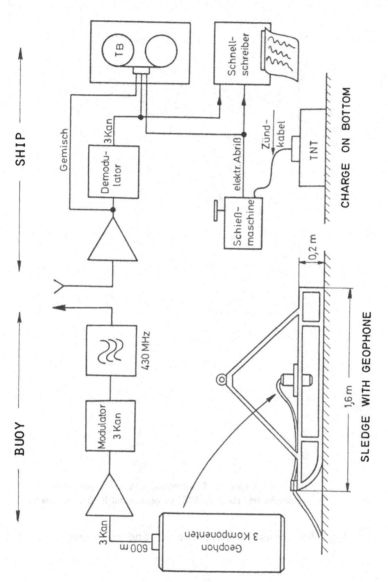

Fig. 3: Block diagram of the geophone circuit. Left: electronics in the buoy. Right: on board of RS PLANET

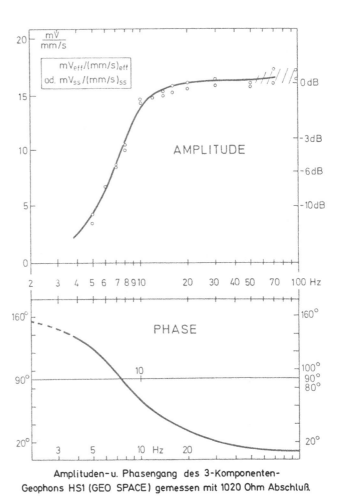

Amplituden- u. Phasengang des 3-Komponenten-
Geophons HS1 (GEO SPACE) gemessen mit 1020 Ohm Abschluß.

Fig. 4: Transfer function of the used geophone HS1

RESULTS

Some remarks

We fired more than 30 shots with charges between 1 and 10 kg, but, surprisingly, only 5 shots (two of them with 3 kg TNT and the others with 1 kg TNT) generated a large amplitude interface wave. Sometimes we were not able to repeat a seismogram, even under the same conditions. My impression is that the exitation of an interface wave is very sensitive to the constitution of the bottom near the explosion point.

Because I was also interested in the P-wave in the bottom I was not able to use greater charges than 10 kg. In this case the radio link would have been exceeded in its dynamic range.

Velocities

In Fig. 5 a composition of 17 seismograms is shown. Line B in this figure corresponds to the direct compressional wave in the sandy bottom with the velocity

$$C_p = 1.8 \text{ km/s.}$$

Line A in Fig. 5 corresponds to the direct water wave with velocity

$$C_{water} = 1.477 \text{ km/s.}$$

Line C in Fig. 5 is due to a refracted wave of velocity 2.3 km/s. The refractor has a depth of (560 ± 20) m below the bottom.

In the following Fig. 6 seven seismic traces are shown. Note the large time scale - up to 25 seconds at only 2.5 km! It is not possible to show the P wave with this scale. The large vertical amplitude results in a direct wave with

$$\text{group velocity 111 m/s,}$$

which I think is a Scholte wave.

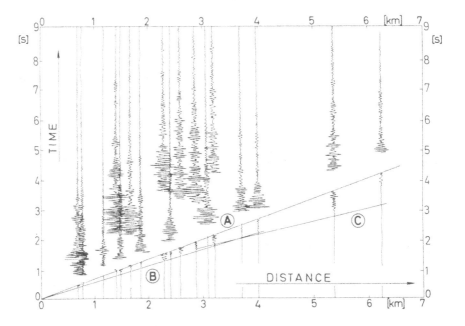

Fig. 5: Arrival of the P wave "B", of a refracted wave "C" and of
        the direct water wave "A". The traces are adjusted in
        distance according to the water wave.

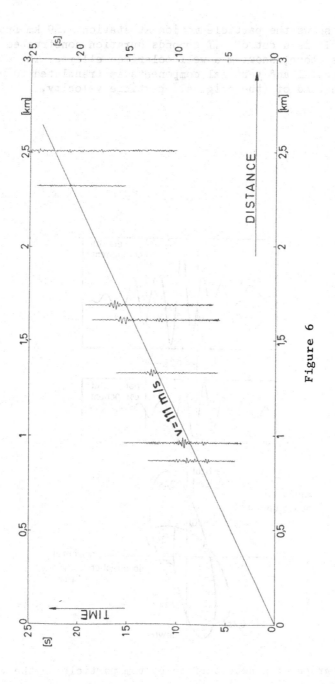

Figure 6

Hodograph

Fig. 7 shows the particle motion at station 1.69 km from the
shot point. It is a cut of .7 seconds duration constructed from
both signals above, showing a well polarized ellipse. The abscissa
of the horizontal and vertical components is translated to units
of length instead of (the original) particle velocity.

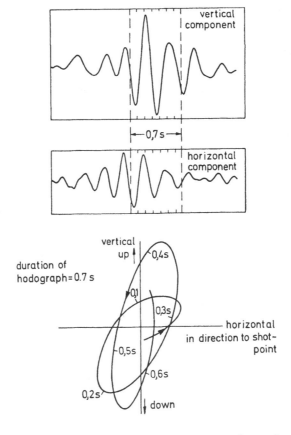

Fig. 7: Hodograph of a near surface bottom particle in the vertical
plane

Spectra

From the shot shown in Fig. 7 the power spectra of the spatial components are given in Fig. 8. The ordinates have the same sensitivity, which is given in relative units of particle motion energy per frequency spacing.The total energy of the three spatial components is related by the ratio 1 to .27 to .29.

Wavelength

From a group velocity of 111 m/s and a peak frequency of 4.5 Hz it follows that the wavelength is of the order

$$\lambda = (25 \overset{+}{-} 4) \text{ m.}$$

This wavelength is similar to the wavelength of the P wave from the same shot. This similarity of wavelengths occurs frequently even when the P wave velocity is 10 to 20 times the velocity of the S-type wave. This could be due to layers in the bottom which act as a "wavelength filter".

Attenuation

To find the attenuation of the interface wave it was necessary to use shots with charges of 3 and 10 kg TNT together. To extrapolate the seismic amplitudes of the 3 kg shots to 10 kg signals the empirical relation of Hirschleber and Menzel[3] between charge W and vertical seismic amplitude was used:

$$\text{Ampl.} = \text{const.} \cdot W^{0.56}.$$

It was then postulated that there is a geometrical spreading of energy proportional to 1/R, where R is the distance from the shot-point. In this manner the total energy of the vertical component of the interface wave of all seismic traces was corrected to the dotted line in Fig. 9. Thus, we find a good fit to a relation of the form $E_R = \text{const.} \cdot e^{-\alpha R}$. The correlation coefficient is here .89. (The minimum H value for four degrees of freedom is .80 for confidence limits of 5% to 95%.) This results in an

$$\text{attenuation coefficient} = -6.8 \cdot 10^{-3} \text{dB/m}$$

which contains absorption and scattering. It is related to the frequency band $(4.5 \overset{+}{-} 1)$ Hz.

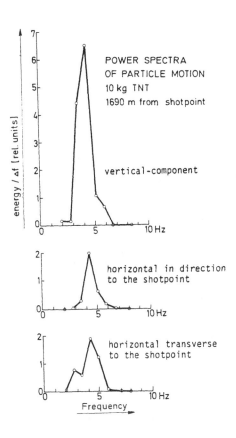

Fig. 8: Energy spectra of spatial particle motion near surface

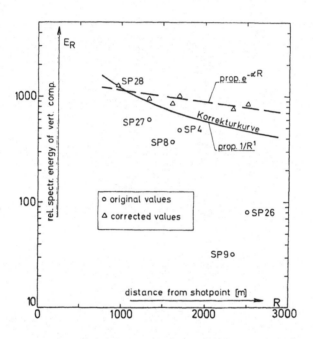

Fig. 9: Decrease of seismic energy of vertical component as a
        function of distance.
        Dotted line: Corrected to charge and geometrical spreading

References

1.   D. Rauch, Propagation of a plane, monochromatic
          Scholte wave along the ocean bottom, SACLANTCEN
          SM-138, SACLANT ASW Research Centre, La Spezia, 1980.
2.   L. Cagniard, "Reflection and Refraction of Progressive
          Seismic Waves", Mc.Graw-Hill, New York, (1962).
3. H.B. Hirschleber und H. Menzel, "Das Amplituden-Ladungs-
          Gesetz für Sprengungen im Kleinen Belt",
          Z. F. Geophysik, 32, (1966).

MODEL COMPUTATIONS FOR LOW-VELOCITY

SURFACE WAVES ON MARINE SEDIMENTS

Heinz-H. Essen

Institut für Geophysik
Universität Hamburg
Bundesstr. 55, 2000 Hamburg 13, Germany

ABSTRACT

In seismology Rayleigh waves, which propagate along the
free surface, are of great importance for determining
the elastic parameters of the upper mantle. With regard
to shear-wave velocities in the upper layers of ocean
sediments the same may apply to Scholte waves, which
propagate along the interface between ocean and sea
floor. In the case of homogeneous halfspaces both, Rayleigh and Scholte waves are nondispersive, however the
dependences of propagation velocities on the elastic
parameters are different. But in presence of layers the
Rayleigh as well as the Scholte wave splits up into a
finite number of dispersive modes. As for Rayleigh
waves this effect seems to be of importance for Scholte
waves, too. Measurements of Schirmer et al. from the
shore belt of the North Sea show two different dispersive modes. In this paper simple theoretical models
are presented in order to investigate the dispersion
of "Rayleigh"- and "Scholte"-modes on ocean sea-floors.
Compared to the earth mantle the sea floor propagates
Rayleigh (and Scholte) waves of much smaller wavelength,
and moreover the ratio of shear to compressional wave-
velocity is lower by a factor of about 10.

INTRODUCTION

Though shear-wave velocities are of importance for
studies of several geophysical problems, there is still
a great lack of measured data, cf. Hamilton (1976).

This fact is caused mainly by experimental difficulties.
In the past years two experiments were carried out by
the University of Hamburg with the objective of determining shear-wave velocities in the uppermost layers of
marine sediments. In both experiments explosive sources
were used and the signal was recorded by bottom mounted
geophones.
Results from the first experiment (1976), which took
place in the North Sea at a water depth of 150 m, are
presented by Dr. Schirmer in this conference. The second
experiment was carried out in 1977 at the coast of the
North Sea, and the water coverage was only 1 m. In order
to obtain maximum energy the explosive charges were
brought into 1 m deep holes during ebbtide and detonated
at floodtide. Due to the simpler experimental conditions
the recorded time series of the second experiment are of
higher quality than those from the first experiment.
As especially the second experiment gave occasion to my
model computations I present a record section for the
vertical component of the recorded bottom velocities,
see Fig. 1.

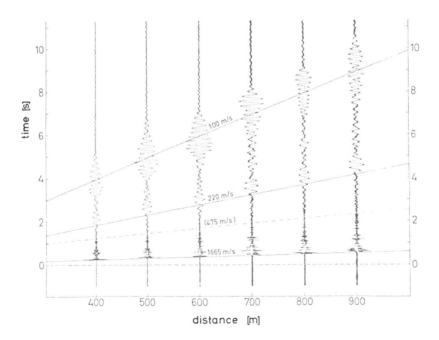

Fig. 1: Vertical component of bottom velocity,
generated by explosive sources on a shoal of
the North-Sea coast,
unpublished measurements of H. Janle, F. Schirmer, and J. Siebert, 1977

I will not discuss the details of the measurements and
data processing but concentrate on the fact that at
least two different wave groups are to be seen. Furthermore the waves seem to be dispersive. For this reason
I analysed some seismograms by multiple filter technique,
as described by Dziewonski et al. (1969). Fig. 2 shows
the Gabor matrix of the vertical component at 900 m distance from the source, cf. Fig. 1. This matrix is obtained by filtering the seismogram with a narrow band-
pass at different center frequencies and computing the
envelopes of the filtered seismograms as function of
frequency. The lines in Fig. 2 represent the contours
of the amplitude, which verify the assumption of at least
two dispersive waves.

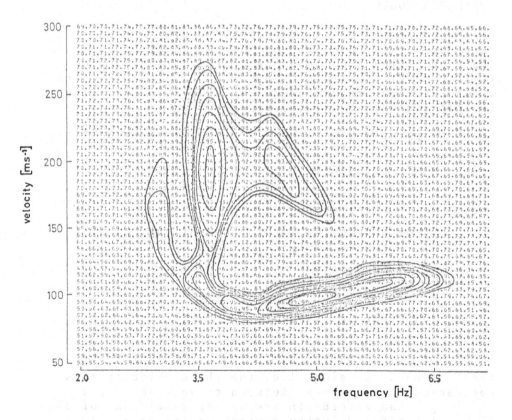

Fig. 2: Gabor matrix, obtained by multiple filter
        technique from the seismogram at 900 m distance
        in Fig. 1. The contours correspond to the iso-
        lines of the amplitude level of the filtered
        seismogram.

THEORETICAL CONSIDERATIONS

The presented experimental results show waves propagating along the water-sediment interface with a velocity much smaller than the sound velocity of water. From theoretical investigations it is known that such waves exist for a non-zero shear-wave velocity in the sedimentary sea floor. As consequence of experience it may be assumed,

$$0 < c_t < c_w < c_l \tag{1}$$

with, c_w = sound velocity of water
c_t = shear-wave velocity $\left\{ \begin{array}{l} \text{in the sedimen-} \\ \text{tary sea floor} \end{array} \right.$
c_l = compressional-wave velocity

TWO-LAYER MODEL

The simplest model, which allows the propagation of interface waves, consists of a water layer of constant depth overlying a homogeneous sedimentary halfspace. In this model the limiting cases of zero and infinite water coverage are included. The dispersion relation in accordance with Eq. (1) becomes,

$$\rho_w \cdot v^4 \cdot \gamma_l \cdot \tanh\left(\frac{\omega}{v} \cdot \gamma_w \cdot h\right) = \rho_s \cdot c_t^4 \cdot \gamma_w \cdot (4 \cdot \gamma_l \cdot \gamma_t - (1 - \gamma_t^2)^2)$$

$$\text{with,} \quad \gamma = \sqrt{1 - \left(\frac{v}{c}\right)^2} \tag{2}$$

and, h = water depth

$\rho_{w,s}$ = density of water and sediment, respectively

$\frac{\omega}{2\pi}$ = **frequency**

v = phase velocity

In the two limiting cases,

$\tanh(...) = 0$ no water coverage (Rayleigh waves)
$\tanh(...) = 1$ high water coverage (Scholte waves),

there is no dispersion, i.e. the phase velocity does not depend on frequency. Between these cases where wavelength and waterdepth are of the same order of magnitude the phase velocity depends on frequency. In Fig. 3 this dependence is shown. A logarithmic frequency scale is chosen in order to cover a broad range. The phase velocity decreases by only 7 %, i.e. it may be stated that the dispersion is weak.

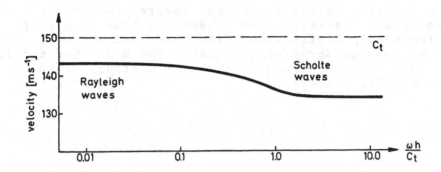

Fig. 3: Phase velocity (v) of guided waves on the
water-sediment interface, cf. Eq. (2).

Model parameters: $c_t = 150$ ms^{-1}

$c_w = 1500$ ms^{-1}, $c_1 = 1700$ ms^{-1}, $\rho_s/\rho_w = 1.9$

MULTI-LAYER MODEL

In disagreement with our experimental results the
simple 2-layer model allows the propagation of one mode
only. This problem can be overcome by assuming the
shear-wave velocity increasing with depth. Then the
different observed waves are trapped modes within the
uppermost sedimentary layers.
For simplicity a sea-floor model is introduced which
consists of a certain number of homogeneous layers of
constant thickness overlying a homogeneous halfspace.
Usually the shear-wave velocity of the halfspace is
the highest one in the model, but still smaller than
the water sound velocity. In this case the dispersion
relation can be solved by means of a matrix formalism,
first described by Haskell (1953). It should be noted
that the conditions of Eq. (1) are different from
usual seismic surface-wave conditions where shear-wave
velocities (of rocks) exceed the sound velocity of
water.
First computations were carried out with a 3-layer model, which are the water layer, one sedimentary layer
and a homogeneous halfspace. For comparison with the
contours in Fig. 2 it is necessary to compute group-
velocity curves because the filtered wave groups propagate with group velocity. It turns out that only the
shear-wave velocities influence the frequency depen-

dence of group velocity, whereas compressional-wave
velocity and density may be chosen arbitrarily within
the reasonable limits known from experience.
Fig. 4 shows group-velocity curves from a 3-layer model,
which are in good agreement with the contours of the
Gabor matrix, cf. Fig. 2.

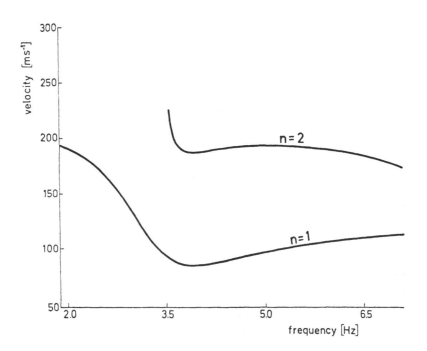

Fig. 4: Group-velocity curves from a 3-layer model.
        The model parameters are chosen with the objec-
        tive to fit the observed contours, cf. Fig. 2.

Model parameters:

layer	depth	density	compressional-wave velocity	shear-wave velocity
1)water	1.0 m	1.0 gcm^{-3}	1500 ms^{-1}	0 ms^{-1}
2)sediment	16.0 m	1.8 gcm^{-3}	1700 ms^{-1}	135 ms^{-1}
3)sediment	∞	1.9 gcm^{-3}	1800 ms^{-1}	230 ms^{-1}

Group velocities obtained by the analysis of further
seismograms, also horizontal components, are in accordance with Fig. 4. But contrary to this good agreement
the computed and observed axial ratios of the horizontal and vertical particle velocity differ considerably
(the computed horizontal component is too small by a
factor of about 2). Numerical tests show that this disagreement may not be overcome within a 3-layer model.
Thus it is necessary to introduce at least one more
layer or on the other hand a gradient of the shear-
wave velocity in the uppermost sedimentary layer.

For the moment I can't present a satisfactory result
but I hope to be successful in future.

CONCLUSIONS

Methods of numerical analysis, developed for seismic
waves generated by earthquakes, may be applied to artificially generated waves on the sea floor, as well.
Dispersion will be observed if the shear-wave velocity
increases with depth. The analysis yields group velocities as well as the axial ratios of horizontal and
vertical particle velocity. The dependence of these
quantities on the shear-wave velocity as function of
depth is described by the theory of surface (or interface) waves, provided that the depths of the first
layers are small compared to the observation range.
Then inversion techniques can be applied in order to
obtain an optimal model.
In this paper only group velocities are analysed, and
no axial ratios. With regard to inversion the simple
"trial and error" method is used. But already at this
stage considerable information on shear waves in ocean
sediments is obtained.

REFERENCES

Dziewonski, A., S. Bloch, and M. Landisman, A technique
        for the analysis of transient seismic signals,
        Bull. Seim. Soc. Am. 59, 427-444, 1969
Hamilton, E.L., Shear-wave velocity versus depth in
        marine sediments: a review,
        Geophysics 41, 985-996, 1976
Haskell, N.A., The dispersion of surface waves on
        multilayered media,
        Bull. Seimol. Soc. Am. 43, 17-34, 1953

EXPERIMENTAL AND THEORETICAL STUDIES OF SEISMIC INTERFACE WAVES

IN COASTAL WATERS

Dieter Rauch

SACLANT ASW Research Centre

La Spezia, Italy

ABSTRACT

A digital three-component ocean-bottom seismometer has been developed at SACLANTCEN to study extremely-low-frequency sound propagation in coastal waters (e.g. below the cut-off of the shallow-water duct). It has been used successfully in conjunction with a variable depth hydrophone during several sea trials off the Italian coast. By exploding small charges (45 to 900 g TNT) at various distances (up to 5 km) it has been demonstrated that the infrasonic energy is mainly transmitted via a seismic interface wave. In general this propagation mode may be termed a modified Scholte wave as it is guided by the acoustically most significant interface between water column (including liquid-type sediments) and solid basement. On a sedimentary sea floor very pronounced wavelets have been observed, at frequencies in the 1.5 to 6 Hz band, group-velocities in the 250 to 70 m/s range, and hodographs in the vertical radial plane. Together with this experimental program theoretical studies have been made using a seismic Fast-Field-Program, which has proved to be a very powerful tool to model infrasonic phenomena in shallow-water environments.

INTRODUCTION

As a result of heavy shipping and the installation of numerous offshore platforms (e.g. oil rigs) the extremely-low-frequency noise-level has increased considerably in many coastal-water areas during the last few years. The pertinent spectra are characterized by pronounced lines generated by the power strokes of the various engines or generators and by manifold structural vibrations and hull resonances. This paper is devoted to the frequency band of about 1 to 25 Hz, which is usually termed "infrasonics" due to the extensive pioneer work done in air acoustics[1]. With regard to the classical aspects of underwater acoustics this infrasonic regime seemed to be of no importance because these frequencies stay usually below the absolute cut-off frequency of the shallow-water duct or otherwise are extremely damped by the bottom interactions of the underwater sound field. Though the pertinent energy cannot be transmitted by the compressional or P-wave modes insonifying the water column, its propagation has been demonstrated quite often by hydrophone records taken near the sea floor.

To recapitulate briefly the underlying physics, Figs. 1 and 2 present the basic propagation conditions for two strongly simplified shallow-water environments.

As an example of very good propagation conditions, Fig. 1 shows the dispersion curves for an isovelocity water layer (with the phase velocity c_w of the compressional wave) on top of a homogeneous crystalline-rock sub-bottom, with the phase velocities c_p and c_s of the compressional (P) and shear (S) waves). The occurring phase velocities of the different modes are plotted against the product of water depth H and radiated frequency f. Starting with number 1, all typical modes of the water layer show a low-frequency or shallow-water cut-off as soon as their phase-velocity reaches the relative high-shear velocity c_s of the

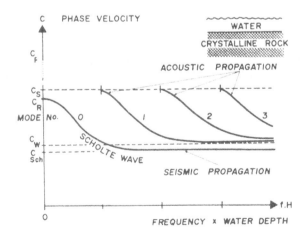

Fig. 1 *Phase-velocities of the lowest modes in shallow water over an extremely "hard" rock-bottom*

sea floor. All these modes we term "acoustic modes" to simplify matters and to make a clear distinction between the different propagation phenomena. With decreasing water depth H or sound frequency f the number of these modes is diminished, and we finally reach a range without any direct acoustic contribution to the elastic wave-field. Only mode Number 0 is not cut off and is thus the only mechanism for extremely-low-frequency propagation in coastal waters.

This particular wave type assumes the Rayleigh-wave velocity c_R of the rock as its low-frequency or shallow-water limit, and gradually approaches the smaller Scholte-wave velocity C_{Sch} in the opposite case. Thus it has to be a pure or modified version of the last-named seismic-wave type that is trapped by the water/rock interface.

The relatively-strong dispersion of the Scholte wave in our liquid/solid system reveals that this interface wave perceives the sea surface and therefore is strongly modified until a relatively high value of water depth or frequency is exceeded.

Based on this hypothetical case we can pass over to a more realistic example of a less-hard sea bed formed by a sedimentary rock or a consolidated sediment. The dotted curves of Fig. 2 show that there are no longer discrete modes in the water layer. The acoustic field consists now of blurred interference patterns that behave like a non-regular and very weak mode structure. In this

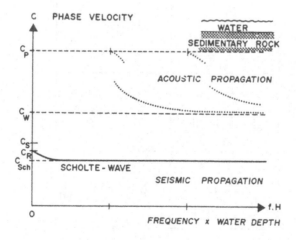

Fig. 2   *Phase-velocities of the lowest modes in shallow water over a relative "soft" rock-bottom*

case, the seismic interface is the only discrete mode of the liquid/solid system and its relatively weak dispersion indicates a minor penetration depth in the water column.

Those more-realistic bottom conditions cannot be handled by the well-known normal-mode models for shallow-water environments because they ask for a very hard ocean floor, at least as a sub-bottom underlying some liquid-like sediment layers. Especially the latter assumption of liquid sediments holds only for top layers with an extremely low-shear rigidity. These two idealized environments demonstrate above all that the transmission of infra-sound can be achieved only through seismic-wave propagation. Among the great variety of seismic wave types the Scholte-wave not only offers the closest relationship with the acoustic field in the water layer but also some other quite interesting propagation features. The basic characteristics of "pure" or "idealized" interface waves are well known from the literature and their more-complicated manifestations under more realistic conditions (layered media) are discussed in some detail by Rauch[2][3], Schirmer[4] and Essen[5]. Favourable characteristics of these waves are their cylindrical-spreading and the strict channelling due to a strong decay of their energy with growing distance from the guiding interface.

CLASS OF INTERFACE	TYPE OF FREE INTERFACE WAVE
VACUUM $c_p'\ c_s'$ SOLID	RAYLEIGH WAVE $c_R' = n'c_s'$ $0.873 < n' < 0.956$
LIQUID c_p $c_p'\ c_s'$ SOLID	SCHOLTE WAVE $c_{Sch} \leq \min \begin{Bmatrix} c_p \\ c_s' \end{Bmatrix}$
SOLID $c_p\ c_s$ $c_p'\ c_s'$ SOLID	STONELEY WAVE $\max \begin{Bmatrix} c_R \\ c_R' \end{Bmatrix} < c_{St} < \min \begin{Bmatrix} c_s \\ c_s' \end{Bmatrix}$

Table 1   Classification of the basic interface-waves, with an estimation of their phase-velocities

To establish a clear nomenclature we follow a suggestion of Cagniard[6] who proposed that these waves should go by the name of the scientists associated with them: Rayleigh, Scholte, and Stoneley (see also[2][3]). This relationship is illustrated in Table 1, together with an estimation of the respective phase velocities under idealized conditions (two adjacent homogeneous half-spaces). These latter data show that these interface waves are the "slowest" contribution to the elastic wave-field. For that reason, radiation losses to bulk waves in the different media or other possible trapped modes are not possible. Real-world interface waves are formed by one or several hybrid propagation modes due to the layered structure of the earth's crust. Hence they have a strong normal dispersion and the particle displacement or velocity vector does not always sweep in a retrograde or anticlockwise sense through the well known elliptical orbits or hodographs in the vertical-radial plane (with respect to interface and source).

I   COMPUTER MODELLING

As mentioned already, these wave types are beyond the scope of the usual underwater acoustic models. Fortunately, a seismic program based on the Thompson-Haskell method and termed "Fast-Field-Program" (FFP) due to the solution technique, was made available to SACLANTCEN by Lamont-Doherty Geological Observatory[7]. The version implemented on the Centre's UNIVAC computer[8] can calculate the complete monochromatic wave-field in a stack of isovelocity liquid and solid layers with options for the depth of the source and receiver. To work out some essential features and trends of the interesting seismic contribution to short- and medium-range transmission, we have chosen a very simple type of sea bed in which a sand layer ($\rho = 2.0$ g/cm^3; $c_p = 1800$ m/s; $c_s = 700$ m/s) with absorption ($\alpha_p = 0.75$ dB/λ; $\alpha_s = 1.5$ dB/λ) and constant thickness L is covering a sedimentary rock ($\rho = 2.2$ g/cm^3; $c_p = 2000$ m/s; $c_s = 1400$ m/s; $\alpha_p = 0.1$ dB/λ; $\alpha_s = 0.05$ dB/λ).

We assume a 10 Hz sound source, S, in the middle or at the bottom of a water column ($\rho = 1.0$ g/cm^3; $c_w = 1500$ m/s; $\alpha_w = 1.8 \ 10^{-6}$ dB/λ) of mostly W = 60 m depth, with the receiver always placed on the ocean floor. In agreement with our strongly simplified splitting of the field, all waves propagating faster than the sound in water are included in the acoustic portion, while all slower types are counted as the seismic portion. Of course this latter combination obscures the fine structure of the field and disregards especially the overlapping and interaction of certain wave-types, but here we can easily tolerate these adulterations. To demonstrate the influence of the sediment layer we keep the source position fixed in the middle of a 60 m water

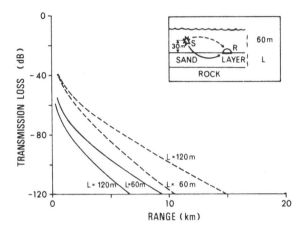

Fig. 3   *Range dependence of the acoustic (dashed) and seismic*
         (solid) transmission-loss   for   a   10 Hz source   over
         different thick sand layers

column and decrease the layer thickness stepwise. In Fig. 3 we
have plotted the transmission loss against range for the acoustic
and the seismic fields in the case of the two deep sand layers of
120 m and 60 m thickness. In both cases the acoustic propagation
is quite bad and the seismic one even worse, but we realize that a
decrease of the layer thickness slightly improves the conditions
for seismic sensing and deteriorates those for acoustic detection.
A more-detailed analysis reveals that the acoustic field comprises
merely one mode-like structure, while there are different weak
contributions   to   the   seismic   field.   Among   the   latter,   the
Scholte-wave at the water/sand interface ($c_{Sch} \cong 600$ m/s) can be
identified clearly, but due to the high absorption in the sediment
it is of no importance.

     Omitting   some   intermediate   steps,   which   confirm   the
above-mentioned   trend,   we   pass   over   immediately   to   three
relatively-thin sand layers of 10 m, 5 mand 0.5 m thicknesses.
The   pertinent   curves   in   Fig. 4   show   impressively   that   the
environment   strongly   favours   the   seismic   propagation.   The
remarkable ranges predicted for it are solely a consequence of a
modified Scholte-wave that is guided along the "sand-coated"
water/rock interface ($c_{Sch} \cong 1150$ m/s).   It should be mentioned
that there is no direct contribution to the seismic field from the
side   of   the   sand/rock   interface,   because   our   choice   of   the
parameters excludes the existence of a "pure" Stoneley wave at
this boundary.

     The excitation of the modified Scholte-wave, and thus the
good seismic propagation, would be improved if the source position
were shifted towards the ocean floor.

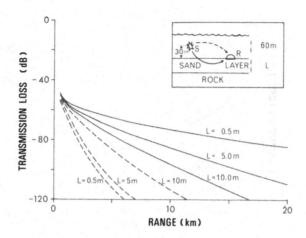

*Fig. 4   Range dependence of the acoustic (dashed) and seismic
(solid) transmission-loss for a 10 Hz source over
different thin sand layers*

For that reason, Fig. 5 compares the seismic transmission loss of
the preceding plot with the corresponding curves calculated for a
bottom-mounted transmitter S of the same source level.   In all
three cases, a decrease in loss of about 10 dB confirms the
expected better coupling of source and sea bed.

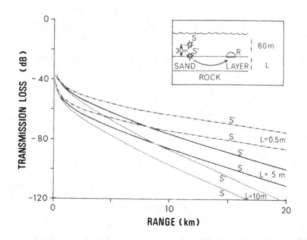

*Fig. 5   Range dependence of the seismic transmission loss
for a 10 Hz source over and on a thin sand layer*

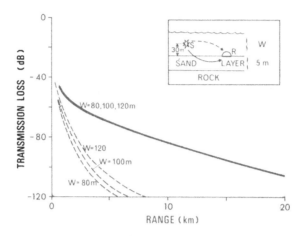

Fig. 6   *Range dependence of the acoustic (dashed) and seismic*
         (solid)   transmission loss   for   a 10 Hz   source   in
         different water columns over a thin sand layer

To show finally that the selected water depth of only 60 m
does not imply a serious limitation to our conclusions, we return
to the intermediate example of Fig. 4 (source S, 30 m above a 5 m
sand layer) and calculate the transmission loss for water columns
of 80 m, 100 m, and 120 m.   From the curves of Fig. 6 we learn
that   the   poor   acoustic   propagation   undergoes   only   slight
improvements   with   increasing   water   depth,   while   the   relatively
good seismic propagation is not affected.   Figure 7, which plots

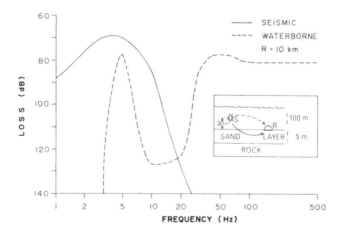

Fig. 7   *Frequency dependence of the acoustic (dashed)   and*
         seismic (solid) transmission loss at a fixed range
         of 10 km for a 10 Hz source over a thin sand layer.

the mean propagation loss against frequency for the intermediate case of Fig. 6 and a fixed range of 10 km, provides another instructive comparison of the waterborne and seismic propagation paths. Above 25 Hz the acoustic propagation is relatively good and the seismic counterpart is of no importance. Decreasing the frequency below this limit we notice the expected strong deterioration of the acoustic path and a drastic improvement of the seismic propagation. The optimal conditions for the latter path occur at about 4 Hz where the seismic field is dominated by a modified Scholte-wave at the sand-coated water/rock interface. This Scholte-wave represents the transition from a highly-attenuated Scholte-wave at the sand/water interface to a modified Rayleigh-wave in the bottom. The acoustic propagation suffers a complete cut-off at 3 Hz, but it previously goes through a local maximum around 5 Hz because the wave no longer "feels" the sand layer. For more-detailed and more experiment-oriented studies, SACLANTCEN will soon operate an improved FFP-version that can calculate synthetic seismograms from broadband sources.

II   FIELD WORK

To tackle these problems experimentally a digital tri-axial ocean bottom seismometer (OBS) was developed at SACLANTCEN using a

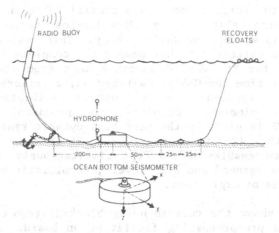

*Fig. 8   Installation of the sensor package on the sea floor
and mooring of its radio buoy in shallow water*

new type of active and lightweight 1 Hz geophones (Teledyne
Geotech S-500). These sensors do not need blocking for trans-
portation and because they can be applied in any orientation do
not have to be precisely levelled for proper operation. The
seismic sensor-set is completed by a variable-depth hydrophone
mounted outside of the OBS-container or floating above it.
Figure 8 shows how this sensor package is installed in shallow
water together with its radio buoy.
Figure 9a presents a simplified block-diagram of the pertinent
electronics which are similar to those described by Blackinton and
Odegard[9].

The output of each of the four basic sensors is preamplified,
low-pass filtered (100 Hz), and fed directly and via three additional amplifier-stages of 18 dB each into a 4 x 4 channel multi-
plexer. Each of the four signal-levels is scanned with a sampling
frequency of 600 Hz, and the optimally-amplified signal is
transmitted to a fast 12-bit A/D converter. The resulting 12-bit
mantissa is then combined with the pertinent 2-bit exponent, with
a parity bit and another independent bit to a basic data-word of
16 bits. After three scanning cycles of the basic sensors we thus
have available twelve additional bits for the transmission of
important parameters and control data. Eleven of those are used
to form a secondary data-word, which is cyclically assigned to one
of the four auxiliary sensors (compass, tilt X, tilt Y and
temperature) and one of the four electronic checkpoints. Thus,
every 24 basic scanning cycles we can not only monitoring the
actual position of the OBS, the nearby water temperature, and the
power supply of four important networks, but also have at our
disposal an additional set of eight bits that is used for the OBS
station's identification code. These digital data-sequences are
transferred to the surface buoy via a coaxial cable and from there
to the receiving ship by an FM-modulated radio-transmitter
(170 MHz, max 15 W) in the surface buoy. This buoy also houses
the rechargeable battery-set that powers the complete system and a
radio receiver for the ON-OFF commands sent from the receiving
platform. Every time the OBS is switched on, a calibration signal
is automatically applied to each geophone calibration-coil for
about one to two minutes. One of the most prominent features of
this digital OBS is given by the very high dynamic range of 120 dB
(66 dB from the 12-bit mantissa and 54 dB from the 2-bit
exponent), which enables us to cope simultaneously with a low-
level seismic background and high-level deterministic signals from
nearby CW-sources or explosions.

Figure 9b shows the corresponding block-diagram of the data-
acquisition and pre-processing facilities on board. Here we are
using a Hewlett-Packard 21 MX computer with disc unit, which gives
us the possibility of getting a printout or a plot of all
essential data channels a few seconds after recording an event.

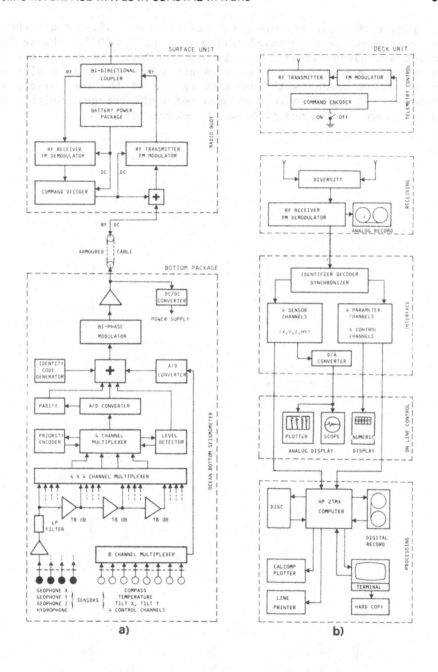

Fig. 9 *Electronic block-diagram for: (a) the OBS and its radio-buoy, (b) the onboard receiving, recording, controlling and pre-processing systems*

Figure 10 indicates two positions off the Italian coast where
the instruments were deployed during a cruise in 1979. The sensor
station was always oriented in such a way that the directions of
the planned acoustic runs (at constant water depth or a sloping
sea floor) coincided with the axis of one of the two horizontal
geophones. As sound sources we used small TNT charges, which were
usually placed on the sea floor and fired electrically.

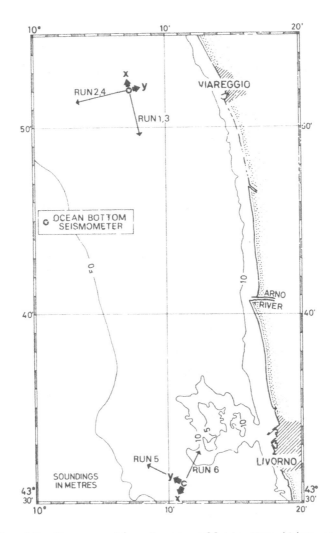

Fig. 10 The operational area off the Versilian coast
         (Italy) with the two chosen OBS-positions

To discuss some of the typical features of the interface waves, we will look first at an event from Run 3 along the 20 m depth profile off Viareggio, where the hydrophone was floating in the middle of the water column. Figure 11 presents a lineprinter plot of the four time-series generated by a charge of 180 g TNT at a distance of about 1.3 km. Beginning at the lower rim of this plot, the first three traces, 0, 1, 2, have been chosen to display the measured particle velocities, u, v and w, while the uppermost trace, 3, was used for the pressure history recorded by the hydrophone. The latter channel was always subject to an additional pre-amplification of 30 dB. Due to this scaling factor we usually had to clip the first arrival, which was formed by the unavoidable, high-frequency water wave. On this and all following plots the time-series were passed through a 10 Hz low-pass filter to stay below the cut-off frequency of the water duct.

The interface wave we are looking for arrives somewhat later than the water wave. As expected from our geometrical consideration, this interface wave is detected only by the vertical geophone GeoZ and the radial one GeoX, which is horizontal and oriented parallel to the direction of propagation. The ripples on the output of the transverse sensor GeoY are probably due to imperfections of the wave guide and lateral inhomogeneities.

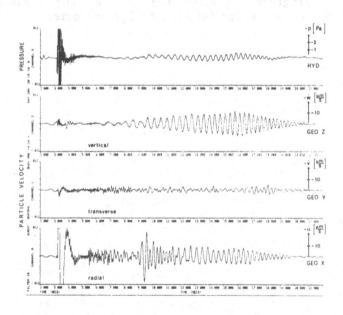

Fig. 11   Line-printer plot of the signals from the four basic sensors (180 g TNT fired at 1.3 km distance)

The wavelet itself consists of a long-lasting, narrow-band signal that shows the expected normal dispersion, indicating that the higher frequencies are left behind. Especially at the output of the radial sensor GeoX we realize that the beginning of the main signal is concealed by another superimposed wavelet that has relatively-high amplitudes and an extremely small bandwidth. In agreement with the theoretical predictions, the hydrophone also senses the low-frequency pressure variations created by the interface wave.

As a final check for identification we have displayed the corresponding velocity vector in the radial/vertical X-Z plane using the radial velocity u as abscissa and the vertical one w as ordinate. On top of Fig. 12 we have therefore again plotted the appropriate sections (windows) of both signals and below we have displayed the resulting hodographs for 14 successive one-second windows.

Due to the above-mentioned superimposed wavelet, the first four windows A-D are still quite irregular and curve B especially seems to be complicated by the superimposition of a horizontal ellipse. This latter feature could indicate the coexistence of a higher mode. Nevertheless, the following ten hodographs (E to N) form very regular vertical ellipses that are circumscribed clockwise or prograde. Towards the tail of the signals these ellipses fade away and merge into the background noise.

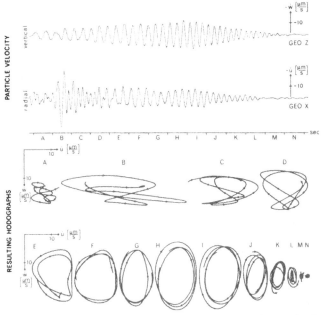

Fig. 12 Radial and vertical particle velocity of the interface wavelet in Fig. 11 with the resulting hodographs

To prove the almost perfect orientation of these ellipses in the vertical/radial plane we have chosen a fixed two-second window from another signal and displayed the pertinent hodographs for different observer positions by rotating the X- and Y-axis in the horizontal plane. Figure 13a explains the geometry of this procedure and Fig. 13b presents the results. In particular, the view against the propagation direction ($\alpha = 0°$) and the view from the top of the station demonstrate that the recorded signal does indeed come close to the ideal case predicted by theory.

This rotation of the horizontal reference axes plays an important role if a given OBS-station has to be aligned with respect to a moving source or to other sensors of an array.

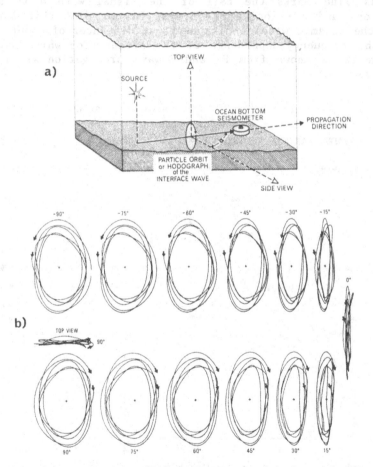

Fig. 13   (a)   Geometry of the computerized hodograph-display
              for   different observer-positions,   (b) Series   of
              side-views and one top-view for the same two-second
              time-window of an interface wavelet

After this unambiguous identification of the wave type we
would like to discuss more carefully some of its propagation
characteristics. For this purpose we have stacked in Fig. 14a the
output of the vertical geophone GeoZ for Run 2, using the same
scale for all signals but increasing the actual charge size with
growing distance according to the indication on the right-hand
side.

As reference quantity, the first dotted line marks the onset
of the unavoidable water wave propagating with a velocity of
1536 m/s. The first continuous line indicates the front of the
seismic signal, which is roughly characterized by 1.5 to 2 Hz
oscillations and a velocity of about 240 m/s, while the second
continuous line marks the tail of the signal with 4 to 5 Hz
oscillations and a velocity of 75 m/s In our plot this latter
tail of the seismic wavelet disappears at distances of about 1 km
due to the frequency-dependence of the attenuation which may be
linear as it is known from P- and S-wave propagation at higher
frequencies.

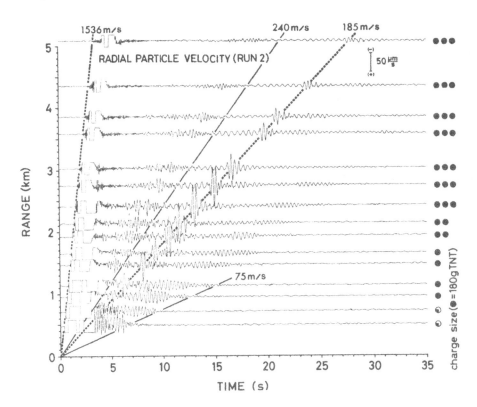

*Fig. 14   (a) Stacking of the vertical particle-velocity for an
acoustic run over a layered sediment-bottom*

The seismic wave-guide under consideration thus offers optimal propagation conditions for frequencies below 3 Hz. The second dotted line emphasizes the above-mentioned superimposed wavelet, which has a propagation velocity of 185 m/s.

Figure 14b shows the stacking of the corresponding output of the radial geophone GeoY. Here the onset of the interface wave is more hidden by some refracted arrivals but the superimposed wavelet is much more pronounced and reveals its very narrow frequency band of 2.5 to 3 Hz.

To understand the impulse-response of the system in question (shallow-water/layered-bottom) we made seismic sub-bottom profiles using the Centre's Uniboom equipment. Along the tracks of our runs off Viareggio the sea floor consists of a quite-regular sediment layer having a total thickness of 20 to 25 m resting on top of sedimentary rock. To get some dispersion curves that could explain our observations in this environment we assumed the bottom

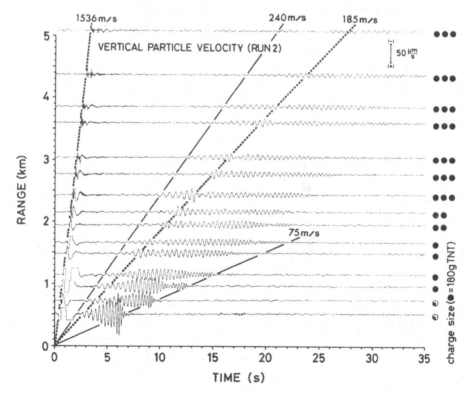

Fig. 14   (b) Stacking of the radial particle-velocity for the same acoustic run as in (a)

parameters indicated on the top of Fig. 15. While the sound speeds in the water-column represent a good approximation to our expendable bathythermograph data (typical summer profile) we had to extract those of the solid layers from core data and other acoustic measurements. The important values for the shear velocity c_s are based on literature data and reasonable assumptions. Handling these given bottom layers with the Centre's FFP-Program we tried to simulate roughly the expected velocity gradients. The dotted and continuous curves of Fig. 15 present the calculated phase velocity for the zeroth and the first mode while the dashed curves give their group velocities.

The group-velocity curve of the zeroth mode predicts a main wavelet characterized by 2 Hz oscillations propagating with about 250 m/s at the front and by approximately 4 to 5 Hz oscillations propagating with about 80 m/s at the tail. From the group-

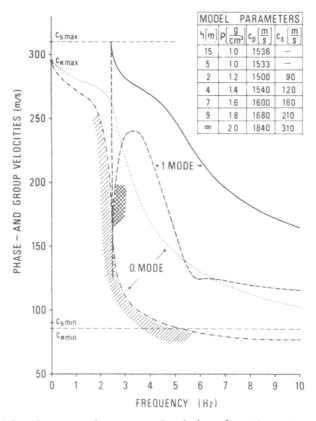

MODEL PARAMETERS			
$h\,[m]$	$\rho\left[\frac{g}{cm^3}\right]$	$c_p\left[\frac{m}{s}\right]$	$c_s\left[\frac{m}{s}\right]$
15	1.0	1536	—
5	1.0	1533	—
2	1.2	1500	90
4	1.4	1540	120
7	1.6	1600	160
9	1.8	1680	210
∞	2.0	1840	310

Fig. 15   Phase- and group-velocities for the only two modes
of the Scholte wave in a model-environment for the
acoustic run of Fig. 14

velocity curve of the higher mode we may expect a more-or-less monochromatic wavelet having a frequency of about 2.5 Hz and a velocity in the range 160 ± 30 m/s. The hatched areas cover roughly our experimental results from the OBS-deployment off Viareggio. It is quite obvious that the chosen bottom-model not only provides a qualitative interpretation of our data but also gives a good quantitative approach. From the fact that the low-frequency limit of the dispersion curve (zeroth mode) is always determined by the Rayleigh-wave velocity of the sub-bottom, and the present wave-guide obviously favours the extremely low frequencies, we learn that optimal detection-conditions would be obtained by digging the OBS into the sediments. Anyway, this technique may be strongly recommended for weak signals as it improves the signal-to-noise ratio by sheltering the OBS from some interfering noise sources in the water-column, such as strong currents or marine life.

A more-detailed analysis of the wavelets recorded during Runs 2 and 4 indicates that the share of the vertical component is about 65% of the total elastic energy and the damping coefficient may be of the order of 0.15 dB/λ.

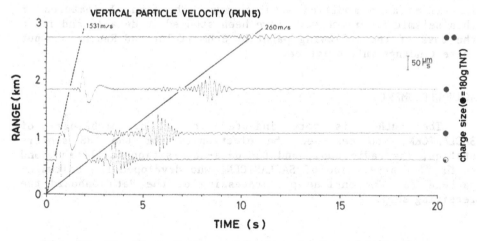

Fig. 16   Stacking of the vertical particle-velocity for an
          acoustic run over an inhomogeneous and rough sedi-
          mentary rock-bottom

The corresponding results for Runs 1 and 3 are only slightly different due to the upslope propagation on their track. With the other deployment off Livorno we intended to study the Scholte-wave propagation along a rock-like sea-floor. Usually such a hard bottom can be expected to be a better approach to the classical half-space problem with no dispersion and a retrograde elliptical particle motion. In Fig. 16 we have stacked a few seismograms recorded by the vertical sensor for Run 5. The expected onset of the forerunning P-wave cannot be identified, but the wavelet of the Scholte wave is indeed much more compressed and reveals a weak normal dispersion for its quite narrow frequency content of 5.5 ± 1.5 Hz. The front of this wavelet is propagating with a velocity of about 260 m/s (solid line), and an inspection of the pertinent hodographs proved their retrograde sense of rotation.

In general, this experiment on a hard bottom was characterized by strongly varying results and the shorter ranges achieved. This was due to the fact that the rock had a quite-complicated internal structure there (inclined and irregular layering) and also a very rough surface (many grooves and outcrops). Therefore, not only the excitation and propagation of the Scholte wave took place under more unfavourable environmental conditions, but also the very-important ground-coupling of the OBS (see e.g.[10] [11]) was made worse by the small-scale roughness of the sea floor.

CONCLUSIONS

By modelling and measuring the transmission of infrasound in shallow water areas the pertinent propagation-mechanism has been identified as a modified Scholte-wave. The typical features of this seismic interface wave have been studied in detail, and it is the task of the following paper by Schmalfeldt and Rauch[12] to put these findings into practice.

ACKNOWLEDGMENT

The author is very indebted to Mr E. Michelozzi of SACLANTCEN, who designed the electronics for the OBS sensor-station, its radio buoy, and the data-acquisition facilities, and to Mr V. Duarte, also of SACLANTCEN, who developed the software-package for the on-line pre-processing of the data onboard the receiving ship.

REFERENCES

1. L. Pimonov, "Les Infrasons", Editions du CNRS, Paris (1976).
2. D. Rauch, Propagation of a plane, monochromatic Scholte wave
    along an idealized ocean bottom, SACLANTCEN SM-138, SACLANT
    ASW Research Centre, La Spezia, Italy (1980).
3. D. Rauch, Seismic interface waves in coastal waters,
    SACLANTCEN SR- in press, SACLANT ASW Research Centre, La
    Spezia, Italy.
4. F. Schirmer, Experimental determination of properties of the
    Scholte wave in the bottom of the North Sea. (This volume).
5. H. H. Essen, Model computations for low-velocity surface waves
    in marine sediments. (This volume).
6. L. Cagniard, "Reflection and Refraction of Progressive Seismic
    Waves", McGraw-Hill, New York, N.Y. (1962).
7. H. W. Kutschale, Rapid computation by wave theory of propaga-
    tion loss in the Arctic Ocean, Lamont-Doherty Geological
    Observatory Tech Rpt 8, Columbia University, Palisades,
    N.Y. (1973).
8. F. B. Jensen and W. A. Kuperman, Environmental acoustic
    modelling at SACLANTCEN, SACLANTCEN SR-34, SACLANT ASW
    Research Centre, La Spezia, Italy (1979). [AD A 081 853]
9. J. G. Blackington and M. E. Odegard, An ocean-bottom seismo-
    graph using digital telemetry and floating-point
    conversion, IEEE Trans Geoscience Electronics GE-15:74-82
    (1977).
10. R. J. Hecht, Background of the problems associated with
    seismic detection of signal sources in the ocean, in:
    "Proceedings of a Workshop on Seismic Propagation in
    Shallow Water held in Washington D.C., July 1978", U.S.
    Office of Naval Research, Arlington, Va. (1978).
11. R. J. Hecht, Investigation of the potentialities of using
    seismic sensors for the detection of ships and other naval
    platforms, Tech.Rpt, Underwater Systems Inc., Silver
    Springs, Md. (1978).
12. B. Schmalfeldt and D. Rauch, Ambient and ship-induced low-
    frequency noise in shallow water. (This volume).

AMBIENT AND SHIP-INDUCED LOW-FREQUENCY NOISE IN SHALLOW WATER

Bernd Schmalfeldt and Dieter Rauch

SACLANT ASW Research Centre

La Spezia, Italy

ABSTRACT

This paper describes some features of extremely low-frequency hydroacoustic and seismic noise (1 to 10 Hz) in coastal waters. The data have been collected by a single tri-axial ocean-bottom seismometer equipped with an additional hydrophone. The following aspects are considered.

    a.   modification of the background power-spectra by ship noise.

    b.   coherence and phase spectra between the seismic components.

    c.   direction-finding based on the azimuthal power-distribution.

The hydroacoustic and seismic power-spectra are very similar in general shape. However, for some significant lines the output of the horizontal geophones indicates a higher signal-to-noise ratio. The horizontally polarized energy may be used to determine the direction to the source once the propagation mechanism has been identified by means of all seismic sensors. It has been proved that ship-induced infrasonic noise is propagating in form of seismic interface waves with their characteristic elliptical particle motion.

INTRODUCTION

The preceding enquiry by Rauch[1] has studied in detail the mechanism of extremely-low-frequency sound propagation in shallow

329

water by means of powerful deterministic events. The present
paper picks up this thread and will apply statistical processing
methods to the infrasonic line-spectra recorded from a passing
merchant ship.

We will try to discriminate a ship's line-spectrum from a
characteristic seismic-noise background and determine the bearing
of this source with respect to a single, tri-axial ocean-bottom
seismometer (OBS). This directional capability is offered by the
unique features of seismic interface waves of the Rayleigh-,
Scholte- or Stoneley-type. Those waves are characterized by an
elliptical particle motion in the vertical-radial-plane (with
respect to interface and source)[1,2,3]. The sign of this phase
shift depends on the excited propagation mode and thus, finally,
on the local bottom conditions.

The receiving station, with its three velocity sensors and a
variable-depth hydrophone, has been described in the previous
paper[1]. The following data samples stem from a deployment off the
Italian coast close to Viareggio, which was also used to explore
the propagation conditions of explosion-generated Scholte-waves
with frequencies from about 1.5 to 6 Hz. At this location the
water depth was about 20 m and the hydrophone was floating in the
middle of the water column. The SACLANTCEN research ship was used
to simulate the passage of ships whose parameters (e.g. the shaft
revolutions and hence the propeller-blade rate) were to be
continuously recorded.

Figure 1a shows a map of the test area and Fig. 1b (on a much
bigger scale) the two positions of the ship we have selected to
examine more carefully. It should be noted that these two positions are characterized by bearings of 295° and 270° with respect
to the given x-axis of the OBS.

I   DATA PROCESSING

For the seismic data processing a new software package was
designed and fitted to a Hewlett-Packard 2116 B-computer.

First, the original sampling frequency of 600 Hz was reduced
to 150 Hz by decimating the data. Then the selected time series
(synchronous scanning of all four sensor channels) was divided
into sub-series of 6.83 s (corresponding to 1024 samples) length
and displayed on the screen for visual inspection. This data
check was needed to exclude from the processing those sub-series
that suffered strong adulterations due to superimposed mechanical
or electronic disturbances (pulling at the OBS by the coaxial
cable to the surface buoy, interferences on the radio link, or

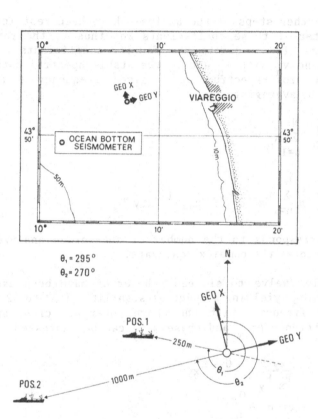

Fig. 1 a) Operational area

     b) Geometry:
         POS.1, POS.2 = Positions of ship during noise
         measurements,   GEO X, GEO Y = Axes of the two
         horizontal geophones (the third (Z) geophone
         axis is downwards))

drop-outs during the playback of the data). After subtracting the mean value from these sub-series, weighting them with a COS-window, and overlapping them by 50%, the pertinent fourier coefficients have been calculated up to 75 Hz by means of a fast Fourier transform. Hence a frequency resolution of 150/1024 = 0.146 Hz was achieved.

All further steps of the analysis have been restricted to the first quarter of these coefficients and thus to the range from 0 to 18.75 Hz. If we denote the fourier coefficients of the sub-series x_n and y_n with X_n and Y_n the stable spectral estimates for power- and cross-spectrum at a given frequency f are simply calculated by averaging

$$P_x = \frac{1}{N} \sum_{n=1}^{N} X_n X^* \quad , \tag{1}$$

$$C_{x,y} = \frac{1}{N} \sum_{n=1}^{N} X_n Y_n^* = L_{x,y} - i Q_{x,y} \quad , \tag{2}$$

where N corresponds to the number of sub-series analyzed and the asterisk denotes the complex conjugate.

Usually twelve to sixteen sub-series have been averaged for our processing, yielding a spectral stability of 32 to 42 equivalent degrees of freedom. With the above power and cross spectra the squared coherence $\rho^2_{x,y}$ and phase $\phi_{x,y}$ can be expressed in the form (see e.g.[4]):

$$\rho^2_{x,y} = \frac{C_{xy}^2}{P_x P_y} \quad 0 < \rho^2_{xy} < 1 \tag{3}$$

$$\phi_{x,y} = \arctan -\frac{Q_{x,y}}{L_{x,y}} \tag{4}$$

Dealing with directional sensors we must be aware that the spectral estimates of Eqs. 1 to 4 represent merely the respective components and thus have to be corrected for the bearing of the incoming wave. Assuming cartesian coordinates and a bearing angle θ with respect to the x-axis we can determine the more meaningful radial and transverse fourier-coefficients $R_{n,\theta}$ and $T_{n,\theta}$ simply by the rotation of the reference axis.

$$R_{n,\theta} = X_n \cos\theta + Y_n \sin\theta \tag{5}$$

$$T_{n,\theta} = -X_n \sin\theta + Y_n \cos\theta \tag{6}$$

If we insert these latter coefficients instead of X_n and Y_n into Eq. (1), for example, we obtain the radial or transverse power spectra for a given frequency f and a chosen bearing θ.

Integrating these spectra over a frequency range Δf and repeating this procedure for all possible directions we get an azimuthal power distribution $P(\theta)$ for any chosen frequency band.

To understand the resulting plots properly, one has to take into consideration the basic directivity pattern of the sensors themselves. As geophones have a dipole characteristic, the best possible azimuthal power pattern for an incoming wave with a perfectly unidirectional horizontal deflection component is a $\cos^2$-pattern. In general the signal bearing θ has an ambiguity of 180° due to the symmetry of such a lobe pattern.

II   RESULTS FOR A DETERMINISTIC EVENT

At first we tested the software by processing some of the deterministic propagation phenomena. For the following demonstration we chose (Fig. 2) the same wavelet of a Scholte wave as that used in the previous paper[1] to show some almost perfect elliptical hodographs circumscribed in a prograde sense (i.e. clockwise when observing the wave propagating from left to right).

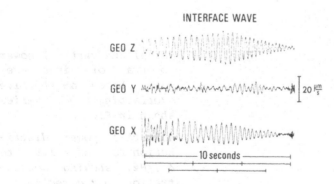

Fig. 2 *Seismograms of interface waves from TNT charge at*
1300 m distance in the-x direction

This wavelet was generated by a charge of 180 g TNT, exploded at a distance of about 1300 m exactly in the (-x) direction from the sensor.

On the basis of five overlapping sub-series with lengths of 3.4 s each, we calculated the radial and vertical power spectra shown in Fig. 3a. These curves confirm that most of the artificially created seismic energy is confined to the range from 2 to

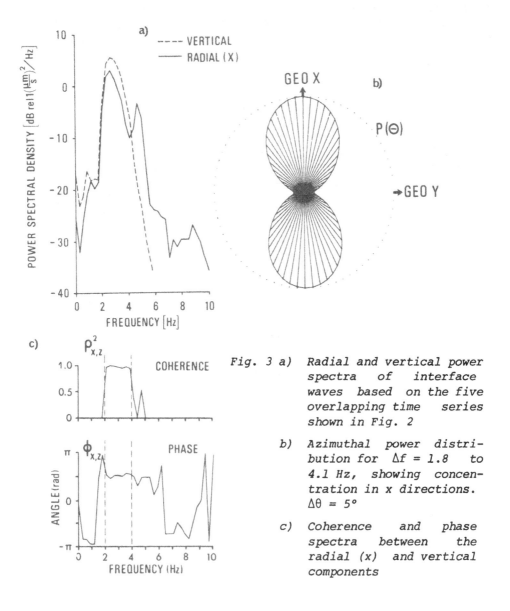

Fig. 3 a) Radial and vertical power spectra of interface waves based on the five overlapping time series shown in Fig. 2

b) Azimuthal power distribution for $\Delta f = 1.8$ to 4.1 Hz, showing concentration in x directions. $\Delta\theta = 5°$

c) Coherence and phase spectra between the radial (x) and vertical components

4 Hz. The resulting power pattern for this frequency band is plotted in Fig. 3b where its lobes are indeed seen to point exactly in the ±x directions. Obviously there is almost no energy carried by the transverse deflection component.

Figure 3c presents the corresponding curves for the (squared) coherence and the phase spectrum between the radial and the vertical geophones. We notice that the range from 2 to 4 Hz is not only characterized by an almost constant coherence close to the value 1 but also by a quite stable phase shift of $+ \pi/2$. This result demonstrates that the radial and the vertical particle velocities are highly correlated and that the former component is leading the latter one. Based on our knowledge of the source position this phase shift indicates without any ambiguity that the pertinent hodographs should be prograde ellipses. Conversely, the bearing of the source can be determined without ambiguity if the propagation mechanism has been identified before.

III APPLICATION TO SHIP-INDUCED NOISE

After having demonstrated some typical features of seismic interface waves in terms of statistical processing methods we would like to return to the two events simulating a passing ship (Fig. 1b).

During these experiments the noise background was also recorded for several two-minute periods while the main engine of the research ship was stopped. Figure 4a shows the power spectra from the three seismic sensors and the hydrophone during one of these periods; in the following paragraphs these spectra are referred to as "ambient-noise spectra". The general shapes of all four spectra are very similar and characterized by the well-known enormous increase of the low-frequency noise-level below 4 Hz. While the steep slope from 1 to 4 Hz (about -27 dB/oct) is a well-reported seismic background noise feature, the more drastic decrease below 1 Hz is partly a consequence of the sensor characteristics (e.g. the response curve of the geophones shows a slope of -18 dB/oct in this range). Nevertheless the pressure spectrum from the water column also indicates that there is indeed a maximum of the level around 1 to 2 Hz. This has also been reported by many other studies (e.g.[5]).

Figure 5a shows that the pertinent azimuthal power-distribution is nearly omnidirectional due to the many-fold seismic and man-made phenomena contributing to it. Furthermore, the hydrophone data point out another increase of energy at 0.15 to 0.30 Hz (see Fig. 4a); this maximum is caused by microseisms and is created by wave action at the sea surface.

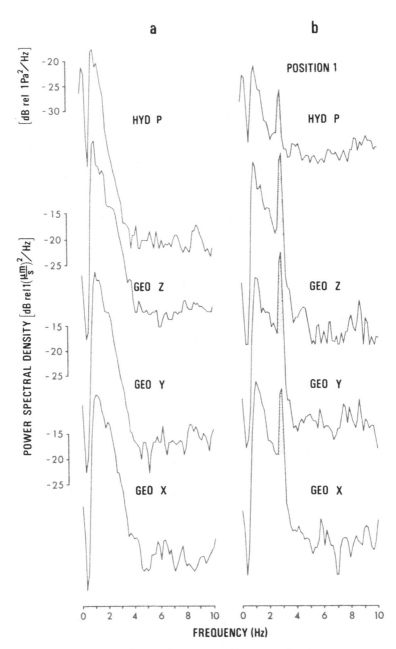

Fig. 4 Spectra from three geophones and one hydrophone
        a) Ambient noise. b) Ship noise, ship at POS.1
        The spectra within the relevant frequency band
        are shaded.

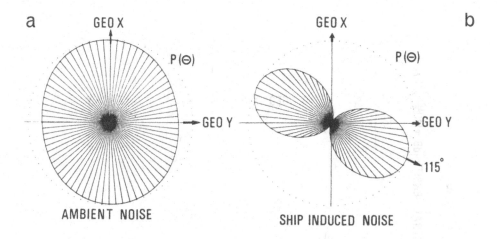

Fig. 5   Azimuthal power distributions for P(θ)
   a)   Δf = 0.6 to 2 Hz ambient noise
   b)   Δf = 2.5 to 3.7 Hz ship noise (ship at POS.1),
         comprising relevant peak of Fig. 4b

Let us now examine the ship noise recorded for position 1 in Fig. 1b, which corresponds to a distance of 250 m from the OBS and a bearing of 295° relative to its x-axis. In Fig. 4b we have plotted the pertinent spectra; these are dominated by a pronounced and slightly broadened line at about 2.9 Hz, which stems from shaft rotations of around 174 rev/min.

Plotting the azimuthal seismic power distribution for this ship-generated line around 2.9 Hz we obtained the almost ideal pattern of Fig. 5b. It is oriented towards 115° and 295° (with respect to the x-axis) and the energy at the perpendicular bearing is reduced to about 5% with respect to the maximum in lobe direction.

To obtain some further information on the propagation mechanism and its direction we have calculated the pertinent squared coherences ($\rho^2_{x,z}$ and $\rho^2_{y,z}$) and phase spectra (ϕ_{xz} and ϕ_{yz}). The results are shown in Figs. 6a and 6b, together with the power spectra of the appropriate horizontal velocity components.

For the line under discussion (Δf = 2.7-3.2 Hz) both coherence spectra show a significantly high value of $\rho^2 \cong 0.9$ and the phase spectra are remarkably stable with the shifts of $\phi_{x,z} \cong -\pi/2$ and $\phi_{y,z} \cong +\pi/2$.

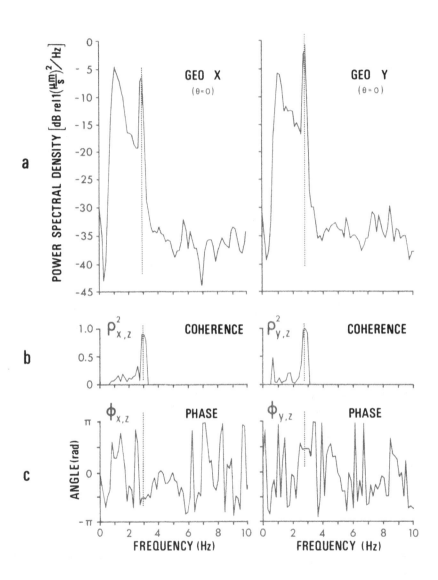

Fig. 6   *Spectra of ship noise from ship at POS.1, 250 m distance.*
         Relevant frequency band indicated by dotted lines at centre.
              a) Power spectrum of x-/y-component
              b) Squared coherence } *between x-/y- and z-component*
              c) Phase spectrum

Together with our findings for deterministic events under completely identical operational and environmental conditions, these results with a merchant-ship source lead to the unambiguous conclusion that we have in fact sensed an interface wave. This wave has a velocity vector with components in the (-x) and the (+y)-directions and thus points in the second quadrant of the underlying coordinate-system. Accordingly, the propagation direction of the interface wave is given by 115° and hence the bearing of the source is 295°.

To make a final check of these conclusions we have rotated the horizontal plane by 115° (x rotated into the direction of the wave propagation). The result is shown in Fig. 7. The coherence and phase shift (Fig. 7a) of the radial relative to the vertical component around the shaft frequency are identical to the results for the deterministic wavelet in Fig. 3c. In contrast, the transverse sensor (Fig. 7b) indicates non-significant coherence and has a phase shift with high variance.

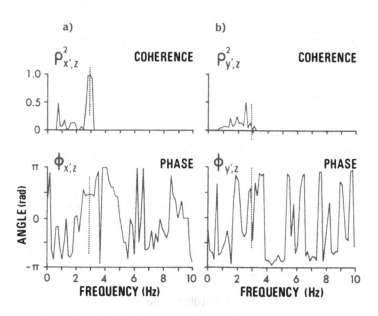

Fig. 7 Coherence and phase spectra after rotation by 115° into propagating direction. Relevant frequency band indicated by dotted lines at centre.
   a) between radial (x') and vertical (z) component
   b) between transverse (y') and vertical (z) component

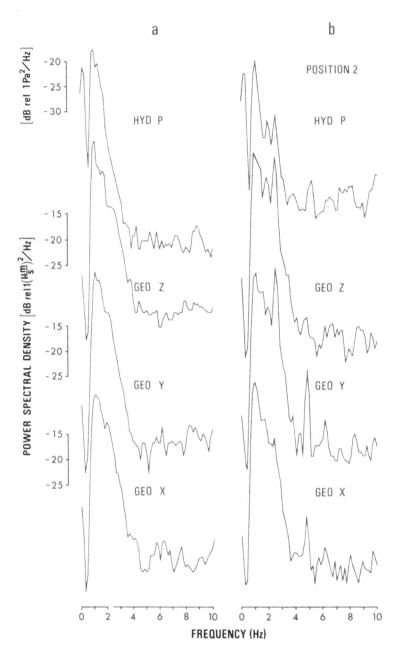

Fig. 8   Spectra from three geophones and one hydrophone
         a) Ambient noise.   b) Ship noise, ship at POS.2
                                    (compare with Fig. 4)
         The spectra within the relevant frequency band
         are shaded.

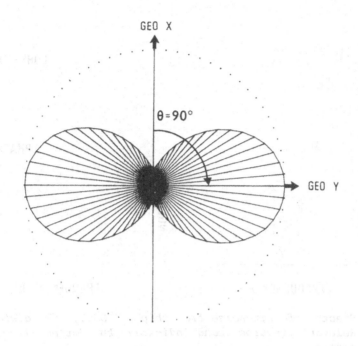

GEO X

θ=90°

GEO Y

Fig. 9 Azimuthal power distribution for Δf = 2.05 to 2.95 Hz ship
noise (ship at POS.2), comprising relevant peak of Fig. 8

After examining the 250 m case at position 1, we can try to
apply our direction-finding method to the more distant position
(POS. 2 in Fig. 1b) with a range of 1000 m and a bearing of 270°.
Figure 8 presents the power spectra for all sensors (which are
shown together with the ambient noise spectra as in Fig. 4) that
reveal a pronounced line around 2.6 Hz corresponding to a 156
rev/min shaft rotation. The seismic sensor's signals at 2.6 Hz
are more clearly visible than the hydrophone signal. The highest
S/N ratios are reached by the sensors Geo Y and Geo Z, since the
ship is in the -y axis direction. Another relatively strong line
at 4.8 Hz is visible on three channels, but it was obviously not
radiated by the ship because the pertinent power pattern pointed
to a different direction.

Figure 9 shows the power pattern of the energy radiated from
the ship at position 2 in the frequency range comprising the
relevant peak. It is to be seen that the main energy arrives at
the OBS from the ±y direction.

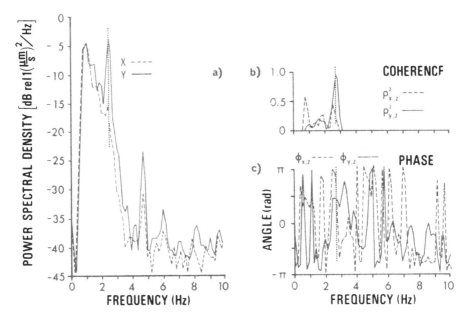

Fig. 10   Spectra of ship noise from ship at POS.2, 1 km distance.
Relevant frequency band indicated by dotted lines at
centre
a) Power spectra of (---) x-, (——) y-component
b) Squared coherences between (---) x- and z-components
(——) y- and z-components
c) Phase spectra between (---) x- and z-components
(——) y- and z-components

Figure 10a repeats and overlaps the power spectra from the
radial (y) and transverse (x) sensor, to aid in visualizing the
level differences. Figures 10b and c plot the corresponding
coherence and phase spectra, which indicate at the 2.6 Hz line a
significant coherence between Geo Y and Geo Z amounting to
$\rho^2_{y,z} \cong 0.8$ and a quite stable phase shift of $\phi_{y,z} \cong + 0.42\pi$. Hence,
the prograde circumscribed elliptical hodograph has been slightly
inclined towards the propagation direction, which is probably due
to damping effects in the sea floor.

CONCLUSIONS

This paper has shown that in shallow-water areas extremely-low-frequency ship noise is transmitted in the form of seismic interface waves. Based on this propagation mechanism, the bearing of the source has been determined by a single tri-axial ocean-bottom seismometer station.

REFERENCES

1. D. Rauch, Experimental and theoretical studies of seismic interface waves in coastal waters, (This volume).
2. H. H. Essen, Model computations for low-velocity surface waves on marine sediments, (This volume).
3. F. Schirmer, Experimental determination of properties of the Scholte wave in the bottom of the North Sea, (This volume).
4. G. C. Carter, C. H. Knapp and A. H. Nuttall, Estimation of the magnitude-squared coherence function via overlapped fast Fourier transform processing, IEEE Trans. Audio Electro-acoustics AU-21: 337-344 (1973).
5. H. Berckhemer, The concept of wide band seismometry, in: "Proceedings of XIIe Assemblée Générale de la Commission Séismologique Européenne, Luxembourg, 21-29 Sep, 1970", pp. 214-220.

DISPERSION OF ONE-SECOND RAYLEIGH MODES THROUGH OCEANIC SEDIMENTS

FOLLOWING SHALLOW EARTHQUAKES IN THE SOUTH-CENTRAL PACIFIC OCEAN BASIN

Emile A. Okal*   and   Jacques Talandier**

* Department of Geology & Geophysics, Yale University
   Box 6666, New Haven, Connecticut 06511, USA

**Laboratoire de Géophysique, Commissariat à l'Energie
   Atomique, Boîte Postale 640, Papeete, Tahiti.

INTRODUCTION

The purpose of this paper is to report the observation of anomalously long wavetrains of high-frequency seismic energy (0.3 - 1 Hz) following shallow events at a number of epicenters in the South-central Pacific basin, and to model their propagation and generation as due to efficient coupling of the oceanic water column with the solid Earth through a well-developed transitional layer of sediments. The following paragraphs are a short summary of the seismic features of the area, and provide a framework for our study.

The French Polynesia seismic network, consisting of 15 permanent stations, is presently the only large aperture seismic array operating in an oceanic intraplate environment. As such, it can provide exceptional insight into phenomena associated with intraplate seismicity. The network has been fully described elsewhere [Okal et al., 1980; Talandier, 1972]. Among its stations, Rikitéa (RKT), and the four stations on Hao atoll are of primary importance for detection capabilities in the region extending between Polynesia and the East Pacific Rise. Solid triangles on Figure 1 show the location of these five stations, superimposed on the bathymetric map of Mammerickx et al. [1975]. For clarity, the other stations in the network, located on Tahiti, Rangiroa and Tubuai, and irrelevant to the present study, are not shown.

The stations are equipped with short-period vertical seismometers, coupled to a variety of electronic filters [Talandier, 1972] which damp ocean-generated noise, and permit routine magnifications of 10^5 at 1 Hz and 2×10^6 at 3 Hz, the latter allowing detection of

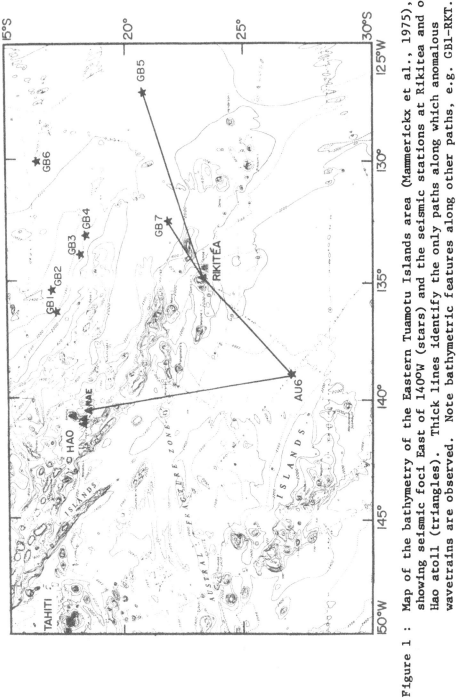

Figure 1 : Map of the bathymetry of the Eastern Tuamotu Islands area (Mammerickx et al., 1975),
showing seismic foci East of 140°W (stars) and the seismic stations at Rikitea and on
Hao atoll (triangles). Thick lines identify the only paths along which anomalous
wavetrains are observed. Note bathymetric features along other paths, e.g. GB1–RKT.

low-amplitude T waves converted at the island shore. The one-second magnification of 10^5 is one order of magnitude above that of WWSSN stations in oceanic environment, and results in considerable boosts in the detection capabilities of the network: most of the area covered by Figure 1 has a detection threshold of $M_L = 3.0$ to 3.5.

Seismicity in the area of the Gambier Islands has been studied extensively by Okal et al. [1980]. Eight epicenters have been identified to the East of 140°W, called GB1-GB7 and AU6 in these authors' classification. They are shown as stars on Figure 1. Over the past 15 years, 116 events have been recorded at these eight locations, with a maximum m_b of 5.5; among the eight epicenters, two (GB4 and GB5) have been particularly active. With 96 events recorded in 1976-79, GB5 is the most active seismic focus inside the Pacific plate (Hawaii excepted). Because of poor station coverage at teleseismic distances, and of the relatively low magnitude of these events, their precise study has been difficult. Since GB5 lies in an area where bathymetric coverage is almost inexistant, the structural interpretation of this seismicity can only be speculative. Nevertheless, Okal et al. [1980] have argued on the basis of the systematic observation of water multiples that these foci must be very shallow, probably within a few km of the water-sediment interface. They have also been able to interpret this seismicity as due to the release of compressional tectonic stress in the oceanic plate.

OBSERVATIONS

Routine recording at RKT of seismic signals from events at GB5 (21°S, 127°W; $\Delta = 7.95$ degrees), GB7 (22°S, 132°W; $\Delta = 2.96$ degrees) and AU6 (26.7°S; 138.9°W; $\Delta = 5.08$ degrees) shows a relatively high-frequency wavetrain lasting up to 20 minutes, whose onset grossly corresponds to the expected arrival time of Rayleigh waves.

Figure 2 is a reproduction of the seismograms obtained from AU6 on November 20, 1979. Focal characteristics for this event are given by Okal et al [1980] as : Latitude 26.75°S, Longitude 138.88°W, Origin Time 06:45:02, Magnitude $m_b = 5.3$. The intermediate and bottom frames on Figure 2 are the continuation of the top one. The trace labeled (a.), which uses a low-gain broad-band amplifier, clearly shows the arrival of P_n, followed by S_n about one minute later. Shortly after S_n, the one-second wavetrain starts arriving with a group velocity of 4 km/s. It lasts for about 13 minutes, before the amplitude dies off to a level comparable to the noise preceding the event. The corresponding group velocity is then 0.6 km/s. This longlasting wavetrain must be opposed to records from events of similar magnitude at similar distances, such as at GB1 or GB4, whose energy dies off to the noise level over a period on the order of 3 to 4 minutes. The wavetrain observed for the AU6

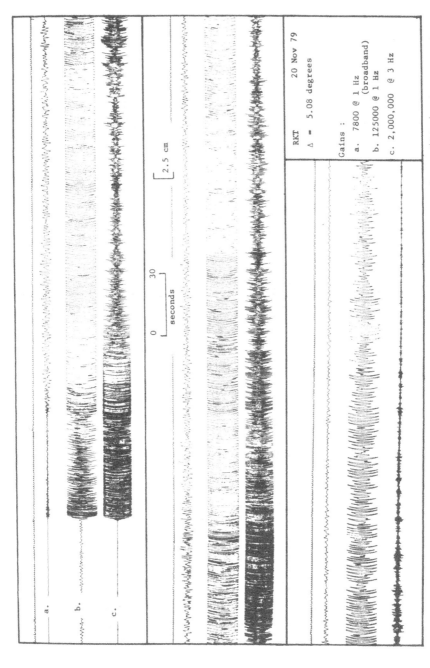

Figure 2 : Seismograms obtained at RKT following event at AU6 on Nov. 20, 1979. The anomalous wave train is best seen on trace (b.), lasting more than 13 minutes. Note low level of signal on high-frequency channel (c.).

Table 1. Summary of observations of anomalous wavetrains

Epicenter	RKT	NAE	Other stations
	(distance in °)		on Hao
GB1	no	no	no
	(5.63)	(4.21)	
GB2	no	no	no
	(6.10)	(5.49)	
GB3	no	no	no
	(5.14)	(6.63)	
GB4	no	no	no
	(5.11)	(7.42)	
GB5	YES	no	no
	(7.85)	(13.22)	
GB6	no	no	no
	(10.69)	(13.90)	
GB7	YES	no	no
	(2.96)	(8.88)	
AU6	YES	YES	no
	(5.08)	(8.47)	

event saturates the high-gain channel with a magnification of
125,000 at 1 Hz (trace b.); from the low-gain records, we can infer
a peak-to-peak ground displacement at this period of 9×10^{-7} m
or 0.9 microns, for this $m_b = 5.3$ earthquake. On the other hand,
channel (c.), whose response is peaked around 10 Hz, with a gain of
2×10^6 at 3 Hz (see Talandier and Kuster [1976] for details), is
saturated only during the arrival of the T phase generated by the
event (at the beginning of the second frame in Figure 2), and shows
little energy present over background noise in the later part of the
seismogram. This, and a direct measurement of the predominant
periods on trace (b.), indicate that most of the energy present
in the observed long-lasting wavetrain (which, for clarity, we will
henceforth call "anomalous wavetrain") is concentrated around one
to two seconds in period.

Observations of a totally similar nature are repeatedly made at
Rikitéa for events occurring at GB5. The group arrival times characterizing the onset of the anomalous wavetrain and its coda correspond
to velocities of 4.0 and 0.7 km/s, and the amplitude recorded at

RKT for major events (e.g. October 31, 1977, m_b= 5.1; January 5, 1978, m_b= 5.5) indicates a peak-to-peak ground motion on the order of 0.4 microns. For the single event at GB7 (November 6, 1978), similar values are obtained, despite a lower magnitude (m_b = 4.5).

Apart from RKT, the only station in French Polynesia having recorded anomalous wavetrains is Naké (NAE), at the southeastern tip of the atoll of Hao (see Figure 1). For the November 20, 1979 event at AU6 (Δ = 8.47 degrees), a wavetrain lasting 18 minutes, with a maximum peak-to-peak ground motion of 0.5 microns, was recorded despite a waveform less regular than at RKT, and a considerably lower signal-to-noise ratio. It is important to note that the signal at the other three stations in the Hao subarray falls down to noise level after about 4 minutes, as in the case of 'normal' earthquakes, and thus no anomalous wavetrain can be identified.

Furthermore, no anomalous wavetrains are observed, either at Rikitéa or Hao, for events located at the other nearby epicenters (GB1-4 and GB6), despite comparable magnitudes; nor are they observed at NAE for GB5 and GB7 events. Finally, no such wavetrains have ever been observed in the Tahiti or Rangiroa stations. Table 1 summarizes observations of anomalous wavetrains in the five Eastern Tuamotu stations.

INTERPRETATION

There can be a priori two causes to the generation of the observed anomalous wavetrains: a source effect or a path effect. We can readily reject a source effect on the basis of three observations:

First, an investigation of teleseismic records of events from GB5 and AU6 has failed to reveal any anomalous waveforms at large distances from the source. Anomalous wavetrains are therefore characteristic of local propagation;

Second, it is clear from GB5 events, which give rise to anomalous wavetrains at RKT but not at NAE, that anomalous wavetrains are observed only along preferential paths;

Finally, the duration of the wavetrain is grossly proportional to the distance traveled, and corresponds to limiting group velocities of 4.0 and 0.6 km/s; a source of long duration would generate a wavetrain of the same duration at all distances; furthermore, it would be totally unrealistic to envision magnitude-5 earthquakes involving a rupture lasting 10 minutes or more: such events have fault dimensions typically on the order of 2 km, and rupture times of no more than 2 seconds [Geller, 1976].

It appears therefore that the observation of anomalous wavetrains is due to a path effect. The possibility that this may represent dispersion involving an unusually steep group velocity

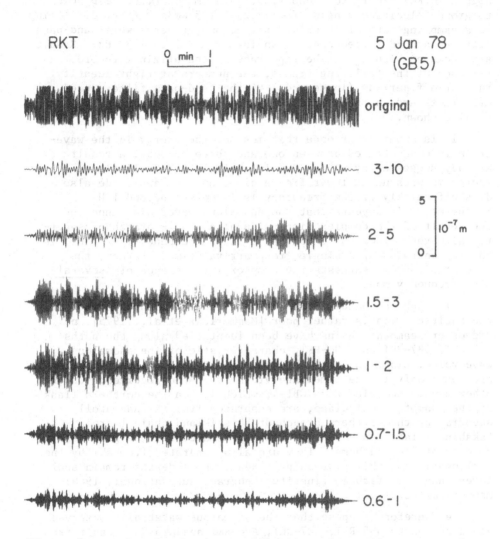

Figure 3 : Selective band-pass filtering of anomalous wavetrain
recorded at RKT for event at GB5 on Jan. 5, 1978. Top:
original seismogram; other traces: seismogram filtered
using cut-off periods given on right-hand side (in s).
Note strong decay of signal for frequencies greater than
1 Hz.

curve was examined next, through selective band-pass filtering of
a high-quality record. We used a stretch of 8 minutes of the
high-gain RKT record for January 5, 1978 (m_b = 5.5), corresponding
to group velocities ranging from 4.2 to 1.3 km/s. After digitizing
at a sampling rate of 10 points per second, records were band-pass
filtered around 6 frequencies ranging from 0.2 to 1.33 Hz. Results
are shown on Figure 3. The top trace is the original record, and
for each of the following traces, the numbers at right identify
the cut-off periods (in seconds) of the band-pass filters used.
All traces are on the same scale, whose ground motion equivalent
is also shown.

It is apparent at once that most of the energy in the wavetrain is concentrated between one and three seconds, a result
already suggested by a time-domain analysis. Very little
energy is present at lower frequencies, and the amplitude also
dies off quickly if the frequency is increased beyond 1 Hz.
Furthermore, it appears that the anomalous wavetrain cannot be
the result of the complex dispersion of a single branch of super-
ficial waves, since the energy at a given frequency (say 1 Hz) is
not concentrated at a single group arrival time. Rather, the
present situation suggests the complex interference of several
high-frequency modes.

Although the bathymetric data available for this portion of
the Pacific Ocean is rather poor [Mammerickx et al., 1975], no
ridges or seamount chains have been identified along the paths
GB5-RKT, GB7-RKT and AU6-RKT, characteristic of the anomalous
wavetrains. Along the path AU6-NAE, substantial bathymetric features
are found only in the immediate vicinity of the receiver. On the
other hand, the epicenters GB1-4, which lie on the northern flank
of the Tuamotu archipelago, are separated from the Hao atoll
subarray by the northern branch of this island chain (Raroia -
Fakahina - Tatakoto - Réao), which forms a continuous plateau at
a depth of 1500 fathoms. They are also separated from RKT by the
prolongation of this plateau past Réao, where depths remain shallower than 1700 fathoms [Institut Géographique National, 1969;
Mammerickx et al., 1975].

We therefore propose that the anomalous wavetrains observed
along the paths GB5-RKT, GB7-RKT, AU6-RKT and AU6-NAE result from
the interference of a number of branches of modes propagating
primarily in the water column and sedimentary transition zone.
Any disruption in this structure, such as the presence of a barrier
of seamounts, makes this propagation impossible and leads to the
disappearance of the wavetrain. A similar situation is encountered
if the propagation path crosses an area of irregular, rugged relief,
such as a succession of fracture zones. This explains why no anomalous wavetrains were recorded in Tahiti from AU6, a path crossing
the Austral Fracture Zone, and whose bathymetry is very irregular.
Also, the low attenuation characteristic of modes of the water

column favors propagation over great distances at sea, but results in comparatively very inefficient propagation over an island structure, once the energy has been converted to more conventional seismic waves at the island shore. The inability of high-frequency seismic energy to propagate in detritic structures such as those present on oceanic atolls as been described in T wave studies [e.g. Talandier and Okal, 1979], and explains the absence of anomalous wavetrains at the three northern stations of the Hao subarray for the AU6 event, despite a good signal at NAE.

We further propose that the sedimentary and transitional layers are fundamental coupling agents between the water and the solid Earth, and are responsible for the efficient excitation of the modes involved in the anomalous wavetrains by seismic sources in the crust. In the next section, we will study the dispersion of normal modes for two very crude models, in order to justify our proposed interpretation, and discuss this matter more quantitatively. It should be kept in mind however, that this modelling can only be tentative and crude, since the wavelenghts involved are on the order of a few km, several orders of magnitude below the accuracy of our present knowledge of the lateral variations of the bathymetry in this part of the Pacific.

NORMAL MODE MODELLING

In this section, we investigate theoretically the dispersion of the first few Rayleigh overtones in the period range 1-3 seconds and for the two models of oceanic structure described in Table 2. Model O involves no sedimentary layers and a sudden transition from water to crust at a depth of 4 km. Model S includes two transitional layers, which represent the so-called transition zone, and the basement layer. In the absence of any data on the shallow structure of the particular area of the Pacific under study, we chose to adapt Eaton's[1962] oceanic model by incorporating a transition layer, inspired from Spudich and Helmberger's [1979] model H. It should be again emphasized that the purpose of this section is to crudely investigate the influence of such transitional layers on Rayleigh overtones at high frequencies, and not to provide a detailed modelling of the South-Central Pacific Basin.

Spheroidal mode solutions were computed in the range ℓ (angular order) = 3000 - 16000, and T (period) = 1 - 3 seconds, using an eigenfunction function program originally written by R.A. Wiggins. The number of overtone branches computed was n = 10 for Model S and n = 6 for Model O. These numbers were chosen so as to include all modes with $T \geq 1$ s for ℓ = 10000. Increments in the value of ℓ between modes computed on a single branch varied substantially depending on the amount of coupling between branches involved [Okal, 1978], from a minimum of $\delta\ell$ = 10 to a maximum of $\delta\ell$ = 1000.

Table 2. Models used in Normal Mode Investigation

Layer	Thickness km	Density g/cm^3	P-wave vel. km/s	S-wave vel. km/s
MODEL 0				
1. Water	4	1.0	1.485–1.495	0
2. Crust	12	2.7–2.8	6.4–6.6	3.5–3.8
3. Mantle	∞	3.0	8.0	4.6
MODEL S				
1. Water	4	1.0	1.485–1.495	0
2. Transit- ional	2	1.9	2.5–2.7	0.8
3. Basement	1	2.0–2.05	4.7–4.8	2.0–2.1
4. Crust	9	2.7–2.8	6.4–6.6	3.7–3.8
5. Mantle	∞	3.0	8.0	4.6

Full solutions, including excitation coefficients, were obtained
for all modes computed. We will first concentrate on the results
concerning angular order (ℓ), period (T) and group velocity (U),
presented on Figure 4. It is immediately apparent that a totally
different dispersion of the modal energy is involved in the two
models. In the pure 'oceanic' model 0 without a transitional layer
of sediments, and apart from the fundamental, slow propagating, and
poorly excited Stoneley branch, most of the group velocity values
are found to lie between 2 and 4 km/s. The only exceptions involve
a few 'crossover' periods, such as 1.4 s, for which systematic
branch coupling occurs. On the contrary model S, which includes
sedimentary layers, has group velocity covering the range 0.6 to
4 km/s more or less uniformly over the period range 1.2 to 2.5 s.
This pattern is in much better agreement than the one derived from
model 0 with our experimental results, which indicate a wide dispersion of group arrival times at all periods. This proves that anomalous wavetrains are due to efficient coupling of seismic energy
into the ocean through the transitional sedimentary layers.

A study of the mode eigenfunctions can give insight into the
difference in excitation of comparable modes in both models. For

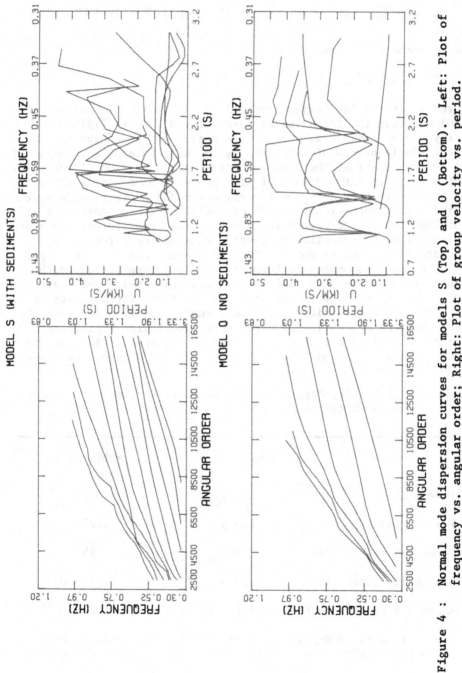

Figure 4 : Normal mode dispersion curves for models S (Top) and O (Bottom). Left: Plot of frequency vs. angular order; Right: Plot of group velocity vs. period.

this purpose, we make use of the excitation coefficients K_0, K_1, K_2 and N_0 [Kanamori and Cipar, 1974; Okal, 1978]. We define

$$K = \sup \{|N_0| \ell^{-0.5}, |K_0| \ell^{-0.5}, |K_1| \ell^{0.5}, |K_2| \ell^{1.5} \} ,$$

and we focus on the quantity

$$E = I K^2/ U^2 ,$$

where I is the Lagrangian integral $I_1 + \ell(\ell+1) I_2$ of the mode involved. From the formalism of Kanamori and Stewart [1976], one easily derives that E is the power spectrum of the energy excited into a single branch of overtones by the most favorable geometry of the moment tensor at the source. A typical comparison of the relevant quantities for a couple of slow propagating modes is given in Table 3. In this example, we investigate modes of very comparable periods and group velocities, which would therefore generate similar contributions to the wavetrain, both in frequency content and group arrival time. The excitation coefficients are computed for a standard earthquake of moment 10^{27} dynes-cm, and a source depth of 7.5 km, corresponding in both cases to a source buried in the crust. The data in this table show that, while values of K are similar in the two models (meaning that surface displacements of the water would be similar), far more energy is channeled into the given overtone branch in the case of the model with transitional layers. Specifically, about 100 times more energy is involved in model S than is in model O. Since the number of overtone branches

Table 3. A Typical Example of the Relative Excitation of Modes in both Models.

	MODEL O	MODEL S
Mode	$_1S_{10000}$	$_1S_{16000}$
Period (s)	1.695	1.678
Group velocity (km/s)	0.974	0.955
Lagrangian Integral	1.06×10^{-4}	1.06×10^{-2}
K at 7.5 km (see text)	1.93×10^{-3}	1.69×10^{-3}
Spectral Energy Density ($I K^2/U^2$)	4.16×10^{-10}	3.32×10^{-8}

covering our area of interest in the T–U plane is also greater in
model S, we come to the conclusion that the presence of a well-
developed transitional layer favors the excitation of a substantial
amount of energy into the seawater. Since the process of conversion
of this energy at the receiving shore is a complex one, depending
on the particularities of the local topography, the energy power
spectrum is probably more representative of the final amplitude
of the seismogram than would be the simple displacement at the top
of the water column, a quantity commonly used in surface-wave
seismology at lower frequencies.

If the source depth is increased, say to 11 km, the difference
in energy excitation between our two models is greatly reduced
(to a value of only 10 in the example studied above). This reflects
the fact that most of the solid Earth is a virtual node for the
water-sediment modes involved in anomalous wavetrains. This property
strongly suggests that the excitation of anomalous wavetrains
requires extremely shallow seismic sources. Although a more quantitative statement will require the systematic use of synthetic
seismograms, and therefore a better knowledge of the shallow structure of the area, these results are in qualitative agreement with
the suggestions by Okal et al.[1980] that events at GB5 are no
deeper than a few km below the water-sediment interface, based on
their study of p(n)wP phases regularly generated by these events.

CONCLUSION

In conclusion, we have shown that anomalous wavetrains (observed only along paths involving a smooth bathymetry, and therefore
a continuous sedimentary layering), whose energy is concentrated
between periods of 1 and 3 seconds, and travels at group velocities
of 0.6 to 4.0 km/s, can be interpreted as due to the interference
of overtone branches of surface modes. The presence of the
transitional layer(s) results in strong coupling of the water column
with the upper crust, and in efficient excitation by shallow sources
located not more than a few km below the sedimentary layers. The
models used in the present study are, of course, very crude, but
we hope they will provide a framework for more quantitative studies
which could greatly help us understand both the mechanism of the
input of seismic energy into the ocean, and the intricate origins
of oceanic intra-plate seismicity.

ACKNOWLEDGMENTS

Normal mode solutions obtained in this study were computed
using a program written by R.A. Wiggins, and on which R.J. Geller
provided useful comments. The use of the DEC-20 facilities at Yale's
Department of Computing Science is gratefully acknowledged. This

study was supported by the Office of Naval Research under Contract N00014-79-C-0292.

REFERENCES

Eaton, J.P., 1962, Crustal Structure and Volcanism in Hawaii, in "The Crust of the Pacific Basin, Geophysical Monogr. Ser.", vol. 6, pp. 13-29, American Geophysical Union, Washington, D.C.

Geller, R.J., 1976, Scaling Relations for Earthquake Source Parameters and Magnitudes, Bull. Seism. Soc. Amer., 66:1501-1523.

Institut Géographique National, 1969, Carte de l'Océanie au 1/2 000 000 [Map], Paris.

Kanamori, H. and Cipar, J.J., 1974, Focal Process of the Great Chilean Earthquake of May 22, 1960, Phys. Earth Plan. Inter., 9:128-136

Kanamori, H. and Stewart, G.S., 1976, Mode of Strain Release along the Gibbs Fracture Zone, Mid-Atlantic Ridge, Phys. Earth Plan. Inter., 11:312-332.

Mammerickx, J., Chase, T.E., Smith, S.M. and Taylor, I.L., 1975, Bathymetry of the South Pacific [Map], Scripps Institution of Oceanography, La Jolla, California.

Okal, E.A., 1978, A Physical Classification of the Earth's Spheroidal Modes, J. Phys. Earth, 26:75-103.

Okal, E.A., Talandier, J., Sverdrup, K.A. and Jordan, T.H., 1980, Seismicity and Tectonic Stress in the South-Central Pacific, J. Geophys. Res., submitted.

Spudich, P.K.P. and Helmberger, D.V., 1979, Synthetic Seismograms from Model Ocean Bottoms, J. Geophys. Res., 84:189-204.

Talandier, J., 1972, "Etude et prévision des Tsunamis en Polynésie Française", Thèse d'Université, Université Paris VI, Paris.

Talandier, J. and Kuster, G.T., 1976, Seismicity and Submarine Volcanic Activity in French Polynesia, J. Geophys. Res., 81: 936-948.

Talandier, J. and Okal, E.A., 1979, Human Perception of T waves: the June 22, 1977 Tonga Earthquake felt on Tahiti, Bull. Seism. Soc. Amer., 69:1475-1486.

VELOCITY SPECTRAL ESTIMATES FROM THE ARRAYS OF THE ROSE PROGRAM

Arthur B. Baggeroer

Ocean & Electrical Engineering
M.I.T.
Cambridge, MA 02139

1. Introduction

The ROSE (Rivera Ocean Seismic Experiment) was a large seismic/
acoustic program conducted near the Clipperton Fracture Zone in
Jan/Feb 1979 off the western coast of Mexico. Ten oceanographic
institutions and Navy laboratories, deployed over seventy ocean
bottom seismometers/hydrophones, two vertical hydrophone arrays and
a 24 channel towed hydrophone array over approximately 120 nm x
120 nm area near the East Pacific Rise. Explosive sources ranging
in weight from .1kg to 1000 kg generated signals that propagated
through the water and seabed and were recorded by these instruments.
Seismically the experiment focused upon determining the characteristics of the crust near an active plate boundary. Acoustically
it concerned analyzing low frequency propagation within a thinly
sedimented seabed.

The acoustic component of ROSE involved three arrays, a 2.5 km
24 channel towed seismic streamer array with digital recording, a
2 km, 12 channel MABS (Moored Acoustic Buoy System) vertical hydrophone array with an analog data recording capsule, and an array of
seismometers of various manufacture distributed over two 12 km legs
close to the mooring point of the vertical array. The approximate
location of the mooring site was 12.26°N, 101.90°W over a seabed
with topographic features of $\pm$300m.

The arrays were employed for several reasons. First, the
complicated multipath of signals interacting with the seabed requires a dense coverage of receivers if one is to resolve the

---

*The program was originally planned to be sited near the Rivera
Fracture Zone. Problems in obtaining permission to operate in
Mexican territorial waters required relocating it south to inter-
national waters north of the Clipperton Fracture Zone.

various paths. Second, the arrays can be steered in the vertical
direction to measure vertical angles, or equivalently the horizontal
phase velocities which is indicative of the mode of propagation
within the seabed. Finally, the combination of the seismometers,
deep moored hydrophones, and the surface towed arrays can be used
to measure changes in the signal waveform as it propagates across
the seabed boundary and through the water column.*

We concentrate on the first two issues - resolving the multipath and its propagation modes. This is done by using the concept
of high resolution velocity spectra analysis. This is a technique
that has been employed extensively in the oil exploration industry
to resolve complicated multipath structure in seismic reflection
profiling.[2] We have adapted this to the refraction and long
range reflection environment and modified the signal processing to
improve upon the spectral resolution that have been commonly used
in the oil exploration industry.[3,4] A velocity spectra is an
estimate of the signal energy that propagates across an array within a specified time window. In the processing of acoustic signals
the spectral estimate is a function of three variables--the time
after the shot instance, T, the apparent phase velocity, Cp, or
vertical angle across the array, and the center frequency, fo, of
the analysis band of the signal. In this form it provides a particularly convenient means of analyzing bottom interacting acoustic
signal propagation generated by an impulsive source. The time
after shot instant localizes the signal epoch, the phase velocity
is indicative of the propagation speed in the refracting medium or
the relection angle and the frequency band can be centered at the
dominant region of the source spectrum to filter extraneous out-
of-band noise and reverberation.

In this paper we present some velocity estimates from the
ROSE data and their acoustic interpretations. At the present
time we are using the towed array data since it was directly
available in digital format. (The vertical array data has been
digitized, it is now being formatted, edited and analyzed using the
same velocity spectra program). We emphasize that these are initial
results presented primarily to be indicative of the power of
velocity spectra estimation in interpreting bottom interacting
acoustic data.

2. Experiment format

The format of the data for this acoustic experiment during
ROSE is illustrated in Fig. 1. A shooting ship, the R/V Kana Keoki
of the Univ. of Hawaii dropped 5 and 25 lb charges at a depth of
35 meters as it steamed away from the vertical array. Simultaneously, the receiving ship, the R/V Conrad of Lamont-Doherty

---

* There is, however, some uncertainty in the true transfer function
  between an ocean bottom seismometer and the seabed.[1]

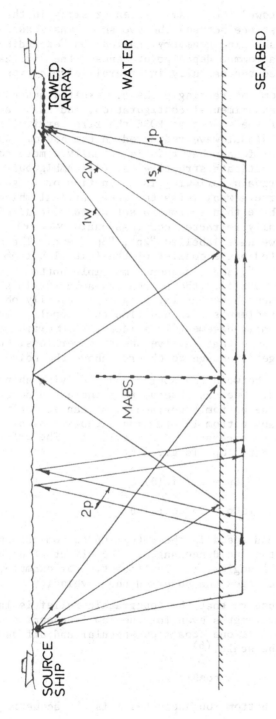

Fig. 1 Geometry and ray paths in ROSE acoustic experiment

Geological Obsv., towed the seismic streamer array in the opposite
direction. The distance between the two ships was measured using
an RF link. This shooting geometry, called an "expanding spread
profile" leads to a common depth point beneath the vertical array,
and it has been used successfully in several seismic experiments.[5]

Several bottom interacting paths are excited and can be observed using this experimental configuration, and these are indicated in Fig. 1. For the seabed at ROSE the first arrival at long
ranges is a compressional wave refracted in the bottom which we
label "1p". We have found that this is actually a more complicated arrival whose multipath structure is still ambiguous. The
sequence of the arrivals following is a function of range. Generally, one can observe a weak refracted shear arrival which we indicate by "1s". There then follows a set of refracted/reflected
paths of the initially refracted compressional waves reflected at
the surface which we have labelled "2p", "3p", etc. The bottom
interacting reflected paths consist of the initial bottom bounce,
which we label as "1w", and a sequence multiple bottom bounces,"2w",
"3w", etc. At the ROSE site the seabed has a very thin sediment
cover and no prominent basement layering as in usually observed
older crust where sediments have had time to accumulate and where
layering of the oceanic basement is evident. Additional paths
exist, especially those that involve shear conversions, but they
have not been energetic enough to observe above the noise.

The bathymetry between the source and receiving ships is
indicated in Fig. 2. One can observe that there are several prominent features in the bottom topography that can strongly influence
the properties of any bottom interacting signals. Slopes over
extended distances in excess of 1° are evident. The effects on the
observed phase velocity, Cp, is given by

$$Cp = Co/\sin(\theta) , \qquad\qquad (1a)$$

$$\text{or} \qquad \Delta Cp \simeq Cp \cot(\theta)\Delta\theta \qquad\qquad (1b)$$

where Co is the sound speed in the water, Cp is nominal phase
velocity and θ is the incidence angle. The effect at high phase
velocities, or small angles, is significant. For example, at
7 km/s one degree changes the observed phase velocity by .5 km/s.

We can also observe that the topographic relief is large
compared to the wavelengths even for the low frequencies of
interest (6-16 Hz). If one separates specular and diffuse scattering according to the scale.[6]

$$2\bar{h} \cos\theta/\lambda \qquad\qquad (2)$$

where $\bar{h}$ is the rms bottom roughness and λ is the acoustic wavelength

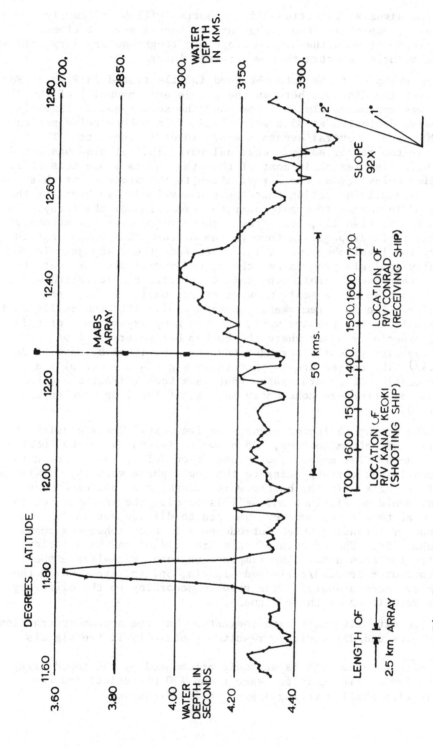

Fig. 2 Bathymetry near vertical array for ROSE acoustic experiment

then the signals reflecting off the bottom will be diffusely
scattered, especially those that are reflected several times.
The effects of both the bottom slope and roughness are very evident
in the velocity spectra that we present later.

An example of the data obtained is illustrated in Fig. 3. For
this shot the distance between the source and channel 1 of the
array was approximately 24.7 kms. and the source was 25 lbs. of
explosive detonated at 35 m which leads to a bubble pulse period
of 200 msecs. Several events are prominent in the data. The
first is the refracted compressional wave, 1p. It also has a coda,
1p', which is present in most of the other shots along this line.
The time delay across the array, often termed moveout, of this
event is small indicating that the phase velocity is high, or the
angle of incidence is small, near the broadside of the array. The
frequency is also low, since the higher frequencies are strongly
attenuated in propagating through the bottom. The next event is
a very low level shear wave, 1s. One needs the array gain in the
velocity analysis program and the redundancy of several shots to
detect this event. Following this is a refracted/reflected wave,
2p. This also has a small moveout corresponding to high phase
velocity. The more remarkable observation is that the amplitude is
comparable to the primary wave, 1p. In many experiments of this
type, especially those where the bottom bathymetry is level, the
2p signal is typically stronger than the 1p signal by several decibels.[7] The conjecture is that these signals are actually a
superposition of several paths that have been refracted off the
bottom and interfere coherently because of the long wavelengths
involved.

The bottom reflected signals arrive next.* These consist of
a series of bottom bounces, two of which 1w and 2w can be localized
in the data. The moveout, or propagation delay, across the array
is much longer corresponding to the lower phase velocity, or larger
vertical angle. If the bottom were smooth, each bottom bounce
signal would be distinct and well isolated. The roughness of the
bottom at this site, however, has led to diffuse scattering and
wavenumber spreading which introduces a lack of coherence among
the channels. The effects of this are evident in the velocity
spectra for this data. Superimposed on these signals are the
sequence of refracted/reflected signals, 3p, 4p, etc., which the
velocity spectra analysis can resolve according to the differing
phase velocities of these signals.

The extensive multipath indicates that the bottom interaction
is complex and difficult to resolve, especially if the signals

*The direct water path is strongly attenuated by the towed array
by the low endfire gain for each group (50 m section) and the
combination shallow towing depth and low frequency.

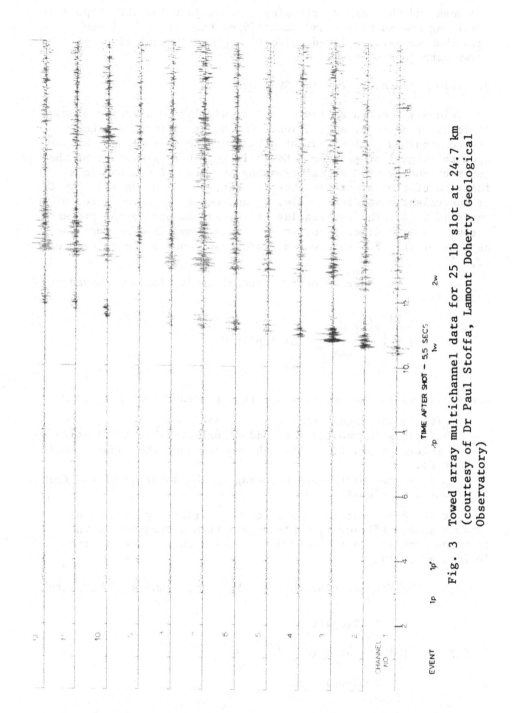

Fig. 3 Towed array multichannel data for 25 lb slot at 24.7 km (courtesy of Dr Paul Stoffa, Lamont Doherty Geological Observatory)

are weak and they are overlapping. It is just for the purpose of resolving the multipath and quantifying the strength of each path that we have employed velocity spectra in analyzing the array data from ROSE.

3. Velocity spectra for the ROSE data

Velocity spectra analysis is very similar to directional wave spectra, or bearing, estimation. The major difference concerns the constraints that transient nature of the shot signal imposes upon the signal processing. Essentially, it is a measure of the coherent energy of a signal crossing an array of sensors as a function of time after the shot instant, T, the phase velocity Cp, or equivalently vertical angle, θ, and center frequency, fo, of the analysis band. The intricacies of the signal processing prevent a comprehensive discussion; so we simply summarize the important aspects of it. The interested reader can consult the references (2,3,4).

Applied to phase velocity spectral estimation the conventional definition is

$$\hat{S}(T,Cp) = \int_{T-T_w/2}^{T+T_w/2} \left| \sum_{n=1}^{n} x_n(t+\Delta T_n(Cp)) \right|^2 dt ; \qquad (3)$$

where $\hat{S}(T,Cp)$ is the velocity spectra as a function of T and Cp;

$x_n(t)$ is the signal of the nth channel;
$\Delta T_n(Cp)$ is the moveout; or additional travel time relative to a reference point, at the nth channel when the phase velocity is Cp.
T_w is window width corresponding to the duration of the transient shot signal.

It is convenient in many instances, primarily because of computational efficiency, to transform this expression to the frequency domain. After some manipulation this leads to the following: Define

(i)   the Fourier transform of the shifted and windowed data

$$X_n(f,T,Cp) = \int_{T+\Delta T(Cp)-T_w/2}^{T+\Delta T(Cp)+T_w/2} dt\, x_n(t) e^{-j2\pi ft} ;$$

(ii)   a matrix, $\underline{R}(f,T,Cp)$, whose elements are given by

$$[\underline{R}(f,T,Cp)] = X_m(f,T,Cp)X_n^*(f,T,C_p)$$

(iii) a steering vector $\underline{E}(f,Cp)$, of phase delays given by

$$\underline{E}^+(f,Cp) = [e^{j2\pi f\Delta T_1(Cp)}, \ldots , e^{j2\pi f\Delta T_n(Cp)}]$$

Using these definitions it can be shown that the frequency domain expression for the velocity estimate is given by the quadratic form

$$\hat{S}(T,Cp) = \int df \; \underline{E}^+(f,Cp)\underline{R}(f,T,Cp)\underline{E}(f,Cp) \tag{4}$$

(The quadratic form makes implementation of this particularly attractive using an array processor.)

Signal processing research has led to several improvements upon this basic quadratic form. First, the integral across frequency has been divided into bands so that we have frequency selective results centered at fo, or

$$\hat{S}(fo,T,Cp) = \int_{fo-W/2}^{fo+W/2} df \; \underline{E}^+(f,Cp) \; \underline{R}(f,T,Cp) \; \underline{E}(f,Cp) \tag{5}$$

where W is the desired bandwidth for the processing.
Second, research on adaptive seismic and sonar arrays has demonstrated that Eq.(5) can be modified to obtain higher resolution by using the expression[4]

$$\hat{S}(fo,T,Cp) = [\underline{E}(fo,Cp)\overline{\underline{R}}^{-1}(fo,T,Cp)\underline{E}(fo,Cp)]^{-1} \tag{6}$$

$$\text{where } \overline{\underline{R}}(fo,T,Cp) = \int_{fo-W/2}^{fo+W/2} \overline{R}(f,T,Cp) \; df$$

(The user should be aware that ultimate implementation using digital data imposes some additional constraints in the processing.) Equation (6) is the form that we use to produce our velocity spectral estimates of the ROSE array data.[8]

Velocity spectra for the data illustrated in Fig. 3 are indicated in Fig. 4a and 4b. Center frequencies of fo = 8 Hz and fo = 12 Hz are displayed. Several acoustic interpretations can be obtained from these spectral estimates.

i) The bottom bounce reflected signals 1w, 2w, 3w, etc. are

spread in phase velocity with each successive bounce having
greater spreading. After the fourth bounce, 4w, it is
difficult to identify the bounce as coherent energy since
it is spread over such a large extent in phase velocity.
This is in sharp contrast to results obtained from flat
bottoms where the reflected signals remain very coherent
over many more bounces. The first bounce, 1w, is actually
spread over a significant angular extent with a spread
of 50 m/sec. Inverting eq. (1) yields

$$\Delta\theta = \tan(\theta) \frac{\Delta Cp}{Cp}$$

so a 50 m/sec spread at Cp = 1550 m/sec leads to an
angular spread of 7°. The linear display versus phase
velocity compresses the near grazing; however, on a
flat bottom with good signal to noise ratio, this array
geometry can resolve phase velocities to ± 5 m/sec.

(ii) The compressional refracted signal, 1p, has a phase
velocity of approximately 7 km/s at this range of 24.7 km
the energy in the refracted signal at 8 Hz and 12 Hz is
of the same order of magnitude as the bottom bounce
reflected signals. It is difficult to determine the
extent of spreading since the array resolution at higher
phase velocities is limited. (Again at high phase velocities small angular changes correspond to large changes.)

The refracted signal, 1p, is followed by the sequence
of refracted/reflected signals, 2p, 3p, etc. separated by
4.5 secs. These paths also have high phase velocities
suggesting that they propagate as refracted signals in
the bottom, traverse the water column two ways by reflecting off the sea surface, and propagate again as refracted
signals in the bottom. The amplitude of the path 2p is
approximately equal to that of the primary path, 1p. In
several experiments of this type where the seabed is flat,
the amplitude of path 2p is actually stronger than that
of 1p by several decibels. One possible explanation is
that path 2p actually consists of the superposition of
an infinite number of paths refracting off the seabed and
interfering constructively because of the long wavelengths
and near normal incidence geometry of the paths. This
would be consistent with the observation that the affect
appears to be more pronounced when the bathymetry is
constant.
Paths 3p, 4p, etc. have lower amplitudes and greater
spreading in phase velocity. Eventually, they overprint
the bottom reflected signals 1w, 2w, etc. To some degree

the array resolve these two paths when they overlap; however, the interaction ofr the later arrivals is quite complex when they overlap.

iii) One of the more interesting observations in this data
set is the presence of a doublet path, 1p', following 1p.
This path appears in the velocity spectra of several
shots and often at more than one frequency. The explanation of this path is still somewhat ambiguous. One possible
one is that there is a low velocity zone beneath the crust
at the seabed. This has been suggested previously from
refraction lines shot to telemetering sonobuoys.[9] The
other notable aspect is that the shot at a range of 35 kms
generates very strong amplitude levels which is consistent
with the synthetic seismograms of the preferred model in
Reference 9. The analysis of this path obviously needs to
be refined, especially to reject the effects of bottom
topography roughness. The velocity spectra indicates
that it appears consistently, so there are some intriguing
acoustics involved.

iv) If we examine Fig. 4b, the velocity spectra at 12 Hz,
just prior to epoch of event 2p we observe the faint suggestion of shear wave propagation, or path 1s. This signal
is very weak, typically 25 dB below the compressional wave,
1p. We can improve our confidence in this by noting that
it occurs in a region where there is no reverberation
from other paths and at other frequencies and shots. The
conversion to shear here is very weak, possibly a consequence of the bottom roughness.

The above observations illustrate some of the observations
that can be made from the ROSE array data using velocity spectra
analysis. By repeating this type of analysis over a sequence of
shots, we can obtain a more synoptic perspective of the characteristics of the acoustic propagation and the bottom interaction.

In Fig. 5 we have plotted the phase velocity for each path as
measured by the array over a sequence of shots where the source-
to-receiver distance increased from 15 to 40 kms. These velocities were obtained by averaging the value for the peak of the
velocity spectra for each path at center frequencies of 8, 12 and
16 Hz.

For the refracted compressional signals we observe that the
phase velocities at short ranges are between 7.5 and 8.2 km/s.
This probably reflects the complex topography and slopes for the
shallow interaction at short ranges. For source/receiver ranges
between 20 and 30 kms the velocities decrease consistently to
7.5 km/s. Between 30 and 40 km. the velocities increase to approximately 8.5 km/s. We suggest that the bottom topography is

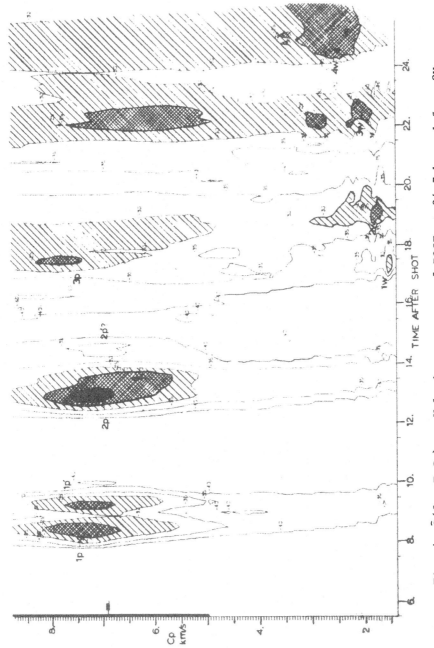

Fig. 4a $\hat{S}(f_o, T, C_p)$: Velocity spectrum of ROSE at 24.7 km and f_o = 8Hz

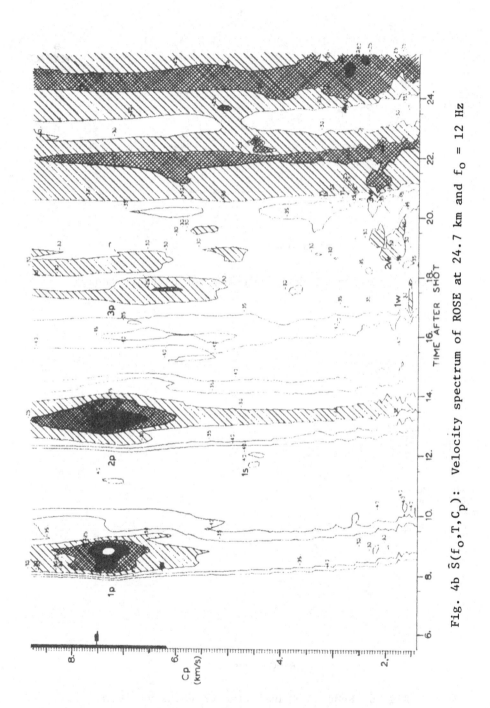

Fig. 4b $\tilde{S}(f_o, T, C_p)$: Velocity spectrum of ROSE at 24.7 km and f_o = 12 Hz

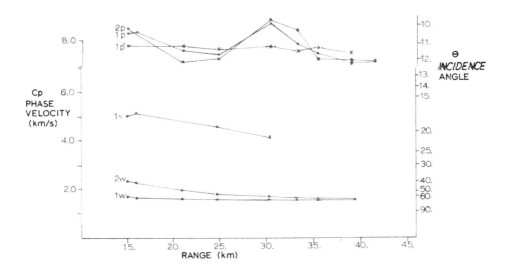

Fig. 5   Estimated phase velocity for paths vs range

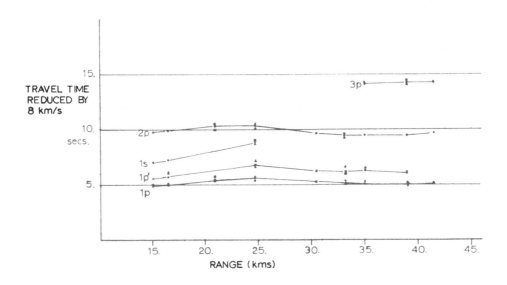

Fig. 6   Reduced travel time of paths vs range

causing this change in the apparent phase velocity. When we examine the bathymetry plot in Fig. 2, we observe that there is an
extended distance in the range of 30 to 40 km where the bottom
slope is approximately 2°. This slope would cause an increase in
the observed phase velocity of 1km/s using ray theory analysis.
Beyond this range the compressional phase velocities return to
approximately 7.5 km/s which is more typical of young oceanic crust
at this range. It is evident that the influence of the bottom
topography roughness such as this need more extensive study to
understand how it modulates local changes in the observed phase
velocity of a signal across an array.

The shear wave phase velocity measurements are sparse and
difficult to make because of the low signal to noise ratio for
this path. Nevertheless, there are enough measurements to argue
that they are present. The velocity spectra indicate a systematic
decrease from 5.2 to 4.2 km/s which is consistent with the observed
compressional wave velocities. Analysis of the data with larger
charge weights for sources are needed for a more definitive analysis
of the shear wave propagation. (A line with 500 lb source weights
was shot during the ROSE.)

The phase velocities of the bottom bounce reflected paths
illustrate the expected behavior of the decrease in phase velocity
to the near grazing geometry as the range increases. Eventually,
they approach incidence angles where the array has reduced sensitivity.

Another way to analyze the phase velocities is in terms of
reduced travel time curves. In Fig. 6 we have plotted the reduced
travel time for each path versus range. The travel time has been
"reduced" by a velocity of 8 km/s, which means that at each range
R, the travel time has been reduced by an amount of R/8. A path
with a velocity of 8km/s would have a constant "reduced" travel
time, while those with a lower velocity would have an increased
"reduced" travel time versus range for ranges less than 25 km,
the compressional signals have a slight positive slope indicating
a phase velocity less than 8 km/s. Beyond this range the slope
decreases indicating a velocity exceeding 8 km/s. This is consistent with the local phase velocities measured for each shot
illustrated in Fig. 5. At the longer ranges, the slopes appear to
increase which would be consistent with the expected behavior of
these refracted signals at long ranges where they typically have
values of approximately 8.3 km/s.

The most difficult behavior to interpret is the observed
energy in the various paths as a function of range. In the shooting and receiving program, great care was given to shooting the
lines with identical charge sizes and constant depth so that uncertainty in the source spectrum variations could be kept to a
minimum. The data indicated in Fig. 7 indicates the peak value of

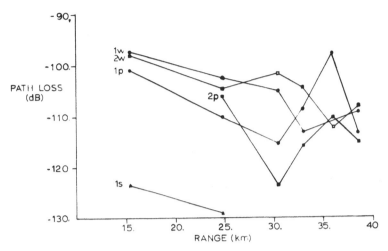

Fig. 7 Path loss and energy partitioning vs range

the velocity spectra at 8 Hz for each of the paths using 25 lb
source charges detonated at 35 m. Each value has been corrected
for the mirror effect of the seasurface and the directivity index.
DI(θ) of each 50 m group in the streamer. The correction is

$$CF(\theta) = -20 \log_{10}|1-e^{-j4\pi\frac{d}{\lambda}\cos\theta}| + DI(\theta) \qquad (7)$$

where θ is the incidence angle d is the array depth

The path gain, or transmission loss, TL(R) was calculated by using the equation

$$TL(R) = \quad SPL -[10 \log_{10}\hat{S}(f,T,Cp) + CF(\theta) + PG \qquad (8)$$

$$-20 \log_{10}|G(f)|]$$

where $\hat{S}(f,T,Cp)$ is the value of the velocity spectrum for the
path. (The velocity spectrum computes values
on a per Hertz basis, so we use an energy
density formulation.);

SPL   is the spectral level of the source per Hertz
in the band of interest (226 dB re 1μPa/Hz @
8 Hz and 1m);

$20 \log_{10}|G(f)|$ is the sensitivity of the array
($\approx$-185 dB re 1v/1μPa)

PG   is a correction factor for digital signal
processing in the velocity analysis program
(-52 dB).

The gain for the bottom interacting paths for ranges between 15 and 25 kms indicate that the reflected paths are slightly more energetic than the refracted compressional paths and that the level of the shear wave is 25 dB below these paths. There is a 5 to 7 dB decrease in path gain in increasing the range from 15 to 25 kms. Beyond this range the amplitude behavior becomes significantly more complex. The most notable phenomenon is the very high level for the refracted path 1p at the 35 km range. The conjecture above was that this may be associated with a low velocity layer; however, this must be tested extensively against a synthetic modelling analysis.

4. Conclusion

The important observation for this paper is that the technique of velocity spectra analysis have produced quantitative measurements of the acoustic levels of the bottom interacting paths We have used but one small part of the ROSE data set for this paper. As the entire data set is processed we will be able to refine our acoustic interpretations significantly, and one of the important analysis methods will involve the use of velocity spectra.

5. Acknowledgements

We want to thank Mr. Richard Gozzo and his associates at the Naval Underwater Systems Center for their work in deploying the MABS array during the ROSE; Dr. Paul Stoffa of the Lamont-Doherty Geological Obsv. for providing the towed array data; and Mr. David Gever of Woods Hole Oceanographic Inst. for processing the data and generating the velocity spectra.

6. References

1. "The Lopez Island ocean bottom seismometer intercomparison experiment," Journal of Geophysical Research (in press).
2. Taner, M. and Koehler, F., "Velocity Spectra-Digital Computer Derivation and Application of Velocity Functions", Geophysics, Vol. 34, No. 6, p. 859-881, Dec. 1969.
3. Capon, J., "High Resolution Frequency Wave Number Spectrum Analysis," Proceedings of the IEEE, Vol. 57, No. 8, p. 1408-1448.
4. Baggeroer, A.B., "High Resolution Velocity Depth Spectra Estimations for Seismic Profiling," Proccedings of the 1974 IEEE Conference on Engineering in the Ocean Environment, Vol. II, p. 201, 1974.
5. Stoffa, P. and Buhl, D., "Two Ship Multichannel seismic experiments for deep crustal studies: expanded spread geometries and constant offset profiles, Journal of Geophysical Research, p.

7645-7660, Vol. 80, No. B13, Dec. 10, 1979.

6. Urick, R.J., <u>Principles of Underwater Sound for Engineers,</u>
    · McGraw-Hill Book Co., 1967.

7. Ewing, J., private conversation

8. Leverette, S.J., "Data Adaptive Velocity/Depth Spectra Esti-
    mation in Wide Angle Seismic Reflection Analysis," Ph.D.
    Thesis, Dept. of Ocean Engineering, M.I.T.-WHOI and Joint
    Program, 1977.

9. Lewis, B.T.R. and Snydsman, "Evidence for a low velocity layer
    at the base of the oceanic crust," <u>Nature</u>, Vol. 266, p. 340-
    344, March 24, 1977.

BOTTOM INTERACTION REPRESENTED BY IMPEDANCE

CONDITIONS IN NORMAL-MODE CALCULATIONS

David F. Gordon

Naval Ocean Systems Center
San Diego, CA 92152

DeWayne White

Naval Ocean Research and Development Activity
NSTL Station, MS 39529

INTRODUCTION

We have found that a normal mode program designed to treat
sound speed profiles of many layers can be extended to model sediment layers below the water. While doing so it is easy to compute
reflection coefficients at the interface between water and sediment.
A second normal mode program has been developed in the process of
developing rigorous control models for use by the Acoustic Model
Evaluation Committee. This program uses a bottom loss table similar
to that used by ray theory.

We have shown that by using exact reflection coefficients from
the first program as inputs to the second program, the second program
will give identical results. Despite this result, the use of a ray
theory table of bottom loss in the second program has several short-
comings. We will demonstrate that a phase shift table is required.
Also, the slopes of both the reflection loss and phase shift tables
are required. The slope requirement places greater stress on the
interpolation methods used in these reflection tables.

Several examples of linear interpolation are shown here.
Besides bottom reflected modes we will consider modes that correspond
to rays that do not touch the bottom.

REFLECTION COEFFICIENTS IN NORMAL MODES

We will first present a few equations to define the normal mode method we use. We will discuss accuracy from the point of view of how certain quantities vary as different approximate inputs are used. Figure 1 gives some essential equations. Equation 1 shows the mode terms that are summed to give the sound pressure at a range r for source and receiver depths z_0 and z. For a given mode to be accurate the eigenvalue, λ_n, must be accurate so that both its phase rate and damping rate will be expressed correctly by the Hankel function $H_0{}^2$. The depth function of mode n is shown in Eq. 2. It is a linear combination of two linearly independent solutions to the depth dependent part of the wave equation, Eq. 3. D_n in Eq. 2 is the normalizing factor and is usually the most touchy part of the depth function when accuracy is in question.

Equations 4 and 5 give the surface and bottom conditions which must be satisfied. We term these generalized impedance conditions. Here we will use a perfectly reflecting condition for the surface. This is done by setting g_0 to 0 in Eq. 4.

We wish to use a specified value of reflection coefficient, R, for any given wave number (or grazing angle), so following the method of Bucker[1] we find that f_1 and g_1 are given by Eqs. 6 and 7. The vertical component of the wave number, ℓ, is given by Eq. 8.

R here is a ratio of sound pressures and we use −20 log in converting its absolute value to or from dB levels. To be useful in mode theory R must be a complex number. That is, a phase shift as well as a loss must be considered in the reflection process.

Useful sound speed models usually consist of layers with different parameters in each layer. A mode depth function is then required to be continuous with continuous depth derivative across each layer interface. These conditions plus the impedance conditions of Eq. 4 and 5 are the boundary conditions for the depth functions.

$$\psi = \pi i \sum_n U_n(z) U_n(z_o) H_o^{(2)}(\lambda_n r) \tag{1}$$

$$U_n = [A f(z) + B g(z)]/D_n \tag{2}$$

$$d^2 U/dz^2 + (k^2 - \lambda^2)U = 0 \tag{3}$$

Fig. 1 Normal mode equations

SURFACE

$$\left[f_o(\lambda)U + g_o(\lambda)dU/dz\right]_{z=0} = 0 \qquad (4)$$

BOTTOM

$$\left[f_1(\lambda)U + g_1(\lambda)dU/dz\right]_{z=z_B} = 0 \qquad (5)$$

$$f_1(\lambda) = i\ell(1 - R) \qquad (6)$$

$$g_1(\lambda) = -(1 + R) \qquad (7)$$

$$\ell = (k_B^2 - \lambda^2)^{\frac{1}{2}} \qquad (8)$$

Fig. 2   Impedance conditions

We will now turn to a specific example. Figure 3 shows a sound speed profile where the two upper layers represent a deep ocean profile. The third layer is a typical sediment layer, 200m deep, with a sound speed gradient of 1.0. The density in the sediment is 1.5, and in the negative gradient half space below is 2. Sound absorption in the sediment is 2.5 dB/km at the top of the sediment and increases to 5.0 at the bottom and remains at 5.0 in the negative gradient half space.

The negative gradient half space plus an outgoing wave condition are, in effect, an impedance condition. However, the formulation differs from that given here and is the usual normal mode model that we have reported elsewhere.[2]

Using this usual normal mode model, the bottom interaction for the profile of Fig. 3 can be determined. For each eigenvalue, which corresponds to a specific phase velocity, we can compute R using Eqs. 5-8. If a table of these reflection losses is now carefully prepared so that a second normal mode program designed to use the impedance

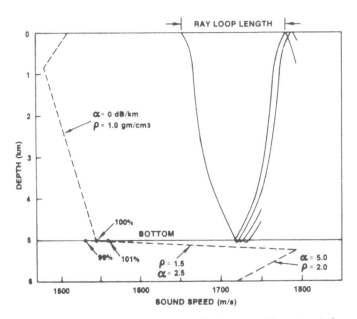

Fig. 3 Sound speed profile including sediment with ray paths

condition of Eq. 5 can read the exact value of P. as computed, then
this second program will determine the exact eigenvalues that give the
table of reflection losses. This second program only uses the two
water layers of Fig. 3 and the reflection loss table. However,
questions of more interest are how closely must the true loss table
be approximated, and can standard ray theory loss tables be used by
the normal mode program. We will attempt to answer at least the first
of these two questions.

Reflection losses will be shown for the profile of Fig. 3 at 25
Hz. Besides this model which has continuous sound speed at the water-
sediment interface we will show losses for sediments in which the
sound speeds are increased and decreased by 1% from that in the water.
These are typical fast and slow sediment values and are designated as
101 and 99%.

Also shown in Fig. 3 is a ray path showing the difference in ray
loop length as the ray splits, part reflecting from the bottom and
part being refracted in the sediment. Multiple refracted paths are
shown but they are weak and only the first is of importance here.
These will be considered later.

Figure 4 shows reflection losses for the three bottom models. These are expressed as decibels given by $-20 \log |R|$. The dots are the actual phase velocities where eigenvalues occur and R is computed. A small arrow is used in this and future figures to mark the bottom sound speed in the water which we will call the bottom point. Eigenvalues with phase velocity greater than this correspond to ray paths which touch the bottom. Those less than this are refracted in the water above the bottom. The curves in this figure pass near zero loss at the bottom point but not necessarily through it. Also shown in Fig. 4 is a three point linear fit to the middle curve. This fit is similar in detail to those often used in ray theory. The accuracy loss due to substituting this coarse fit for the actual loss points will be discussed later. Also an 11 point linear fit will be compared. The small x's along the curve show the end points of this fit.

Figure 5 shows the phase of the reflection coefficient R. It is the angle of the complex number, R. At zero grazing angle it is near 180° and then decreases steadily. The three point and 11 point approximations to this curve are also shown.

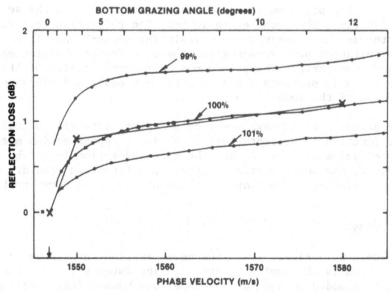

Fig. 4 Reflection loss for three sediment speeds at 25 Hz

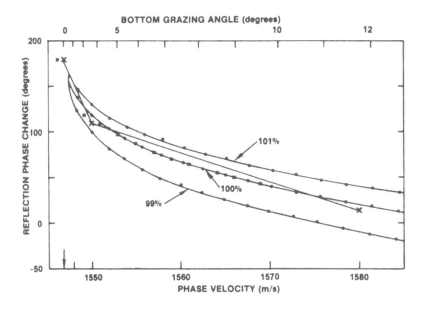

Fig. 5 Reflection phase shift for three sediment speeds at 25 Hz

Some feeling for the nature of the reflection loss curve can be gained by comparing losses for the reflected and sediment refracted path. Figure 6 shows the Rayleigh reflection coefficient and an estimate of the absorption suffered by the refracted path in the sediment. This is 2.5 times the path length in km. The reflection loss for the 100% bottom is shown again on this scale for comparison. It is clear that the refracted path accounts for the main features of the mode reflection loss curve. The pronounced difference in size of these two arrivals probably prevents phase interference beats and gives the smooth shape of the mode curve.

The accuracy of eigenvalues when an approximate fit is made to the loss function is predictable. If a linear approximation misses the decibel value of R by 1% then the mode absorption is off by about 1%. However, the normalization of the depth functions presents a fundamental problem. The slope of the depth function curve must be known.

NORMALIZATION

Equation 9 in Fig. 7 shows the normalization terms from Bucker.[1] The first term is the familiar normalization integral. The other two terms must be added because of the impedance boundaries. Because we are using a perfectly reflecting surface, the first of these is zero. The last term arises from the bottom impedance condition.

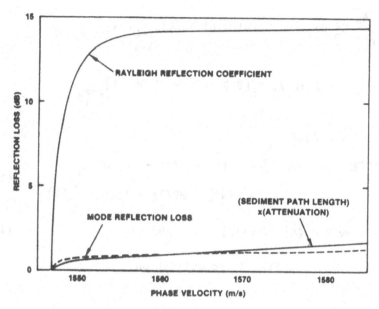

Fig. 6 Sources of bottom reflection loss for 100% sediment sound speed

U and V are both depth functions. U satisfies the surface impedance condition and all the interface conditions below it. It does not necessarily satisfy the bottom impedance condition. V satisfies the bottom impedance conditions and all interface conditions above it except the surface. If one of these satisfies both impedance conditions, then it is an eigenfunction. That is, an eigenvalue has been found. In this situation U and V differ only by a constant. Thus the ratio of U_n and V_n in Eq. 9 can be found by evaluating both at any convenient depth.

Equation 10 evaluates the rest of the normalizing term in terms of the functions of Eq. 5 or in terms of R and its derivative. This derivative can be expressed in terms of the slopes of our linear fits to the reflection loss and reflection phase shift by Eq. 11.

Before showing the success of these fits, we should mention that $\partial R/\partial \lambda$ can be computed by the usual normal mode program to show that exact values of D_n do result. However, one must take care. If the normal mode program with impedance conditions is to evaluate depth functions as a U - that is from the surface downward - then the derivative of R must be evaluated as a function of V - that is from the bottom upward. Derivatives of the coefficients in each layer must be evaluated in turn up through the sediment layers. The derivative of R at the eigenvalue and R itself is then sufficient information to exactly duplicate the mode from just the water column above.

$$D_n = \int_0^{z_B} U_n^2 \, dz - [U \, \partial U'/\partial\lambda - U' \, \partial U/\partial\lambda]_{\substack{z=0 \\ \lambda=\lambda_n}} / 2\lambda_n \qquad (9)$$

$$+ U_n^2 / 2\lambda v_n^2 [V \, \partial V'/\partial\lambda - V' \, \partial V/\partial\lambda]_{\substack{z=z_B \\ \lambda=\lambda_n}}$$

where $U' = \partial U/\partial z$

$$[V \, \partial V'/\partial\lambda - V' \, \partial V/\partial\lambda] = W^2 (f_1 \, \partial g_1/\partial\lambda - \partial f_1/\partial\lambda \, g_1)$$

$$= -iW^2 [2\ell \, \partial R/\partial\lambda + \lambda/\ell (R^2 - 1)] \qquad (10)$$

$$\partial R/\partial\lambda \simeq -v^2 R/\omega \, (\Delta|R|/\Delta v \cdot 1/|R| + i \, \Delta\theta/\Delta v) \qquad (11)$$

Fig. 7 Normalization

Figure 8 shows the accuracy of the linear fits to the reflection loss and phase. For the first 10 bottom reflected modes the percentage difference between the depth function for the entire water and sediment column is compared with that for the water column plus reflection loss table. The normalizing factor, D, accounts for most of this variation. In the upper curve, the derivative of R was set to zero. In the three point curve the first steep segment included the first two modes and apparently overestimated the slope. The 11 point fit shows that errors can be controlled if the slope of the loss and phase curves are carefully measured. The greatest difficulty appears to be for modes close to the bottom point where the loss and phase curves are steep.

It is likely that higher order interpolation or spline fits to the loss and phase curves could improve the error. However, if one is using a given ray theory bottom table there is a definite question as to whether the exact fit with stepwise slopes is intended or a smooth interpolation of the function is implied.

A second method of normalization should be mentioned briefly. The eigenvalue condition is expressed as a set of linear equations giving the boundary and interface conditions. Eigenvalues are found by finding the zeros or roots of the determinant of this set of equations. It turns out that the normalizing factor, D, can be written as a function of the derivative of the determinant with respect to the eigenvalue. However, evaluating the derivative of

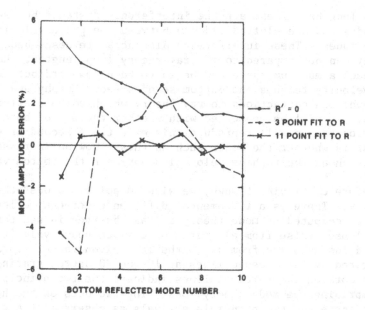

Fig. 8 Errors due to approximations to R' in mode normalization

a determinant is laborious. We have tried using numerical derivatives which must be determined anyway to find the roots of D. However, these are not sufficiently accurate.

A method of computing the derivative of the determinant that
requires less computing time for profiles of four or more layers has
been published by Otsubo.[3] We believe this is a viable alternative
method for finding D. However, this method also requires the
derivative of the reflection loss, R.

BOTTOM EFFECTS CORRESPONDING TO RAYS VERTEXING ABOVE THE BOTTOM

We have shown that modes can be responsive to a bottom loss
table for real bottom grazing angles. In the remainder of this
paper we will show that similar tables can effect modes that correspond to rays that refract above the bottom. In these tables we
will plot losses versus phase velocity rather than grazing angle
because the grazing angles become complex. Figure 9 shows the wave
effects we hope to preserve.

Figure 9 shows ray loop length as a function of phase velocity.
To the right of the bottom point the ray splits and we show the loop
length of the reflected path and the first sediment refracted path.
Near the top of the figure a few points show mode values equivalent

to loop lengths. These are the interference distance between adjacent modes and are plotted mid way between the phase velocities of
the two modes. These interference distances are described elsewhere[4]
but they can be compared to the ray theory loop lengths. Note that
they reach a maximum range and begin to turn downward before the
phase velocity reaches the bottom sound speed. Murphy and Davis[5]
have developed corrections to ray theory which compute effects such
as this. We will be concerned with whether the use of bottom
reflection tables will reproduce this effect. A second effect of
interest is whether the modes such as these show an increase in
attenuation as their phase velocities approach the bottom velocity.

Before continuing though, we wish to point out one other feature
on Fig. 9. There is a fundamental difference between reflection
losses as computed by mode theory and as observed in sea tests. Mode
bottom losses arise from all paths of a given phase velocity while
observed losses arise from all paths at a given range. This is
illustrated by the crossed lines at 10 and 20 degree grazing angles.
The two dots on the vertical lines indicate the two principle arrivals comprising the mode theory loss. The two dots on the horizontal
line indicate the two principle arrivals as observed at a given range.

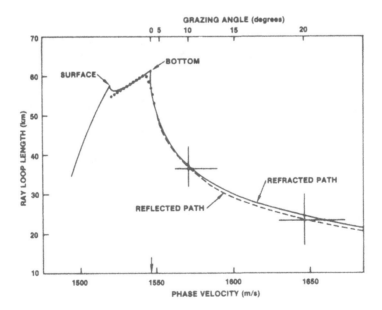

Fig. 9 Loop lengths for ray and mode theory

As can be seen, the discrepancy between the two sets of paths becomes
more severe at larger grazing angles. We suggest that beyond a grazing angle of 20 degrees a correction for this effect should be
considered.

Returning to the pre-bottom modes, Fig. 10 shows reflection
losses for an extended range of phase velocities. To the right of
the reflection point are the same reflection losses shown previously
but on a compressed scale. To the left is the loss for pre-bottom
modes. Most of these modes have no appreciable amplitude at the
bottom depth. They are trapped well up in the water channel and
their bottom loss is probably meaningless. However, for modes with
phase velocities greater than 1540 m/s bottom effects become important. Even here, though, we see modes with large negative losses,
or actual gains. It is not clear what these gains mean, but we know
that by providing them and the corresponding phase shifts, very
accurately as impedance conditions, the identical modes will be produced by the appropriate mode program. As in the first section of
this paper, we will demonstrate the effects of an approximation to
these losses and phase shifts.

The x's in Fig. 10 are points to which linear fits were made
to the 100% line. They were shifted some by a feeling that the
curve should go through zero at the bottom point. It seems apparent
now that that requirement is detrimental to the fit.

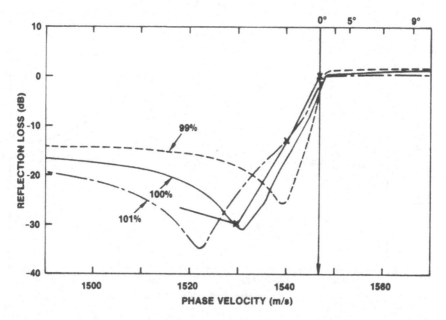

Fig. 10 Reflection loss for three sediment speeds at 25 Hz

Figure 11 shows the phase shifts or angles of the reflection coefficient. To the right of the reflection point again are the phase shifts shown previously. The x's again show linear segments fit to the 100% curve. The effect of these linear fits is shown in Fig. 12.

In Fig. 12 a portion of the ray loop length curve is again shown. Mode interference lengths are shown for the three sediment speeds. To the left of 1540 m/s they and the x's are essentially identical. The x's were determined by the impedance condition program using the linear fits to R of the last two slides. The two bad values arise from the displacement of a single mode in phase velocity since it forms a pair with the mode on either side of it. This mode has a phase velocity just less than the bottom point. As before it is the mode closest to the bottom that requires the most careful fitting of R.

Figure 13 shows the effect on the mode attenuation of the linear fits to R. Attenuation for the sediment model increases smoothly as the bottom point is approached. Our linear fit approximates this effect but again the mode nearest to the bottom point is seriously in error.

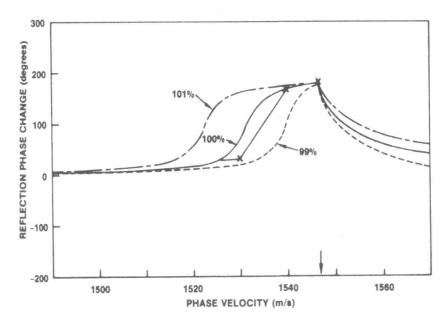

Fig. 11 Reflection phase shift for three sediment speeds at 25 Hz

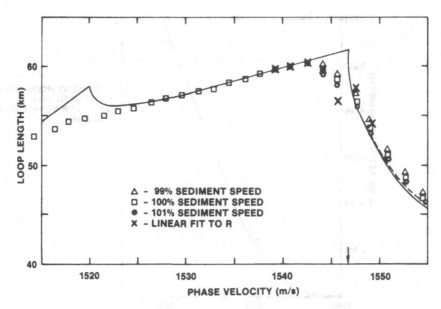

Fig. 12   Loop lengths for ray and mode theory

In the past we have arbitrarily set R to zero sound pressure at the bottom for pre-bottom modes. This gives the zero attenuation to the left of the bottom point as shown by the square points in the figure.

The above is our first attempt to use reflection tables for non-reflecting modes. The results do not indicate whether it is a viable technique or not. The fact that the losses are actually gains removes any simple physical intuition that might aid one in designing a technique. It is apparent that further investigation is necessary.

CONCLUSION

Reflection coefficients can be computed for sediment bottom models using normal mode programs. These coefficients are complex numbers giving a loss and a phase shift. They differ from observed bottom loss when multi-paths are present because they add arrivals of the same phase velocity rather than the same range. This difference is significant at grazing angles greater than 20 degrees for the example used here.

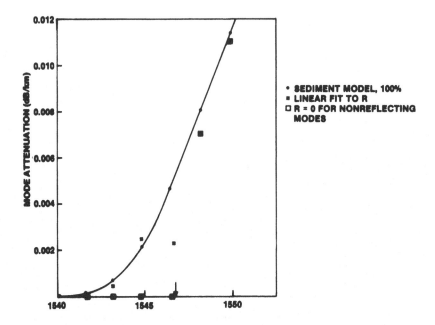

Fig. 13 Mode attenuation for eigenvalues near the bottom sound speed

A second normal mode program which requires that an impedance condition be given at the bottom will give identical eigenvalues if the reflection coefficients determined by the first program are entered as impedance conditions. However, to normalize the depth functions the derivative of the reflection coefficient is also required.

Tables of reflection losses (with phase shifts) can be used as impedance conditions for the normal mode program. However, since slopes are also required, the way that interpolation in the table is done to obtain slopes can change the size of depth functions by several percent.

Tables of reflection losses can be continued to phase velocities less than the bottom sound speed. Examples show that more realistic mode eigenvalues can result from this. However, the reflection losses themselves are actually gains and as such are not readily understood.

ACKNOWLEDGEMENT

This work was supported by the U.S. Navy Acoustic Model Evaluation Committee (AMEC).

REFERENCES

1. H. P. Bucker, "Sound propagation in a channel with lossy
    boundaries," J. Acoust. Soc. Am., 48:1187 (1970).
2. D. F. Gordon, "Underwater Sound Propagation Loss Program,
    Computation by Normal Modes for Layered Oceans and
    Sediments," NOSC Technical Report 393 (17 May 1979).
3. Hisayasu Otsubo, Kazuhiko Ohta, and Shunji Ozaki, "Normal-
    mode solution in the ocean with absorbing bottom sedi-
    ments which have a sound-speed gradient," J. Acoust.
    Soc. Jpn (E) 1:47 (1980).
4. K. M. Guthrie, "The connection between normal modes
    and rays in underwater acoustics", J. Sound &
    Vibration, 32:289 (1974).
5. E. L. Murphy and J. A. Davis, "Modified ray theory for
    bounded media," J. Acoust. Soc. Am., 56:1747 (1974).

CYCLE DISTANCE IN GUIDED PROPAGATION

D. E. Weston

Admiralty Underwater Weapons Establishment
Portland
Dorset, England

ABSTRACT

The concept of cycle distance is reviewed, drawing initially
on recent studies by C T Tindle, S G Payne and the author. Note
that in underwater acoustics the bottom loss and the cycle distance
together control an important part of the attenuation, and this is
the central theme of the work cited. For a given mode it is not
easy to find a definition of cycle distance that is both precise
and generally applicable; two new attempts are described here.
The most successful of these is in terms of group velocity; and
its implications, including those for bottom interaction, are
still being examined.

INTRODUCTION

The idea of cycle distance is well-known in underwater
acoustics, but the intention here is to look at this concept from
a number of different viewpoints, some of them novel. Because of
the time restriction and because work is still in progress we
will only attempt a brief look at five approaches, only two of
these being new. But it should be stressed that this list is
incomplete and that there is much more to be said.

The central feature of the paper is the quest, at a given
mode and a given frequency, for a definition of cycle distance
which is both precise and unique.

This question arose in the course of some recent studies[1-5] into the relationships among attenuation, bottom loss, cycle distance etc. We return to this point later, a secondary objective of the paper being to draw attention to these studies.

DERIVATION FROM RAY TRACING

When acoustic energy is trapped in a sound channel a good description of the propagation may be obtained by tracing rays. It is common to assume that the sound velocity profile does not vary with range, and that the sound velocity changes only gradually with depth. By following a ray to the point where it comes back to its initial depth, travelling in its original direction, a geometric cycle distance D_g may be directly calculated.

$$D_g = 2 \int \cot \theta \; dz = 2h \int \gamma^{-1} \; dz. \tag{1}$$

In these integrals over depth z we have θ as the grazing angle, h as the horizontal wave number and γ as the vertical wavenumber. These simple and familiar relations usually work well. But they are not exact, because the ray concept itself is not in general an exact one. For example no account is taken of the reflected energy at intermediate points along the path, which results from small velocity or density changes.

DERIVATION FROM DISPLACEMENT

The ray tracing formulae can be improved by paying attention to the turning points or reflection points at the depth extremes of the path. According to wave theory some energy will usually penetrate deeper than the nominal reflection point. The theory for an acoustic beam[6] shows that it will in effect suffer a lateral displacement Δ equal to -dφ/dh, where φ is the phase change on reflection. Thus a Δ correction term should be added to equation (1) for the lower reflection and perhaps for the upper reflection as well.

If reflection occurs at the sea bed it would be natural to define Δ for that depth. But it can be defined for any depth, and it is legitimate to take the equation (1) integral down to some lesser depth and to add on an enhanced value for Δ. In the limit the explicit integral contribution vanishes to give

$$D = \Delta_d + \Delta_u. \tag{2}$$

The terms Δ_d and Δ_u refer to the values looking down and looking up from the same horizon or very thin stratum. Choosing this horizon at the surface can make Δ_u vanish, and produce an even simpler result.

Although Δ is precisely defined it does depend on the local impedance of the material in the stratum from which it is measured, so the new result in equation (2) is not the perfect solution we are seeking.

DERIVATION FROM MODAL INTERFERENCE

From considerations of the interference between neighbouring normal modes a formula for "modal" cycle distance may be written[7]

$$D_n = -2\pi/(dh/dn). \tag{3}$$

Here n is the mode number, an integer, so that equation (3) cannot be exact.

An interesting connection with the ray-trace distance may be obtained[8,3,7] by starting with the mode eigenvalue equation, written using the WKB approximation as

$$2n\pi = 2\int \gamma dz + \phi_1 + \phi_2. \tag{4}$$

The subscripts 1 and 2 refer to the two turning or reflection points. Differentiate with respect to $-h$, and use the relation between h and γ.

$$-\frac{2\pi}{dh/dn} = 2h\int \frac{dz}{\gamma} - \frac{d\phi_1}{dh} - \frac{d\phi_2}{dh}, \tag{5}$$

or

$$D_n = D_g + \Delta_1 + \Delta_2. \tag{6}$$

This repeats in symbolic form the comments early in the previous section on the enhancement of D_g by the Δ terms. Computer plots checking this equality appear in refs. 1 and 4, and show that the predictions of equation (3) are usually but not always well-behaved.

DERIVATION FROM ATTENUATION

The work in refs. 1-5 was centred round the value for the intensity attenuation coefficient 2α, and whether or not it may be predicted by dividing the bottom loss B by the cycle distance. For our present purposes the relationship will be turned round, and we try out a definition of cycle distance as

$$D = B/2\alpha. \tag{7}$$

This approach has the attraction that α is a well-defined quantity for a given mode. Unfortunately B is not so well-defined, even if one thinks of it more generally as a loss per cycle. It is easy to choose compatible units, but one finds that the value of B depends on the stratum from which it is measured.

In addition it is really necessary to model an attenuating mode using so-called inhomogeneous plane waves[2,5], and the appropriate value of B leads to the geometric cycle distance D_g. If one uses the different B value appropriate for ordinary or homogeneous plane waves the cycle distance turns out to include the Δ correction or corrections[2-5].

Thus there are several trade-offs in the assumed values for B and D, and although equation (7) can be very useful it certainly does not define D in a unique way. Understanding of the above points is important for an overall understanding of bottom interaction.

DERIVATION FROM GROUP VELOCITY

A connection with group velocity can be established in several ways. For example start with equation (3), but consider dependence on the continuous variable "frequency" rather than the discrete variable "n". We will use a more general approach here, looking at the propagation of energy in a wave packet, and initially thinking of the propagation as one-dimensional. As the wave packet travels along, so the individual crests will slide either forward or backward relative to the packet envelope. When the slide extends to one complete wavelength the appearance of the packet will be exactly the same as at the beginning. Let us call the distance travelled the "repetition distance" L. The travel times to this distance, predicted using the phase velocity c_p or the group velocity c_g, will differ by one wave period T so that

$$\frac{L}{c_g} - \frac{L}{c_p} = T. \tag{8}$$

Note that the argument so far will work for almost any form of wave. It is also possible to put forward a more specific rays-with-phase or WKB argument to reach this result for a two-dimensional mode.

Equation (8) gives

$$L = \frac{c_p c_g T}{c_p - c_g}.$$ (9)

To find the cycle distance for higher modes it is necessary to put

$$D = nL.$$ (10)

Here at last is a well-defined cycle distance, unique for a given mode, since all the quantities in equations (9) and (10) are well defined. In view of the importance of the group velocity c_g it is perhaps surprising that the distances, related to it through equations (9) and (10), have received so little attention.

The cycle distance from equation (10) is commonly close to the other versions discussed above, but it is not in general identical. The meaning and significance will be discussed later.

CONCLUSION

The various definitions of cycle distance all have their advantages, but the last discussed is both novel and unique. It may be written in several forms, eg

$$D = \frac{n c_p c_g T}{c_p - c_g}.$$ (11)

This formula is general enough to allow application to a very wide variety of wave motions.

ACKNOWLEDGEMENT

The ideas discussed here have their roots in my earlier work at the Applied Research Laboratories of the University of Texas with C T Tindle, S G Payne and other colleagues.

REFERENCES

1.   C. T. Tindle, The equivalence of bottom loss and mode
     attenuation per cycle in underwater acoustics, J. Acoust.
     Soc. Am. 66, 250-255 (1979).
2.   D. E. Weston and C. T. Tindle, Reflection loss and mode
     attenuation in a Pekeris model, J. Acoust. Soc. Am. 66,
     872-879 (1979).
3.   C. T. Tindle and D. E. Weston, Connection of acoustic beam
     displacement, cycle distances, and attenuations for rays
     and normal modes, submitted to the Journal of the
     Acoustical Society of America.
4.   C. T. Tindle, D. E. Weston and S. G. Payne, Cycle distances
     and attenuation in shallow water, submitted to the Journal
     of the Acoustical Society of America.
5.   D. E. Weston, Oblique reflection of inhomogeneous acoustic
     waves, submitted to the Journal of the Acoustical Society
     of America.
6.   L. M. Brekhovskikh, "Waves in Layered Media", Academic,
     New York (1960).
7.   I. Tolstoy and C. S. Clay, "Ocean Acoustics", McGraw Hill,
     New York (1966).
8.   E. L. Murphy and J. A. Davis, Modified ray theory for bounded
     media, J. Acoust. Soc. Am. 56, 1747-1760 (1974).

COMPUTATION OF AVERAGED SOUND-PROPAGATION LOSSES AND FREQUENCY/ SPACE COHERENCE FUNCTIONS IN SHALLOW WATERS

Robert Laval and Yvon Labasque

Société d'Etudes et Conseils AERO

3, avenue de l'Opéra, Paris, France

ABSTRACT

In the frequency range covered by active sonars (above 1 kHz) sound propagation in shallow waters is characterized by fast fluctuations of the propagation-loss term as a function of range, source depth, receiver depth, and frequency. These fluctuations may be interpreted as the result of interferences between a very large number of modes. A method is presented that allows the propagation losses to be decomposed into two parts:

    1. An averaged propagation-loss term, which takes the form of a slowly-varying function of range, depth and frequency.
    2. A fluctuation term, which will be assimilated to a random function and will be characterized by its coherence functions in the range, depth and frequency domains.

Assuming that the wavelength is much smaller and the horizontal range much larger than the water depth, some approximations may be introduced that allow the above functions to be expressed by continuous integrals; these integrals can be solved numerically through a rather simple computer program. The method is illustrated by some applications to a number of realistic cases in order to show the relative influence of the various parameters characterizing the bottom and the sound-velocity profile.

1.   INTRODUCTION

In the frequency range covered by active sonars (above 1 kHz)
sound propagation in shallow waters is characterized by fast fluctuations of the propagation-loss term as a function of range,
source depth, receiver depth and frequency. These fluctuations may
be interpreted as the result of interferences between a very large
number of modes, or a very large number of rays. Experimental
studies of these fluctuations have shown that they were following
a Rayleigh amplitude distribution, which does agree with the hypothesis of a gaussian process resulting from an addition of a large
number of incoherent vectors.

Under the above assumptions, a logical way of approaching the
shallow water sound propagation study would consist in splitting
the problem into two distinct parts:

•The "macroscopic" study of the mean energetic propagation
losses, which should be characterized by slowly varying functions
of range, depth and frequency.

•The "microstructure" study, which should take the form
of a statistical description of the fast fluctuations of the sound
field in space, in frequency and possibly in time.

The experimental studies of sound propagation in shallow water
are generally conducted accordingly to this principle.

On the other hand, the propagation models currently available
are not conceived in this way.

Models based on mode theory are organised in order to compute
the propagation losses as a function of range, for a given frequency and a given source depth. This range profile is generally
computed with all the details corresponding to the fluctuations
along the longitudinal axis; the computation has to be started
again for each new frequency value, and/or each new combination of
source and receiver depth.

Ray models become quite heavy in case of a very large number
of eigenrays, and the propagation losses computations are complicated by a myriad of caustics.

A method for computing directly the averaged propagation
losses and the various coherence functions of the microstructure
is presented here.

2.   PRINCIPLE OF A "PSEUDO-STOCHASTIC" MODELLING

The sound propagation properties can be characterised in the
most general way by a transfer function:

$$\Xi(s,r,f,t)$$

which is a complex function describing the amplitude and phase of
the harmonic sound pressure received at a point r at the time t
when a monochromatic wave of level unity is transmitted by a source
situated at point s. If Ξ is computed by a propagation model

assuming a time invariant horizontally stratified medium, it will
be a function of four variables only:

$$\Xi(x,z_s,z_r,f)$$

where x is the horizontal distance between s and r, z_s the source
depth and z_r the receiver depth.

If Ξ exhibits very fast fluctuations as a function of the four
above variables, one can try to split it as a product of two functions:

$$\Xi(x,z_s,z_r,f) = H_D(x,z_s,z_r,f) \; \widetilde{H}(x,z_s,z_r,f)$$

where x is a slowly varying envelope function of x, z_s, z_r, f, and
$\widetilde{H}$ is a much faster varying function of the same variables.

The possibility of such a decomposition of Ξ into a product of
two functions, the variation scales of which are very different is
at the basis of the method presented here. Its main field of application can be found in shallow waters in the frequency domain above
0.5 to 1 kHz where the number of modes is very high, and for propagation ranges greater than 1 or 2 km, where the sound field has no
more a "coherent structure".

The functions H_D and $\widetilde{H}$ can be normalised in such a way that
$|H_D|^2$ represents the "mean propagation losses", which is the mean
value of $|\Xi|^2$. The mean value of $|\widetilde{H}|^2$ will then be equal to 1.

Ξ being a deterministic function, the concept of mean value
should be interpreted as the result of an averaging within a limited domain of the x, z_s, z_r, f space (in other words a "smoothing"
operation). In the foregoing reasoning it will be easier to consider $\widetilde{H}$ as a particular realisation of a stochastic process (we shall
call it a "pseudo-stochastic process") and to replace the concept
of space-frequency averaging by the one of ensemble averaging.

We shall then write:

$$|H_D|^2 = E\{|\Xi|^2\} \quad \text{and} \quad E\{|\widetilde{H}|^2\} = 1$$

H_D will then be a deterministic function representing the macroscopic properties of the transmission channel.

$\widetilde{H}$ will be considered as a random function representing the
"microscopic" structure of the channel.

The distribution of $\widetilde{H}$ should first be described. It may be
admitted for instance, that $\widetilde{H}$ is a complex gaussian random variable
with zero mean; its amplitude distribution is then a Rayleigh one,
and its phase is equidistributed.

The second order moments should then be given, in the form
of coherence functions of space and frequency:

$$\Gamma(\delta x) = E\{\tilde{H}(x,z_s,z_r,f)\ \tilde{H}^*(x+\delta x,z_s,z_r,f)\}$$

$$\Gamma(\delta z_s) = E\{\tilde{H}(x,z_s,z_r,f)\ \tilde{H}^*(x,z_s+\delta z_s,z_r,f)\}$$

$$\Gamma(\delta z_r) = E\{\tilde{H}(x,z_s,z_r,f)\ \tilde{H}^*(x,z_s,z_r+\delta z_r,f)\}$$

$$\Gamma(\delta f) = E\{\tilde{H}(x,z_s,z_r,f)\ \tilde{H}^*(x,z_s,z_r,f+\delta f)\}$$

The coherence functions which have been defined for the differential variables δx, δz_s, δz_r and δf are remaining identical to themselves only within a limited domain of the x, z_s, z_r, f space; in other words $\tilde{H}$ is not a stationary random function. However, $\tilde{H}$ may be considered as "locally stationary" with respect to a given variable, say x, provided the coherence function $\Gamma(\delta x)$ remains practically identical to itself within an interval of x much larger than the "effective correlation distance" in the x direction (the "effective width" of $\Gamma(\delta x)$).

When the local stationarity conditions are verified for all the variables, the Γ functions can be Fourier transformed with respect to the differential terms, which gives corresponding "scattering functions"[1,2].

For the spatial variables they will be:

$$R(u) = F.T[\Gamma(\delta x)] \quad \text{and} \quad R(v) = F.T[\Gamma(\delta z_r)]$$

The new variables u and v are the "spatial frequencies" along the longitudinal horizontal direction x, and the vertical direction z_r, R(u) and R(v) are the power spectra of the spatial frequencies along the same directions.

If the wave received in the vicinity of the receiver can be decomposed into elementary real plane waves, θ_r being the direction of propagation of one of these with respect to the horizontal, u and v can be written:

$$u = \frac{f}{c_r}\cos\theta_r = 1/\lambda_r \cdot \cos\theta_r, \quad v = \frac{f}{c_r}\sin\theta_r = 1/\lambda_r \cdot \sin\theta_r$$

where c_r is the sound propagation velocity and λ_r the wavelength at the receiver.

If $D(\theta_r)$ is the "angular energy density" of the received wave:

$$R(u) = \left|\frac{d\theta}{du}\right| D(\theta_r) = \frac{\lambda_r}{\sin\theta_r} D(\theta_r) \quad \text{with} \quad \theta_r = \text{arc } \cos(\lambda_r u)$$

$$R(v) = \left|\frac{d\theta}{dv}\right| D(\theta_r) = \frac{\lambda_r}{\cos\theta_r} D(\theta_r) \quad \text{with} \quad \theta_r = \text{arc } \sin(\lambda_r v)$$

If $D(\theta_r)$ is known, $R(u)$ and $R(v)$ can be computed, and their Fourier transform will give $\Gamma(\delta x)$ and $\Gamma(\delta z)$.

This computation is simplified if the small angles approximation can be used (which is always the case in shallow waters applications). Then:

$$R(u) = \frac{1}{\lambda_r \theta_r} D(\theta_r) \quad \text{with} \quad u = 1/\lambda_r (1 - \frac{\theta_r^2}{2})$$

and its Fourier transform becomes:

$$\Gamma(\delta x) = \int D(\theta_r) e^{-2\pi i(1 - \frac{\theta^2}{2}) \frac{\delta x}{\lambda}} .d\theta$$

$$= e^{-2\pi i \frac{\delta x}{\lambda}} \int D(\theta) e^{\pi i \theta^2 \frac{\delta x}{\lambda}} .d\theta$$

The harmonic term $e^{-2\pi i \frac{\delta x}{\lambda}}$ results from the main propagation effect in the x direction. It is the envelope function:

$$\Gamma'(\delta x/\lambda) = \int D(\theta) e^{\pi i \theta^2 \frac{\delta x}{\lambda}} .d\theta$$

which effectively represents the coherence function in the x direction. If $D(\theta)$ is independent of frequency, then $\Gamma'(\delta x/\lambda)$ is also independent.

The Fourier transform $\Gamma(\delta z_r)$ of:

$$R(v) = \lambda D(\theta_r) \quad \text{with} \quad v = \lambda \theta_r$$

can be written:

$$\Gamma(\frac{\delta z_r}{\lambda}) = \int D(\theta) e^{-2\pi i \theta_r \frac{\delta z}{\lambda_r}} .d\theta$$

Expressed as a function of $\delta z_r/\lambda$ it is the Fourier transform if $D(\theta)$.

Generalized coherence or scattering functions depending upon several variables[3,4] could also be introduced, such as the function:

$$\Gamma(\delta z, \delta f) = FT[R(v, \tau)]$$

which has to be used to evaluate the performances of a space-time processor associated with a vertical array, or:

$$\Gamma(\delta x, \delta t) = FT[R(u, \tau)]$$

applying to an "end-fire" horizontal array.

3   COMPUTATION METHODS DERIVED FROM MODE THEORY

The classical expression of normal modes theory[5,6] is recalled.

$$p(f,x,z_s,z_r) = \rho_0^2\, f\sqrt{\frac{\pi}{2x}} \sum_{n=1}^{N} \frac{U_n(z_s,f)U_n(z_r,f)}{\sqrt{K_n}}\, e^{\,i(K_n x - 2\pi f t) - \frac{\alpha_n}{2}x}$$

where:

$p(f,x,z_s,z_r)$ is the pressure received at range x and depth z_r by a so called "unity source" of frequency f and depth z_s.

α_n is the attenuation (in Neper/m) associated with the mode n which depends upon both the water and the bottom properties.

K_n is the wave number associated with the mode n.

$U_n(z,f)$ is the "profile" of the mode number n for the frequency f. It is a pseudo-periodic function of z, with n zero-crossings between surface and bottom.

The pressure p_0 created at 1 m by the so called "unity source" is:

$$p_0(f) = \rho_0^2\, \frac{f}{2}$$

The acoustic pressure can finally be written in the form:

$$p(f,x,z_s,z_r) = p_0(f)\, \Xi(f,x,z_s,z_r) e^{2\pi i f t}$$

where Ξ is the "transfer function":

$$\Xi(f,x,z_s,z_r) = \sqrt{\frac{2\pi}{x}} \sum_{n=1}^{N} \frac{U_n(z_s,f)U_n(z_r,f)}{\sqrt{K_n}}\, e^{\,iK_n x - \frac{\alpha_n}{2}x}$$

If the total number N of the modes giving an effective ener-getic contribution is large enough, and if the ($K_n x$) terms, repre-senting the phases of the different modes, are distributed on a much larger interval than 2π, then Ξ effectively fluctuates rapidly as a function of x and f (and generally also of z). The conditions which have been defined in Ch. 1 are then verified, and it should be possible to calculate the mean propagation loss and the various coherence functions which have been defined.

A first step towards the calculation of the mean energy con-sists in making an incoherent addition of the energies of the dif-ferent modes, which gives a propagation loss term H^2:

$$H^2 = \frac{2\pi}{x} \sum_{n=1}^{N} \frac{U_n^2(z_s,f)U_n^2(z_r,f)}{K_n}\, e^{-\alpha_n x}$$

or:

$$H^2 = \frac{1}{fx} \sum_{n=1}^{N} \nu_n(f) U_n^2(z_s,f) U_n^2(z_r,f) e^{-\alpha_n x}$$

if we introduce the phase velocity $\nu_n(f) = 2\pi/K_n f$.

In order to go further, we shall express the mode profile functions $U_n(z,f)$ in the form of a pseudo periodic function of z modulated in amplitude and phase:

$$U_n(z,f) = W_n(z,f) \sin[2\pi\xi_n(z,f)z + \phi_n]$$

where:

$W_n(z,f)$ represents the amplitude and $\xi_n(z,f)$ the local frequency of the U_n profile.

W_n and ξ_n are varying much more slowly than U_n as a function of z.

The profile $U_n(z,f)$ may be considered as the result of an interference between two wave fronts propagating in the x direction with the phase velocity $\nu_n(f)$. These wavefronts would deviate from the vertical plane of a distance $V_n(z,f)$ for the first one and $-V_n(z,f)$ for the second.

$W_n(z,f)$ represent the amplitude of this wave then:

$$U_n(z,f) = W_n(z,f) \sin\left[2\pi f \frac{2V_n(z,f)}{\nu_n(t)} + \phi_0\right] = W_n(z,f) \sin[\Phi_n(z,f)]$$

$\Phi_n(z,f)$ is the local phase of the pseudo-sinusoid in the z direction.

V_n and W_n are here functions of 3 variables: the continuous variables z and f and the distontinuous one, which is a series of integer numbers. They can also be expressed as a function of the three continuous variables z, f, and the phase velocity ν:

$$V(z,\nu,f) \quad \text{and} \quad W(z,\nu,f)$$

then:

$$V_n(z,f) = V[z,\nu_n(f),f] \quad \text{and} \quad W_n(z,f) = W[z,\nu_n(f),f]$$

Expressing V and W as a function of ν is particularly interesting in the high frequency domain where ray-theory is valid (i.e. in the domain where diffraction effects can be neglected); V and W are then independent of the frequency and function of z and ν only. It may be considered, indeed, that the two elementary wave fronts with phase velocity ν and amplitude W, the angle with the vertical of which being $\pm\theta(z)$, are perpendicular in each point with the upgoing and the downgoing ray, the phase velocity of which is ν. It is well known that in a horizontally stratified medium the horizontal component of the phase velocity along a given ray is constant, and equal to the sound velocity at the depth where the ray is horizontal.

The wave front profile perpendicular in each point with the ray having a phase velocity ν is given by:

$$V(z,\nu) = \int_{z_{min}}^{z_{max}} tg[\theta(z,\nu)]dz$$

where:

θ is the angle of the wave front with the vertical.

z_{min} is the minimum depth of the corresponding ray, which is 0 if the ray is reflected by the surface, or the depth of its upper vertex in the opposite case (then $\theta(z_{min}) = 0$).

z_{max} is the maximum depth, either the depth H of the bottom if the ray is reflected by the bottom, or the depth of the lower vertex in the opposite case (then $\theta(z_{max}) = 0$).

On the other hand, it can be shown that W is also independent of f, and is equal to:

$$W^2(z,\nu) = \frac{4\sqrt{2}}{L(\nu)\sin[\theta(z,\nu)]}$$

where $L(\nu)$ is the length of a total arch of the ray of phase velocity (ν), i.e. the distance between two surface reflections or two upper vertexes. This arch length is equal to:

$$L(\nu) = 2 \int_{z_{min}}^{z_{max}} cotg\theta \, dz$$

The profile $U_n(z,f)$ can then be expressed as a function of the continuous functions $V(z,\nu)$ and $W(z,\nu)$ if the phase velocity $\nu_n(f)$ corresponding to the mode n at the frequency f is known.

$V(z,\nu)$ and $W(z,\nu)$ are only depending upon the sound velocity profile $c(z)$ and they are easy to compute.

For a given frequency f the $\nu_n(f)$ terms does correspond to the eigenvalues of ν satisfying the boundary conditions of the profile U:

$$U(z,\nu) = W(z,\nu)\sin\left[4\pi f \frac{V(z,\nu)}{\nu} + \phi_o\right] = W(z,\nu)\sin\Phi(f,z,\nu)$$

for the depth z_{min} and for the depth z_{max}.

• For z_{min}, the condition is relative to the phase ϕ_0:

$\phi_0 = \pi$ if $v \geqslant c(0)$ (sound velocity at the surface) in which
case the corresponding ray is surface
reflected .

$\phi_0 = \pi/2$ if $v < c(0)$ in which case the corresponding ray is
refracted in its upper part.

• For z_{max}, the conditions does apply to the phase Φ_m:

$$\Phi_m = \frac{4\pi f \ V(z_{max}, v)}{v} + \phi_0$$

must be equal to:

$\Phi_m = \pi/2$ if $v < c(H)$ (sound velocity at the bottom).

$\Phi_m = \phi_H(f, \theta)$ if $v \geqslant c(H)$ in which case the corresponding ray
is bottom reflected.

$\phi_H(f, \theta_H)$ is the phase term of the complex bottom reflection
coefficient $R(f, \theta_H)$ for the frequency f and the grazing angle θ_H,
which is related to v by the SNELL's law:

$$\cos\theta_H = c(H)/v$$

The boundary conditions of U will be satisfied for the values
of v verifying the following equation for the integer values of n:

$$v(f) = \frac{4f \ V(z_{max}, v)}{n - n'} \quad \text{where} \quad n' = \frac{1}{\pi} (\phi_m - \phi_0)$$

ϕ_0 and ϕ_m depending of v.

The attenuation coefficients $\alpha_n(f)$ can also be expressed as a
continuous function of v. In fact $\alpha(f, v)$ is a sum of two independent coefficients:

$$\alpha(f, v) = \alpha(f) + \beta (f, v)$$

$\alpha_0(f)$ is the attenuation coefficient in the water. We shall
admit that it is independent of mode number n, or the phase velocity v, and depends only upon the frequency.

$\beta_n(f)$ is the attenuation coefficient related to the losses in
the bottom. Within the ray theory approximation, its influence is
limited to the domain of v corresponding to the rays which are
touching the bottom. For such rays the averaged bottom losses are
equal to the bottom reflection loss $|R(f, \theta_H)|^2$ divided by the arch
length $L(v)$. It will then be admitted that:

$$\beta(f, v) = \frac{Ln[|R(f, \theta_H)|^2]}{L(v)}$$

We can then write:

$$e^{-\alpha x} = e^{-\alpha_o x} [|R|^2]^{-x/L} \quad \text{if} \quad \nu \geqslant c(H)$$

$$e^{-\alpha x} = e^{-\alpha_o x} \quad \text{if} \quad \nu < c(H)$$

The following step in the process of simplification is to replace the discontinuous summation on the integer values of n by a continuous integral of n, which should then be considered now as continuous variable:

$$H^2 = \frac{1}{fx} \int_o^N \nu W^2(z_s,\nu) W^2(z_r,\nu) \sin^2\left[\frac{4\pi f \ V(z_s,\nu)}{\nu} + \phi_o\right]$$

$$\cdot \sin^2\left[\frac{4\pi f \ V(z_r,\nu)}{\nu} + \phi_o\right] e^{-\alpha_o x} |R|^{2x/L} \, dn$$

The variable ν is now expressed as a continuous function of n and f. H^2 can be expressed in function of ν by changing the variable:

$$n = \frac{4fV(z_{max},\nu)}{\nu} + n'(f,v) \quad \text{where } n' = \frac{1}{\pi}(\phi_m - \phi_o)$$

then:

$$dn = \left[4f \frac{\partial(\frac{V_m}{\nu})}{\partial \nu} + \frac{1}{\pi} \frac{\partial \phi_m}{\partial \nu}\right] d\nu$$

with:

$$\frac{\partial(\frac{V_m}{\nu})}{\partial \nu} = \frac{\partial}{\partial \nu}\left[\int_{z_{min}}^{z_{max}} \frac{tg\theta}{\nu} \, d\theta\right]$$

and:

$$\cos\theta = \frac{c(z)}{\nu}$$

which gives:

$$\frac{\partial(\frac{V_m}{\nu})}{\partial \nu} = \frac{1}{4\nu^2} \int_{z_{min}}^{z_{max}} \cot g\theta \, dz = \frac{L(\nu)}{8\nu^2}$$

where $L(\nu)$ is the arch length.

Then:

$$H^2(f,x,z_s,z_r) = \frac{1}{fx} \int_{\nu_{min}}^{\nu_{max}} \nu W^2(z_s,\nu) W^2(z_r,\nu) \sin^2\left[\frac{4\pi fV(z_s,\nu)}{\nu} + \phi_o\right]$$

$$\times \sin^2\left[\frac{4\pi fV(z_r,\nu)}{\nu} + \phi_o\right] \times \left[\frac{fL(\nu)}{2\nu^2} + \frac{1}{\pi}\frac{\partial \phi_m}{\partial \nu}\right] e^{-\alpha x} \, d\nu.$$

It can be noted that, for $\nu < c(H)$ $\phi_m = \frac{\pi}{2}$ and $\frac{\partial \phi_m}{\partial \nu} = 0$. Furthermore: $\alpha(\nu,f) = \alpha_o(f)$

The integral can be decomposed into a sum of three distinct integrals, and H^2 may be written:

$$H^2 = H_o^2 + H_B^2 + H_\phi^2$$

where:

H_o^2 results from the computation of the integral between the limits ν_{min} and $c(H)$, corresponding to the phase velocities of the rays which are not touching the bottom.

H_B^2 results from the computation of the integral between the limits $c(H)$ and ν_{max} corresponding to the phase velocities of the bottom reflected rays, by keeping $fL(\nu)/2\nu^2$ in the last bracket, and dropping the $\frac{1}{\pi}\frac{\partial \phi_m}{\partial \nu}$ term.

H_ϕ^2 results from the computation of the integral between the same limits as for H_B^2, by keeping the term $\frac{1}{\pi}\frac{\partial \phi_m}{\partial \nu}$ only.

The above integrals can also be epxressed as a function of θ_r which is the angle at the receiver of the ray which has ν for phase velocity. Then:

$$\nu = \frac{c(z_r)}{\cos\theta_r} \quad \text{and} \quad \frac{\partial\nu}{\partial\theta} = \nu \, tg\theta_r$$

All computation being done, H^2 will take the following form:

$$H^2 = 2 \int_{\theta_{min}}^{\theta_{lim}} D_o(\theta_r)d\theta_r + 2 \int_{\theta_{lim}}^{\theta_{max}}[D_B(\theta_r) + D_\phi(\theta_r)]d\theta_r$$

The functions D above being symmetric in θ_r, the integrals are only taken for the positive values of θ_r; θ_{lim} is the angle at the depth z_r of the ray tangent to the bottom.

With some reserve (which are related with the stationarity conditions of the process) D_o may be considered as the "angular energy density" within the angle interval of the rays which are not touching the bottom, and $[D_B(\theta_r) + D_\phi(\theta_r)]$ as the angular energy density of the bottom reflected rays.

The various functions D may be explicited in the form:

$$D_o(\theta_r) = \frac{2}{x} e^{-\alpha_o x} \frac{1}{L(\nu)|tg\theta_s|} \cdot 2\sin^2\phi_s \cdot 2\sin^2\phi_r$$

$$D_B(\theta_r) = \frac{1}{2x} e^{-\alpha_o x} |R(f,\theta_H)|^{-\frac{2x}{L(\nu)}} \frac{1}{L(\nu)|tg\theta_s|} \cdot 2\sin^2\phi_s \cdot 2\sin^2\phi_r$$

$$D_\phi(\theta_r) = \frac{1}{\pi f x} \; e^{-\alpha_o x} \; |R(f,\theta_H)|^{x/L(\nu)} \; \frac{1}{L^2(\nu)|tg\theta_s||tg\theta_r|}$$

$$\cdot \; \nu \; \frac{\partial \phi_H(f,\theta_H)}{\partial \theta_H} \; 2\sin^2\phi_s \cdot 2\sin^2\phi_r$$

where:

$$\phi_s = \frac{4\pi f V(\nu,z_s)}{\nu} + \phi_o$$

$$\phi_r = \frac{4\pi f V(\nu,z_r)}{\nu} + \phi_o$$

All the terms of the above expressions can be expressed as a function of θ_r, from the knowledge of the sound velocity profile $c(z)$ and the complex bottom reflection coefficient $R(f,\theta_H)$. This last term can be computed itself from the elastic properties of the bottom.

The numerical computation of the three D functions can be programmed on a computer. From the functions D, it is possible to compute:

• The expression of the mean propagation-losses H^2 by integration over θ_r.

• The expression of the vertical coherence by a Fourier transformation:

$$\Gamma(\frac{\delta z}{\lambda}) = \frac{1}{H^2} \int D(\theta) e^{-2\pi i\theta \frac{\delta z}{\lambda}} \; d\theta$$

• The expression of the horizontal coherence, by applying the transformation:

$$\Gamma(\frac{\delta x}{\lambda}) = \frac{1}{H^2} \int D(\theta) e^{-\pi i\theta^2 \frac{\delta x}{\lambda}} \; d\theta$$

It should be noted that the terms $\sin^2\phi$ oscillate very rapidly as a function of θ_r and θ_s, except when the source and/or the receiver are in the immediate vicinity of the boundaries (surface or bottom). Except for this last case, the terms in $2\sin^2\phi$ (the mean value of which is 1) can then be dropped out.

Without the $\sin^2\phi$ terms, the above expressions for $D_o(\theta_r)$ and $D_B(\theta_r)$ are identical as the ones obtained by SMITH, after a different type of approach.

The contribution of the $D_\phi(\theta_r)$ term is very small in the case of a simple half space liquid bottom. It may not be negligible in case of a complicated multilayered bottom, where the phase of the reflection coefficient varies very fast with θ_H.

4   APPLICATION OF THE METHOD

A simple computer program has been written for the computation of the various functions which have just been defined.

Some examples of application are now presented in the following context:

    •The computations do not take into account the $\sin^2\phi$ terms. The results are then not valid in the immediate vicinity of the surface.

    •The water depth has been fixed to 100 m and the frequency to 3.5 kHz. With the hypothesis which have been adopted for the bottom, however, only the attenuation in the water is frequency dependent; the results can then be easily transposed to another frequency.

    •Two velocity profiles have been chosen:

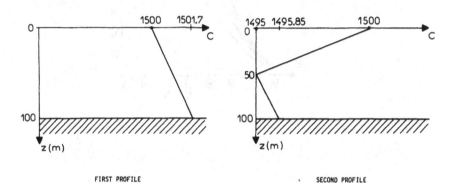

FIRST PROFILE   SECOND PROFILE

    •The bottom is represented by a single liquid layer. Two different characteristics have been selected from [6] and [9]:

	Bottom 1 Coarse sand	Bottom 2 Silty sand
Bottom density / Water density	2.03	1.83
Bottom velocity / Water velocity	1.20	1.10
Losses dB/λ	1.03 dB/λ	1.07 dB/λ

Four cases can be defined by combining the above characteristics:

    Case 1 (profile 1, bottom 1)   Case 3 (profile 2, bottom 1)
    Case 2 (profile 1, bottom 2)   Case 4 (profile 2, bottom 2)

For each of them, the following curves have been traced:

    •Angular energy density in dB at the receiver for a source depth of 25 m, a receiver depth of 50 m and a range of 50 km

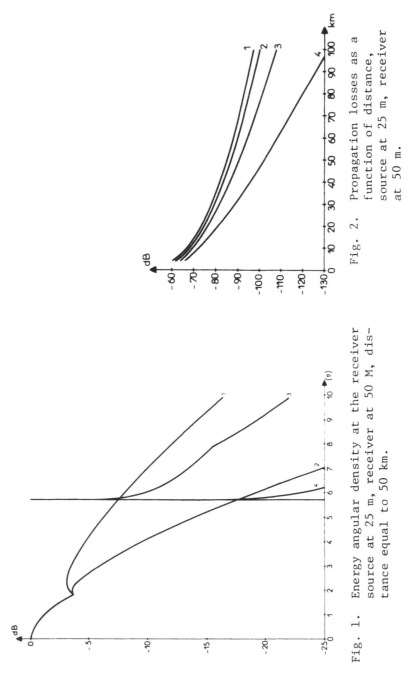

Fig. 2. Propagation losses as a
         function of distance,
         source at 25 m, receiver
         at 50 m.

Fig. 1. Energy angular density at the receiver
         source at 25 m, receiver at 50 M, dis-
         tance equal to 50 km.

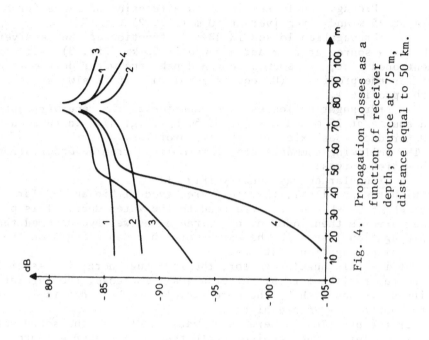

Fig. 4. Propagation losses as a function of receiver depth, source at 75 m, distance equal to 50 km.

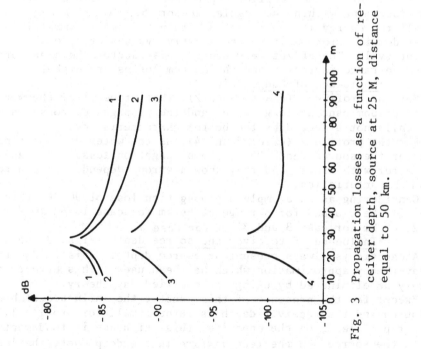

Fig. 3 Propagation losses as a function of receiver depth, source at 25 M, distance equal to 50 km.

(Fig. 1). The curves are normalized to 0 dB for the maximum value.

•Propagation losses in dB as a function of range for a source at 25 m and a receiver at 50 m (Fig. 2).

•Propagation losses in dB as a function of the receiver depth for a source at 25 m and a range of 50 km (Fig. 3). It should be mentioned that, for such a source depth, the water duct does not exist for the profile 2 (H_0^2 equals zero) while it is always present for the profile 1.

•Propagation losses as a function of the receiver depth for a source at 75 m and a range of 50 km (Fig. 4). In this case, the water duct does exist for the two profiles.

The following comments can be formulated from an observation of these few curves:

•Angular energy density (Fig. 1)

With the profile 1, the energy received within an angular sector -1.9° to +1.9° does correspond to the water duct. This part of the curve is then independent of range, as we have assumed that the propagation losses in the water were only depending upon the horizontal propagation distance.

Outside this angular sector, the incoming energy is associated with bottom reflected rays. After a small increase, the angular density decreases with θ_r the more rapidly as the range is longer and the bottom losses are higher.

For the profile 2, there is no water duct. As the sound velocity is smaller at the receiver depth than at the source depth, no energy is coming within the angular sector -5.7° to +5.7°.

All the energy is concentrated within two small symmetric angular domains for positive and negative values of θ_r outside this interval. The effective width of these sectors becomes narrower as the range increases and the bottom losses are higher.

•Influence of Range (Fig. 2)

For the profile 1 (cases 1 and 2), the influence of the water duct is predominant at long range, and the propagation losses are practically not affected by the bottom characteristics.

For the profile 2 (cases 3 and 4), as the water duct does not exist for the source depth 25 m, the propagation losses decrease much faster with range, and they show a strong dependence upon the bottom characteristics.

Considering as an example a propagation loss of 90 dB, this value will be reached for a range of 66 km for case 1, 62 km for case 2, 49 km for case 3, and 31 km for Case 4.

•Influence of receiver and source depth (Figs. 3 and 4)

A caustic is always present at source depth. This is due to the ray-theory approximation which has been used. This effect could probably be eliminated by using the modified ray theory.

Except for the presence of this caustic the influence of both the source and the receiver depth is rather small for profile 1.

For profile 2, on the opposite, this influence is fundamental. If both the source and the receiver are in the deep duct, the losses are practically the same as for the profiles 1, and they are not much affected by the bottom characteristics.

If the source, the receiver or both are in the upper part, the propagation losses are considerably higher, and they are essentially depending upon the bottom characteristics.

The above results are in good agreement with the orders of magnitude which can be obtained by using normal-modes theory or ray-tracing models.

REFERENCES

1. R. Laval. Time Frequency-space Generalised Coherence and Scattering Functions. NATO Advanced Study Institute on Signal Processing, Portovenere, 1976, D. Reidel Publishing Company.
2. R. Laval. Cohérence spatio-temporelle et fonction de diffusion généralisée. 6ème Colloque GRETSI, avril 1977, conférence n° 8.
3. R. Laval. Dégradation du gain associée aux incohérences de la propagation ou aux déformations aléatoires d'une antenne linéaire. 7ème Colloque GRETSI, juin 1979, conférence n° 81.
4. R. Laval and Y. Labasque. Medium Inhomogeneities and Instabilities: Effects on Spatial and Temporal Processing. Nato Advanced Study Institute on Signal Processing. Kollekolle, August 1980 (in preparation). To be published by D. Reidel Publishing Company.
5. S.N. Wolf. Measurements of Normal-mode Amplitude Functions in Nearly Stratified Medium. Oceanic Acoustic Modelling. Proceeding of a Conference at SACLANTCEN, 8-11 September 1975, n° 17: Field Calculations. SACLANTCEN, 15 October 1975, pp. 34-1 to 34-9.
6. W.A. Kuperman, F. Ingenito, Contributions to Propagation Loss in Shallow Water. Oceanic Acoustic Modelling. Proceedings of a Conference at SACLANTCEN, 8-11 September 1975, n° 17, Pt 4: Sea Bottom. SACLANTCEN, 15 October 1975, pp. 15-1 to 15-10.
7. L.M. Brekhovskikh. Waves in Layered Media. Academic Press, 1960, 561 pp.
8. P.W. Smith, Jr. Spatial Coherence in Multipath or Multimodal Channels. J. Acoust. Soc. Am., Vol. 60, n°2, August 1976.
9. E.L. Hamilton. Acoustic Properties of Sea Floor: A Review. Oceanic Acoustic Modelling. Proceedings of a Conference at SACLANTCEN, 8-11 September 1975, n° 17, Pt 4: Sea Bottom. SACLANTCEN, 15 October 1975, pp. 18-1 to 18-96.

INITIAL DATA FOR THE PARABOLIC EQUATION

David H. Wood and John S. Papadakis

Naval Underwater Systems Center, New London Laboratory
Code 3122
New London, CT 06320, U.S.A.

ABSTRACT

When the Helmholtz equation is replaced by the related parabolic equation, we obtain both a profit and a puzzle. The profit is the relative numerical ease of solving the parabolic equation. In this paper then consider the puzzle: What do we use for the required initial data for the parabolic equation? We adopt the position that the solution of the parabolic equation should correspond at long ranges to the solution of the Helmholtz equation at least in the case when the environment does not change with range. By considering the solution of the parabolic equation that reduces to the modal decomposition of arbitrary data at the initial range, we recommend that $\Sigma\phi_n(z_o)\phi_n(z)$ be used for initial data. Here the ϕ_n are the mode functions, z is the variable depth at the initial range, and z_o is the source depth. The fewer terms in above sum, the less variation in the solution of the parabolic equation, and therefore the greater the profit from numerical ease. This would normally suggest including only those modes whose effects can be accurately computed. If, on the other hand, phase errors are to be corrected by a technique such as Polyanskii's it does no harm to include all modes with the weights given in the above sum.

The idea behind the parabolic equation method is that one solves a parabolic equation instead of solving the Helmholtz equation. This solution is then an approximation to the sound field. The profit is that the parabolic equation is much simpler to solve numerically than the Helmholtz equation. However, to solve the parabolic equation, we have to start with two kinds of information: 1) we have to specify the environment to be modeled, and 2) we have to give initial data over depth at a fixed range. But what is this initial data to be?

There are two basic schools of thought on this question. One school says that the solution of the parabolic equation is to be an approximation to the sound field, so the proper data is a sample of the sound field over depth at a fixed range. This is usually calculated using a normal mode program. The other school says that what we are seeking is the impulse response, or Green's function, so the correct data is a delta function at the source point. This is usually simulated by a sharp Gaussian pulse.

This concerns me. When you see calculations from a parabolic equation model in the literature, the author seldom says what he used for initial data. Different data will, of course, give different answers. The sample of the sound field will tend to be highly oscillatory, the Gaussian data will be smooth. How could they lead to the same solution? It does no good to protest that we are only using two different approaches to the same physical problem and will therefore get the same answer. All the physics of the environment are specified before the initial data. The data must be redundant. It is also futile to object that one merely seeks a reasonable approximation to the field and that either approach is approximately consistent with the physics to be modeled. If we always use the wrong data, we may bias the model. This is not a reasonable approximation.

What data we choose depends on what answer we seek. We adopt the position that the solution of the parabolic equation should correspond at long ranges to the solution of the Helmholtz equation at least in the case when the environment does not change with range. In such an environment, both the parabolic equation method and the method of normal modes apply. I want to now compare their results. To do this, I first form an expansion of the initial data in terms of the depth dependent normal modes,

$$P(0,z) = \Sigma \ a_n \ \phi_n(z) \ .$$

This is analogous to expanding the initial data in a Fourier series. In fact, for a uniform medium, it happens to be a Fourier series. Once we have the above expansion, it is easy to write down the form of the solution of the parabolic equation:

$$P(r,z) = \Sigma \ a_n \ \phi_n(z) \ \exp(i\kappa_n r)(i\kappa_o r)^{-\frac{1}{2}} \ .$$

Here, κ_o is a parameter at our disposal and κ_n is the eigenvalue corresponding to the nth mode. On the other hand, the form of the Green's function for the Helmholtz equation is, at long ranges,

$$H(r,z) = \Sigma \ \phi_n(z_o)\phi_n(z) \ \exp(ik_n r)(ik_n r)^{-\frac{1}{2}} \ .$$

Here k_n is the wavenumber corresponding to the nth mode. Finally, we would make these solutions (of two different equations) correspond if we choose

$$a_n = \phi_n(z_o)$$

$$\kappa_n = k_n$$

$$\kappa_o = k_n \ .$$

Unfortunately, we cannot satisfy all these conditions because only κ_o and the coefficients a_n are at our disposal. The very nature of the parabolic approximation is that κ_n is only approximately equal to k_n. This is referred to as the "small angle approximation". (Actually, by tuning the parameter κ_o, we can make κ_n and k_n equal for any one value of n. We would then have a "selected angle approximation.") In practice, we pay little attention to the fact that the numbers k_n do not equal the fixed parameter κ_o because this only leads to amplitude errors, whereas the approximation is much more sensitive to the phase errors, $\kappa_n \simeq k_n$.

In any case, let us assume that we have chosen a nominal κ_o and, of course, the data

$$P(0,z) = \Sigma \ \phi_n(z_o)\phi_n(z).$$

All of the summations up to this point have been infinite summations. If we use an infinite summation at this point we are in agreement with those who would use a delta function for data because

$$\delta(z-z_o) = \sum_{n=1}^{\infty} \ \phi_n(z_o)\phi_n(z)$$

is an approximation (or exact if there is only point spectra).

However, it is important to remove those terms that correspond to non-propagating modes in the Helmholtz equation because of the relationship

$$\kappa_n = \kappa_0 - k_n^2 / (2\kappa_0).$$

The values of k_n that correspond to non-propagating modes are essentially pure imaginary numbers that lead to exponential decay and in effect reduce the normal mode sum to a finite number of terms. The above relationship shows that the κ_n for these same terms will be essentially pure real and will not give exponential decay. This means that our data should contain only those modes that would propagate in the Helmholtz equation if we are to obtain agreement between the parabolic equation solution and the normal mode solution.

We are forced to drop more terms from our data when we consider that due to the small angle approximation, only a few modes in the parabolic equation will have acceptable phase errors. What is the use of retaining terms that are to be added together using meaningless (to the Helmholtz euqation) errors?

Our final recommendation is that the initial data for the parabolic equation should be the above mode sum using only those terms for which the small angle approximation is acceptably accurate.

There are two computational advantages that go with the data we recommend. The first is that since the resulting solution of the parabolic equation will have relatively few terms, it will be less oscillatory. This means that a larger step size can be used, making the computation faster. The second advantage is that only a few normal modes need to be computed, again a savings in computer time - however, still not as fast as using a Gaussian pulse.

I would like to add one additional remark, however. We have identified the desired data, but it may be possible to substitute a more conveniently computed approximation. For example, perhaps a sophisticatedly filtered Gaussian pulse could be used. On the other hand, if the solution of the parabolic equation were processed to correct the phase errors, as in Polyanskii (1974), one would want the data to be the above mode sum including all the propagating modes.

REFERENCE

E.A. Polyanskii, Relationship between the solutions of the Helmholtz and Schrodinger equations, Sov. Phys. Acoust. 20(1974),90.

THE SEAMOUNT AS A DIFFRACTING BODY

Herman Medwin and Robert P. Spaulding, Jr.*
Physics and Chemistry Department
Naval Postgraduate School
Monterey, CA 93940, U.S.A.

ABSTRACT

Diffraction over a seamount has been studied by experiments with a three-dimensional physical model of DICKENS seamount. When laboratory diffraction and scattering data are scaled to the ocean and added to absorption and refraction losses at sea, there is excellent agreement with ocean experiments reported by Ebbeson et al. (J. Acoust. Soc. Am. 64, S76 (1978). Both the magnitude and the frequency dependence of the shadowing at sea are predicted over a decade of frequencies.

I. INTRODUCTION

Major topographical features such as seamounts can cause large losses which are functions of sound frequency and the type of interruption of the sound path.[1,2] It is the thesis of this work that when complete ray blockage appears to occur, the ray description is useful for propagation from source to seamount, and for propagation from seamount to receiver. However, a correct calculation of the total propagation loss requires that one consider: (1) wave scattering at the rough surface upslope of the seamount; (2) three dimensional diffraction over the crest of the seamount; (3) a new ray path from the crest of the seamount to the receiver.

The description of the diffraction loss is aided by defining a new universal concept, "Diffraction Strength", DS, for forward diffraction somewhat in analogy to Target Strength as used in studies of backscatter. The far field diffraction strength of a seamount is a

*LT., U.S. Navy

function of the angle of incidence, angle of diffraction, and the
gross wedge angle along the soundtrack at the crest. It is not a
function of number of wavelengths of source/receiver from the seamount
or of the medium surrounding the real obstacle or its acoustically
scaled model.

The sum of the propagation loss from source to seamount, the
rough surface forward scattering loss at the upslope of the seamount,
the diffraction loss over the seamount, and the additional propagation
loss from the seamount to the hydrophone has been compared with
measurements at sea[3,4] and good agreement is found for a wide range of
frequencies.

II. EXPERIMENT

The physical model of DICKENS Seamount was constructed of approximately 60, 1/8 inch sheets of plywood. This terraced base was then
smoothed by application of a layer of plaster. The completed model
which is approximately 2mx2mx0.2m high, represents an area of 16km x
16km of the actual seamount (Fig 1). The geometrical scaling is 5" =
1 km or 1:7874. The model was used in the large Naval Postgraduate
School anechoic chamber in order to minimize background noise.

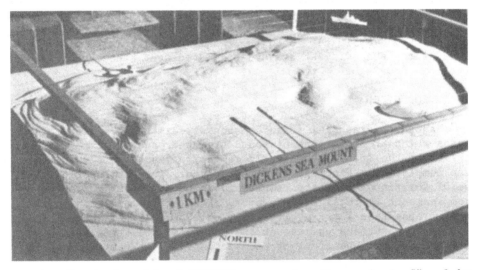

Fig 1. Physical model of DICKENS seamount. The scale is 5" = 1 km.

The diffraction over the seamount was evaluated by using a Bruel
and Kjaer (B&K) 1" microphone as a source and a B&K 1/2" microphone as
receiver. The electrical input signal was a single cycle of a triangular wave form of duration 100 µs and rise time 0.5 ns provided by a
Wavetek Model 175 Arbitrary Waveform Generator which was triggered by
an Interface Technology Model RS648 Timing Simulator. The RS648 was

also the command source which initiated A/D conversion for the Ocean
Physics Environmental Effects Analyzer (OPHELEA), at preset times.
OPHELEA was set to A/D convert at 320K samples/s. Each ping provided
a time series of 128 points which we call a block of data. In order
to improve the signal to noise ratio 1000 blocks were stored in memory
and then averaged for each run. The FFT of this time-averaged pulse
produced a frequency spectrum to 160kHz at 2.5kHz resolution. Because
of the poor high frequency sensitivity of the microphone, data are
reported only from 10-100 kHz. The block diagram of the experimental
arrangement is in Fig. 2.

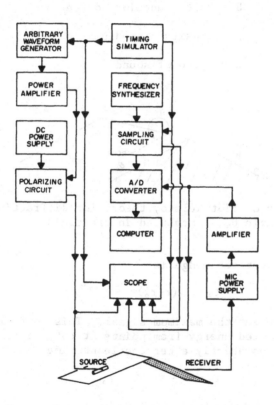

Fig. 2.   Block diagram of electronics. The model is indicated as a
          wedge.

III. BASIS FOR PHYSICAL MODELING OF DIFFRACTION

     Our guiding light in the modeling procedure is the Biot-Tolstoy
theory of pulse diffraction by an infinite rigid wedge.[5] A modifica-

tion of that work[6] gives the diffracted pressure for a source that is a delta function in time as well as space, Fig 3,

$$p(t) = \frac{-S\rho c}{4\pi\theta_w} \{\beta\} \, [rr_o \sinh y]^{-1} \, \exp(-\pi y/\theta_w)$$

Where $\beta = \dfrac{\sin\,[(\pi/\theta_w)\,(\pi\underline{+}\theta\underline{+}\theta_o)]}{1-2\,\exp(-\pi y/\theta_w)\,\cos\,(\pi/\theta_w)\,(\pi\underline{+}\theta\underline{+}\theta)]\,\exp(-2\pi y/\theta_w)}$ (1)

$$y = \text{arc cosh } [\frac{c^2 t^2 - (r^2 + r_o^2 + z^2)}{2rr_o}]$$

S = Delta function source (m^3/sec)

ρ = density of medium

c = speed of sound

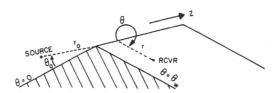

Fig 3. Geometry of Biot-Tolstoy theory for diffraction by a wedge.
Wedge angle θ_w is measured in the fluid.

The least time over the wedge is

$$\tau_o \equiv [(r+r_o)^2 + z^2]^{\frac{1}{2}}/c$$ (2)

at which time we get the maximum signal. This is followed by the principal diffracted energy from points at both sides of this crestal intercept at times shortly after the least time

$$\tau = t - \tau_o << \tau_o$$ (3)

Then $p_s(\tau) = S\rho\beta/[4\pi\sqrt{2}\,\theta_w\,(\tau\tau_o rr_o)^{\frac{1}{2}}]$ when $\tau/\tau_o <<1$ (4)

Eq.(4) may be rewritten

$$p_s(\tau) = A\tau^{-\frac{1}{2}}$$ when $\tau/\tau_o <<1$ (5)

where $A = S\rho\beta/[4\pi\theta_w)\,(2\tau_o r_o\,r)]^{\frac{1}{2}}$

The wedge angle θ_W is measured in the fluid angles; $\theta_W, \theta_o, \theta$ and ranges r_o, r are defined in Fig 3.

Transforming (5) to the frequency domain we have the spectrum due to the pulse diffraction for times small compared with the least time

$$P_o(f) = (A/2)(1+i)f^{-\frac{1}{2}}. \tag{6}$$

Since the major energy will move along or close to the least time path over the seamount, one can expect from (5) and (6) that the spectral pressure amplitudes will be proportional to $(r_o^2 rf)^{-\frac{1}{2}}$. For modeling purposes we therefore define the diffraction strength, DS,

$$DS = 20 \log_{10} \left[\frac{P_D(\theta, \theta_o \theta_W\ r, \ r_o, \ z)}{P_o} \frac{r_o}{R_o} (r/\lambda)^{\frac{1}{2}} \right] \tag{7}$$

where P_o is the reference pressure at the reference distance R_o (=1m) and λ is the wave length.

The final step in the modeling procedure is to define the diffraction loss, DL, by comparison of the diffracted pressure with the pressure that would have existed by direct propagation to the field position

$$DL = 20 \log_{10} \frac{P_o}{P_D} \frac{R_o}{R_D} \tag{8}$$

where P_D is the diffracted pressure at the direct source to receiver range R_D. In terms of diffraction strength (7) we will find the diffraction loss from

$$DL = 20 \log_{10} \left[\frac{r_o}{R_D} (r/\lambda)^{\frac{1}{2}} \right] - DS \tag{9}$$

If the diffraction strength, DS, is to be a useful modeling parameter applicable in air or water, it should be a constant for a given angular geometry, independent of the medium, although it may be a function of the nondimensional distance (r/λ). Even better, if DS is to be extrapolated to ranges or wavelengths beyond the model limits it would be convenient if DS is independent of r/λ. That both requirements are true for a <u>plane wedge</u> is shown by Fig 4. The symmetrical wedge angle chosen for Fig 4 has slopes of 14° with the horizontal (θ_W = 208°) which is common for seamounts and in fact might be characterized as a zero order approximation to DICKENS seamount along the sound track. Fig 4 shows that DS is independent of r/λ and independent of the medium, water or air, for a rigid "infinite" wedge.

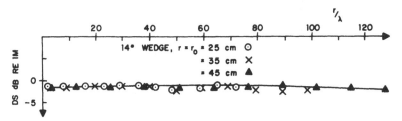

Fig 4.   Diffraction strength for a symmetrical plane rigid wedge with
         θ_w = 208° and z = o.   The solid line gives the theoretical
         value for r = r_o = 3.5 km in water; data points are for air
         model experiment at three ranges.   The model was constructed
         of 6.35 aluminum plates, 1.52m long x 0.6m wide.

     As the next order of approximation to DICKENS seamount a "contour
wedge" was constructed.   This wedge model was made of two sheets of
aluminum contoured to the topography along a sound track over the seamount (track 6 of Ref 1).   The gross wedge angle of the contour is
again approximately θ_w = 208°.   What is found in this approximation
(Fig 5) is that the DS of the contour wedge is less than that of the
plane wedge, and that there is a slight dependence of DS on range
because of destructive interference from secondary scatterers down-
slope from the wedge crest.   Most importantly for extrapolation purposes, DS does have an asymptotic, far field, value (-7.5dB) for
larger r/λ.

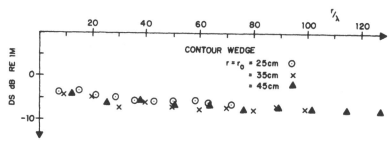

Fig 5.   Diffraction strength for a wedge model with sides contoured to
         track 6 across DICKENS seamount.   Three ranges in air. z = o.

For the remainder of this section we consider the acoustical
behavior of the physical scale model of the seamount (Fig 1). In
effect we are using the scale model as an analogue computer to solve
the wave equation for diffraction by a seamount. An acoustical survey
was conducted for the six sound paths shown in Fig 6. When DS is
plotted against r/λ, five of the paths reached their (different) far
field asymptotic values of DS for $r/\lambda > 30$ (Fig 7). Path 1 is distinguished by being far from the peak of the seamount; its asymptotic
value is not reached even at $r/\lambda = 130$, so that modeling in this case
would be sensitive to r/λ within the range studied. The far
field values of DS for paths 2 through 6 show a variation of about 6
dB. Therefore the value of DS is a function of the sound path across
the seamount; the diffraction loss is also path-dependent.

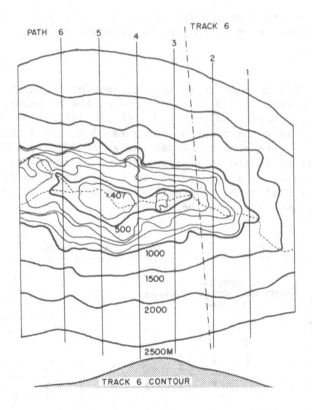

Fig 6.   Depths in meters below sea level for DICKENS seamount. The
         short dashed line represents the crest. Six laboratory acous-
         tical survey paths at spacing 25 cm (approx. 2 km at sea) and
         the ocean sound track 6 of ref. 1 are shown. The contour
         along track 6 is at the bottom, not to scale.

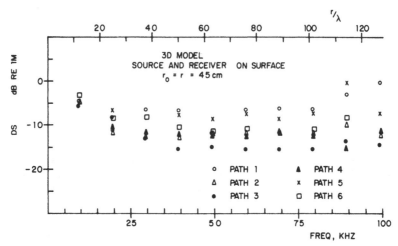

Fig. 7.   Diffraction strength for six sound paths across DICKENS
          seamount. r_0 = r = 45 cm.   Source and receiver on seamount
          model.

     The azimuthal dependence of the diffraction strength has been
studied with central path 3 as reference and the Z coordinate for the
receiver position varied on both sides of the least time central path
(z=o), Fig 8.   Since the receiver range is 45 cm the z displacements
of 25 and 50 cm correspond to offset angles of 29° and 48° relative to
the crest crossing at the least time path.   The asymptotic diffraction
strength is decreased by about 3 dB and 6 dB, respectively, for these
offset receiver positions.   The diffraction strength is therefore a
function of azimuth angle of the receiver.

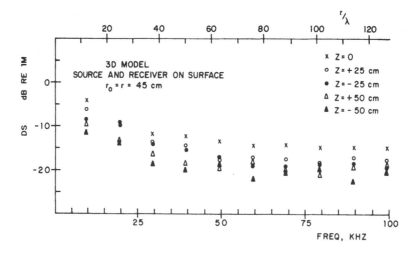

Fig. 8.   Diffraction strength for four azimuthal positions of
          receiver, offset from the least time trajectory of Path 3.

For the remainder of this section we concentrate on sound track 6, for which comparison will be made with the sea trials in Section IV. The effect of range is seen in Fig 9. The diffraction strength for track 6 is essentially independent of the non-dimensional range for a far field that starts at $r/\lambda = 40$. The asymptotic value of DS is about 5dB less for the 3D model than for the 2D contour model.

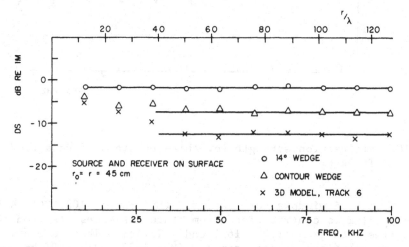

Fig. 9. Diffraction strength for three models of the seamount. Solid line is theory for 14° plane wedge and asymptote for contour wedge and track 6 of physical scale model.

An important consideration is the determination of the part of the crest and scattering surface that contributes to the diffraction strength. In a proposed parabolic equation solution to the scattering by a seamount[7] the cylindrical symmetry of that method effectively assumes that it is sufficient to know the seamount contour at the sound path. In effect our contour wedge also represents this two dimensional approximation. The contribution to the total diffraction from parts of the true surface that are off from the least time path can also be studied in our three dimensional scale model by opening the data window in steps to receive those components.

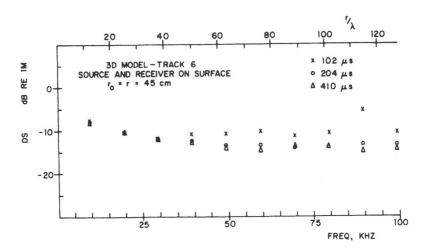

Fig 10. Diffraction strength for three openings of data window.
         Track 6.

     Three data windows were used for this test: 102, 204, 408 μs,
corresponding to contributions from ridge distances off from the model
least time path by $\pm$ 11.2 $\pm$ 16.1 and $\pm$ 23.5cm. This is equivalent to
ridge contributions at sea from $\pm$ 0.88 $\pm$ 1.27 and $\pm$ 1.85 km off from
track 6. Fig 10 shows that for the near field DS measured at the
least time path is not dependent on the off-axis diffraction elements.
However, the more important far field diffraction strength can be too
large by at least 4 dB if only the sound path profile is considered
($\Delta\tau \leq 102\mu s$). This supports the interpretation based on the asymptotic values in Fig. 9. We conclude that a three dimensional wave solution is essential for correct modeling of the effect of DICKENS seamount.

     Anticipating the need for DS as a function of propagation direc-
tion (Section IV), Fig 11 shows the far field values for diffraction
at three angles measured with respect to the horizontal along track 6.
The diffraction strength is indeed a function of θ, being greater
above the horizontal and less as the receiver penetrates more deeply
into the shadow region.

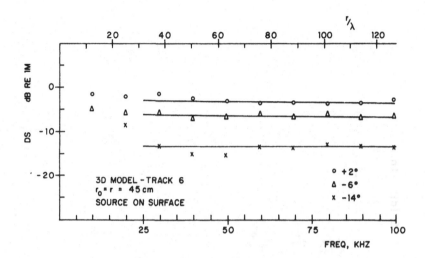

Fig 11.   Diffraction strength for three elevation angles of propagation measured with respect to the horizontal, track 6.

IV.   COMPARISON WITH OCEAN TRIALS

Results of ocean trials of shadowing by DICKENS seamount have
been reported by Ebbeson et al[1]; Fig 12, from their paper shows at
position A, 79 km, a propagation loss of 97 ± 5 dB compared with a
FACT prediction of 82 dB. We believe that this discrepancy of about
15 dB, and its frequency dependence, can be understood by a new
description of this ray-wave propagation which incorporates the wave
phenomena of forward scatter at the upslope of the seamount followed
by diffraction and reradiation over the crest. The proposed ray pattern up to and away from the seamount is shown in Fig 13. The crestal
rays of angle +2° and -6° with respect to the horizontal are emphasized because they are the ones that reach the hydrophone at its depth of
329m, range 79km. A more appropriate true scale of the wave interaction phenomena is presented in Fig 14 where it is seen that the incoming rays meet the seamount at grazing angles of 23 to 17 degrees,
ranges 10.8km to 17.6km, depths 2700m to 1000m.

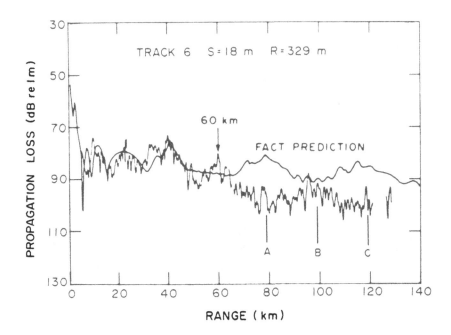

Fig 12.   Measured and FACT-predicted propagation loss as reported by
           Ebbeson et al[1]. Range A, 79 km, with source 19 km from
           seamount is case of apparently complete ray blockage and
           maximum shadowing.

    In the previous section we found that the physical model is
useful for the study of the diffraction component. From Fig 11 the
far field diffraction strength for the +2° ray is −3.5 ± 0.3 dB, and
for the −6° ray it is −6.5 ± 0.3 dB. Then using eq (9) the diffraction losses at 230 Hz, $r = \overline{r_0} = 3.54$ km are calculated to be 24.5 ±
0.3 dB and 27.5 ± 0.3 dB respectively. From eq (9) we see that the
diffraction loss is proportional to $\lambda^{1/2}$; therefore the diffraction
loss increases 3 dB/octave increase of frequency.

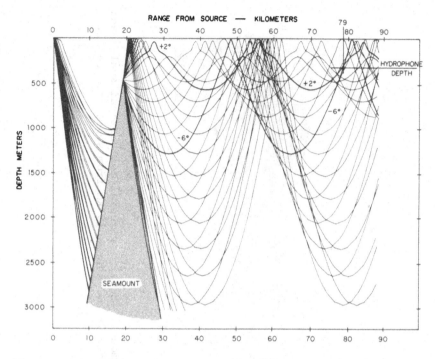

Fig 13.   Proposed ray pattern for range 79 km, source 19 km from sea-
          mount.   Track 6.

        The forward scatter loss (or gain) of the incoming energy is
needed to complete the picture of the wave interaction.   From Fig 14,
we see that the energy that starts to diffract over the crest is the
forward scattered reverberation that was incident at the seamount from
range 10.8 km to 17.6 km.   This situation has been modeled in the
laboratory by placing a receiver at the 17.6 km position (18 cm below
the crest) and aiming the source at the midpoint of the upslope.   The
source was a low Q, 10cmx10cm mylar transducer with 3 dB down points
at 5.5° at 50 kHz.   A separate experiment consisted of direct radiation to the receiver, without the seamount, in order to provide a
reference incident signal.   The computer was used to subtract the
incident signal from the complex superposition of incident plus reverberant signal, in order to determine the reverberant signals by
itself.   The rms values of the incident and reverberant signals were
then determined by direct calculation from sampled time series of the
same duration.   The forward scattering loss was defined by

$$\text{dB scattering loss} = 20 \log_{10} \frac{\text{rms reverberation pressure}}{\text{rms incident pressure}}$$

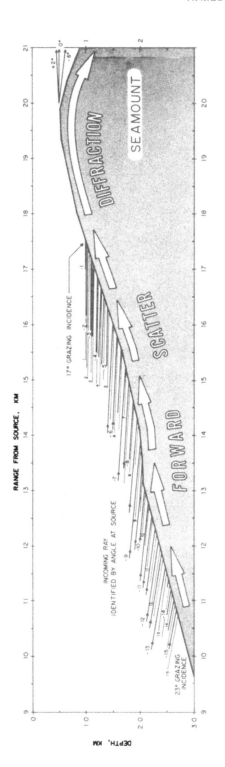

Fig 14. Wave interaction region at upslope of seamount. Equal horizontal and vertical scales and correct angles of incidence. Track 6, range 79 km, source 19 km from seamount. Ref 1.

In fact, because the upslope of the seamount acts as a converging device, surface reverberation results in a <u>gain</u> rather than a loss of signal. The gain was 8.4 $\pm$ 0.7 dB, measured at 50 and 100 kHz, approximately independent of frequency. We note that the forward scatter is in the large Rayleigh roughness, incoherent, regime so that the independence of frequency is to be expected.

The complete bookkeeping for the ocean propagation loss at 230 Hz is shown in Table 1.

Table 1

Contributing Losses at 230 Hz, source depth 18 m,
Receiver depth 329 m, range 79 km, Track 6

Item	+2° Ray Loss, dB	−6° Ray Loss, dB
Source to upslope seamount (3.5 km below crest)	77.3 $\pm$ 1.5	77.3 $\pm$ 1.5
Upslope scattering loss	−8.4 $\pm$ 0.7	−8.4 $\pm$ 0.7
Diffraction loss to 3.5 km beyond seamount crest	24.5 $\pm$ 0.3	27.5 $\pm$ 0.3
Seamount (3.5 km) to receiver	3.8 $\pm$ 0.5	10.1 $\pm$ 0.5
TOTAL (each Path)	97.2 $\pm$ 3.0	106.5 $\pm$ 3.0
TOTAL LOSS	96.7 $\pm$ 3.0 dB	

The total loss of 96.7 $\pm$ 3.0 dB in Table 1 is obtained by assuming that the energies in the two ray paths add incoherently. This compares well with the ocean experimental result 97 $\pm$ 5 dB for a CW source.

More recent explosive shot analysis by the group at DREP[9] has provided an additional comparison. By interpolation between their results at 200 and 250 Hz we find an experimental propagation loss of 97.6 dB to compare with our value 96.7 $\pm$ 3 dB.

The frequency dependence is of great interest. If we focus our attention on the shadowing loss, defined as the difference between the experimental propagation loss and the theoretical loss calculated from the FACT model in the absence of the seamount, we can extend our results to other frequencies. The three contributors are the upslope scattering loss, the diffraction loss which from eq (9) is proportional to $f^{\frac{1}{2}}$, and the additional loss for the new path from the seamount to the hydrophone. For the experimental conditions of Table 1 our prediction of the shadowing loss is the sum:

$$\text{Additional Refraction Loss} = -2.3 \pm 2 \text{ dB} \tag{10}$$

$$\text{Diffraction Loss} = +24.5 \pm 0.3 + 10 \log_{10} (f/230)$$

$$\text{Forward Scattering Loss} = -8.4 \pm 0.7$$

$$\overline{\text{Shadowing Loss} = +13.8 + 10 \log_{10} (f/230) \pm 3.0 \text{ dB}}$$

The comparison of this prediction with ocean experiment is shown in Fig 15. The agreement is quite satisfactory from about 50 Hz to 500 Hz, with the exception of an unexplained rise at 80 Hz.

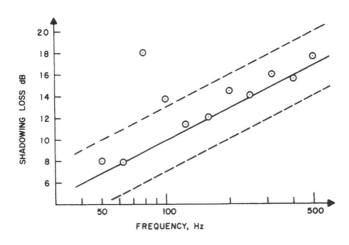

Fig 15.   Frequency dependence of shadowing loss at sea, position A, track 6. Solid and dashed lines are prediction and error estimate of eq. (10). Circles are difference between ocean experimental and FACT-predicted propagation losses, corrected for the displaced convergence zone.

V. CONCLUSION

The prediction of shadowing losses due to a seamount that blocks
a propagation path requires a wave, rather than a ray, description of
the interaction. A rigid three-dimensional scale model of DICKENS
seamount has been used in air to predict the forward non-specular
scatter at the upslope and the forward diffraction over the crest of
the obstacle at sea. When these scaled losses are added to absorption
and refraction losses from source to seamount and from seamount to
receiver, the total is in good agreement with the measured ocean
transmission losses for the frequency range 50 to 500 Hz. There are
no parameters or coefficients that need be selected or adjusted for
this comparison to be made.

The three dimensional physical scale model is a relatively inex-
pensive device to aid in the prediction of shadowing losses caused by
obstructions at sea.

Acknowledgment

The comparison of our predictions with the results of ocean
experiments would not have been possible without the generous cooperation of Gordon Ebbeson and the Defense Research Establishment Pacific,
Victoria, B.C., Canada, who provided the detailed bathymetry of
DICKENS seamount, ray traces, and propagation losses measured in their
sea trials. Supplementary ray traces were calculated by the Fleet
Numerical Oceanographic Center, Monterey. Computer services were
effectively performed by Mrs. B. J. Savage. Financial assistance was
provided by Manager, ASW Systems Project Office and the Office of
Naval Research, U.S. Navy.

REFERENCES

1. G. R. Ebbeson, J. M. Thorleifson and R.G. Turner, "Shadowing of
       sound propagation by a seamount in the northeast Pacific", J.
       Acoust. Soc. Am. 64, S76 (1978).
2. D. A. Nutile and A. N. Guthrie, "Acoustic shadowing by seamounts",
       J. Acoust. Soc. Am. 66, 1813-1817 (1979).
3. R. P. Spaulding, Jr., "Physical modeling of sound shadowing by
       seamounts", M.S. Thesis, Naval Postgraduate School, Monterey,
       CA 93940, (1979).
4. H. Medwin and R. Spaulding, "Shadowing by seamounts", J. Acoust.
       Soc. Am. 66, S76, (1979).
5. M. A. Biot and I. Tolstoy, "Formulation of wave propagation in
       infinite media by normal coordinates with an application to
       diffraction", J. Acoust. Soc. Am. 29, 381-391 (1957). See also

I. Tolstoy "Wave Propagation", McGraw-Hill Book Co., Inc., New York, (1973).

6. H. Medwin, "Shadowing by finite noise barriers", submitted for publication in J. Acoust. Soc. Am. (1980).

7. F. D. Tappert, "The parabolic approximation method", Ch. V in Lecture Notes in Physics, 70, Wave Propagation and Underwater Acoustics, ed. by J. B. Keller and J. S. Papadakis, Springer-Verlag, New York (1977).

8. C. S. Clay and H. Medwin, "Acoustical Oceanography", Wiley, New York (1977). (Ch. 10, Scattering and Reflection of Sound at Rough Surfaces").

9. G. R. Ebbeson, Private communication.

PROPAGATION OF SOUND FROM A FLUID WEDGE INTO A FAST FLUID BOTTOM

Alan B. Coppens and James V. Sanders

Department of Physics and Chemistry
Naval Postgraduate School
Monterey, California 93940

I. INTRODUCTION

The ocean waters lying above the continental shelf often can be
modeled with a fair degree of accuracy by a fluid wedge overlying a
bottom. In those cases where the bottom consists of unconsolidated
sediments, it is also an admissible approximation to treat the bottom
as a fluid. See Fig. 1. If the speed of sound in the bottom exceeds
that in the water layer (a fast bottom), the wedge can support trapped
normal-mode propagation. Let acoustic energy from a distant source
propagate up the slope towards the apex of the wedge. As the apex is
approached and the local thickness of the wedge decreases, the modes
carrying the energy will experience the condition for cutoff at varying distances from the apex. We can estimate the depth at which a
given normal mode would be cut off by calculating the depth H for
which the equivalent mode would just be cut off in a layer of constant
depth. If the wedge angle is β , the lowest mode of propagation
would experience cutoff when the distance from the apex decreases to a
dump distance $X = H/\beta$, the next mode at the distance 3X, the third at
5X, and so forth. Thus, as the acoustic energy propagates towards the
apex, successively lower modes will be cut off and the energy contained in each of them transmitted into the bottom.

For the purpose of developing some additional parameters useful
in qualitative descriptions of the sound field in and below the wedge,
let us treat the problem entirely, but incorrectly, as an exercise in
ray tracing. As a ray of sound propagates from large distances
towards the apex of the wedge, upon successive reflection its angle θ
of elevation (and depression) increases by an amount 2 β. After the
ray attains critical incidence θ_c at the dump distance X, it will continue to bounce between surface and bottom losing energy to the bot-

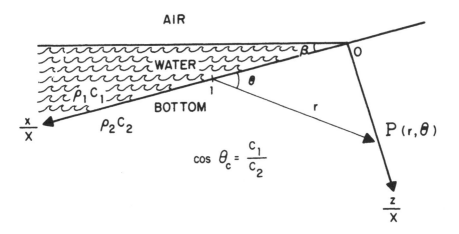

Fig 1. Geometry used to describe the transmission of sound from a
        fluid wedge into a fluid bottom.

tom on each bounce as it advances towards the apex. Upon experiencing
a reflection normal to the bottom, at the turn-around distance
$x_t = X \sin \theta_c$, it will retrace its path, propagating away from the
apex back towards deeper water. For media posessing similar specific
acoustic impedances, the ray will be almost completely dissipated well
before the turn-around distance is reached. The interval $X(1-\sin \theta_c)$
between dump and turn-around distances over which the majority of
energy is lost by the ray can be considered to be the length of the
aperature of an apparent source on the interface radiating into the
bottom.

    The propagation of sound in a fluid wedge has been studied by a
number of investigators. In the case of a fluid wedge in which both
surfaces are acoustically impenetrable, a mathematical examination of
the problem has been performed by Bradley and Hudimac[1] using both
image theory and normal mode theory. In the case of more practical
interest for underwater acoustics, the top boundary may still be
approximated by being considered impenetrable, a pressure release surface, but the bottom surface of the wedge overlying the acoustically-
penetrable ocean bottom can no longer be considered to be perfectly
reflecting. From a mathematical point of view, this generalization of
the boundary conditions has an important repercussion: The wave equation is no longer separable and standard normal mode theory can no
longer be applied. In this more realistic case, there have been three
basic approaches that have been reported in the literature:

(1) Perturbation methods. Tien, Smolinsky, and Martin[2] applied
the method of Marcuse[3] to the optical case of a thin film wedge
overlying a substrate possessing a greater speed of light. The wedge
is replaced with an asymmetrically-stepped thin layer, the individual
steps being small with respect to the layer thickness at each distance
from the apex. The conversion of energy from the bound normal mode in
the layer to the radiation modes in the substrate is calculated at
each step. The computer-generated results confirm in some detail the
experimentally-observed features of the beam formed in the substrate.
However, this theory predicts that all the energy in the normal mode
will be converted to radiation in the substrate before the normal mode
reaches the wedge thickness at which it would achieve the condition of
cutoff.

2) Ray tracing methods. Tien et al[2] also developed an analysis
based on replacing the normal mode approach with a ray approach. At
the depth where the mode would be cut off, it is assumed that a ray
reflects from the bottom of the wedge at the critical angle. This ray
is allowed to reflect from the surface and it reattains the bottom at
a grazing angle which is the critical increased by twice the wedge
angle. The intervening region of the wedge is filled with a large
number of downward-traveling rays whose angles of incidence on the
bottom are uniformly increasing between θ and $\theta + 2\beta$. Then, for each
ray the transmission into the bottom is calculated. This process is
repeated progressively toward the apex of the wedge. There is no
attempt to preserve phase information, and the power of the beam in
the bottom at each angle of depression is simply the sum of the
powers transmitted by the appropriate rays. Calculations performed
with the help of a computer again provide good agreement with their
experimental observations as far as the angle of depression and beam
width of the beam in the bottom were concerned, but this approach predicts that there should be no transmission of sound into the bottom
until the cutoff depth is reached.

This same method was applied to acoustic waves in a fluid wedge
overlying a solid bottom by Odom, Sigelmann, Mitchell, and Reynolds.[4]
Comparison of the predictions of this model with experiment showed
good agreement with the angle of depression of the beam but beam
widths were greatly dissimilar.

A different approach based on the representation of normal modes
by ray bundles has been developed by Kuznetov.[5] The results predict
that the beam in the bottom will be contained within angles of depression of β and 2β. This is in disagreement with all experiments except Kuznetsov's own.

(3) Parabolic equation method. Jensen and Kuperman[6] have investigated this problem using an approach based on the parabolic
equation. Utilization of a specialized high -speed computer has yield-

ed sets of equal-level contours mapping the field in the fluid bottom
as well as in the overlying wedge. The results show unambiguously the
existence of a set of narrow beams propagating in the bottom, all having the same apparent angle of depression at large distance from the
wedge. The transition of the angle of depression of the beam and its
beam width between near-field and far-field behaviors for each of the
beams is also discernible. The only significant limitation of this
method appears to be that inherent in the approximations necessary for
appilicablity of the parabolic equation itself: The method works as
long as the angles of elevation and depression of the rays describing
the normal modes do not become too large. This would appear to limit
accurate application of this method to situations for which the critical angle does not exceed about 20° or so, which corresponds to the
condition $1 > c_1/c_2 > 0.94$.

II. THEORY

Our approach to the problem[7-8] is to combine two fundamental,
classical, and well-established techniques. For the sound within the
wedge, it is possible to determine its amplitude and phase by the
method of images. The wedge is replaced by a set of mirror images
within each of which there is an image of the source. The source and
each of the images radiates spherical waves of the appropriate phase,
and the phase coherent summation of these waves yields the total pressure and phase to be found at any field point in the wedge. For a
source which is at a large distance from the apex and is not too close
to either the surface or the bottom, the successive reflections of
sound from surface and bottom can be accounted for by multiplying the
sound fields radiated by the images by the plane-wave reflection coefficients corresponding to the relevant reflections encountered by each
wave as it propagates towards the field point. Furthermore, for sufficiently large distances between the source and the dump distance of
interest, the curvature of the surfaces of constant phase measured in
the horizontal plane can be assumed negligibly small so that the
intersections of the surfaces of constant phase and the bottom can be
assumed to be a set of lines parallel to the apex. Under these conditions, the pressure distribution on the bottom can be treated as a
collection of infinite line sources, each parallel to the apex. Now,
having a semi-infinite region (the bottom) with a known pressure distribution on its surface, the resultant field in the bottom can be
calculated by the Green's function technique. All attempts to pursue
this approach by analytical techniques using reasonable approximations
have yielded results of insufficient accuracy. As a consequence, we
have come to depend on numerical methods both to calculate the pressure distribution on the interface and to evaluate the Green's function integral. An analytical evaluation has not been completely abandoned, but has been relegated a lesser priority.

Two computer programs have been written. The first, and simplest, performs the calculations for the situation wherein the source

is an infinite distance away from the apex and lies halfway between
surface and bottom. The trigonometric approximations which can be
made in this limit reduce the computer running time by about one order
of magnitude below that of the second program. Placing the source
midway between surface and bottom also insures that only the odd normal modes will exist. This succeeds in separating the field in the
bottom into a number of relatively strong beams emanating from the
dump distances $x/X = 1, 5, 9, \ldots$, with weak beams emanating from dump
distances $x/X = 3, 7, 11, \ldots$ only if there is significant mode conversion from the odd normal modes into the even ones possessing nodes
midway between surface and bottom at large distances from the apex.
This suppression of half the set of available normal modes aids in
reducing mutual interference between adjacent beams. This has been
done in order to facilitate the study of the properties of an individual beam, rather than having to deal with the interpretive complexi-
ties that could arise from more prominent phase coherent mutual interference. Representative predictions of this program are shown in
Figs. 2 and 3.

The second computer program relaxes the restriction on the location of the source. The source may be positioned at any distance from
the apex and at any depth within the wedge. The underlying physical
restrictions on the validity of the method of images when applied with
plane-wave reflection coefficients and the application of the Green's
function integral in cylindrical geometry still apply: (1) the source
must be several wavelengths away from either surface or bottom of the
wedge, and (2) the source must be many times the largest dump distance
of interest away from the apex of the wedge.

The results of our approach appear to be valid. The numerical
evaluations yield a narrow beam of sound propagating into the bottom
at a rather shallow angle of depression, emanating from the immediate
vicinity of the dump distance for the particular normal mode which has
attained cutoff. It is also observed that the pressure amplitude predicted by the image theory at the turn-around distance is substantially smaller than the amplitude at the dump distance. This effect is
seen in Fig. 2 where $x_t/X = 0.44$. In addition, it can be clearly seen
that the ratio $P(x_t)/P(X)$ decreases significantly as the wedge gets
smaller. This is not inconsistent with the idea that the normal mode
retains its identity as it approaches the apex within the dump distance and that it can be modeled in some respects by rays for small
wedge angles.

Furthermore, the apparent angle of incidence on the bottom of the
signal in the wedge as predicted by the image theory deviates less and
less from values just slightly in excess of the critical angle as the
wedge angle decreases. This in turn suggests that there is less energy lost at each reflection, and that the energy transmitted into the
bottom has smaller angles of depression. The beam of sound in the
bottom therefore has an angle of depression approaching zero as the

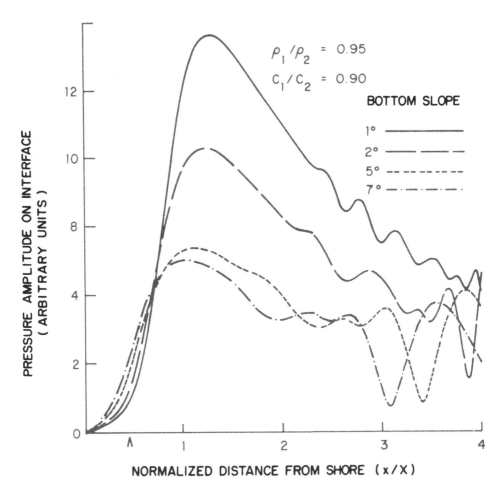

Fig. 2. Predicted pressure amplitude along the wedge-bottom inter-
         face.

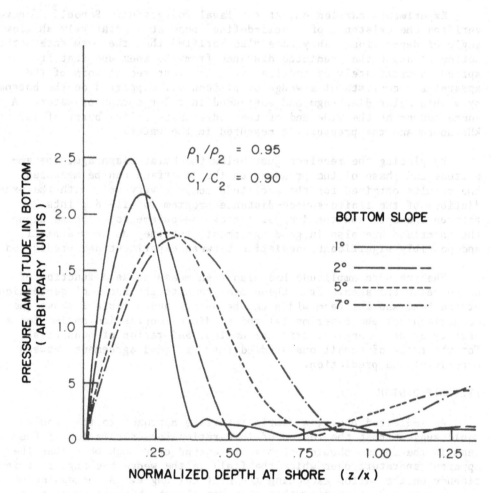

Fig. 3. Predicted pressure amplitude beneath the apex.

wedge angle approaches zero. The aperature of the apparent source
increases with X (for constant θ_c and frequency) so that the width of
the beam should decrease with β_c.

III. EXPERIMENT

     Experiments carried out at the Naval Postgraduate School[9-12] have
verified the existence of a well-defined beam at a relatively shallow
angle of depression. They have also verified that the beam enters the
bottom at about the predicted distance from the apex and that it
spreads approximately cylindrically.   The most recent form of the
apparatus consists of a wedge of silicon oil supported on the bottom
by a thin Mylar diaphragm and suspended in a large tank of water. A
sound source at the wide end of the wedge emits a long burst of 100
kHz sound and the pressure is measured in the water.

     By placing the receiver just below the Mylar diaphragm, the am-
plitude and phase of the pressure on the interface can be measured.
The results obtained for the amplitude compare very well with the predictions of the finite-source-distance program for field points
between the apex and the dump distance. Measurements of phase along
the interface are also in good agreement, but they do show a small,
and possibly significant, deviation between experiment and prediction.

     The pressure amplitude has also been measured as a function of
depth below the apex. From these measurements the angle of depression
of the beam and the beam width can be determined. Table I shows the
comparison between experimental and predicted depression angles for a
variety of wedge angles, critical angles, and ratios of densities.
For the range of conditions studied there is good agreement between
experiment and prediction.

IV. DISCUSSION

     As indicated above, a simple intuitive approach to the problem
would suggest that the region of integration for calculation of the
beam in the bottom should not have to extend over much more than the
apparent aperature over which the field in the wedge has angles of incidence on the bottom exceeding the critical angle. A succession of
computer runs demonstrated that this was almost, but not quite true
enough for accurate prediction. Looking at the beam formed from the
lowest normal mode in the wedge, and integrating from the apex of the
wedge out to distances of X, 2X, 3X, and 4X revealed that the shape of
the beam stabilizes only if the integration extends to or beyond 3X.
For integration over the shorter distances, the beam appears
"smeared", in that the amplitude is lower, the angle of depression
somewhat larger, and the lower portion of the beam (below the apparent
angle of depression) tails off to greater depths than found for the
larger integration distances. Since the predicted beams were essen-

Table I.   Predicted and experimental angle of depression of the beam
           in the bottom as measured beneath the apex.

β	c_1/c_2	θ_c	ρ_1/ρ_2	θ_o EXPERIMENTAL	PREDICTED
2.5°	0.88	28°	0.87	14°	14°
2.2°	0.79	38°	1.02	14°	16°
2.6°	0.92	23°	0.92	15°	12°
2.3°	0.83	34°	0.78	13°	16°
3.9°	0.83	34°	0.73	17°	20°

tially identical for the 3X and 4X integrations, it appears that the
apparent aperture of the beam can be assumed to be less than 4X for
the lowest mode, and corresponding larger for the higher modes.
Again, application of simple physical arguments leads to results which
are qualitatively acceptable but rather inaccurate when compared to
the results of the more rigorous approach.   It must be noted, however,
that this effect may be the result of phase coherent interference
between the beams formed by the first and third normal modes.

      The computation time goes directly as the number of steps that
must be chosen along the interface. However, the increment Δ in x/X
must be sufficiently small that the difference in phase of the integrand of the Green's function integral is not appreciable compared to
2π between successive steps. If it is desired to have no less than
20 points as the phase traverses 2π, then a useful approximate criterion for Δ is

$$\Delta \leq 1/5 \tan \theta_c \tan \beta / (1-\cos\theta_o) \sim 0.4\ \theta_c\ \beta/\theta_o^2 \qquad (1)$$

where θ_o is the apparent angle of depression in the bottom measured
from the dump distance.

      A matrix of cases was obtained with the short program. The
values of parameters were restricted to the sets β = 1, 2, 5, 7, 10°;

c_1/c_2 = 0.85, 0.90, 0.95; ρ_1/ρ_2 = 0.50, 0.70, 0.80, 0.95. Pressure amplitude and phase as functions of distance along the interface and pressure amplitude as a function of depth directly beneath the apex were calculated. We are presently analyzing the results of these runs for parametric dependences of the properties of the pressure distribution on the interface as a normal mode approaches cutoff and begins to radiate into the bottom, and of the properties of the resulting beam radiated away from the interface. Our analysis is not complete yet, but we have some preliminary results which reveal the relative importance of the wedge angle, the ratio of densities, and the critical angle. These three parameters contain all the relevant geometrical and acoustic information. The angle of depression of the beam in the bottom is found to be well estimated by

$$\Theta_o = 1.15\beta^{0.333} \theta_c^{0.773} (\rho_2/\rho_1)^{0.27} \tag{2}$$

where all angles are in radians. This value is the effective angle of depression of the beam for the lowest normal mode, as measured downward from the apex perpendicular to the bottom, assuming that the beam originated at the dump distance. Since this location is still probably within the near field of the beam, the angle is a lower bound to the value to be expected at greater distances away from its apparent aperture. Notice the relatively weak dependence of Θ on the densities and wedge angle. This, most likely, explains in good part the small range of values that have been observed for this angle not only in our acoustic experiments but also in the optical experiments. The distance in the wedge at which the maximum value of the pressure amplitude on the interface occurs is given by

$$(x/X)_{P_{max}} = 0.8 \ (\theta_c^{0.20}/\beta^{0.17}) \ (\rho_1/\rho_2)^{0.23} \tag{3}$$

where all angles are in radians. This quantity is very close to one for most values of parameters, but notice that its value increases as the wedge angle decreases: the location of the maximum pressure migrates outward from the apex as the wedge is narrowed, migrating from $x < X$ to $x > X$ as β decreases; see Fig. 2.

Table II lists the depression angles computed or measured by Tien et al and the (near-field) depression angles computed from the above empirical formula. The optical experimental system had c_1/c_2 = 0.975. Since the relative densities of the wedge and substrate have no relevance in the optical problem, the acoustical problem reduces to the optical if we set ρ_1/ρ_2 = 1. Even though the values of the parameters lie outside the matrix we chose, the agreement is rather good. The fact that our predictions tend to run a bit low may be because of our being relatively close to the aperture of the beam, in the near field.

This work is supported by the Earth Physics Program of the Office of Naval Research under contract No. N00014-80-WR-00117.

Table II. Angle of depression of the beam for the optical case.

β°	Tien et al. θ_0 Calculated	Tien et al. θ_0 Experimental	Eq. 2
2°	10°		7.1°
1	7.4		5.7
0.477		5.2°	4.4
0.4	5		4.2
0.22	4		3.5
0.2		3.6	3.3
0.11	3		

REFERENCES

1. D. Bradley and A.A. Hudimac, The Propagation of Sound in a Wedge Shaped Shallow Water Duct, Naval Ordnance Laboratory NOLTR 70-235 (1970).

2. P.K. Tien, G. Smolinsky, and R.J. Martin, Radiation Fields of a Tapered Film and a Novel Film-to-Fiber Coupler, IEEE MIT-23, 79 (1975).

3. D. Marcuse, Mode Conversion Caused by Surface Imperfections of a Dielectric Slab Waveguide, Bell Syst. Tech. J. 48, 3187 (1969).

4. R.I. Odom, Jr., R.A. Sigelmann, G. Mitchell, and D.K. Reynolds, Navy Contract N66001-77-C-0264 Tech. Rept. #209 (1978).

5. V.K. Kuznetsov, Emergence of Normal Modes Propagating in a Wedge on a Half-Space from the Former into the Latter, Sov. Phys. Acoust. 19, 241 (1973).

6. F.B. Jensen and W.A. Kuperman, Sound Propagation in a Wedge-Shaped Ocean with a Penetrable Bottom, J. Acoust. Soc. Am. 67, 1564 (1980).

7. A.B. Coppens, Theoretical Study of the Propagation of Sound into a Fast Bottom from an Overlying Fluid Wedge, J. Acoust. Soc. Am. 31, Suppl. 1 (1978).

8. A.B. Coppens, The Theory: Transmission of Acoustic Waves into a Fast Fluid Bottom from a Converging Fluid Wedge, ONR/NRL Workshop in Seismic Wave Propagation in Shallow Water (1978).

9.   J. N. Edwards, A Preliminary Investigation of Acoustic Energy
      Transmission from a Tapered Fluid Layer into a Fast Bottom,
      Naval Postgraduate School Technical Report, NPS-33JE76121
      (1976).
10.   G. Netzorg, Sound Transmission from a Tapered Fluid Layer into a
      Fast Bottom, M.S. Thesis, Naval Postgraduate School (1977).
11.   J. V. Sanders, A.B. Coppens, and G. Netzorg, Experimental Study of
      the Propagation of Sound into a Fast Bottom from an Overlying
      Fluid Wedge, (A) J. Acoust. Soc. Am. 31, Suppl. 1, (1978).
12.   J. V. Sanders, The Experiment: Transmission of Acoustic Waves
      into a Fast Bottom from a Converging Fluid Wedge, ONR/NRL
      Workshop in Seismic Wave Propagation in Shallow Water (1978).
13.   M. Kawamura and I. Ioannou, Pressure on the Interface Between a
      Converging Fluid Wedge and a Fast Fluid Bottom, M.S. Thesis,
      Naval Postgraduate School (1978).

RANGE-DEPENDENT BOTTOM-LIMITED PROPAGATION MODELLING

WITH THE PARABOLIC EQUATION

F.B. Jensen and W.A. Kuperman

SACLANT ASW Research Centre

La Spezia, Italy

ABSTRACT

It is demonstrated that mode-coupling effects are included in the parabolic equation method. Propagation over sloping bottoms is then studied demonstrating various mode-cutoff and mode-coupling phenomena. Finally, propagation over a seamount is considered, which includes not only up- and down-slope propagation but also diffraction over the top of the seamount. Some of the theoretical results are compared with experimental data.

INTRODUCTION

Recently there has been increasing interest in the physics of sound propagation in shallow water. Most of the work done in the past has been concerned with range-independent environments since it was the intention of the researchers to isolate the factors that governed bottom-limited propagation without the additional complication of a changing environment. Thus it was found that propagation in shallow water was similar to wave-guide propagation and normal-mode theory turned out to be in good agreement with experiments[1-5]. A natural progression of this research was to investigate the effect of range-varying environments: changing sound-speed profile, changing bottom type, and changing water depth.

The simplest extension of normal-mode theory to a range-dependent environment is to assume that modes adapt to the slowly changing environment; this is the "adiabatic normal-mode approxi-

451

mation"[6-11]. The next step towards increasing complexity is to include "coupling" between modes[12], but so far no practical numerical solution scheme has been devised for this more general method. Simultaneous with the extension of mode theory to range-dependent environments, a new theoretical technique was introduced to describe deep-water range-dependent propagation: the "Parabolic Equation" (PE) method[13]. It was then demonstrated that this technique was also applicable to shallow water environments, particularly due to new developments in computer technology[14-16].

The purpose of this paper is to further demonstrate the applicability of the PE method to bottom-limited propagation problems. We first construct an environment where the implications of mode coupling are clear and we compare them with the PE results, demonstrating that these effects are indeed included in the PE representation. In the rest of the paper we consider propagation in different range-varying bathymetries.

I   THE PARABOLIC EQUATION AND NORMAL-MODE THEORY

In this section we construct an environment where we know what coupled mode theory would predict. We then explain why the adiabatic mode theory does not work whereas the PE result agrees with mode-coupling theory.

Consider the environment shown in Fig. 1. Propagation is from left to right with both source and receiver at a depth of 52 m. The bottom properties change abruptly at a range of 20 km, where we introduce a 10-m thick sediment layer with a lower sound speed than that of the water column. Thus we create here a sound

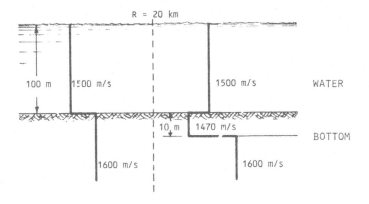

Fig. 1 Hypothetical environment for mode coupling study

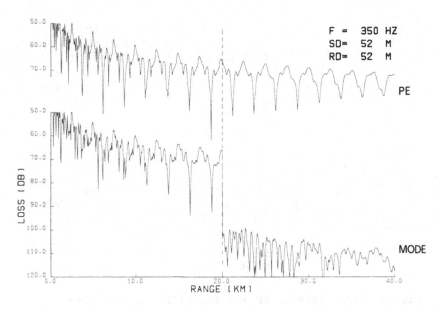

Fig. 2 *Comparison of PE and adiabatic mode results at a frequency of 350 Hz.*
(F: Frequency, SD: Source Depth, RD: Receiver Depth)

channel in the upper part of the bottom. Figures 2 and 3 show computed propagation losses from both PE[17] and adiabatic mode theory[18] for two different frequencies: 350 and 700 Hz. Before we discuss the results, we must mention that the abrupt change of environment depicted in Fig. 1 is such that the adiabatic approximation is not expected to work. The adiabatic theory works only for gradual changes with range, and the change under consideration should therefore rather be treated by a "sudden approximation"[19]. The reason for using the adiabatic approximation in this example will become apparent from the ensuing discussion.

In Fig. 2 we see that the adiabatic mode result for a frequency of 350 Hz shows an abrupt drop-off at a range of 20 km whereas the PE result shows no such change. At a frequency of 700 Hz (Fig. 3) nothing special seems to happen at a range of 20 km for either of the methods. In order to investigate what actually should happen, let us examine the relevant normal modes for this environment. Figure 4 shows two sets of modes for the 350 Hz case; Fig. 4a gives the first 6 (out of a total of 16) modes for the environment on the left of Fig. 1 (segment 1) and Fig. 4b shows

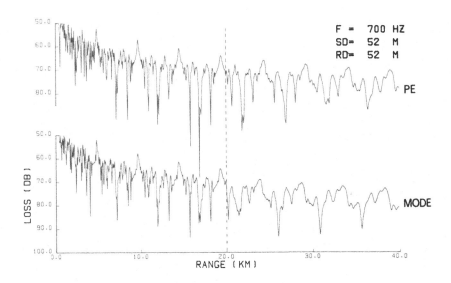

Fig. 3   Comparison of PE and adiabatic mode results
at a frequency of 700 Hz

the first 6 modes for the environment on the right of Fig. 1
(segment 2).   Adiabatic theory requires that modes adapt, or in
other words, that mode number one remains mode number one; mode
number two remains mode number two, etc.   With the chosen
source/receiver depths of 52 m only odd-number modes are excited
in segment 1 (Fig. 4a) and only even-number modes have non-zero
values at the receiver depth in segment 2 (Fig. 4b).   Thus at a
range of 20 km adiabatic theory requires that energy contained in
modes 1, 3, 5...etc. of segment 1 propagate in the corresponding
modes of segment 2, which, however, is not possible since these
modes cannot be excited in segment 2.   Hence we understand why the
adiabatic theory gives the sharp drop-off in propagation loss when
passing from segment 1 to segment 2.

Let us now investigate what a mode theory that allows for
energy exchange between modes would predict for this specific
case.   Mode-coupling theory (and also in this particular example,
the "sudden approximation") states that modes will couple to other
modes that have a similar shape as determined by a maximum value
of the "overlap integrals" between modes[19].   From Fig. 4 we see
that mode number one in segment 1 should therefore couple entirely
into mode number two of segment 2 since these modes have identical
shapes in the water column.   Similarly any mode in segment 1
should couple into the next higher mode in segment 2.   Furthermore, from the above discussion there should be virtually no

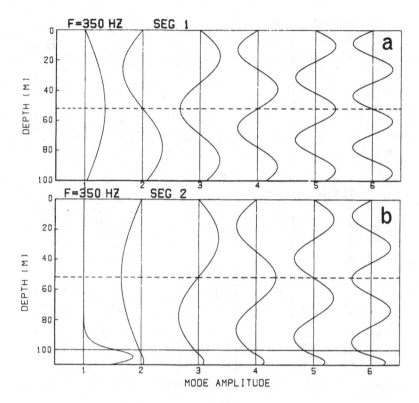

*Fig. 4 First six normal modes for each of the segments
at a frequency of 350 Hz*

coupling into mode number one in segment 2, which has the extra
mode occurring due to the change in environment. Hence there is
just a re-numbering of modes and we therefore expect nothing
special to happen as we go from segment 1 to segment 2. This in
turn means that the mode-coupling result for this particular
example should be identical to a range-independent normal-mode
result. Figure 2 shows that the PE calculation is not affected by
the sudden change in bottom properties at a range of 20 km, and it
can be shown that the PE result is virtually identical to the
range-independent mode result. Hence we may conclude that the PE
method correctly handles this simplified but drastic mode-coupling
problem.

Figure 5 shows the results for 700 Hz (32 modes) where in
this case the modes are shifted by two: mode number one couples
into mode number three, etc. We see from Fig. 3 that the PE
calculation gives the expected result of nothing drastic happening
in the propagation loss. However, this time the adiabatic calcula-

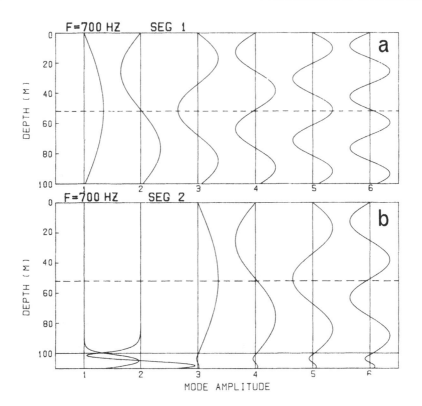

Fig. 5   First six normal modes for each of the segments
at a frequency of 700 Hz

tion looks better and that is because modes three, five, etc. are
excited both in segment 1 and segment 2. The mean level, and to
some extent the interference pattern of the adiabatic calculation,
seem to be in agreement with the PE result, but from the
discussion above, it is more accidental than due to a correct
handling of the physics in the adiabatic approximation. Figure 6
summarizes the discrepancy between PE and adiabatic mode theory.
At higher frequencies we get agreement in predicted propagation
loss at a range of 40 km because we here have more modes and the
possibility of hitting a sequence of nulls as shown in Fig. 4 is
therefore decreased.

In this section we have constructed a hypothetical case where
we believe we understand what mode coupling theory would predict
and we have shown that the PE results are in agreement with
coupling theory. We furthermore believe that the PE computation
is both faster and simpler than the mode-coupling calculation for
realistic range-dependent ocean environments.

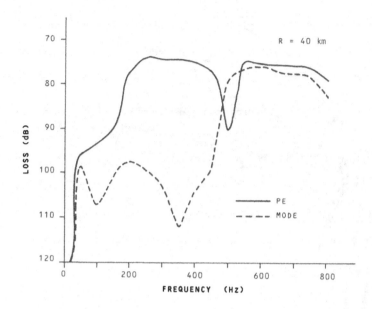

Fig. 6 *Comparison of PE and adiabatic mode results*
        at a range of 40 km

II   UP-SLOPE PROPAGATION

Having gained confidence in the PE method and in its ability
to handle mode-coupling effects in range-dependent propagation, we
now proceed to study propagation over sloping bottoms at a
sufficiently low frequency that phenomena such as mode cutoff and
mode conversion can be investigated in some detail[17]. In this
section we consider up-slope propagation for the environment given
in Fig. 7. The water/bottom interface is indicated on the contour
plot by the heavy line starting at 350 m depth and moving towards
the surface beyond a range of 10 km. The bottom slope is 0.85°.
The frequency is 25 Hz and the source depth is 150 m. The water
is taken to be isovelocity with a speed of 1500 m/s, while the
bottom is characterized by a speed of 1600 m/s, a density of
1.5 g/cm^3, and an attenuation of 0.2 dB/wavelength.

Before interpreting the contour plot, let us have a look at
the simplified sketch in the upper part of Fig. 7. Using the
ray/mode analogy, a given mode can be associated with up- and
down-going rays with a specific grazing angle. The sketch indicates a ray corresponding to a given mode. As sound propagates up
the slope, the grazing angle for that particular ray (mode)
increases, and at a certain point in range the angle exceeds the

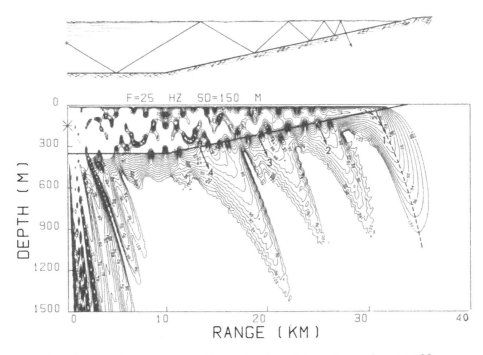

Fig. 7 Up-slope propagation showing discrete mode cutoffs

critical angle at the bottom, meaning that the reflection loss becomes very large and that the ray essentially leaves the water column and starts propagating in the bottom. The point in range where this happens corresponds to the cutoff depth for the equivalent mode.

To emphasize the main features in Fig. 7, we have chosen to display contour levels between 70 and 100 dB in 2 dB intervals. Thus high-intensity regions (loss < 70 dB) are given as blank areas within the wedge, while low-intensity regions (loss > 100 dB) are given as blank areas in the bottom. In this particular case four modes are excited at the source. As sound propagates up the slope we see four well-defined beams in the bottom, one corresponding to each of the four modes. This phenomenon of energy leaking out of the propagation channel as discrete beams has been confirmed experimentally[20], and a detailed study of this phenomenon using the PE method has been reported by the authors elsewhere[15]. Finally, it should be mentioned that the above is a particular example of mode coupling, where discrete modes trapped in the water column couple into "continuous" modes which account for sound radiation into the bottom.

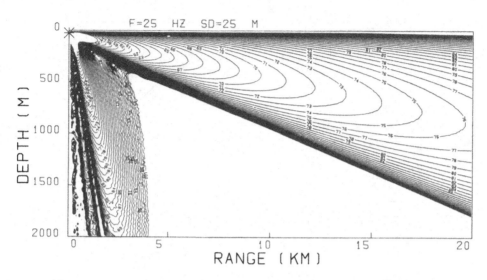

Fig. 8   Down-slope propagation over a constant 5° slope

III DOWN-SLOPE PROPAGATION

We now consider the problem of down-slope propagation as illustrated in Fig. 8. Water and bottom properties are as in the former example. The initial water depth is 50 m and the bottom slope is 5°. The source depth is 25 m and the frequency is 25 Hz. Figure 8 displays two regions of high sound intensity: First we have the sound propagating directly into the bottom (continuous modes) corresponding to sound hitting the bottom at grazing angles greater than the critical angle. Second we have the sound trapped in the water column (discrete modes) corresponding to sound hitting the bottom at small grazing angles. Only one mode is excited at the source while as many as 22 modes can exist at a range of 20 km where the water depth is 1800 m. The smoothness of the contour lines in the water column (no interference pattern) indicates that no additional modes are generated when propagating down the slope. In Fig. 9 we have plotted the field as a function of depth from the PE result at two ranges and we compare it with a mode calculation of the local first modes (for this plot we use normalized modes expressed in arbitrary decibels). We see in this example exact agreement between mode and PE results, indicating that, for this constant slope example, modes adapt and hence adiabatic mode theory is applicable.

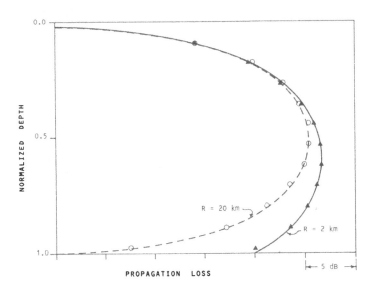

Fig. 9 Comparison of PE results (dots) with local mode
calculations (lines) at two different ranges

Figure 10 shows a PE calculation over the same 5° slope but this time the bottom abruptly becomes flat at a range of 5 km. Beyond 5 km we see a more complicated contour pattern indicating interference between modes and hence the initial first mode has generated (or coupled into) higher modes after the vertex. In this case as many as six modes can exist in the flat part beyond 5 km. That mode coupling is associated with the sharp change of the vertex is shown in Fig. 11 where we gradually go from a 5° slope to a flat bottom. Here we see that the first mode is gradually adapting itself. Thus, for this down-slope case, mode coupling seems to be associated with abrupt changes in bottom slope rather than with the slope itself.

IV   PROPAGATION OVER A SEAMOUNT

Recently an experiment has been performed in the northeastern Pacific studying sound propagation over a seamount[21]. The experimental data in turn have been compared with a small-scale laboratory experiment with good results[22]. In this section we apply the PE model to the same problem, which includes up- and down-slope propagation as studied in Sections II and III as well as profile diffraction effects and diffraction over the top of the seamount. Some aspects of propagation over seamounts have previously been studied using the PE model[23].

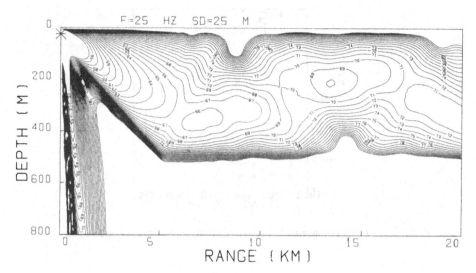

Fig. 10   Down-slope propagation over a 5° slope followed
          by a flat bottom

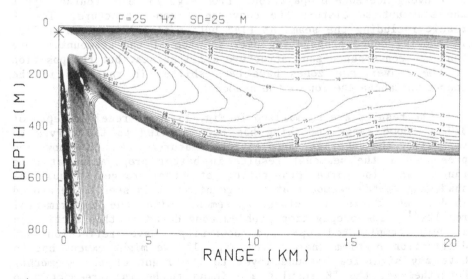

Fig. 11   Down-slope propagation over a smoothly changing slope

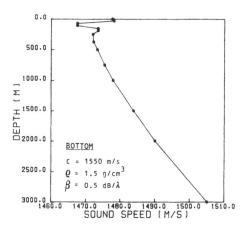

Fig. 12   Environmental input for seamount calculations

A detailed description of the environment is given in[22]. Figure 12 shows the sound-speed profile and the inputted bottom parameters. The computed propagation loss versus depth and range for a source depth of 18 m and a frequency of 230 Hz is shown in Fig. 13. Here the seamount (dashed lines) has not yet been put in. By appropriately selecting the contour intervals we essentially get a ray-type picture showing the characteristic features of convergence-zone propagation. From Fig. 13 we of course expect the seamount to disturb the convergence-zone structure. The PE result including the seamount (14° bottom slope) is given in Fig. 14 showing sound propagating over the top of the seamount. Note that the convergence-zone structure still exists, but the position of the convergence zones have been shifted in range as if the source was above the top of the seamount.

The results are summarized in Fig. 15 for a receiver depth of 329 m. Some range smoothing has been applied to the curves to emphasize the average propagation features. We see how the presence of the seamount results in better propagation at some ranges and in worse propagation at other ranges. Thus the shadowing of the seamount at a range of 80 km is seen to be around 15 dB, which is in close agreement with the experimental results[21]. The propagation problem considered in this section is a very complicated up- and down-slope wave reflection and diffraction problem where, from Fig. 13, we might expect not to have any significant sound level to the right of the seamount. Nevertheless, the PE results are found to be in agreement with experimental data.

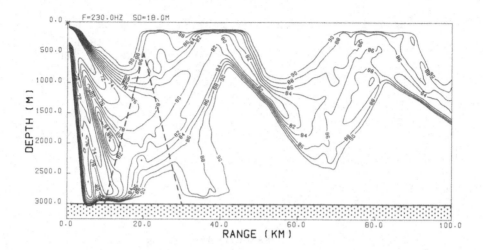

Fig. 13 *Loss contours showing convergence-zone propagation without the seamount*

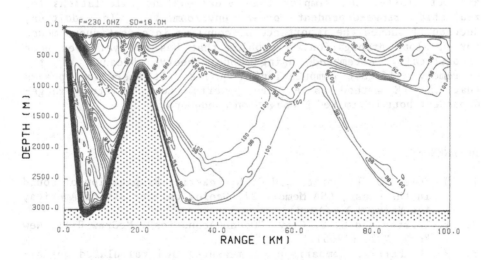

Fig. 14 *Loss contours showing propagation over the seamount*

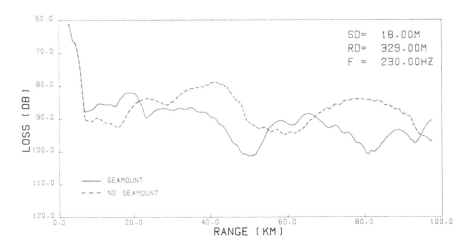

Fig. 15   Comparison of propagation loss levels with and
          without the seamount

CONCLUSIONS

The phenomenon of mode coupling seems to be adequately
included in the parabolic equation. Furthermore, PE computations
are both faster and simpler than mode-coupling calculations for
realistic range-dependent ocean environments. In addition,
Section II showed the importance of coupling into continuous modes
for up-slope propagation, a problem that has not yet even been
theoretically attempted using mode theory, because of the
extremely difficult computations required. It therefore appears
that the PE method is a very powerful tool to study range-
dependent bottom-limited propagation phenomena.

REFERENCES

1. M. Ewing, J. L. Worzel and C. L. Pekeris, Propagation of sound
   in the ocean, GSA Memoir 27, Geological Society of America,
   New York, N.Y. (1948).
2. I. Tolstoy and C. S. Clay, "Ocean Acoustics", McGraw-Hill, New
   York, N.Y. (1966).
3. R. H. Ferris, Comparison of measured and calculated normal-
   mode amplitude functions for acoustic waves in shallow
   water, J. Acoustical Society America 52:981-988 (1972).

4. F. Ingenito, Measurements of mode attenuation coefficients in shallow water, J. Acoustical Society America 53:858-863 (1973).

5. F. Ingenito, R. H. Ferris, W. A. Kuperman and S. N. Wolf, Shallow water acoustics: summary report, NRL Report 8179, U.S. Naval Research Laboratory, Washington, D.C. (1978).

6. A. D. Pierce, Extension of the method of normal modes to sound propagation in an almost stratified medium, J. Acoustical Society America 37:19-27 (1965).

7. D. M. Milder, Ray and wave invariants for SOFAR channel propagation, J. Acoustical Society America 46:1259-1263 (1969).

8. A. O. Williams, Normal-mode methods in propagation of underwater sound, in: "Underwater Acoustics", R. W. B. Stephens, ed., Wiley-Interscience, London, U.K. (1970), pp. 23-56.

9. A. Nagl, H. Uberall, A. J. Haug and G. L. Zarur, Adiabatic mode theory of underwater sound propagation in range-dependent environment, J. Acoustical Society America 63: 739-749 (1978).

10. S. R. Rutherford and K. E. Hawker, An examination of the influence of range dependence of the ocean bottom on the adiabatic approximation, J. Acoustical Society America 66:1145-1151 (1979).

11. S. R. Rutherford, An examination of multipath processes in a range-dependent ocean environment within the context of adiabatic mode theory, J. Acoustical Society America 66: 1482-1486 (1979).

12. F. S. Chwieroth, A. Nagl, H. Uberall, R. D. Graves and G. L. Zarur, Mode coupling in a sound channel with range-dependent parabolic velocity profile, J. Acoustical Society America 64:1105-1112 (1978).

13. F. D. Tappert, The parabolic approximation method, in: "Wave Propagation and Underwater Acoustics, Lecture Notes in Physics 70", J. B. Keller and J. S. Papadakis, eds., Springer-Verlag, New York, N.Y. (1977): pp. 224-287.

14. F. B. Jensen and H. R. Krol, The use of the parabolic equation method in sound-propagation modelling, SACLANTCEN SM-72, SACLANT ASW Research Centre, La Spezia, Italy (1975).

15. F. B. Jensen and W. A. Kuperman, Sound propagation in a wedge-shaped ocean with a penetrable bottom, J. Acoustical Society America 67: 1564-1566 (1980).

16. W. A. Kuperman and F. B. Jensen, Sound propagation modelling, in: "Underwater Acoustics and Signal Processing", L. Bjørnø, ed., Reidel Publishing Co, Dordrecht, The Netherlands (1980).

17. F. B. Jensen and W. A. Kuperman, Environmental acoustic modelling at SACLANTCEN, SACLANTCEN SR-34, SACLANT ASW Research Centre, La Spezia, Italy (1979). [AD A 081 853]

18. F. B. Jensen and M. C. Ferla, SNAP: the SACLANTCEN normal-mode acoustic propagation model, SACLANTCEN SM-121, SACLANT ASW Research Centre, La Spezia, Italy (1979). [AD A 067 256]

19. L. I. Schiff, "Quantum Mechanics", McGraw-Hill, New York, N.Y. (1968): p.292.

20. A. B. Coppens and J. V. Sanders, Propagation of sound from a fluid wedge into a fast fluid bottom. (This volume).

21. G. R. Ebbeson, J. M. Thorleifson and R. G. Turner, Shadowing of sound propagation by a seamount in the Northeast Pacific, J. Acoustical Society America 64:S76 (1978).

22. H. Medwin and R. Spaulding, The seamount as a diffracting body. (This volume).

23. F. D. Tappert, Selected applications of the parabolic equation method in underwater acoustics, in: "Proceedings of International Workshop on Low-Frequency Propagation and Noise", Maury Center for Ocean Sciences, Washington, D.C. (1977).

A LOW-FREQUENCY PARAMETRIC RESEARCH

TOOL FOR OCEAN ACOUSTICS

T.G. Muir, L.A. Thompson,
L.R. Cox, H.G. Frey

Applied Research Laboratories
The University of Texas at Austin
Austin, Texas, 78712, U.S.A.

INTRODUCTION

We conduct research in underwater sound for the primary purpose of supporting and advancing the state of the art in sonar. Efforts in oceanography and marine geophysics are considerably enhanced by this ultimate Naval motivation.

Unfortunately, many of the procedures, tools, and assumptions used in underwater acoustics research are not always compatible with sonar applications. The explosive shot, for example, is a popular and useful tool for propagation measurements, but its signal is an omnidirectional transient with a time-bandwidth product near unity, quite unlike a sonar signal, be it passive or active. Explosive shot data, averaged in one-third octave bands, tells us little about the spatial and temporal coherence of the medium and even less about the dynamics of sonar signal processing.

It is essential to conduct dedicated underwater acoustics work with versatile research tools of immediate significance to sonar and the issues confounding it. Future systems should require longer range, better accuracy, and higher probabilities of detection and classification. This means future sonars must have improved directivity, bandwidth, and signal processing.

As research tools, nonlinear sources offer a means of studying these requirements, especially with respect to quantifying both the limitations of the medium as well as the payoffs of new approaches and techniques. In the following pages, we report the development of one such research tool and indicate some of the issues to which it may be addressed.

PARAMETRIC ARRAYS

The fundamental processes of a parametric array are sketched in Fig. 1. A source simultaneously emits two high frequency primary waves at frequencies f_1 and f_2. These beat together in amplitude modulation. Nonlinear interaction occurs in a zone encompassed by the primary beams out to the range where the primary waves are absorbed. Each elementary volume in this irradiated zone becomes a nonlinear oscillator producing vibrations at the sum and the difference of the two original frequencies. The difference frequency radiation is of practical significance because of its high directivity, which is achieved at low operating frequencies. Large bandwidths, useful in signal processing, can also be achieved by this technique. The main advantage of the parametric technique is that it produces a radiation pattern at the difference frequency which is highly directive and is completely devoid of undesirable minor lobes. The main disadvantage of the technique is that it is one of low conversion efficiency. These topics will be addressed in the discussion of the present system.

Those unfamiliar with nonlinear acoustics may find one of the survey and review papers useful.[1,2]

THE PRESENT SYSTEM

Research and development on parametric arrays has progressed to the point where large systems for applications in ocean acoustics can be considered. The system described in this paper was intended as a research tool for a wide variety of measurements in shallow water over the continental shelf. A photograph of the transducer is shown in Fig. 2. It is a circular piston array 2.3 m in diameter, operating at primary frequencies in the neighborhood of 12.5 kHz, with difference frequencies in the neighborhood of 500 Hz to 5 kHz. Through nonlinear processes in the water, the 12.5 kHz radiation also produces energy at its harmonic frequencies. The transducer array consists of 720 mass loaded elements. Alternate elements are driven at slightly different frequencies centered around the primary center frequency to generate the two frequencies in the water which then interact to produce the difference frequency radiation. The circular holes in the transducer backplane are provided to house low frequency elements designed to receive echoes from the parametric difference frequency radiation as well as to transmit signals linearly in the low frequency band. These devices were not installed when the photograph was taken. These are also mass loaded elements with a resonance at 2400 Hz. The arrays are mounted on a stainless steel backing plate. Located behind the array is a canister to house matching transformers and miscellaneous electronics fixtures. The array and the canister with elements installed weighs approximately 11 kg.

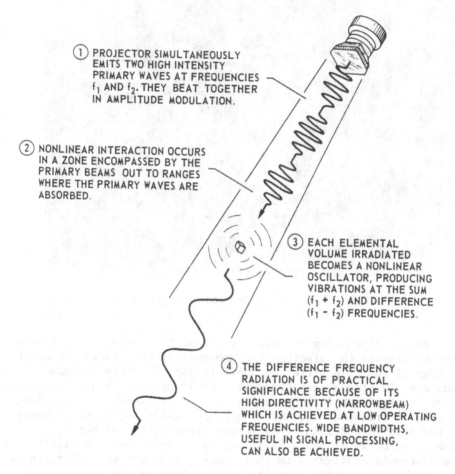

① PROJECTOR SIMULTANEOUSLY
EMITS TWO HIGH INTENSITY
PRIMARY WAVES AT FREQUENCIES
f_1 AND f_2. THEY BEAT TOGETHER
IN AMPLITUDE MODULATION.

② NONLINEAR INTERACTION OCCURS
IN A ZONE ENCOMPASSED BY THE
PRIMARY BEAMS OUT TO RANGES
WHERE THE PRIMARY WAVES ARE
ABSORBED.

③ EACH ELEMENTAL
VOLUME IRRADIATED
BECOMES A NONLINEAR
OSCILLATOR, PRODUCING
VIBRATIONS AT THE SUM
$(f_1 + f_2)$ AND DIFFERENCE
$(f_1 - f_2)$ FREQUENCIES.

④ THE DIFFERENCE FREQUENCY
RADIATION IS OF PRACTICAL
SIGNIFICANCE BECAUSE OF ITS
HIGH DIRECTIVITY (NARROWBEAM)
WHICH IS ACHIEVED AT LOW OPERATING
FREQUENCIES. WIDE BANDWIDTHS,
USEFUL IN SIGNAL PROCESSING,
CAN ALSO BE ACHIEVED.

Fig. 1.   Process in a parametric transmitting array.

Fig. 2.   Transducer array.

    Before discussing the performance of this array it is instruc-
tive to discuss the electronic components of the system. A block
diagram of the transmitter is shown in Fig. 3. This diagram
indicates 20 staves each consisting of 18 primary frequency elements
driven at f_1, with 20 identical staves of 18 elements each driven
at primary frequency f_2. As previously stated, adjacent elements
in the array are driven at different frequencies, either f_1 or f_2.
Each stave of 18 elements is driven by a two kW power amplifier and
the system is configured so that there are 20 two kW amplifiers at
f_1 and 20 two kW amplifiers at f_2 all driven in parallel. The
signal generator for each primary frequency consists of a micro-
computer controlled oscillator that permits automatic selection of
a wide variety of pulse shapes and pulse types. For example, a
primary frequency can be sequentially varied in ten steps from pulse
to pulse, starting at a low frequency and progressively stepping to
higher frequencies to provide the experimenter with sequentially
changing difference frequencies in the water. Capabilities for FM
chirp and CW pulses are also provided. Swept bandwidth can be
varied from a few hundred hertz out to a FM slide of 5 kHz.

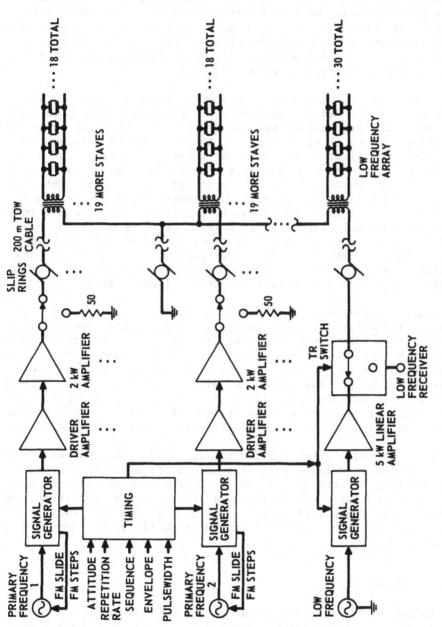

Fig. 3. Transmitter block diagram.

Sequential operation enables experiments to be done essentially simultaneously at different frequencies to permit the frequency dependence of an experiment to be determined very quickly. For linear transmissions the same kinds of signals can be applied to the low frequency array through a separate 5 kW low frequency driver system as shown at the bottom of Fig. 3. Provision is made for alternately exciting parametric and linear sources in the afore-mentioned stepped frequency pulse sequence.

A sketch of the transmitter system is shown in Fig. 4. The electronics is mounted in two separate mobile enclosures. To con-serve funds the transmitter was constructed of components from several existing U.S. Navy fleet sonars. It is built around a vacuum tube type power amplification system. The motor generator hut shown at left in the figure houses devices to convert 440 Vac to 5400 Vdc plate supply voltage; these are the main motor generator sets. This hut also houses auxiliary motor generators that provide the voltages needed for screen, grid, and filament supplies. The transmitter hut houses various controller devices as well as three stacks of power amplifiers consisting of 40 two kW power amplifier modules which supply a total of 80 kW of electrical power. This hut also contains auxiliary equipment, including a dummy load as well as the 5 kW power amplifier used to drive the low frequency linear elements.

Initial experimentation will be performed using a fixed platform, the Stage I facility of Naval Coastal Systems Center, Panama City, Florida. This facility is depicted in Fig. 5. It is located approxi-mately 20 km offshore in approximately 34 m deep water. The platform is about 30 m on each side. The transmitter electronics huts will be mounted on deck, and the transducer will be mounted on the rail-guided elevator seen near the left side of the near face of the platform. The fixed platform tests will be followed by mobile tests aboard a research vessel with the system configured as a variable depth sonar.

The tests will allow limited but controlled shallow water studies to be done first to isolate and assess essential parameters without the mitigating effects of ship motion and/or changing geographical conditions. Once the key features of the study are identified, spe-cialized experiments will then be conducted aboard the research vessel to ascertain how the findings from the fixed platform study stand up underway.

SYSTEM PERFORMANCE

Before moving this equipment to Florida, several measurements of its acoustic properties were made in fresh water at Lake Travis near Austin, Texas. Representative radiation patterns are shown in Fig. 6. On the left is the pattern for a 13 kHz primary radiation obtained from linear operation of the f_1 array. Next to this pattern

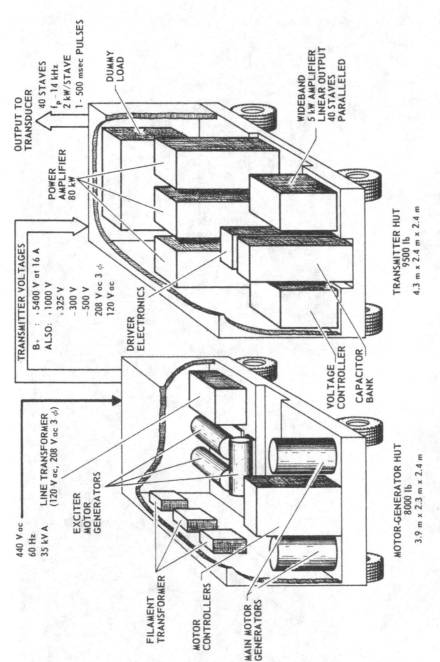

Fig. 4. Transmitter electronics equipment.

Fig. 5.   Stage I installation.

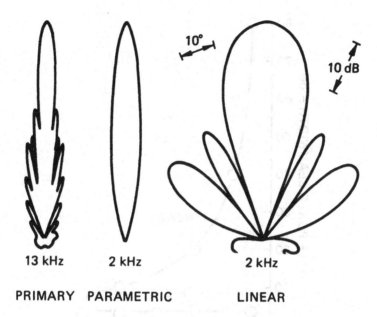

13 kHz   2 kHz   2 kHz

PRIMARY   PARAMETRIC   LINEAR

Fig. 6. Beam patterns.

is a 2 kHz difference frequency pattern obtained by driving the system
at 12 and 14 kHz and allowing the 2 kHz to be generated in the water
through parametric interaction. Notice the narrow beamwidth as well
as the complete absence of minor lobes. The pattern on the right is
that obtained from 2 kHz linear emission by the array of conventional
low frequency elements. Notice the considerably broader major lobe,
as well as the existence of minor lobes in this radiation pattern.
One of the major advantages of nonlinear acoustics is illustrated in
the comparison of the two patterns at 2 kHz: with a parametric beam
one realizes superdirective radiation without minor lobes, which is
ideal for viewing targets as well as for suppression of reverberation.

A comparison of half-power beamwidths versus frequency for the
parametric and linear arrays is shown in Fig. 7. The linear radiation
has a half-power beamwidth around 80° at 500 Hz which decreases with
increasing frequency to about 8° at 5 kHz, a decade change in frequency. On the other hand the parametric radiation has a relatively
constant beamwidth, shown here to be approximately 4°, over the same
frequency range. These data were taken at a nearfield range of only
80 m. In the farfield the parametric beamwidth should decrease to
approximately 2.5° and remain constant over the 500 Hz to 5 kHz

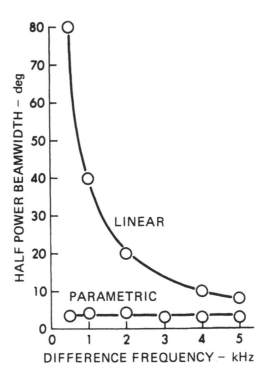

Fig. 7. Beamwidth characteristics.

frequency range. The fact that the narrow beam parametric radiation
can be designed to have a constant beamwidth over a decade of frequency change makes it an excellent research tool for conducting a
wide variety of experiments. Among these are measurements of reverberation, bottom penetration, scattering, and mode selection in the
shallow water waveguide.

Measurements of source level versus frequency of the system are
shown in Fig. 8. The primary frequency source level is in the neighborhood of 250 dB re 1 μPa at 1 m over the frequency range 10 to 16
kHz. Difference frequency source levels, extrapolated from nearfield
measurements at 80 m, vary from about 88 dB to 112 dB re 1 μPa at 1 m
over the same frequency range. This variation occurs because the
parametric interaction process improves in efficiency with increase
in difference frequency. These data are from measurements made
within the growth region of the parametric interaction volume.

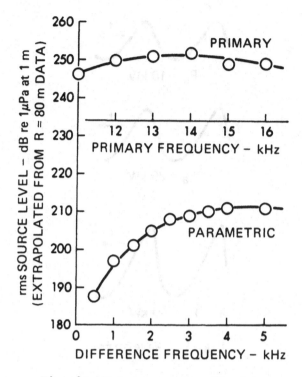

Fig. 8.   Frequency response data.

Values extrapolated from farfield measurements should be considerably higher, probably at least 10 dB.

Characteristic waveforms for radiation from this high intensity parametric source are shown in Fig. 9. Shown are primary frequency signals as a function of radiated acoustic power. With only 10 kW radiated power, a near sinusoidal waveform is produced. At 20 kW, the waveform is distorted, and at 40 kW radiated power shock waves are evident. Steep discontinuities occur at the shock front and asymmetry between the positive and negative pressure portions of the wave occurs. This asymmetry is the subject of studies in nonlinear acoustics centered around the role of diffraction in the distortion process. Shock waves are the hallmark of nonlinear sonar. The

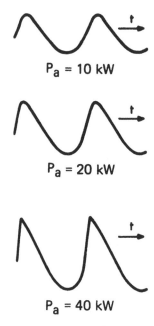

P_a = 10 kW

P_a = 20 kW

P_a = 40 kW

Fig. 9. Waveforms.

reason that shocked waveforms are desired is because they contain
useful harmonic radiations. A Fourier analysis of the waveforms of
Fig. 9 will show the existence of a series of harmonic radiations
which are useful both in sonar research and in sonar applications.

The utility of harmonic radiation is demonstrated in Fig. 10
which compares the radiation pattern of the system operated at the
fundamental, or first harmonic, shown on the left, with the pattern
at the second harmonic, center, and at the third harmonic, right.
The patterns become progressively narrower and minor lobe suppression
increases with increase in harmonic number. Theoretically it has
been shown that the harmonic patterns vary as the fundamental pattern
raised to the nth power, n being the harmonic number of the particular harmonic in question.

How useful are these harmonics? That, of course, depends on
their amplitudes, which are, in fact, much higher than the amplitude
of the difference frequency radiation. Amplitude data are shown in
Fig. 11 for the fundamental, second, and third harmonic of the
present experiment when 40 kW of acoustic power is radiated.   The

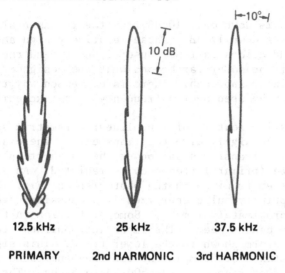

Fig. 10. Harmonic beam patterns.

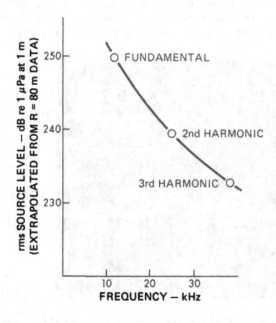

Fig. 11. Harmonic source levels.

second harmonic is down only 10 dB from the fundamental while the third harmonic is down 17 dB. These relatively high amplitudes make the harmonic radiations extremely useful and extend the usable frequency range of the sonar far beyond what the designer could accomplish by any linear approach. There is no reason why the present system could not be used over a frequency range exceeding two decades.

An interesting feature of a nonlinear parametric sonar is its ability to produce highly directed transients. These are known as self-demodulation transients and occur when a short pulse centered at some discrete (primary) frequency is radiated. An example for the present system is given in the photograph of Fig. 12. The upper trace shows a primary pulse energized in a Gaussian envelope with a duration of approximately 1 msec. Some distortion in this high frequency waveform can be seen. The important nonlinear feature is the demodulated waveform shown in the lower trace. This signal was received at a distance of 80 m in the low frequency band with a hydrophone having a flat response from 500 Hz to 5 kHz. The form of the low frequency demodulated wave is proportional to the second time derivative of the Gaussian envelope. For the demodulated signal in Fig. 12 the peak to peak excursion represents a source level in excess of 210 dB re 1 µPa, when these data are extrapolated from the measurement range of 80 m back to 1 m. If the measurement had been made in the farfield of the parametric array, the signal would have a higher source level. The demodulated signal contains a spectrum of frequency components; the fundamental in this example is about 2.5 kHz. Since the demodulated waveform is a wideband, highly directed entity, it is extremely useful in experiments relating to the philosophy of the impulse response and convolution techniques in signal processing.

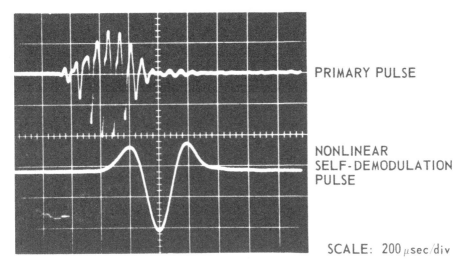

PRIMARY PULSE

NONLINEAR
SELF-DEMODULATION
PULSE

SCALE: 200 µsec/div

Fig. 12.   Self-demodulation of a primary radiation pulsed in a
           Gaussian envelope (ARL:UT research tool).

APPLICATIONS

Figure 13 is a list of research topics that will be addressed with the parametric system described in this paper. It is beyond the scope of this report to go into detail on any of these topics; however it will be useful to briefly discuss them here to indicate what is expected from this work.

The first topic is mode selection and the influence of thermal structure. A number of previous model studies have shown that the narrow beam parametric array can selectively excite modes in the shallow water waveguide. These studies will be continued at sea in the presence of solar induced thermal structure in the water to evaluate how the model studies translate into the real world.

The next topic is bottom loss comparisons, between continuous wave pings and explosive shots. Many scientists use explosive shots in their research while sonars use pings, and there is significant difference in the resulting signals. We will utilize the present

¤ MODE SELECTION AND THE INFLUENCE OF THERMAL STRUCTURE

¤ BOTTOM LOSS COMPARISONS (PINGS vs SHOTS)

¤ MODE CONVERSION OVER SLOPED BOTTOMS

¤ DOPPLER DIVERSITY IN FORWARD AND BACKSCATTERED SIGNALS

¤ BIOLOGIC ATTENUATION AND SCATTERING

¤ BOTTOM PENETRATION BY PARAMETRIC AND LINEAR BEAMS

¤ REMOTE SENSING OF SEDIMENT PARAMETERS

¤ ATTAINABLE RANGE RESOLUTION AND BANDWIDTH

¤ SPATIAL AND TEMPORAL COHERENCE IN PROPAGATING SIGNALS

¤ SIGNAL PROCESSING ASPECTS OF MODE SELECTION

Fig. 13. Summary of current shallow water acoustics research
Physical Acoustics Division, ARL:UT.

parametric system to compare between shots and pings and explore the relationships in the different types of data acquired. This will help to interpret shot data and lead hopefully to meaningful predictions and extrapolations with respect to sonar.

The third topic, mode conversion over sloped bottoms, is presently an active topic in acoustic propagation. With the present parametric system single propagation modes can be selectively excited. This will permit examination of how they are excited and how they are converted over sloped bottoms within a context of adiabatic mode propagation.

Doppler diversity is important to understanding sonar performance. Measurements of the frequency shifts wrought by currents and wind driven surfaces on the highly directive, pure tone signals that are generated by the parametric system will be made to quantify the Doppler diversity of the medium with much higher resolution than heretofore available.

The topic of biologic attenuation and scattering is an old one that will be examined with coherent signals over exceedingly wide frequency ranges, allowing for new contributions in this area.

Bottom penetration of parametric and linear beams has been studied in model tank experiments in the laboratory; results were recently published in the Journal of Sound and Vibration.[3] In these studies we observed the steep penetration of transmitted beams into sediments at angles below the critical grazing angle. Theoretical work to resolve the resulting philosophical difficulties has been done[4,5] and expanded experiments in support of these theoretical formulations will be carried out.

Remote sensing of sediment parameters is of great interest to the oceanographic community. With a narrow beam parametric system, it is possible to construct acoustic refractometers that can measure not only the critical angle in the bottom but also sound velocity in the sediment as well as its reflectivity. This should provide us with a tool to remotely sense the acoustic impedance of the bottom.

The next topic is attainable range resolution and bandwidth. Multimode excitation and intermodal interference in shallow water propagation limit the range of frequencies that can be used. When there are frequency limitations, there are also bandwidth limitations and consequently range resolution limitations. The parametric system will permit us to examine the maximum range resolution attainable in shallow water propagation.

Spatial and temporal coherence in propagating signals is important in signal processing. The medium affects the quality of the signal, which is measured by its spatial and temporal correlation functions. We will study these functions to determine how the medium affects a signal which might be designed or preferred for use in signal processing.

Signal processing aspects of mode selection refers to the reduction of signal degradation due to the medium by mode selection at the receiver. This is usually accomplished by using a vertical receiving array. With the parametric system described, we will study aspects of (vertical) mode selection of propagating signals with a view toward enhancing their coherence for signal processing.

ACKNOWLEDGMENTS

This work was supported by the U.S. Navy Office of Naval Research.

REFERENCES

1. T. G. Muir, "Nonlinear Acoustics: A New Dimension in Underwater Sound," in Science, Technology, and the Modern Navy, E. I. Salkovitz, Ed. (U.S. Government Printing Office, Washington, 0-242-122, 1977).

2. L. Bjørnø, "Parametric Acoustic Arrays," in Aspects of Signal Processing, G. Tacconi, Ed. (Reidel Publishing Co., 1977).

3. T. G. Muir, Claude W. Horton, Sr., and Lewis Thompson, "The penetration of highly directional acoustic beams into sediments," J. Sound Vib. 64 (4), 539-551 (1979).

4. J. N. Tjötta and S. Tjötta, "Reflection and Refraction of Parametrically Generated Sound at a Water/Sediment Interface," in Bottom Interacting Ocean Acoustics, W. Kuperman and F. Jensen, Eds. (Plenum Press, New York, 1980).

5. H. O. Berktay and A. H. A. Moustafa, "Transmission of a Narrow Beam of Sound across the Boundary between Two Fluids", in Bottom Interacting Ocean Acoustics, W. Kuperman and F. Jensen, Eds. (Plenum Press, New York, 1980).

TRANSMISSION LOSS VARIABILITY IN SHALLOW WATER

John E. Allen

U.S. Naval Oceanographic Office
NSTL Station
Bay St. Louis, MS   39522

ABSTRACT

Bathymetry, sediment type and sediment thickness information is required to explain the variability of transmission loss in shallow water. Bathymetry and sediment charts may indicate uniform sediment type and relatively uniform bathymetry in shallow water. This uniformity leads to transmission loss estimates with little, if any spatial variability. Incorporating sediment thickness into model predictions explains much of the variability which is observed in experiments. Data and model results are presented.

INTRODUCTION

Recent efforts to model the transmission of sound in both deep and shallow water have realized the importance of the geology of the ocean bottom. Although the sound speed profile continues to be the single most important environmental parameter in determining the interaction of sound with the geologically controlled ocean bottom, the bottom far surpasses the oceanographic controlled sound speed profile in complexity and lack of knowledge.

Several geoacoustic models are now available which can reproduce measured results. However, for the models to work, the environment must provide a set of range and bearing independent environmental parameters. The ability to predict the results of sound interaction with the ocean bottom in areas where the environment is not constant will require considerable expertise in understanding the interrelationship of geology and acoustic

propagation. Several examples of acoustic results are shown
which demonstrate the importance of range and depth related
geological parameters before predicting the propagation conditions in an ocean environment. These acoustic measurements were
taken under similar bathymetric and sound speed conditions.

Bathymetric blockage by seamounts in the water column form a
broadband sound barrier to sound energy propagating through the
ocean medium. Recent surveys by the U.S. Naval Oceanographic
Office point to a similar occurrence of sound blockage in sediments
by buried geological structures. This not easily observable
geological parameter of subbottom blockage effectively blocks
sound propagation in sediments as seamounts block the sound in
the sea.

DISCUSSION

Sound Speed in Water

The most important single environmental parameter in determining the effect of the ocean bottom on sound propagation is the
sound speed profile of the water mass, figure 1. A sound speed
profile with a negative gradient will cause the sound energy to
be refracted downward thereby causing a bottom limited condition
where energy is either reflected at the sediment water interface
or refracted by the sediments, figure 1(a). Positive sound speed

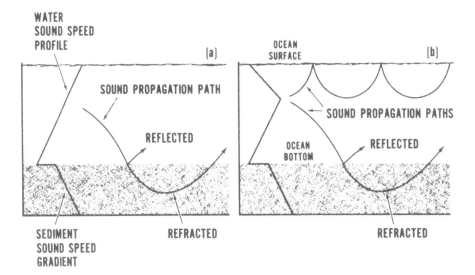

Figure 1.   Sound Speed Profile

profiles or shallow ducts, such as figure 1(b), result in the
sound energy being propagated through the water with less interaction with the ocean bottom. The effect of seasonal changes in
the sound speed profile in shallow water can cause severe
differences in the role the ocean bottom plays in contributing to
the transmission loss of sound in shallow water. Because of the
severe differences, the examples of transmission loss variability
presented in this paper will be restricted to areas where the
sound speed profiles and water depths are similar.

Sediments, Layering and Ocean Bottom Roughness

Figure 2 shows a sound speed profile and bathymetric profile
for a shallow water region. The bottom topography is gently
sloping from 110 to 150 meters along the station track. The
sound speed profile for the area has a slightly negative gradient
from surface to bottom. The surface sediment in the area is a
layer of sand and gravel varying in thickness from .1 to 3 meters.
The material under the layer is mud. The transmission loss
contours shown in figure 3 are derived from measured values at
each of the 1/3 octave bands shown and at 18 ranges from 50 to
125 kilometers. The transmission loss contours show the
optimum propagation frequencies from 50 to 315 hertz. Frequencies
above and below this preferred zone do not propagate as well.
The transmission loss results are low compared to other data
observed at nearby locations.

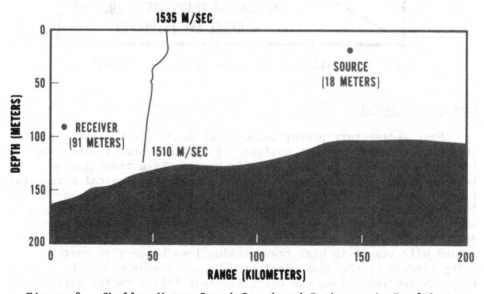

Figure 2.   Shallow Water Sound Speed and Bathymetric Profile

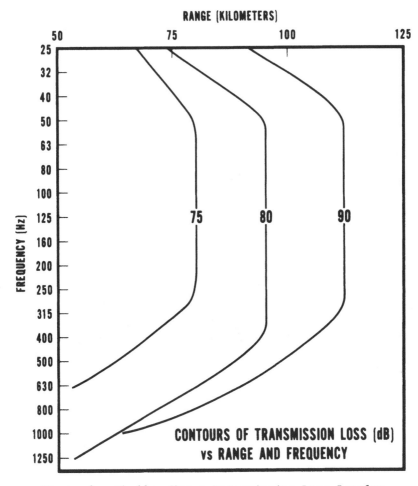

Figure 3.   Shallow Water Transmission Loss Results

Blockage of Sound

Many sedimentary properties control sound propagation through
sediments. However recent analysis of shallow water transmission
loss data has shown the relationship between high transmission
loss and subbottom blockage of sound by buried geological structures.
Figure 4 explains the effect of buried geological structures on
low frequency transmission loss. In figure 4 two parallel ridges
and a sediment filled valley are overlain by thin sediment cover.
Measurements of transmission loss across the ridge and valley
regime will result in high transmission loss because of energy
being blocked by the geological structure. Measurements of
transmission loss along the deep, sediment filled valley will
exhibit results characterized by lower loss and preferred propagation of low frequencies as shown in figure 3. The sediment

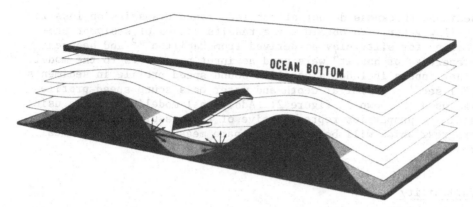

Figure 4.   Diagram of Buried Subbottom Structures Showing
            Preferred Propagation Direction

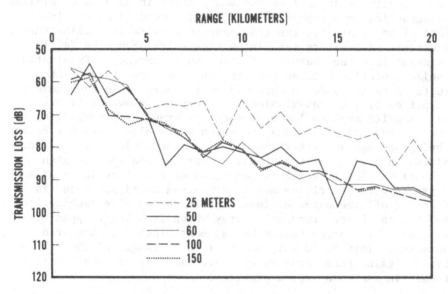

Figure 5.   Influence of Sediment Thickness on Transmission Loss

thickness over the ridge and valley structure of the basement
will control the effect of the buried structures on transmission
loss results.

    Figure 5 shows results from modeling shallow water trans-
mission loss using the U.S. Naval Research Laboratory Shallow
Water Normal Mode Model.[1]   This model shows the influence of
sediment thickness on transmission loss at 100 hertz.   From this
we can conclude that buried structures below about 100 meters of

sediment thickness do not affect long range transmission loss in
shallow water. To obtain these results values of sediment properties for silty clay as derived from Hamilton[2,3] and basement
properties of basalt[4] were used as input parameters to the model.
Other inputs included a positive sound speed profile in sediments
of 1 sec^{-1}. A similar depth and water mass sound speed profile
was used as shown in figure 2. Additional modeling efforts using
sediment properties representative of the measurement area are in
progress which will be used to compare measured and theoretical
results.

Variability

     Surface sediment distribution charts and bathymetry charts
may indicate an environmental condition in which little, if any,
spatial variability in transmission loss should occur. Figure 6
is a composite of several measurements taken in an area of similar
oceanographic, topographic and sedimentary conditions nearly
identical to figure 2. The transmission loss results illustrate
the extreme variability seen in an area which should be fairly
consistent from one measurement location to another. Examination
of seismic profile records confirm the existence of basement
structures buried under a thin sediment cover. This environment
is typified by the transmission loss contours shown in figure
6(a). Results show no low frequency preferential propagation.
Several frequency inversions are shown where high frequency energy
is being propagated better than the low frequencies. The surface
sediment in this area is responsible for the low loss at high
frequencies. The material is comprised of a surface layer of sand
and gravel with underlying mud as discussed earlier. Sound is
reflecting off the water sediment interface and experiencing less
loss than the lower frequency energy which is being trapped or
scattered in the ocean bottom by basement blockage. Measured
transmission loss could only be observed to ranges of 50-60 kilometers. Again, these measurements are taken in areas of similar
bathymetric and sound speed profiles.

     Figure 6(b) shows the variability that can occur at one
measurement location. The two contoured plots represent two
different directions of measurements from the same receiver. Note
that in one, the maximum range for measurement of transmission
loss is approximately 45 kilometers. The transmission loss in the
other direction shows the preferred propagation of low frequency
energy with measured results extending beyond 120 kilometers.
Examination of the bathymetric and subbottom profile records
clearly show the relationship between sediment covered topographic
blockage and the measured results.

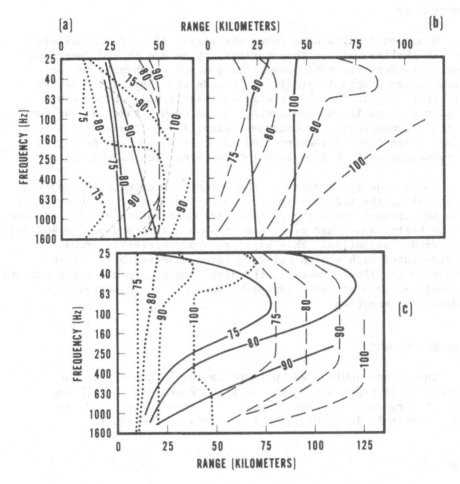

Figure 6.   Transmission Loss Variability in Shallow Water

Figure 6(c) shows three measurements of transmission loss.
The wide variation of the results are clearly observable if the 80 dB
contour interval is used as a comparison from group to group.
Also shown is the characteristic preferred propagation frequencies.
However, these frequency bands range from as low as 32-50 hertz to
50 to 315 hertz.   Propagation of energy is very good and the 80 dB
transmission loss contour extends to a range of 120 kilometers in
40 to 65 hertz band.   Also shown on figure 6(c) is a frequency
inversion of results from 500 to 1250 hertz at approximately 45
kilometers.   Results such as figure 6(c) are considered to be
representative of thick, uniform sediments without blockage.
Surface sediment conditions do not seem to be as important, as in
the results shown in 6(a).

CONCLUSIONS

Measured results have shown the variability of transmission loss in shallow water. This variability is due to the range and depth dependent sediment properites and not to changes in the water column. Seismic profiling records show buried geological structures within the thin sedimentary layer which block low frequency sound in the sediments. Measurements taken at the same location demonstrate the extreme variability of transmission loss. Transmission loss varied from 100 dB at 40 kilometers to values of transmission loss less than 100 dB beyond 120 kilometers.

The frequency response of the bottom can be clearly observed by examining the transmission loss as a function of frequency. Some measurements show the preferential propagation of low frequency (< 400 hertz) and other measurements show high frequency (> 400 hertz) preferred propagation. This high frequency preferred propagation is associated with a high velocity layer observed by a shallow subbottom profiler. However, the high frequency preferred propagation can only be seen at short ranges because of the suspected high sediment attenuation.

ACKNOWLEDGEMENT

The author would like to thank Carlos Belgodere for the geological interpretation used in this report and Steve Ewing for the modeling results. Appreciation is extended to Dr. W. Jobst of the Acoustic Projects Division for reviewing the paper.

REFERENCES

1. F. J. Miller and F. Ingenito, Normal Mode Fortran Programs for Calculating Sound Propagation in the Ocean, NRL Memo Report 3071, Naval Research Laboratory, Washington, DC (1975)
2. E. L. Hamilton, Sound Attenuation in Marine Sediments, NUC TP 281, Naval Undersea Research and Development Center, San Diego, CA (1972)
3. E. L. Hamilton, Variation of Density and Porosity with Depth in Deep-Sea Sediments, Journal of Sedimentary Petrology 46:280-300 (June 1976)
4. D. P. Mackensie, The Mohorovilic Discontinuity, in: "The Earth's Crust and Upper Mantle," J. P. Hart, ed., American Geophysical Union, Monograph No. 13, Washington, DC (1969)

INFLUENCE OF SEMICONSOLIDATED SEDIMENTS

ON SOUND PROPAGATION IN A COASTAL REGION

Suzanne T. McDaniel and John H. Beebe

Applied Research Laboratory
The Pennsylvania State University
P. O. Box 30
State College, PA 16801

ABSTRACT

In 1976, acoustic measurements were performed in shallow water off the coast of Jacksonville, Florida. Seismic refraction and Stonely wave experiments, performed as part of the measurement program, revealed that the seabed at the measurement site consisted of a semiconsolidated sediment overlain by a thin layer of sand. Compressional and shear wave velocities determined for the semiconsolidated sediment layer were 2400 m/sec and 670 m/sec, respectively. Using a normal-mode model that includes the effects of sediment rigidity, the attenuation of the first acoustic normal mode was computed and compared with experimental results for frequencies of 40 Hz to 800 Hz.

I. INTRODUCTION

In November of 1973, the Applied Research Laboratory (ARL) participated in a shallow-water field exercise conducted off the coast of Jacksonville, Florida by the Naval Research Laboratory. Analysis of the experimental data revealed that low-frequency acoustic energy was strongly attenuated. To investigate the cause of the large low-frequency losses, ARL returned to the same area in 1976 to perform a more complete set of acoustic and environmental measurements.

Acoustic experiments were performed to measure the attenuation of the first normal mode for frequencies of 40 to 800 Hz for propagation along two constant-depth paths at the same site. Environmental measurements to determine seabed properties included seismic

refraction, shear velocity and dispersion experiments. A bottom
model was formulated based on the results of the seabed measurements
and on the analyses of core and grab samples. Using this bottom
model, the attenuation of the first normal mode was predicted and
compared with the experimental results.

A most interesting result of the acoustic experiments was the
large difference in the measured low-frequency attenuation of the
first normal mode for the two propagation paths. The seabed experiments revealed that the bottom at the measurement site consisted of
a thin layer of sand overlaying a semiconsolidated sediment. Model
results show that the difference in the low-frequency loss measured
for the two propagation paths can be accounted for by variations in
the thickness of the surficial sand layer.

II. EXPERIMENTAL PROCEDURES

Figure 1 shows the locations at which experiments were performed
and where grab and core samples were obtained. The acoustic data

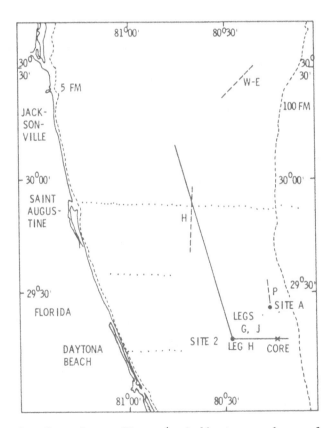

Fig. 1. Experiment Sites (• indicates grab samples).

presented in this paper were acquired along Legs G and H using a receiving array deployed in 32 m of water at Site 2. Seismic refraction and shear-velocity data were also acquired along these two legs. Previous seismic refraction experiments[1,2] performed in the area are shown by the dashed lines HE and WE. A short-range seismic experiment was also performed by ARL at Site A in 1973. At that time a long-range seismic experiment (P) was also performed.

Explosive charges were used as a source for all the experiments. The charges were deployed either by an aircraft (Leg J) or from a support ship. The air dropped charges were .82 kgm (equivalent TNT) SUS charges detonated at 18 m. Charges deployed from the USNS BARTLETT (T-AGOR 13) were 7 gm (equivalent TNT) PETN for the acoustic experiments and the equivalent of .45 kgm of TNT for the seismic refraction and shear-velocity measurements at Site 2. Small 7 gm charges were employed for the short-range seismic refraction experiment conducted from the USNS HAYES (T-AGOR 16) in 1973.

The receiver for the acoustic measurements consisted of a vertical array of eight hydrophones equally spaced to cover the water column. For the seismic refraction and shear-velocity measurements, a bottom-mounted sensor package containing two hydrophones and three mutually orthogonal geophones was used as a receiver. Signals received by the sensors were cabled to a surface buoy and telemetered to the support ship for recording and later analysis in the laboratory. The short-range seismic refraction experiments at Site A employed an array of hydrophones on the bottom which was hardwired to the support ship.

III. ACOUSTIC MEASUREMENTS

Experiments were performed to measure the attenuation of the first acoustic normal mode for propagation along two legs at Site 2. The bathymetry along the two propagation legs is shown in Fig. 2; sound-speed profiles were isovelocity varying by less than 1 m/sec throughout the water column. During the acoustic measurements, a

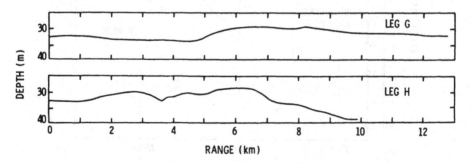

Fig. 2. Propagation Path Bathymetry.

sea state of 1 was recorded. Charges were detonated at a depth of 17 m which was close to a node of the second normal mode as is shown in Fig. 3. This source depth was selected to provide strong excitation of the first mode and to minimize excitation of the second mode. The signals received at hydrophones 3 through 6 were examined and, except ·at very short ranges and high frequencies, only the first mode was found to be present.

Data from three hydrophones were analyzed to determine the frequency dependence of the attenuation of the first normal mode. Total energy in one-third octave bands was determined for those shots free of contamination from higher-order modes. The energy levels were corrected for cylindrical spreading; the results shown for three frequencies in Fig. 4 are data from hydrophone 5. A straight-line, least-squares fit to the data is also shown in Fig. 4. Almost identical results were obtained for hydrophones 3 and 6.

As is evident in Fig. 4, at 63 Hz, where acoustic energy penetrates deeply into the bottom, a much higher attenuation was measured for Leg H than for Leg G. The difference in attenuation decreases with increasing frequency; at frequencies above 250 Hz, where little penetration into the sediment occurs, no difference in attenuation was measured for the two legs. Based on these findings, the observed differences in low-frequency attenuation are attributed to differences in the sub-bottom structure along the two legs.

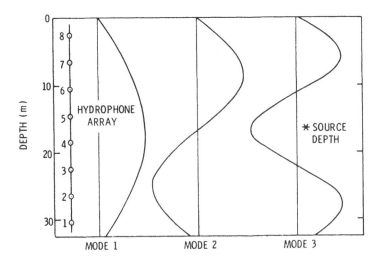

Fig. 3. Mode Excitation and Reception.

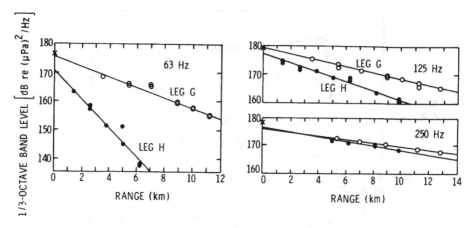

Fig. 4. Attenuation of the First Acoustic Normal Mode.

IV. SEABED MEASUREMENTS

Acoustic propagation in shallow-water is strongly influenced by
the character of the ocean floor. The attenuation due to bottom
interaction depends on the complex compressional and shear modulii
of the seabed. To determine these parameters, a variety of environmental measurements were performed. The compressional and shear-
velocity structure of the seabed was determined from the analysis of
dispersion data and the results of seismic refraction and Stonely
wave experiments. The analysis of core and grab samples provided
data from which absorption coefficients were derived.

Seismic Refraction

Previous seismic refraction experiments in the area showed a
thick (80-300 m) surficial layer of unconsolidated sediment having
a compressional velocity of roughly 1720 m/sec. Long-range seismic
data acquired along Legs G and H yielded much the same results.
The short-range seismic experiment performed at Site A in 1973
revealed, however, a quite different bottom structure. In this
experiment, small explosive charges were detonated at a range of
approximately 32 m from one end of a line array of hydrophones
deployed on the bottom. The spacing between hydrophones was 15 m.

Figure 5 shows typical data acquired using the bottom array.
In this figure, energy propagating through the seabed and arriving
before the water-borne arrival is evident. The data were analyzed
(using standard techniques) to determine the thickness of the surficial sediment layer and the velocities of the two uppermost
layers.

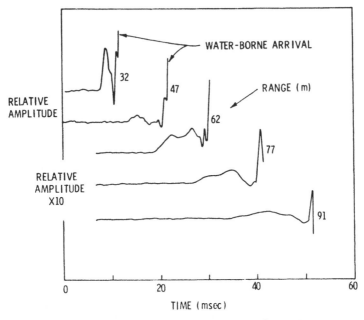

Fig. 5.   Short-Range Seismic Refraction Data.

The first arrival in Fig. 5 is a refracted arrival propagating
with a velocity of 2400 m/sec in a sub-bottom layer. A second
arrival having a velocity of 1700 to 2000 m/sec is identifiable in
Fig. 5 at ranges of 62 and 77 m. This arrival corresponds to propagation in the surficial sediment layer. The estimated thickness of
the unconsolidated sediment layer was 5 m--substantially lower than
the values obtained from the long-range experiments.

Dispersion Analysis

To verify that the same bottom structure was present at Site 2
as that deduced from the short-range seismic refraction experiments
at Site A, a dispersion analysis was performed. Data acquired on
Legs G and J were analyzed to determine the group velocity of the
first acoustic normal mode as a function of frequency. The results
of this analysis are shown in Fig. 6.

Also shown in Fig. 6 is the dispersion predicted for the bottom
models of Table 1, where H is the layer thickness, ρ is the density,
and C_p is the compressional velocity. The agreement between the
measured and predicted dispersion confirms the bottom structure
obtained from the short-range seismic experiment--a thin layer of
unconsolidated sediment over a higher velocity semiconsolidated
sediment.

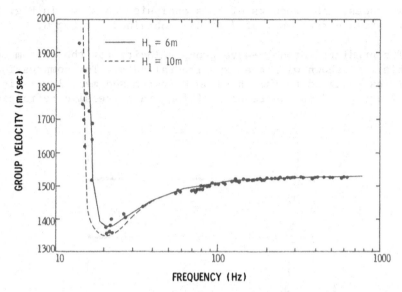

Fig. 6.   A Comparison of Measured and Predicted Dispersion.

Table 1.   Compressional Sound-Velocity Structure

Layer	H (m)	ρ (gm/cm^3)	C_p (m/sec)
water	32	1.0	1530
1	6, 10	1.8	1720
2	∞	2.0	2400

Stonely Wave Experiment

The seismic refraction and dispersion measurements described in
the two previous paragraphs provided a clear picture of the compressional velocity structure of the near-surface sediment layers.   To
determine shear velocity as a function of depth in the sediment,
experiments were conducted to measure the velocity of Stonely waves.
These waves were generated using .45 kgm charges of TNT detonated
on the bottom and were detected by bottom-mounted geophones.

Raw data acquired on Leg H is shown in Fig. 7.   The low-frequency late arrival has been identified as a Stonely wave.
Filtering the signals revealed two clearly dispersed modes of propagation, as is shown in Fig. 8, in which the large-amplitude first
mode is preceded by a lower-amplitude second mode.   The data were
analyzed to determine group velocity as a function of frequency for

the two modes. The results of this analysis are shown in Fig. 9.
No data above a frequency of 7 Hz were obtained for Leg G.

The predicted Stonely-wave group velocity for the bottom model
of Table 2 is shown with the experimental data. The compressional
velocities assigned to the third and fourth sediment layers in
Table 2 are based on the results of long-range seismic refraction

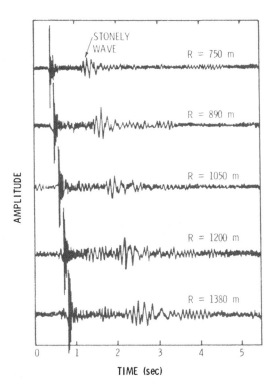

Fig. 7. Vertical Geophone Signals--Stonely Wave Experiment.

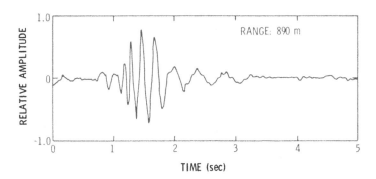

Fig. 8. Filtered Stonely Wave Data--Vertical Geophone.

Table 2.   Shear Velocity Structure

Layer	H (m)	ρ (gm/cm^3)	C_p (m/sec)	C_s (m/sec)
water	32	1.0	1530	0
1	3	1.8	1720	200
2	40	2.0	2400	670
3	60	2.0	2400	900
4	∞	2.3	3570	2100

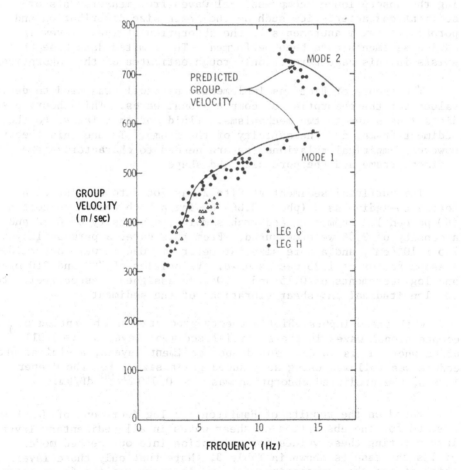

Fig. 9.   A Comparison of Measured and Predicted Stonely Wave Group Velocities.

experiments. As the predicted Stonely-wave group velocity is
insensitive to compressional velocities, the values are presented
for completeness only. The predictions were also found to be insensitive to the shear velocity assumed for the surficial sediment
layer--but were strongly dependent on the first layer thickness.
The maximum layer thickness that provided agreement with the Leg H
data was only 3 m, however, the Leg G data could accommodate a
layer 10 m thick.

Absorption

Whereas the velocities of compressional and shear waves could be
derived from seabed experiments, no direct measurement of the absorption of these waves (at the frequencies of interest) was possible.
A strong theoretical and empirical base exists, however, for predicting the absorption of compressional waves from measurements of
sediment characteristics such as the grain size distribution and
porosity. Few measurements of the absorption of shear waves in
marine sediments have been performed. The limited data base that
exists in this case permits only rough estimates of the absorption.

The theory of Biot[3] as implemented by Stoll[4] was used to derive
values for the absorption of compressional waves. This theory predicts losses due to two mechanisms: fluid motion relative to the
sediment frame, and inelasticity of the frame. To use this theory,
however, empirical relationships are needed to characterize the
sediment frame and the pore size ahd shape.

The surficial sediment at Site 2 was found to consist of a
coarse-to-medium sand (phi = 0.86) having a high carbonate content
(60 percent). From core and grab samples, a porosity of 0.38 and
a density of 2.06 were measured. From this data, a permeability of
$7.5 \times 10^{-7} cm^2$, and a pore size parameter of .034 mm were determined.
A shape factor of 1.25 was assumed. Velocities of 700 and 210 m/sec
and log decrements of 0.15 and 0.20 were assigned, respectively, to
the longitudinal and shear vibrations of the sediment frame.

With these inputs, Biot's theory predicted an absorption of
compressional waves in the surficial sediment layer of $\alpha = 0.011 f^{1.76}$
dB/km where f is in Hz. For deeper sediment layers, a similar procedure was followed using an assumed grain size. For the deeper
layers, the predicted absorption was $\alpha = 0.010 f^{1.00}$ dB/km.

Based on the results of Hamilton[5], a log decrement of 0.30 was
assumed for the absorption of shear waves in all sedimentary layers.
Incorporating these values for absorption into our seabed model
yields the results shown in Table 3. Note that only those layers
which affect the propagation of acoustic energy at the frequencies
of interest have been included in Table 3.

Table 3. Attenuation

Layer	Thickness (m)	ρ (gm/cm^3)	C_p (m/sec)	k_p*	n_p*	C_s (m/sec)	K_s*	n_s*
water	32	1.0	1530	0	0	0	0	0
1	3-10	2.0	1720	.011	1.76	200	10	1.00
2	∞	2.0	2400	.010	1.00	670	5	1.00

*$\alpha = K f^n$ dB/km, f in Hz.

V. COMPARISON OF THEORY AND EXPERIMENT

Using the bottom models formulated in Section IV, the attenuation of acoustic energy in the Site 2 environment was predicted as a function of frequency. For this purpose, a normal-mode model that treats a bottom consisting of an arbitrary number of visco-elastic layers was used. The model is an extension of Haskell's[6] matrix method to include absorption, and to treat an arbitrary sound-speed profile in the water column.

The model uses trial values of the modal wavenumber to compute the complex ratio $(\partial\phi/\partial z)/\phi$ at the ocean bottom where ϕ is the velocity potential. The real part of this ratio is used in an iterative search routine to find the real part of the modal wavenumber. The imaginary part is then used to compute the mode attenuation-- the imaginary part of the wavenumber.

Predictions of the attenuation of the first acoustic normal mode were made using the bottom model of Table 3. The shear velocity of the uppermost sediment layer, and the absorption of shear waves in both layers were found to have no effect on the predicted attenuation. The attenuation predicted by the model was due to two causes: at low frequencies, conversion of compressional waves to shear waves at the interface between the first and second sediment layers; and at high frequencies, absorption by the surficial sediment.

Figure 10 presents the measured attenuation of the first mode as determined in Section II for Legs G and H. Shown for comparison is the predicted attenuation for two assumed first layer depths. As is evident, differences in the depth of the surficial sediment layer suffice to account for the large differences in measured low-frequency attenuation.

At frequencies above 250 Hz, where losses are due to absorption in the surficial sediment layer, the modeled attenuation is in

general agreement with the data and, hence, supports the frequency
dependence for sediment absorption derived from Biot's theory. We
note that previous investigators[7],[8] have measured the same frequency
dependence for absorption in sands. At 800 Hz, the predicted attenuation is roughly one-half of that measured; the lack of agreement
may be due to a failure of the method used to isolate the first
mode.

VI. SUMMARY AND CONCLUSIONS

    Experiments to measure the attenuation of the first acoustic
normal mode were performed in shallow water off the coast of
Jacksonville, Florida. The acoustic measurements were supported by
a variety of environmental experiments performed to ascertain the
seabed structure. From the seabed measurements, it was determined
that the bottom consisted of a semiconsolidated sediment (having
compressional and shear velocities of 2400 and 670 m/sec, respectively) which was overlain by a thin surficial layer of coarse
sand. Differences in the thickness of the surficial layer were
found to account for the large differences in low-frequency loss
obtained for two propagation paths·at the measurement site.

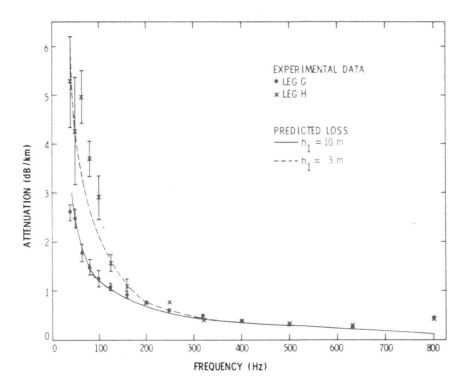

Fig. 10. A Comparison of Measured and Predicted Mode Attenuation.

A theory due to Biot was used to predict sediment absorption from core and grab sample data. At high frequencies where sediment absorption was the dominant loss mechanism, the measured mode attenuation agreed well with the predictions. These findings, along with the results of earlier studies, indicate that the absorption of compressional waves in sands is strongly dependent on frequency.

ACKNOWLEDGEMENTS

The authors wish to thank I. P. Sechrist for programming and computational assistance. The consultations provided by H. Kutshale of the Lamont-Doherty Geological Institute, who made available the program KUPVEX for the Stonely wave computations, are gratefully acknowledged. The authors also wish to thank O. Pilkey of Duke University for the analysis of grab samples. This research was performed at the Applied Research Laboratory under the sponsorship of the Naval Sea Systems Command.

REFERENCES

1. J. B. Hersey, E. T. Bunce, R. F. Wyrick, and F. T. Lietz, "Geophysical Investigation of the Continental Margin Between Cape Henry, Virginia, and Jacksonville, Florida," Geol. Soc. Am. Bull. 70, 437-466 (1950).

2. J. L. Worzel and M. Ewing, "Explosion Sounds in Shallow Water," Geol. Soc. Am. Mem. 27 (1948).

3. M. A. Biot, "A Generalized Theory of Acoustic Propagation in Porous Dissipative Media," J. Acoust. Soc. Am. 34, 1254-1265 (1962).

4. R. D. Stoll, "Acoustic Waves in Saturated Sediments," in: Physics of Sound in Marine Sediments, Loyd Hampton, ed., Plenum Press, New York (1974).

5. E. L. Hamilton, "Attenuation of Shear Waves in Marine Sediments," J. Acoust. Soc. Am. 60, 334-338 (1976).

6. N. A. Haskell, "The Dispersion of Surface Waves on Multilayered Media," Bull. Seism. Soc. Am. 43, 17-34 (1953).

7. F. Ingenito, "Measurements of Mode Attenuation Coefficients in Shallow Water," J. Acoust. Soc. Am. 53, 858-863 (1973).

8. W. R. Caswell, "Mode Attenuation Measurements in Shallow Water off the Coast of Panama City, Florida," J. Acoust. Soc. Am. Suppl. 1 65, S18(A) (1979).

GEOACOUSTIC MODELS OF THE SEABED TO SUPPORT

RANGE-DEPENDENT PROPAGATION STUDIES ON THE SCOTIAN SHELF

John H. Beebe and Suzanne T. McDaniel

Applied Research Laboratory
The Pennsylvania State University
P. O. Box 30
State College, PA   16801

ABSTRACT

Joint long-range propagation experiments were performed at two sites in the Western North Atlantic by the Applied Research Laboratory (ARL) of The Pennsylvania State University, and the Defence Research Establishment Atlantic (DREA) of Dartmouth, Nova Scotia. One result of this program was the development of geoacoustic models of the seabed along the propagation tracks between the shallow (ARL) and deep (DREA) receiving arrays. The seabed models are based on profiling performed with the Huntec Deep-Tow System, grab sample data supplied by the Bedford Institute of Oceanography of Dartmouth, Nova Scotia, and dispersion and seismic refraction analysis. Each track is divided into a number of regions, with each region characterized by a geoacoustic model of two to four layers. Compressional wave attenuation values are predicted for each layer, from Huntec and grab sample data, by the Biot-Stoll Sediment model and by historical data. Using the seabed models as inputs to an adiabatic, normal-mode propagation model, transmission loss is predicted and compared with experimental losses measured at the ARL array.

INTRODUCTION

Joint long-range propagation experiments were performed at two sites in the Western North Atlantic during September 1978. The participating laboratories were the Applied Research Laboratory of The Pennsylvania State University (ARL) and the Defence Research Establishment Atlantic (DREA) of Dartmouth, Nova Scotia. In late August prior to the joint experiment, the propagation tracks were

profiled by CFAV SACKVILLE (AGOR 113) using the Huntec Deep-Tow
System. This paper documents geoacoustic models of the seabed,
based primarily on the bottom and sub-bottom profiles obtained with
this system, for the tracks between the ARL and DREA positions at
both sites. Surficial sediment and bedrock data obtained from the
Bedford Institute of Oceanography (BIO) of Dartmouth, Nova Scotia,
and dispersion and seismic refraction analysis performed by ARL
were also used to formulate the seabed models. The models so
obtained were used as inputs to a range-dependent, normal-mode
propagation model. Using the propagation model, transmission loss
was calculated as a function of range from the ARL array for frequencies of 25, 80, and 250 Hz at Site 1 and 250 and 800 Hz for
Site 2. Experimental data was analyzed to determine propagation
loss in one-third octave bands at the same frequencies and compared
with the predicted values.

GEOACOUSTIC MODELS

Huntec Processing

    Parrott, et al.[1] indicate that the processed Huntec data has
proven to be a very reliable indicator of surficial sediment type.
The Huntec system uses a depth-compensated electro-dynamic source
which produces a repeatable pulse with an energy content of about
500 joules with a peak power of one kilowatt. The width of the
primary pulse is 63 µsec and the broadband frequency spectrum is
centered at 5 kHz. Sediment classification is possible due to the
repeatable pulse and a statistical analysis of the returned energy
found in two selected time windows, r_1 and r_2. The first time
window r_1 (40 µsec) encompasses the first bottom return and measures the coherent fraction of the energy reflected from the ocean-
sediment interface. The second time window r_2 (960 µsec) gives a
measure of the incoherently scattered reflection from non-normal
directions and from shallow-sub-bottom reflectors. The reflectivity
values r_1 and r_2 plotted along with the sub-bottom profile represent the processed Huntec data. The original (real-time) profile is
also available for analysis.

Layer Identification and Velocities

    Using processed Huntec data and surficial sediment maps[2,3],
sediment types of gravel and sand, sand and gravel, silt, and clay
were identified along the propagation tracks. The intermediate
layers were, in most cases, identified from the original analog
Huntec record. The identification of sediments below the water-
bottom interface was based on the appearance of the fine structure
on the record.

    The bedrock layers were identified in many cases on the analog
records. In those cases where the depth-to-bedrock could not be

determined, geological cross sections[2,3,4] were used to estimate the bedrock interface depth.

The velocities for the surficial sediment layers were obtained from the regression equation of Anderson[5] which uses mean grain size M_ϕ (in phi units) as the independent variable. The velocity (in m/sec) at a temperature of 20°C and a pressure of one atmosphere was calculated from the expression:

$$v = -68.156\,M_\phi + 3.053\,M_\phi^2 + 1874.4 . \tag{1}$$

These velocities were then corrected to the bottom temperature and depth. Values for the mean grain size obtained from grab sample analyses or estimated from Huntec reflectivity measurements provided an input to Equation (1).

Equation (1) is not applicable to very coarse sands and gravels. In cases where such sediments were present, other methods including seismic refraction and dispersion analyses or use of a sediment model were used to determine velocities. The sediment model used was based on the classical theory of Biot[6] as recently implemented by Stoll[7].

Velocities for intermediate layers were based on the identification of sediment type, the association of a mean grain size with the sediment type, and a correction for depth below the sediment-water interface. Bedrock velocities were based on geological age. Tertiary and Cretaceous formations were assigned velocities of 2000 and 2300 m/sec, respectively. A third bedrock type, which consists largely of granite, was present in the test area. The granite has been identified[4] as Cambrian to Devonian with a velocity of 4800 to 5500 m/sec.

Compressional Wave Attenuation

Propagation model studies required estimates of compressional wave attenuation for each layer. Grain size distribution data was obtained from Dr. Lewis King at BIO for sample stations nearest the tracks. The mean grain size M_ϕ and standard deviation σ_ϕ (in phi units) were computed and used for calculations of permeability, and a pore size parameter. The permeability and pore size parameter were then used in the sediment model. At Site 2, the mean grain size was determined from Huntec data and an appropriate value for σ_ϕ was assumed. Attenuation values were then predicted by the sediment model.

For deeper layers, the variation of attenuation with depth has been discussed by Hamilton[8], where two types of variation were noted. For sands, he recommends a reduction in the one-meter depth attenuation by $D^{-1/6}$, where D is the desired depth in the appropriate

unit. Silts and clays have been shown to increase in attenuation to
a depth of 100-200 m and then decrease thereafter. The appropriate
increase was predicted by linear interpolation between the predicted
K value at 1 m (K in dB/km·Hz from the equation $\alpha = K f^n$ is
obtained from grain size or porosity data) and a K value of .21 at
100 m. This procedure has been followed where appropriate, that is,
for cases in which a silt layer was overlain by a clay or silt
layer.

For many of the sections, the second layer was a till layer.
An estimate of the attenuation was made by assuming that the till
was a poorly sorted medium sand with $M_\phi = 1$ and $\sigma_\phi = 4$.

Attenuation for the bedrock layers was obtained from data[9] for
the Cretaceous strata and estimated for the Tertiary strata. King[4]
states that the Wyandot Formation composed of chalk dominates the
Cretaceous unit. White[9] lists attenuation for "Chislehurst Chalk"
as $\alpha = .009 f$ dB/km, thus, this value was used where required. The
Tertiary and latest Cretaceous strata appear to be dominated by the
Banquereau Formation[4]. This formation contains sediments ranging
from muds, to silts and sands, and according to King, the top of
the section consists of medium-grained, well-sorted sands. A value
of $\alpha = .05 f$ was assigned to the Tertiary bedrock. Attenuation for
various granite samples[9] ranged from .005 f to .235 f. A value of
.10 f was used. Shear attenuation for the granite was estimated as
.50 f; for other bedrock types, shear was assumed to contribute very
little to losses.

Site 1 Bottom Model

The sediment structure determined for the propagation track at
Site 1 is shown in Figure 1. The acoustic parameters assigned to
designated sections are presented in Table 1. Three sections re-

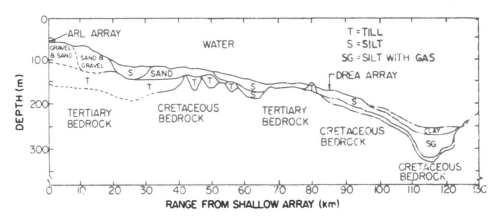

Figure 1. Sediment Structure—Site 1.

Table 1. Seabed Parameters—Site 1

Range (km)	Water Depth (m)	Layer	M_ϕ (phi)	σ_ϕ (phi)	Thickness (m)	Vel (m/sec)	Density (g/cm³)	K (dB/km·Hz)	n
0	57	1	1.2	0.6	50	1900	2.06	.0090	1.78
		2			50	1950	2.10	.0065	1.30
		3			∞	C 2000	2.20	.0500	1.00
						S 900			
6	67				SAME AS ABOVE				
14	76	1	1.2	0.6	50	1850	2.06	.0090	1.78
		2			30	1950	2.10	.0065	1.30
		3			∞	C 2000	2.20	.0500	1.00
						S 900			
22	106	1	6.3	2.7	30	1600	1.56	.0560	1.01
		2			50	1900	2.00	.0065	1.30
		3			∞	C 2300	2.30	.0090	1.00
						S 1200			
28	118				SAME AS ABOVE				
40	126	1	3.0	1.7	28	1650	1.96	.0370	1.14
		2			24	1900	2.10	.0065	1.30
		3			∞	C 2300	2.30	.0090	1.00
						S 1200			
53	137	1	7.0	2.3	13	1513	1.52	.0570	1.00
		2			21	1850	2.10	.0065	1.30
		3			∞	C 2000	2.20	.0500	1.00
						S 900			
62	150	1	5.8	1.7	25	1550	1.52	.0570	1.00
		2			10	1850	2.10	.1020	1.00
		3			∞	C 2000	2.20	.0500	1.00
						S 900			
74	161	1	4.2	1.7	10	1580	1.56	.0560	1.01
		2			∞	C 2000	2.20	.0500	1.00
						S 900			
90	214	1	4.2	1.7	18	1580	1.58	.0560	1.01
		2			6	1800	2.10	.0870	1.01
		3			∞	C 2000	2.20	.0500	1.00
						S 900			
112	265	1	8.0*	2.0*	7	1480	1.54	.0560	1.00
		2			10	1800	2.10	.0720	1.01
		3			6	1850	2.10	.0870	1.01
		4			∞	C 2300	2.30	.0090	1.00
						S 1200			
120	236	1	8.0*	2.0*	10	1480	1.54	.0560	1.00
		2			40	1550	1.67	.1120	1.00
		3			6	1850	2.10	.0870	1.01
		4			∞	C 2300	2.30	.0090	1.00
						S 1200			

*M_ϕ deduced from Huntec Reflectivities; σ_ϕ assumed.

quired modeling techniques somewhat different from those discussed
in previous sections and, hence, are considered in some detail.
The data is summarized for the other sections.

In Table 1, grab samples near the array (0-6 km) indicate a
surficial sediment layer consisting of gravel and sand which is
probably very thin. Velocity calculations for the mean grain sizes
obtained from the grab samples predict very high velocities for this
section. A somewhat lower velocity, 1900 m/sec, has been assumed.
This assumption appears justified based on dispersion analysis of
data from this site.

The intermediate layer for this region (0-6 km) consisted of
glacial till, estimated to be around 50 m thick. The layer was
assumed to be semiconsolidated with a velocity (1950 m/sec) higher
than that of the first layer. The bedrock for this region was from
the Tertiary period[4] with an estimated velocity of 2000 m/sec.

Seismic refraction analysis, performed near the array confirmed
this interpretation to some extent. The most recent analysis predicted a thick layer with a velocity of 1820 m/sec.

Results for the region near 14 km were obtained in a similar
manner. However, a lower velocity was assigned to the surficial
sediment layer on the basis of Huntec reflection data.

The region at the opposite end of the track (near 120 km in
range) lay in the center of a sedimentary basin. It consisted of a
thin clay layer overlying a pocket of silt with gas-seeps diffused
throughout. Grab sample data was not obtained for this region so
the mean grain size indicated in Table 1 was assumed for the clay
layer. The silt layer with diffused gas was treated as an unconsolidated gas-sand which has been shown[10] to have lower velocity
and higher attenuation than a normal sand. The silt for this layer
was modified by assuming a mean grain size as for normal silt, but
with the fluid parameters in the sediment model modified to represent seawater with a ten percent concentration of methane gas. The
resulting velocity and attenuation changed as expected, and were
further modified according to the depth below the seabed to give the
values in Table 1. For the remaining sections, surficial sediments
and intermediate and bedrock layers were characterized by the procedures noted earlier.

Site 2 Bottom Model

The geoacoustic model for Site 2 was based largely on Huntec
data, along with the consideration of grab sample analyses, seismic
refraction results and historical data. Figure 2 and Table 2 present the bathymetry and sedimentary structure for the propagation
path at Site 2. Out to a range of 12 km from the ARL array, the

bottom consisted of granite bedrock overlain with a thin layer of
sediment. The depth of the sediment layer was variable, with bedrock occasionally projecting through the sediment. Layer depths,
where observable on the profiling records, are shown. At a range
of 12 km from the ARL array, the sediment cover thickened and
changed in character. At ranges greater than 23 km, the bottom
consisted of sandstone overlain with successive layers of silt and
clay.

Seismic refraction experiments in the vicinity of the ARL array
at Site 2 revealed two distinct arrivals with velocities of approximately 5500 and 3300 m/sec. These two arrivals have been identified,
respectively, as the compressional and shear velocities of the granite
basement at this site. This interpretation is in agreement with the
findings of Keen and Loncarevic[11]. The velocity of arrivals that may
correspond to propagation in the surficial sediment ranges from 1700
to 2000 m/sec.

Grab samples have been analyzed[3] in the general vicinity of
Site 2, however, none of them are within 10 km of the propagation
path at this site. No grab samples suitable for analysis were
obtained during the field exercises.

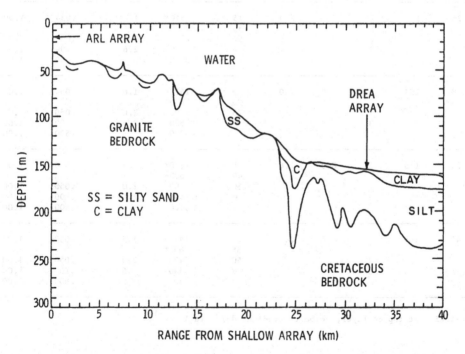

Figure 2. Sediment Structure—Site 2.

Table 2.   Seabed Parameters—Site 2

Range (km)	Water Depth (m)	Layer	M_ϕ (phi)	σ_ϕ (phi)	Thickness (m)	Vel (m/sec)	Density (g/cm³)	K (dB/km · Hz)	n
0	32	1	-3.0	0.5	6	1780	2.0	1.680	0.90
		2			∞	C 5500	2.6	.010	1.00
						S 3300		.050	1.00
1.5, 5	44				SAME AS ABOVE				
8, 12	56				SAME AS ABOVE				
17	83	1	3.0	1.7	10	1640	1.8	.037	1.14
		2			∞	C 5500	2.6	.100	1.00
						S 3300		.500	1.00
20	109				SAME AS ABOVE				
23	131				SAME AS ABOVE				
26	151	1	7.0	2.0	15	1510*	1.6	.054	1.00
					35	1600*			
		2			50	2400	2.0	.050	1.00
		3			∞	C 5500	2.6	.100	1.00
						S 3300		.500	1.00
28	153	1	7.0	2.0	2	1510*	1.6	.054	1.00
					20	1600*			
		2			75	2400	2.0	.050	1.00
		3			∞	C 5500	2.6	.100	1.00
						S 3300		.500	1.00
33	158	1	8.0	2.0	5	1490*	1.5	.050	1.00
					40	1600*			
		2			50	2400	2.0	.050	1.00
		3			∞	C 5500	2.6	.100	1.00
						S 3300		.500	1.00
43	168	1	8.0	2.0	15	1490*	1.5	.050	1.00
					55	1600*			
		2			50	2400	2.0	.050	1.00
		3			∞	C 5500	2.6	.100	1.00
						S 3300		.500	1.00

*Velocity at top of layer, velocity gradient dc/dz is 1 sec^{-1}.

Because of the lack of grab samples along the propagation path
and the failure of the seismic refraction analysis to provide the
surficial sediment velocity at Site 2, the assignment of acoustical
properties to the sedimentary layers at this site was based almost
entirely on Huntec data. A mean grain size for the surficial sediments was determined from the Huntec reflectivity measurements.
This grain size (listed in Table 2) and an assumed standard deviation
were used as discussed previously to predict sediment velocities and
attenuations. Deeper sedimentary layers were assigned acoustical
properties on the basis of identification of sediment type.

EXPERIMENTAL AND PREDICTED TRANSMISSION LOSS

Signals were received on a vertical hydrophone array from 456 g
(one pound) charges detonated at selected depths along the propagation tracks. The data obtained were analyzed to determine transmission loss in one-third octave bands. Source levels were established
using the Gaspin and Shuler[12] model and the depth scaling technique
of Hughes[13]. Data from the bottom hydrophone of the array (1.5 m
above the bottom) is presented for both sites.

The adiabatic normal-mode model was formulated as a modification of the basic model of Miller and Ingenito[14] and predicted
transmission loss, without mode coupling, for a sloping, fluid or
solid bottom. Representative sections from the geoacoustic models
developed previously were selected as inputs to the range-dependent
model for calculation of transmission loss. Predicted and experimental values were then compared.

Site 1 Comparisons

As indicated in Figure 3, the ARL array was placed in the lower
half of the water column in water 57 m deep. The sound velocity
profiles, shown in Figure 3, are downward refracting out to approximately 20 km and then have a sound channel centered near a depth of
60 m for the remainder of the shot track. The charges were detonated
at 38 m out to 20 km and at both 38 and 79 m at longer ranges. As
can be seen, both shot depths were near the sound channel axis. This
had a profound effect on the measured transmission loss as shown in
Figure 4. In this figure, measured and predicted data are shown for
both source depths at one-third octave frequencies of 25, 80, and
250 Hz. In general, the downward-refracting sound-velocity profile
caused losses in excess of cylindrical spreading at ranges less than
20 km due to bottom interaction. At ranges greater than 20 km, the
sound channel dominated and losses were near cylindrical spreading
in magnitude. The change in source depth had no apparent effect on
the loss at the two lower frequencies, but it appears that the shallow source suffered greater losses at 250 Hz for ranges of 40 to
90 km. Beyond 90 km, at the lowest frequency, an increase in atten-

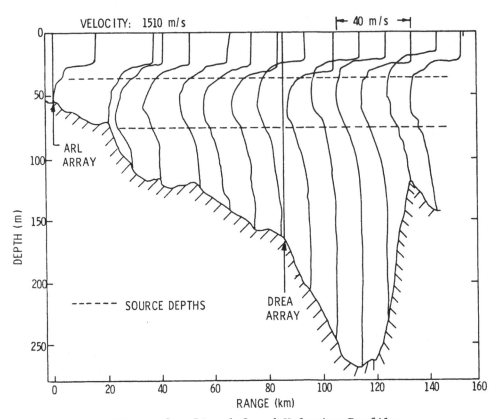

Figure 3.   Site 1 Sound Velocity Profiles.

uation was apparent for both source depths. This behavior appeared
to be related to changes in the seabed structure occurring near this
point in range; that is, a pocket of silt with gas seeps was covered
by a thin clay layer in the region around 110 km.

Although eighteen distinct regions were originally defined in
the geophysical model, the twelve indicated in Table 1 are considered
an adequate representation of the bottom for propagation modeling.
The seabed parameters listed are used as inputs to the adiabatic
normal-mode model. This model assumes that each normal mode propagates independently; that is, no energy is exchanged between modes
due to changes in bottom slope, bottom roughness, sound-velocity
profile, etc., with range. The model does include the option of a
bottom with high shear velocity, and calculates mode attenuation due
to scattering, and volume attenuation due to the absorption of both
compressional and shear waves in the sediment. At Site 1, a thick
layer of unconsolidated sediment is present so the bottom is considered a fluid bottom. Thus, normal mode runs are made at the
ranges along the propagation track in Table 1, giving the mode

amplitudes, wave numbers and mode attenuation coefficients. The
transmission loss for unit source strength is obtained by a local
summation of the normal modes. The loss for intermediate ranges
is obtained by interpolation of mode amplitudes, wave numbers and
attenuation coefficients.

In comparing experimental and predicted transmission loss, one
fact should be noted; the bottom models were obtained independently
of the experimental data, i.e., no attempt was made to refine the
bottom models. If we consider the 25 Hz data in Figure 4, it is
apparent that excellent agreement is obtained for most of the propagation track. However, at ranges greater than 90 km or so the model
failed to predict the increased attenuation due to the gas-seeps in
the basin region. Apparently, the changes made to the sediment model
to account for the diffuse gas did not provide enough additional
attenuation at this frequency.

At 80 and 250 Hz, the predictions were in relatively good
agreement with the experimental data. The attenuation at 80 Hz
provided a better match in the region of diffuse gas. At 250 Hz,
the agreement was somewhat poorer beyond 20 km but was still good
at ranges less than this. The increased scatter in experimental
data at 250 Hz appeared to result from the increased sensitivity
to variations in the sound-speed profile occurring at higher frequencies.

Site 2 Comparisons

At this site, the ARL array was deployed in 32 m of water. As
shown in Figure 5, there was a weak sound channel in deeper water,
but within the first 15 km of the propagation path the sound velocity
profile was downward refracting. Above this depth, a mixed layer of
near-constant sound speed existed. The charges were detonated at ·
18, 38, and 79 m as the water depth permitted. Charges were detonated at ranges of 4 to 43 km from the shallow array.

In Figure 6, the experimental and predicted transmission loss
is plotted. The experimental loss is shown for one-third octave
bands centered at 250 and 800 Hz for the three source depths noted.
At frequencies lower than 250 Hz, the signal-to-noise ratio was
inadequate for analysis. Losses at even the closest range of 4 km
exceeded 100 dB for a frequency of 100 Hz, as compared to Site 1
where the losses were only 50 dB. No real dependence on source
depth was noted.

The adiabatic normal-mode model was run using the parameters
shown in Table 2. As this data shows, the array was located above
a bottom consisting of a thin layer of sand over granite bedrock;
because of this, the solid version of the normal-mode program was

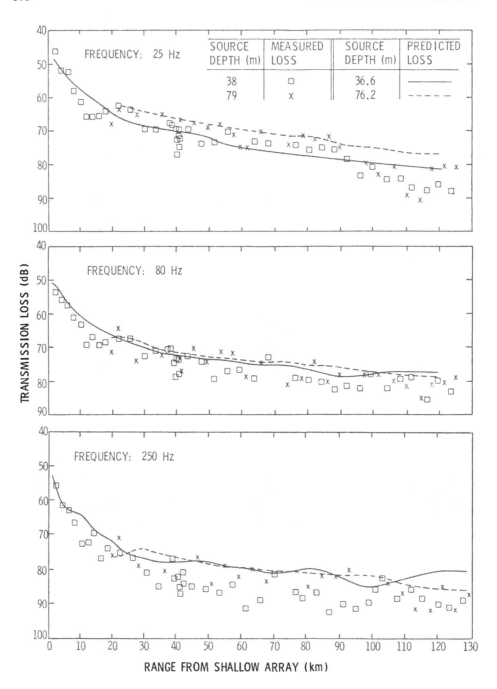

Figure 4. Transmission Loss Comparison—Site 1.

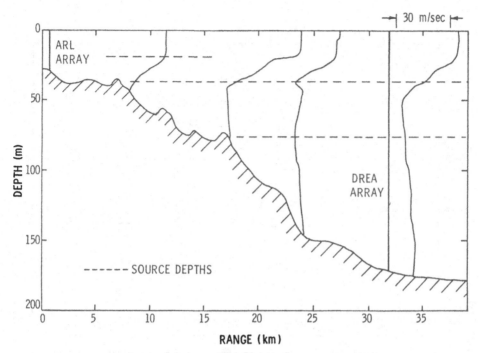

Figure 5. Site 2 Sound Velocity Profiles.

used. The predicted losses for the three source depths shown in
Figure 6 indicated that agreement between measured and predicted
losses was somewhat poorer at this site than at Site 1. For example, at the low frequencies where losses were not shown, the losses
were estimated to be in excess of 100 dB; the model predicted only
50 dB of loss at 25 and 80 Hz. As noted earlier, a weak sound
channel and downward refracting sound-velocity profile were present
over the propagation track. These effects caused the bottom to be
even more dominant than at Site 1. Generally, the bottom can cause
increased losses due to scattering, and compressional and shear
wave attenuation. Estimates of the scattering losses at low frequency have dismissed this mechanism as the source of the large,
low-frequency losses. The high losses must then be due to compressional and shear attenuation in the bottom. Coefficients for this
attenuation have been estimated from the data in the literature
referenced previously, but apparently larger attenuation coefficients
should have been used in order to match the experimental transmission
loss at the low frequencies. At 250 Hz, agreement was somewhat
better at ranges less than 20 km; at longer ranges, the measured
losses were considerably greater. The losses at 800 Hz were dominated by scattering and small variations in the mixed layer and sound
channel. These effects apparently caused mode coupling—a mechanism
unaccounted for in the present propagation model.

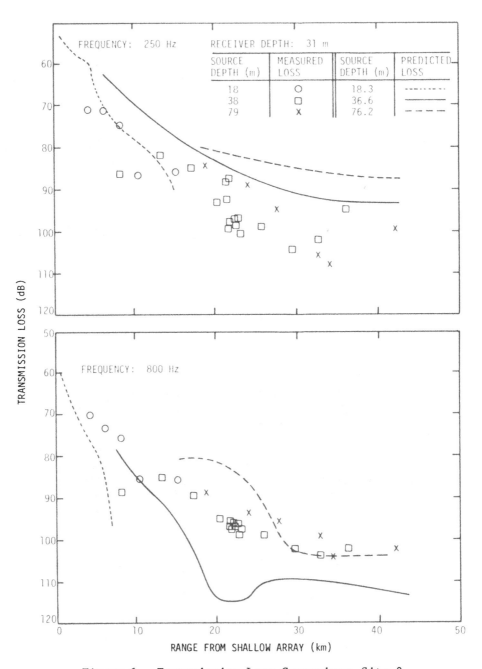

Figure 6. Transmission Loss Comparison—Site 2.

SUMMARY

    Geoacoustic models for the seabed have been formulated for two
sites in the Western North Atlantic. These models were based on
sub-bottom profiles obtained with the Huntec Deep-Tow System, grab
sample data, and dispersion and seismic refraction analysis. Perhaps the most interesting aspect of the models was their sharp
contrast and the consequent effect on propagation. The receiving
array at Site 1 was placed above a bottom consisting of a thick
layer of gravel and sand, a thick till layer and then a semi-consolidated type of bedrock (Tertiary) with an estimated velocity of
2000 m/sec. Along the track, the surficial sediments became more
fine and generally consisted of fairly thick layering above the
relatively low velocity bedrock. In contrast, at Site 2 over half
of the track consisted of a thin sand layer (< 10 m) over a high
velocity (5500 m/sec) granite bedrock.

    The sound-velocity profiles found along the early part of each
track caused downward refraction of the source energy and resulted
in bottom dominance of the propagation loss. These early losses at
Site 1 were approximately 20 dB at 25 and 80 Hz while at Site 2,
the losses were near 100 dB for the same frequencies. The sound
channel found at 15-20 km and beyond resulted in minimal interaction
with the bottom at Site 1 but somewhat more interaction at Site 2.
For example, an examination of the 250 Hz data in Figures 4 and 6
revealed that beyond 20 km at Site 1, the losses increased an additional 10 dB or so, while at Site 2 the additional attenuation was
on the order of 20 dB.

    Model predictions at Site 1 showed good agreement for 25 and
80 Hz except for the basin region at 25 Hz where the predicted
attenuation was not large enough. The losses at 250 Hz showed good
agreement for the first 20 km but were not large enough beyond this
range, indicating that the frequency dependence of the bottom
attenuation may have been incorrect. No real source depth dependence was noted for any of the frequencies analyzed. At 800 Hz, the
onset of mode coupling was noted, indicating the limits of applicability for the present propagation model.

    At Site 2, very large, 100 dB, losses were measured at frequen-
cies below 250 Hz but were not predicted by the propagation model.
The dominant low-frequency mechanism was determined to be the attenuation of compressional and shear waves in the bottom. Hence, it
was assumed the coefficients used for these mechanisms were in error.
Scattering began to dominate at 800 Hz and since the propagation
model was not designed to handle scattering or mode coupling, large
errors resulted.

ACKNOWLEDGEMENT

The authors would like to thank the personnel of the Defence
Research Establishment Atlantic for their assistance and cooperation, and, in particular, Dr. Richard Hughes for his work in providing the source levels for the explosive charges. Appreciation
is also expressed to Dr. Lewis King and Dr. Gordon Fader of the
Bedford Institute of Oceanography for providing grab sample data,
and to personnel of Huntec, Ltd. for their cooperation in interpreting the Huntec profiling data. The work performed by L. A. Rubano
and I. P. Sechrist of the Applied Research Laboratory is greatly
appreciated.

This research was performed under the sponsorship of the
Naval Sea Systems Command.

REFERENCES

1. D. R. Parrott, D. J. Dodds, L. H. King, and R. G. Simpkin,
    "Measurement and Evaluation of the Acoustic Reflectivity
    of the Sea Floor," Huntec Report H7903-02/SB/DRP, presented
    at the First Canadian Conference on Marine Geotechnical
    Engineering; March 5, 1979.
2. L. H. King, "Surficial Geology of the Halifax-Sable Island Map
    Area," Marine Sciences Branch, Department of the Environment,
    Ottawa, Canada (1976).
3. G. Drapeau and L. H. King, "Surficial Geology of the Yarmouth-
    Browns Bank Map Area," Marine Science Paper 2, Department of
    the Environment, Ottawa, Canada (1972).
4. L. H. King and B. MacLean, "Geology of the Scotian Shelf,"
    Marine Science Paper 7, Department of the Environment,
    Ottawa, Canada (1976).
5. R. S. Anderson, "Statistical Correlation of Physical Properties
    and Sound Velocity in Sediments," from Physics of Sound in
    Marine Sediments, Loyd Hampton, ed., Plenum Press, New York
    (1974).
6. M. A. Biot, "Generalized Theory of Acoustic Propagation in
    Porous Dissipative Media," J. Acoust. Soc. Am., 34, 1254-
    1264 (1962).
7. R. D. Stoll, "Acoustic Waves in Saturated Sediments," from
    Physics of Sound in Marine Sediments, Loyd Hampton, ed.,
    Plenum Press, New York (1974).
8. E. L. Hamilton, "Acoustic and Related Properties of the Sea
    Floor--Sound Attenuation as a Function of Depth," Naval
    Undersea Center Report, NUC TP 482; July (1975).
9. J. E. White, Seismic Waves--Radiation Transmission, and Atten-
    uation, McGraw-Hill, New York (1975).

10. G. H. F. Gardner, L. W. Gardner, and A. R. Gregory, "Formation
      Velocity and Density--the Diagnostic Basics for Stratigraphic
      Traps," Geophysics, 39, 770-780 (1974).
11. C. Keen and B. D. Loncarevic, "Crustal Structure on the Eastern
      Seaboard of Canada: Studies on the Continental Margin,"
      Canadian Journal of Earth Sciences, 3 (1966).
12. J. B. Gaspin and V. K. Shuler, "Source Levels of Shallow Under-
      water Explosions," Naval Ordnance Laboratory NOLTR 71-160
      (1971).
13. R. C. Hughes, "Low Frequency Underwater Explosive Source
      Levels," J. Acoust. Soc. Am., 60, Suppl. 1, S72 (1976).
14. J. F. Miller and F. Ingenito, "Normal Mode FORTRAN Programs
      for Calculating Sound Propagation in the Ocean," Naval
      Research Laboratory Memorandum, Report 3071; June (1975).

PROPAGATION LOSS MODELLING ON THE SCOTIAN SHELF:

THE GEO-ACOUSTIC MODEL

D.M.F. Chapman and Dale D. Ellis

Defence Research Establishment Atlantic
P.O. Box 1012, Dartmouth, Nova Scotia, Canada
B2Y 3Z7

ABSTRACT

A model of the acoustic environment of the ocean bottom sediments has been prepared for a selected area on the Scotian Shelf where both summer and winter propagation loss data have been collected. This geo-acoustic model is part of the input to a shallow water propagation loss model described in the accompanying paper, "Propagation Loss Modelling on the Scotian Shelf: Comparison of Model Predictions with Measurements". An attempt has been made to provide an independent bottom model, based on geophysical evidence, including continuous records from a high-resolution seismic profiling system, seismic refraction experiments, grab samples, and consultation with marine geologists familiar with the area. The final criterion for acceptance of the model parameters is the agreement between propagation loss model predictions and experimental measurements for both summer and winter conditions.

INTRODUCTION

The ability to predict acoustic propagation loss in shallow
water environments depends upon understanding the mechanisms of
propagation and attenuation of acoustic waves and on having accurate
knowledge of the acoustic properties of the medium, especially the
sub-bottom sediments. The nature of shallow water acoustics is such
that the medium can be treated as a range-dependent, stratified,
lossy waveguide; consequently, the energy trapped in the waveguide
is the significant contribution to the acoustic field at all but the
shortest ranges. Computer codes have been developed which reliably
compute the normal mode contribution to the field for a specified
acoustic environment; the problem of predicting propagation loss then
becomes one of determining an appropriate model of the medium. Such
a geo-acoustic model would include the sound speed profile in the
water, the bathymetry, and the thicknesses of the sub-bottom sediment
layers with associated sound speeds, densities, attenuation coefficients, all possibly changing in range. Attenuation mechanisms other
than volume attenuation, such as surface and bottom roughness, and
conversion of energy from compressional to shear waves, may be
included in the propagation model and would require inclusion as
parameters of the geo-acoustic model.

Some of the information which contributes to the geo-acoustic
model is available from direct measurement, using expendable bathythermographs (XBT's), ship's echo sounder, visual estimates of sea
state, etc. The acoustic properties of the sub-bottom are less
accessible to the researcher, often requiring the execution of other
experiments to deduce the relevant parameters. The interpretation
of these experiments to produce a geo-acoustic model for the purpose
of propagation loss modelling is perhaps the most difficult task the
modeller faces, and the one most susceptible to error. This difficulty is not alleviated by the sensitivity of propagation loss predictions to the acoustic properties of the bottom.

To expect to predict propagation loss in an extensive shallow
water domain with varied bottom structure is impractical, due to the
heavy computational load and the immense data library that would be
required. In areas whose bottom and sub-bottom have been thoroughly
surveyed, and for which propagation loss data have been collected,
such a modelling effort allows the researcher to gain a physical
understanding of the importance of the factors influencing propagation loss. This may suggest simpler models which function with
less detailed knowledge of the medium, but whose predictions are no
less accurate. These models might be more suitable for incorporation
into operational systems for which reliable propagation loss predictions are necessary.

Over the last several years, the Defence Research Establishment
Atlantic (DREA) has engaged in shallow water acoustics research,

concentrating on the Scotian Shelf area of the eastern Continental
Shelf off the coast of Nova Scotia and the Grand Banks south of
Newfoundland. As well as collecting propagation loss data for a
variety of sediment types, bottom slopes, and seasons, the research
group has developed a range-dependent (adiabatic modes) propagation
loss model so that the data may be interpreted in the light of our
understanding of propagation and attenuation mechanisms, and our
knowledge of the acoustic environment.

In February 1978, a propagation experiment using explosive
sources was conducted over a 175 km track on the Scotian Shelf. The
water depth varied between 55 m and 75 m over a sandy bottom, with
the winter season contributing a slight upwardly refracting profile.
At a separate time during the cruise, a seismic profiler was towed
over the same nominal track to probe the sub-bottom. In July 1979,
a similar propagation experiment was conducted over the same track,
with the summer season contributing a downwardly refracting sound
speed profile. In July 1979 and February 1980, seismic refraction
experiments were conducted in the vicinity of the site of the
receiving array for the propagation experiments.

This paper describes the geo-acoustic model which was developed
at DREA for those propagation tracks, based on the geophysical data
as much as possible. The accompanying paper describes DREA's range-
dependent propagation model PROLOS and compares model predictions
with the experimental data. The role of the geo-acoustic model in
the propagation modelling process is illustrated in Fig. 1. The

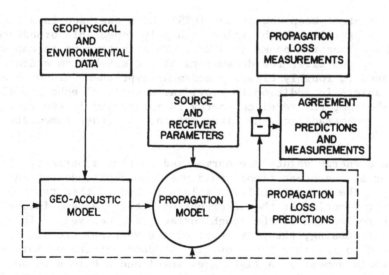

Fig. 1. The role of the geo-acoustic model in the modelling of
         propagation loss.

initial geo-acoustic model is based on the geophysical and environmental data available to the researcher, some of which may be assumed
from previous experience or historical data. This geo-acoustic model
serves as an input to the propagation loss model, along with the
various source and receiver parameters. The predictions of the model,
when compared with the experimental data, may provide occasion for a
judicial adjustment of certain parameters of the geo-acoustic model
in order to improve the degree of agreement. In some cases, the
propagation model itself may have to be altered, perhaps to account
for some attenuation mechanism previously overlooked. The final
criterion for acceptance of the sub-bottom portion of the geo-
acoustic model is that both the summer and winter data be fitted
using the same bottom model, with only the environmental portion
changing.

THE GEOPHYSICAL DATA

     In this section the sources of the geo-acoustic model are pre-
sented with a discussion of the methods used to interpret them. A
discussion of the actual parameters used in the model is found in
the next section.

Seismic Profile Records

     Since seismic profile records were taken over the entire
propagation track, showing layer depths and sediment types (when
properly interpreted), these comprise the basis of the geo-acoustic
model.

     The Huntec[1] Deep Tow System (DTS) which was used is a high
resolution profiler which employs a highly repeatable broadband
pulse whose spectrum peaks at 5 kHz. This instrument is capable of
penetrating the first few decameters of the sub-bottom sediments and
can be used to roughly classify sediment type when "ground truth"
is available. In addition to providing a vertical echo profile of
the sub-bottom, the received energy is integrated in two time
windows, one straddling the first return, the other immediately
afterwards.

     These energy values are normalized so that a perfect, flat
reflector at the ocean floor would yield an amplitude reflectivity
of unity. The two reflectivity values r_1 and r_2 then represent the
coherent reflection at the water/bottom interface, and incoherent
off-axis reflection by the rough surface, respectively. In addition
to this processing, the grey level of the vertical profile is
calibrated to indicate the acoustic impedance of the medium, so that
a clay would appear as a light grey level and a hard sand would
appear dark. Fig. 2 is a reproduction of a processed record from
the Huntec DTS which was selected for the variety of features

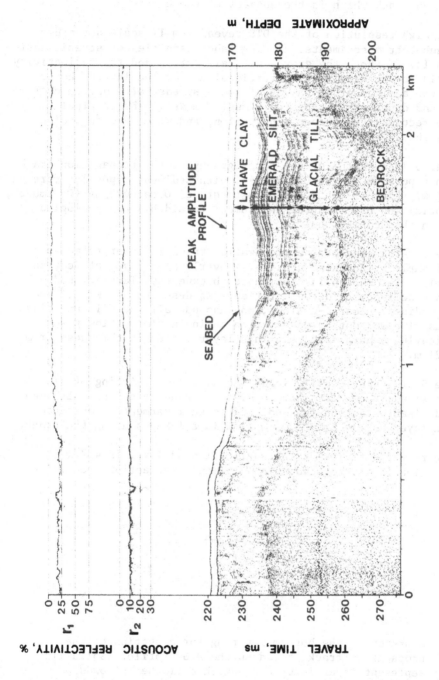

Fig. 2. A processed record from the Huntec DTS, showing the seismic profile and the reflectivity values r_1 and r_2. The sediment classifications are due to King.[2] This sample is not from the propagation track.

encountered. The profile shown is not representative of the
propagation track which is the subject of these papers.

The high resolution of the DTS reveals small-scale structure
in the sub-bottom sediments, showing inhomogeneities or stratification
within a layer. Both the grey scale calibration and the reflectivity
values aid in identifying both surficial and buried layers, hence
guiding the assignment of values of sound speeds, attenuation coefficients, and densities in the geo-acoustic model. The original
analogue records show deeper penetration, and were used to assign
layer depths.

Fig. 3 is a sketch of the stratigraphy of the propagation track
based on a portion of the DTS record, with sediment types interpreted
by Dr. Lewis King of the Bedford Institute of Oceanography (Dartmouth,
Nova Scotia) who has used the instrument to aid his own geological
surveys on the Scotian Shelf.

The stratigraphy of the propagation track is interpreted as a
layer of Sable Island sand and gravel overlying a layer of Scotian
Shelf drift all over an indeterminate thickness of Tertiary bedrock,
which is a semi-consolidated material. As described by King,[2] the
sand is medium to coarse-grained, and has had all the silt and clay
fractions winnowed out by wave action during a time of lower sea
level, leaving a hard surface. This layer varies in thickness from
3 m to 21 m.

The Scotian Shelf drift is a glacial till consisting of all
sediment sizes, from clay through silt and sand to gravel, although
the predominant fraction is sand. It is an impermeable substance.
The till layer varies from 2 m to 13 m in thickness along the track.

The r_1 reflectivity of the surficial sand layer is 40% ± 5%.
If this value is interpreted as an amplitude reflection coefficient,

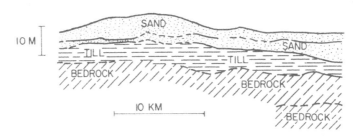

Fig. 3. A sketch of the bottom layering for a section of the
         propagation track, based on the DTS profile. Solid lines
         represent layer interfaces which could be followed
         continuously; dashed lines represent intermittently
         observable interfaces.

this gives a sound speed of approximately 1750 m/sec, using a density of 2.06 g/cm^3.

The Nova Scotia Research Foundation (Dartmouth, Nova Scotia) has conducted some seismic surveys in the area using a sparker source, which gives greater penetration with the expense of lower resolution. Examination of these records confirmed the identity of the Tertiary bedrock layer, since the cross-bedding of this layer was observable.

Seismic Refraction Experiments

Seismic refraction experiments were conducted at the receiving array site using 1 lb charges detonated on the bottom on a 20 km line centred at the array. The lower hydrophones were used as receivers, with the gain set to achieve a good S/N ratio for the refracted arrivals, with the waterborne arrival saturating the recording system. Ranges were determined from radar ranges and the times of the refracted arrivals measured relative to the waterborne arrival. Fig. 4 is a plot of arrival time vs range for the refracted arrivals identified from the records. The slopes of the linear fits to this data were used to calculate the layer velocities, using a sound speed in water of 1460 m/sec.

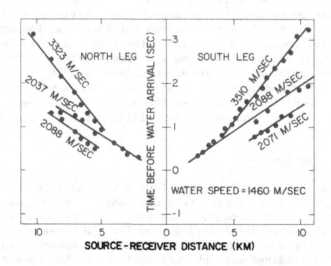

Fig. 4. The results of the seismic refraction survey. The data points represent the refracted arrivals identified from the raw data.

Table 1. Seismic Refraction – Officer and Ewing[3]

Layer	Sound Speed (m/sec)	Sediment Type
1	1680	Unconsolidated
2	2480	Semi-consolidated
3	3910	Consolidated
4	5600	Basement

A previous seismic refraction survey by Officer and Ewing[3] in a nearby area is summarized in Table 1. The DREA experiment apparently did not penetrate to the basement of 5600 m/sec speed and was too coarse to identify an arrival associated with the unconsolidated layer. There is evidence from the DREA experiment that the semi-consolidated layers of Tertiary bedrock have speeds between 2040 m/sec and 2090 m/sec, and that there is a deeper, consolidated sediment with a speed about 3400 m/sec.

The seismic refraction survey is not definitive of sound speeds, but it does suggest that, for the purposes of propagation loss modelling, Tertiary bedrock effectively forms the acoustic basement and should be assigned a sound speed of about 2050 m/sec.

Grab Sample Data

Grab samples of the surficial sediments on the Scotian Shelf have been collected by King.[2] For a previous experiment conducted jointly by DREA and the Applied Research Laboratory at Pennsylvania State University, Beebe[4] has calculated the sound velocity for several samples of Sable Island sand and gravel using the regression equation due to Anderson.[5] He obtained values of 1756, 1779, and 1799 m/sec with a standard error of 46 m/sec. These values must be corrected for the difference between the actual bottom temperature (3°C) and 20°C, to give values of 1690, 1713, and 1733 m/sec, respectively.

Bathymetry and Thermometry

The source ship, in addition to laying the explosive charges for the propagation experiments, recorded continuous bathymetric profiles using its echo sounder. When corrected for sound speed, these provide the water depth information for the geo-acoustic model. At points where the source and survey tracks were known to coincide, the echo sounder readings were checked against the bottom-surface ghost of the DTS record.

The source ship also measured water temperature profiles with XBT's approximately every 15 km along the track. The water

temperature at the surface was measured separately to validate the
XBT records. These measurements, along with the pressure and
salinity information, were used to calculate sound speed profiles
along the track. DREA uses a trilinear salinity profile for the
Scotian Shelf which is a representative average of measurements taken
in many locations in all seasons. The sound speed profiles and
bathymetry are presented in the second paper, along with the data
and modelling results.

THE GEO-ACOUSTIC MODEL

    The DREA range-dependent propagation loss program PROLOS
requires acoustic profiles of the medium as input at several range
points along the track. The sub-bottom layers did not vary in
thickness as much as the water depth, so the latter was used to
determine the range points. In particular, all significant local
maxima and minima of the bottom topography were included. At each
point, the layer thicknesses were estimated, and the closest XBT
record was used to compute the water sound speed. The acoustic
properties of each layer were assigned according to sediment type.
Since the actual tracks of the summer and winter runs were not
identical, a different bathymetric profile was used for each run,
with the corresponding XBT records, but the sub-bottom model was substantially the same. At higher frequencies, the adiabatic modes
approximation performs better, numerically speaking, if there are
more range points along the track. When this was necessary, the
extra points were determined by simply interpolating the original
model.

Layer Depths Including Bathymetry

    The sediment layer thicknesses were read directly off the Huntec
DTS analogue record in units of time, and converted to units of
distance using the sound speed values estimated for each sediment
type.

    The water depth was read directly from the echo sounder records
from the source ship. One consideration in using this water depth
in the model is the effect of a bottom slope perpendicular to the
propagation track. Harrison[6] has shown that such a transverse slope,
using a ray picture, has the effect of refracting the ray into deeper
water, requiring that the actual ray path between a source and
receiver in the same depth of water travel up-slope on a hyperbolic
path. This horizontal refraction relative to the line-of-sight path
depends on the horizontal angle of the ray so that, strictly speaking,
each normal mode at each range point would need a separate bathymetric
profile to account for the effect.

    A sample calculation shows that, for a range of 100 km over a
bottom at water depth 70 m with a transverse slope of .06°, a ray of

horizontal angle 5° would travel 500 m up-slope, representing a
decrease in water depth of .5 m.   A similar calculation for a ray of
horizontal angle 30° gives a transverse travel distance of 9.4 km,
representing a depth decrease of 9.4 m.   Since the high order modes
corresponding to large horizontal angles would suffer high attenuation at such ranges, we feel that a correction for the transverse
slope effect is unnecessary in this case.

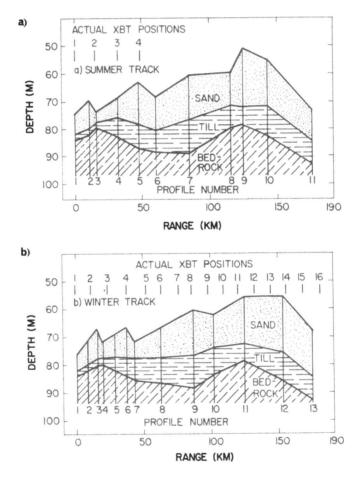

Fig. 5.   Schematic diagrams of the bottom models showing range points,
          layer types and thicknesses, and the actual positions at
          which XBT's were dropped.
          (a) Summer; (b) Winter.

Figs. 5a and 5b show the layer depths, including bathymetry, used for the summer and winter runs, respectively, with the actual ranges of the XBT's. The layer thicknesses at each range point of the model are given in Table 2 along with the XBT's used for velocity calculations. During the summer run, the XBT apparatus malfunctioned, and XBT's are available only for the first portion of the track. For the remainder, the last reliable XBT was used, since the independently measured surface temperature did not vary much.

Table 2. Layer Thicknesses

Profile No.	Range (km)	XBT No.	Layer Thickness (m)		
			Water	Sand	Till
		Summer			
1	0	1	75.0	7.0	2.0
2	11	2	69.5	10.5	2.0
3	16	2	74.0	3.0	2.0
4	32	3	68.5	7.5	7.0
5	48	4	63.0	15.0	9.0
6	60	4	68.0	12.5	8.0
7	86	4	61.0	16.0	12.0
8	116	4	59.5	11.5	7.5
9	123	4	50.5	21.5	6.0
10	142	4	55.0	16.5	10.5
11	175	4	73.0	11.0	8.5
		Winter			
1	0	1	76.5	6.0	2.0
2	8	2	71.5	8.5	2.0
3	16	2	66.5	8.5	2.0
4	20	3	71.5	6.0	3.0
5	29	3	69.0	6.0	3.0
6	38	4	66.0	11.0	7.0
7	44	4	72.0	11.0	7.0
8	63	6	66.5	10.5	9.5
9	87	8	60.0	16.5	12.0
10	102	9	61.5	12.0	10.0
11	125	11	55.0	17.0	6.0
12	153	14	55.0	20.0	10.5
13	175	16	67.0	17.0	8.0

Acoustic Properties of the Sediments

Table 3 lists the sound speed, density, and attenuation coefficient for each of the sediments encountered, as determined from the data available. No need was seen to include range-dependence in these values. The layer thicknesses were judged to be small so possible gradients of these acoustic parameters have been disregarded.

The attenuation law used is linear in frequency [$\alpha = kf$ with k in dB/km//Hz]. A power law of the type proposed by Stoll[7] was considered, but abandoned on the grounds that it predicts a frequency dependence of propagation loss at long ranges which disagrees with the trend in the data.

The values of many of the acoustic parameters are identical with those determined by Beebe[4] for similar sediment types encountered in the 1978 ARL/DREA joint experiment. The DREA survey estimates a lower sound speed for the sand, and a higher speed for the bedrock. The k-value for sand was assigned after investigations by Dodds,[8] who estimates attenuation from the frequency dependence of echoes from reflectors buried in the sediment, using the Huntec DTS as a source. The acoustic properties of till are the least known, due to its highly variable composition (a consequence of its geological origin). No samples of the till in this area have been available for study, and the acoustic parameters assigned are broad estimates. A k-value for till was adapted from Beebe's value by calculating the attenuation predicted by the power law at 160 Hz, and extrapolating linearly from that point.

Table 3. Acoustic Properties of the Sediments

Type	Sound Speed (m/sec)	Density (g/cm^3)	Attenuation (dB/km//Hz)
Sand	1750[a]	2.06[b]	.26[c]
Till	1900[b]	2.1[b]	.030[b]
Bedrock	2050[a]	2.2[b]	.050[b]

[a]Determined from DREA survey.

[b]After Beebe.[4]

[c]After Dodds.[8]

Other Mechanisms of Attenuation

In the course of attempting to model propagation losses which agreed with experimental data, it became necessary to consider including mechanisms of attenuation other than volume absorption in the sub-bottom, especially those mechanisms which would contribute greater loss at high frequency. Volume absorption in the water was included following Thorpe.[9] The attenuation due to scattering at rough boundaries was included via the Kirchhoff approximation, following Kuperman and Ingenito.[10] This latter attenuation mechanism requires environmental input in the form of the r.m.s. value of the deviation of the rough surface from its mean value. The bottom roughness was estimated to be .1 m for deposits of Sable Island sand. The sea surface roughness was estimated to be .5 m for the summer experiment, and 1 m for the winter experiment, from visual estimates of sea state and wind speed measurements.

CONCLUSION

Tables 2 and 3 constitute the original geo-acoustic model which was determined from the geophysical data available. The PROLOS program was executed with this model, and the predictions compared with the experimental propagation loss data. Deferring the detailed discussion of this comparison to the second paper, it can be said that it became necessary to review the values assigned to the acoustic properties of the sub-bottom sediments.

The values for the uppermost, sand layer were considered reliable and alteration of these would significantly change the character of the model predictions, affecting all frequencies. The bedrock layer is the deepest, and the model predictions are least sensitive to the values assigned to it, as only the very lowest frequencies penetrate to this depth. The values of the till layer were most suspect, and consequently were altered to gauge their effect.

King[11] suggests that the till under the experimental site may have a slightly different geological history than that encountered by Beebe.[4] Specifically, it may not have had a heavy overburden of glacial ice during deposition, which would result in a less compacted material, implying that the sound speed should be lowered from Beebe's value. This hypothesis is supported by the fact that, on the DTS records, the impedance contrast at the sand/till interface appeared considerably less than that of the till/bedrock interface. These considerations led to a new value of 1850 m/sec for the sound speed in till. The original attenuation value for till was less than that of the bedrock. In retrospect, this was considered unlikely, since the bedrock is more compacted, so the k-value of till was increased to an arbitrary value of .08 db/km//Hz.

The combined effect of these changes was negligible, having a noticeable effect only at 25 Hz, at which frequency the propagation loss increased by about 2 dB at long ranges.

This concludes the discussion of the geo-acoustic model used for propagation loss modelling in the experimental area on the Scotian Shelf. The sources of the model have been presented, and the method of determining the model parameters has been demonstrated, as well as some of the arguments used to alter the original model to obtain better agreement between the experimental data and the model predictions.

The second paper will present the experimental data, and a comparison of predictions with data. In brief, the agreement between the data and the predictions of the independently constructed geo-acoustic model is reasonable: the agreement is best in the frequency range 25-100 Hz but the model seems to predict lower losses in the 200 Hz - 800 Hz range, especially for the summer conditions. For the winter conditions, the model predicts an optimum propagation frequency which varies from about 100 Hz at 75 km to about 200 Hz at 150 km. This effect is present in the data. The absence of a predicted optimum propagation frequency at long ranges for the summer conditions (which exists in the data) will be discussed. A better overall fit to the winter data might be obtained with a higher attenuation value for sand. Otherwise, no substantial improvement in the agreement is expected from further alteration of the parameters of the sub-bottom portion of the geo-acoustic model.

ACKNOWLEDGEMENTS

The authors thank Dr. Lewis King of the Bedford Institute of Oceanography for his interpretation of the DTS seismic records, and Jack Dodds of Huntec ('70) Ltd. for supplying his attenuation measurements. We also thank John Beebe of the Applied Research Laboratory at Pennsylvania State University for supplying his values for the acoustic properties of some of the sediment types encountered on the Scotian Shelf.

REFERENCES

1. Huntec ('70) Ltd., 25 Howden Road, Scarborough, Ontario, Canada, M1R 5A6.
2. Lewis H. King, "Surficial Geology of the Halifax-Sable Island Map Area, Marine Sciences Paper 1", Dept. of Energy, Mines and Resources, Ottawa (1970).
3. C. B. Officer and M. Ewing, "Geophysical Investigations in the Emerged and Submerged Atlantic Coastal Plain, Part VII", Bull. Geol. Soc. Am. 65: 653 (1954).
4. J. Beebe, "Long-Range Propagation Over a Mixed-Sediment, Sloping Bottom", J. Acoust. Soc. Am. Suppl. 1, 67: S30 (1980).

5. R. S. Anderson, "Statistical Correlation of Physical Properties and Sound Velocity in Sediments", in: 'Physics of Sound in Marine Sediments', Loyd Hampton, ed., Plenum Press, New York (1974).

6. C. H. Harrison, "Three-Dimensional Ray Paths in Basins, Troughs, and Near Seamounts by Use of Ray Invariants", J. Acoust. Soc. Am. 62: 1382 (1977).

7. R. D. Stoll, "Acoustic Waves in Saturated Sediments", in: 'Physics of Sound in Marine Sediments', Loyd Hampton, ed., Plenum Press, New York (1974).

8. J. Dodds, personal communication.

9. W. H. Thorp, "Analytical Description of the Low-Frequency Attenuation Coefficient", J. Acoust. Soc. Am. 42: 270 (1967).

10. W. A. Kuperman and F. Ingenito, "Attenuation of the Coherent Component of Sound Propagating in Shallow Water With Rough Boundaries", J. Acoust. Soc. Am. 61: 1178 (1977).

11. Lewis H. King, personal communication.

PROPAGATION LOSS MODELLING ON THE SCOTIAN SHELF:

COMPARISON OF MODEL PREDICTIONS WITH MEASUREMENTS

Dale D. Ellis and D.M.F. Chapman

Defence Research Establishment Atlantic
P.O. Box 1012, Dartmouth, Nova Scotia
Canada, B2Y 3Z7

ABSTRACT

Propagation loss data using explosive sources have been obtained over a 180 km track on the Scotian Shelf during both summer and winter conditions. The water depth varies between 55 and 75 meters over a sandy bottom. A sub-bottom model is discussed in the accompanying paper, "Propagation Loss Modelling on the Scotian Shelf: The Geo-acoustic Model". In the present paper the bottom properties determined from the geo-acoustic model are used as part of the input to the DREA range-dependent normal mode program. The predictions from the normal mode calculations are compared with the summer and winter propagation loss measurements for a number of ranges, depths and frequencies. The difference between the calculations and measurements can be used to refine the geo-acoustic model.

INTRODUCTION

For several years, the Shallow Water Acoustics Group at DREA has conducted extensive surveys of propagation loss on the eastern seaboard of Canada. More recently, efforts have been made to provide a predictive capability in the absence of historical data, using a range-dependent normal mode computer code, PROLOS, developed at DREA for propagation loss modelling in shallow water. In modelling the acoustic environment, emphasis has been placed on generating geo-acoustic models from independently acquired geophysical data, rather than on "fitting"a bottom model to the observed propagation loss data. Provided that the researcher has knowledge of the seafloor properties, and has confidence in his models, this method would ultimately provide propagation loss predictions in areas for which experimental data has not been collected. Admittedly, this approach entails extensive geophysical surveys in the areas of interest, but much of this work is already being undertaken by other research and exploration agencies, the information having broad application in marine science and technology.

The first paper[1] described the method used in generating a geo-acoustic model for a selected area on the Scotian Shelf whose seafloor properties have been investigated in detail, and where propagation loss experiments have been conducted under both summer and winter conditions. This paper will present the experimental data, the predictions of the propagation loss model based on the independent geo-acoustic model, and the results of a limited study to gauge the sensitivity of the model predictions to certain parameters of the bottom model.

The propagation track was 180 km long over a sandy bottom in water of depth ranging from 55 m to 75 m. A study of the propagation loss data over a long track reveals features which are not apparent at shorter ranges. In particular, in this area, the data shows a pronounced frequency of optimum propagation in the vicinity of 100 Hz at ranges over 50 km. It will be demonstrated that this effect is reproduced in the model predictions for the winter conditions, in which case the model agrees fairly with the data up to 400 Hz. The effect is absent in the model predictions for summer, for which the model only agrees with data below 100 Hz. It is suspected that the failure to predict the frequency dependence of propagation loss in summer is not a deficiency in the geo-acoustic model, but is due to an attenuation mechanism not presently incorporated in the propagation model.

The sound speed profiles, bottom model, and a general description of the environment is presented in the next section. A brief description of the experimental procedure and analysis

techniques follows, with the propagation loss data. The general
features of both the winter and summer data are discussed. The
model predictions are presented and compared with the data, and
the results of the limited sensitivity study are discussed. The
concluding section presents an evaluation of the geo-acoustic
model based on the comparison of data and model predictions, and
outlines alterations to both the propagation model and the geo-
acoustic model which are expected to improve the degree of
agreement.

ENVIRONMENTAL DESCRIPTION

Figure 1 shows the sound-speed profiles and bathymetry for the
winter and summer propagation loss runs. In winter the water is
almost isospeed, although slightly upward-refracting; typical
sound speeds are 1453 m/s at the surface and 1460 m/s at the bottom.
In summer there is a warm surface layer, below which is a negative
thermocline to a depth of about 35 m; between the thermocline and
the bottom the sound speed is almost constant. Note that the
summer sound-speed profiles are more variable, and that due to an
equipment failure the profiles beyond 50 km could not be trusted
so have been omitted. For these ranges the last profile shown in
Figure 1b was used in the normal-mode calculations. The surface
temperature remained constant for the remainder of the track so no
great changes in the profile are expected.

The summer and winter tracks are nominally the same but due
to variations in the path of the source ship and the position of
the receiving array the bathymetry is slightly different. The
bottom has an incline which is less than 0.1° perpendicular to the
propagation track.

A geo-acoustic model of the bottom, based primarily on
interpretation of the Huntec sub-bottom profiler[2] has been
presented in the first paper. Table 1 lists the sound speeds,
densities and attenuation coefficients for the sand, glacial till
and tertiary bedrock. Table 2 of the first paper lists the
thicknesses of the bottom layers at a number of ranges selected
for use with the range-dependent normal-mode program. At each
range the sound-speed profile used for the normal-mode calculation
is also indicated.

A r.m.s. bottom roughness of 0.1 m was estimated using
information obtained from bottom photographs of the area and
analysis of Huntec records over a similar sandy bottom.[3]

During the winter experiment a r.m.s. wave height of 1 m was
estimated from visual observations of the sea state and measurements
of wind speed. During the summer experiment the wave height varied

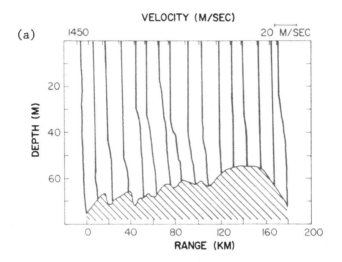

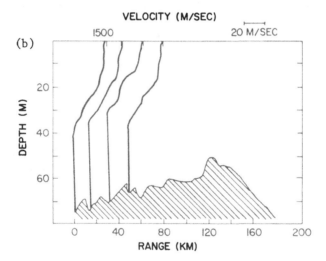

Fig. 1.   Sound-speed profiles and bathymetry: (a) winter,
(b) summer. The tic marks at the top refer to a sound
speed of 1450 m/s in winter or 1500 m/s in summer. The
point at which the profiles intersect the bottom
indicates the range at which they were measured.

due to a rising wind. For modelling purposes a r.m.s. roughness of 0.5 m was used.

DESCRIPTION OF EXPERIMENT AND DATA

Along the 180 km track one-pound (0.45 kg) explosive charges were dropped from a source ship at 1.8 km intervals at a depth of 37 m. The detonation depths were checked later by cepstrum analysis of the received charge.

The broadband energy from the charges was received by a vertical array of hydrophones spanning the water column. The analogue signals from 10 hydrophones were digitized at a sampling rate of 5 kHz, multiplexed and telemetered to the receiving ship CFAV QUEST. On board, the data were recorded both on high-density recording tape and, also in real time, on disk using a PDP-11/34 computer. Between shots the information retained on disk was transferred to magnetic tape. At a later time the shots were analyzed in one-third octave bands from 25 to 1625 Hz.

Measured propagation loss as a function of range is shown in Figure 2 for a selection of frequencies in winter and summer. Losses in summer are generally greater than the losses in winter, although the same overall trends are shown in both seasons. In all cases the losses are considerably greater than cylindrical spreading (by 0.1 to 0.5 dB/km). At ranges less than about 30 km the lower frequencies propagate best, but as the range increases the optimum frequency rises to a broad maximum in the neighbourhood of 100 Hz. More details of the frequency dependence are shown in Figure 3.

The effect of receiver depth is shown in Figure 4. In winter the depth dependence of the propagation loss shows a broad minimum at mid-depth (30 m). The depth dependence was found to become more pronounced as range and frequency increased. At 100 Hz the losses are about 10 dB greater at the surface and bottom than they are at mid-depths. In summer the optimum depth is relatively close to the bottom (about 60 m). For receivers at 30 m and 20 m losses are 5 and 15 dB greater respectively; in the latter case the extra losses are due to shielding by the thermocline.

The differences between the winter and summer results are mainly due to the sound-speed profiles. In summer the mixed surface layer and thermocline decrease the surface interaction and increase the bottom interaction. In winter there is no high-speed layer to shield the surface, so losses due to surface scattering are more important, particularly at the higher frequencies.

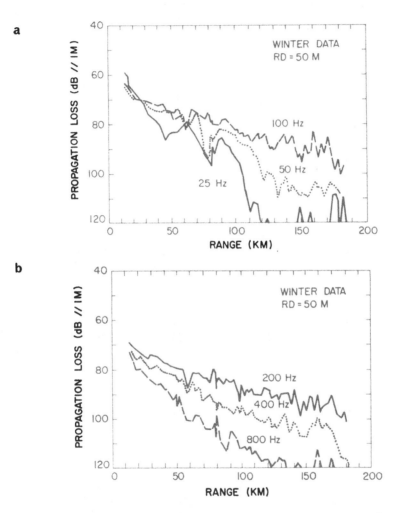

Fig. 2. Measured propagation loss vs range for a selection of
         frequencies: (a) and (b) winter; (c) and (d) summer. The
         source depth is 37 m; the receiver depth is 50 m.

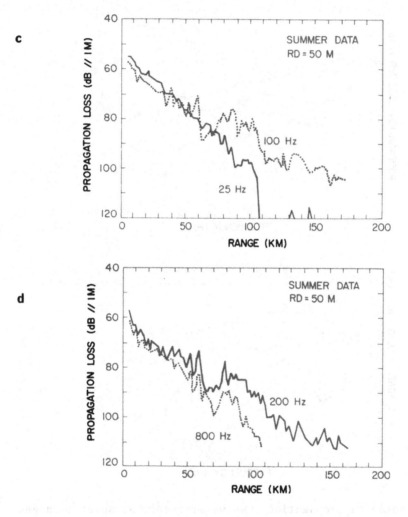

Fig. 2 (Cont'd)

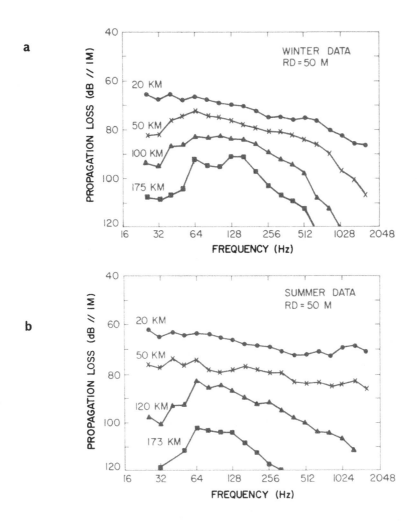

Fig. 3. Measured propagation loss vs frequency at several ranges:
(a) winter, (b) summer. The source depth is 37 m; the
receiver depth is 50 m.

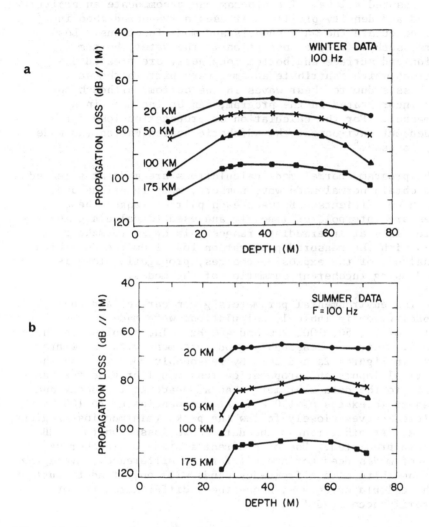

Fig. 4. Measured propagation loss vs receiver depth at several
ranges: (a) winter, (b) summer. The source depth is
37 m; the frequency is 100 Hz.

MODEL PREDICTIONS

The DREA adiabatic, range-dependent normal-mode program[4] PROLOS was used to calculate the propagation loss for comparison with the measured values. The program can accommodate an arbitrary sound speed and density profile. It uses a two-ended shooting method to calculate the wave numbers and mode functions. Loss mechanisms, such as volume absorption in the water, bottom attenuation and surface and bottom roughness, are treated as perturbations which contribute an imaginary part to the wave number. Losses due to shear waves in the bottom, although not presently incorporated in the program, can be included in a similar manner. For the calculation of propagation loss for a range-dependent environment the adiabatic approximation (no mode coupling) is used.[5]

In the program, normal-mode calculations are made at selected ranges to obtain normal-mode wave numbers, mode functions and attenuation coefficients. Between each pair of ranges these quantities are interpolated linearly and used to calculate the propagation loss at intermediate ranges. In order to make a comparison with the measured propagation loss from the one-third octave analysis of the explosive charges, propagation loss is calculated using incoherent summation of the modes.

Using the environmental parameters given earlier and the bottom model given in Table 1, calculations were made at frequencies of 25, 50, 100, 200 and 400 Hz. The results are shown in Figure 5. There is reasonably good agreement with the winter data shown in Figures 2a and 2b. By reasonably good we mean the geo-acoustical inputs and propagation loss model provide the same quality fit to the data over more than a decade in frequency and out to ranges in excess of 150 km. At frequencies up to 100 Hz the predicted curves closely follow the peaks (minimum loss regions) in the data. At other ranges the data shows losses up to 15 dB greater than predicted. Use of coherent instead of incoherent summation of modes does not account for the difference. Averaging over the one-third octave frequency bands will only tend to further smooth the calculated curves, hence these differences are not satisfactorily accounted for.

At ranges less than 30 km the low frequencies propagate best, but as the range increases the optimum frequency of propagation increases to between 100 and 200 Hz, in qualitative agreement with the data. Both the computed and measured long-range losses at 400 Hz are greater than at 200 Hz due to the increased importance of the surface roughness as a loss mechanism at higher frequencies.

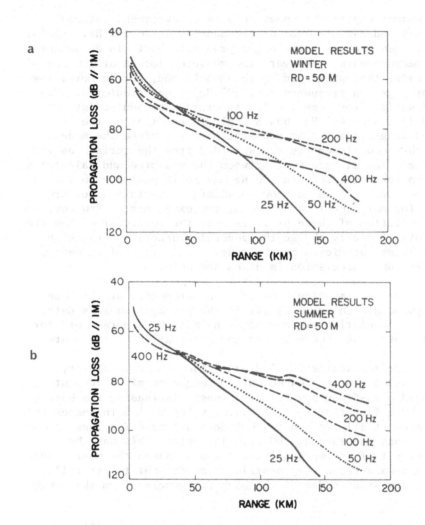

Fig. 5.   Calculated propagation loss vs range for a selection of
frequencies: (a) winter, (b) summer. The source depth is
37 m; the receiver depth is 50 m. In summer at ranges
less than 30 km the propagation loss increases with
frequency between 25 and 400 Hz, so to avoid confusion
the intermediate curves have not been drawn.

Generally a better overall fit to the data would occur if the calculated losses were increased at all frequencies, particularly those above 100 Hz. This point will be addressed later.

The summer experiment shows reasonable agreement between measurements and calculations at frequencies up to 100 Hz. Again, at short ranges the low frequencies propagate best, in agreement with the measurements. However, the measured losses at 200 and 400 Hz are greater than predicted by the model, and, furthermore, the optimum propagation frequency near 100 Hz is not predicted. The effect of surface roughness, which in winter accounted for the increased losses at 400 Hz, has a negligible effect on the calculated long-range summer propagation loss, since the modes carrying the acoustic energy are shielded from the surface by the thermocline. The disagreement between the measured and calculated propagation losses at 200 and 400 Hz may be in part due to the sound-speed profiles which show considerable variation but are available for only the first 50 km of the experiment. However, the expected variation of these profiles over the remainder of the track will not significantly change the general character of the high-frequency losses predicted by the model. This lack of agreement between data and calculation is discussed below.

To assess the importance of certain parameters of the model and perhaps obtain an improved fit to the propagation-loss data, a number of calculations were made using alternate parameters for the bottom roughness, sound speeds and attenuation coefficients.

Since surface scattering increases the losses in winter, it was hypothesized that increased bottom roughness might account for the measured high-frequency summer losses. Increasing the bottom roughness by a factor of 10 to a r.m.s. value of 1 m increases the summer losses at all frequencies but does not make the loss at 400 Hz greater than the loss at 200 Hz. In winter, this has the opposite effect of increasing low frequency losses the most, since the weakly upward-refracting profile is sufficient to partially shield the low-order modes of the high frequencies from the rough bottom.

Considering the bottom properties, those for the till have the greatest uncertainty since it is a highly variable material and no direct measurements of its properties were made. The values listed in Table 1 were obtained by Beebe[6] in his analysis of a nearby site on the Scotian Shelf. It can be argued that the till in the vicinity of our track was less compacted. If we then assign a slightly lower sound speed of 1850 m/s and a higher attenuation coefficient of k=0.08 dB/km//Hz, the results are almost identical to the results shown in Figure 5a. (Only the 25 Hz curve is noticeably affected, with an increased loss of less than 2 dB at 140 km.)

Although the propagation loss is not sensitive to the properties of the till, it is sensitive to the properties of the sand. Increasing the attenuation coefficient from 0.26 to 0.4 dB/km//Hz increases the predicted losses at 25 Hz by more than 30 dB at 180 km in both winter and summer. The effect decreases with frequency to less than 2 dB in summer and 0.2 dB in winter at 400 Hz. (It is noted that the 30 dB increase in losses at 25 Hz is much too large, lending credence to Dodds'[8] value of 0.26 dB/km//Hz for the attenuation coefficient for sand.) Alternatively, the calculated losses can be increased by decreasing the sound speed in the sand. Reducing the sound speed from 1750 to 1700 m/s increases losses from at most 7 dB at the lower frequencies to less than 2 dB at 400 Hz. Neither decreasing the sound speed in the sand nor increasing the attenuation coefficient in the sand predicts significantly increased losses at the high frequencies where they are most needed, although judicious choices might improve the fit to the data at low frequencies.

In all the calculations discussed above the bottom attenuation was assumed to vary linearly with frequency.

Beebe[6] has used a bottom attenuation of the form $\alpha = kf^n$ based on a model by Stoll.[7] Using his attenuation values for sand ($\alpha = 0.009\ f^{1.78}$ dB/km) and till ($\alpha = 0.0065\ f^{1.3}$ dB/km) and our values for sound speeds, calculations of propagation loss were made for both winter and summer. For the winter the optimum frequency is 200 Hz for ranges greater than 120 km. However, the calculated losses particularly at 25 Hz are less than those measured. This model predicts poorest propagation in the 100 to 200 Hz region for the summer data instead of an optimum frequency shown in the data. Thus, this choice of non-linear attenuation appears to degrade considerably the agreement between the measurements and calculations.

DISCUSSION AND CONCLUSIONS

The significant feature of this report is that a model of the bottom independently obtained from geo-acoustic data has been used to model propagation losses. The bottom model, although an over-simplification of the sub-bottom structure, results in reasonable agreement between measured and predicted propagation loss at frequencies up to 100 Hz. In winter the model predicts an optimum propagation frequency near 100 to 200 Hz in general agreement with the measurements.

The main difficulty is accounting for the increased losses in the summer data at frequencies greater than 100 Hz. A number of variations of the bottom properties were discussed in the previous section. The results of computations with these variations

indicate that modifications to the bottom model to give increased
losses will primarily affect the low frequencies where the agreement
is presently reasonable. Increased losses at high frequencies are
necessary to fit the data - a greatly increased bottom roughness
gives increased losses at high frequencies in summer, but not
enough to make the losses at 400 Hz exceed the losses at 200 Hz.
What is needed is a mechanism, not presently accounted for by the
model, that results in attenuation of the low-order modes at the
higher frequencies.

Shear waves are probably not the solution since they would
primarily affect the low frequencies where the agreement is already
reasonable. As discussed in the previous section, a bottom
attenuation coefficient that increases non-linearly with frequency
fails to account for the observed trends. However, it is possible
that some form of non-linear attenuation coefficient or increased
bottom roughness combined with shear waves could predict the
observed trends.

We can speculate that the high-frequency losses obtained in
summer are related to the variability of the sound speed in the
thermocline. Scattering from the thermocline analogous to boundary
scattering may occur. Variations in the depth of the thermocline
from profile to profile may cause coupling between modes, allowing
the acoustic energy from the low-order modes to be coupled into the
higher modes which are then attenuated rapidly. Both scattering and
mode coupling effects become increasingly important as the frequency
increases and would increase the high-frequency losses leaving the
low-frequency losses less affected. The winter results would be
less affected since the sound-speed gradients in both range and
depth are considerably less. Moreover, the iso-speed water in
winter increases the effective water depth seen by the low-order
modes, so the mode coupling caused by changes in water depth is
less than in summer.

Effort must continue to establish the failure of the model
to predict the increased losses at 400 Hz in the summer data.
The next step is to expand the limited sensitivity study to
establish what modifications should be made to the modelled values
to obtain a better overall fit to the data and to establish from
the literature and through measurement whether such adjustment of
the bottom parameters can be justified.

However, it is extremely encouraging that the independently
generated geo-acoustic inputs yield reasonably good agreement over
a wide range of frequencies particularly for the winter data.
Moreover, our preliminary analysis suggests that the faults may not
lie with the bottom model, but rather with an additional loss
mechanism not accounted for by the model. This additional loss
would improve the agreement between the predictions and measurements.

Table 1. Acoustic Properties of the Sediments

Type	Sound Speed (m/sec)	Density (g/cm^3)	Attenuation (dB/km//Hz)
Sand	1750[a]	2.06[b]	.26[c]
Till	1900[b]	2.1[b]	.030[b]
Bedrock	2050[a]	2.2[b]	.050[b]

a. Determined from DREA survey.
b. After Beebe[6].
c. After Dodds[8].

REFERENCES

1. D.M.F. Chapman and D.D. Ellis, "Propagation-loss modelling on the Scotian Shelf: The geo-acoustic model", elsewhere in the proceedings.
2. Huntec ('70) Ltd., 25 Howden Road, Scarborough, Ontario, Canada M1R 5A6.
3. N. Cochrane, Memorial University of Newfoundland, informal communication.
4. D.D. Ellis and B.A. Leverman, Defence Research Establishment Atlantic, informal communication.
5. A. Nagl, H. Uberall, A.J. Haug and G.L. Zarur, "Adiabatic mode theory of underwater sound propagation in a range-dependent environment", J. Acoust. Soc. Am., 63:739 (1978).
6. J. Beebe, "Long-range propagation over a mixed-sediment, sloping bottom", J. Acoust. Soc. Am., Suppl. 1, 67:S30 (1980).
7. R.D. Stoll, "Acoustic waves in saturated sediments", in: Physics of Sound in Marine Sediments", Loyd Hampton, ed., Plenum Press, New York (1974).
8. D.J. Dodds, Huntec ('70) Ltd., informal communication.

SEA FLOOR EFFECTS ON SHALLOW-WATER ACOUSTIC PROPAGATION

Tuncay Akal

SACLANT ASW Research Centre

La Spezia, Italy

ABSTRACT

The acoustic propagation in the 24 Hz to 8 kHz band has been measured in a large number of coastal-water areas using explosive charges and digital techniques. When displaying the losses as a function of range and frequency a marked optimum frequency is observed in most of the cases. The acoustic characteristics of the sea floor over which the propagation measurements were made have been compared with the loss levels and the optimum frequencies and have shown, among other things, a relationship between the optimum frequency and sediment properties.

INTRODUCTION

Over the last decade, SACLANTCEN has been conducting an extensive experimental research programme on the environmental acoustics of shallow waters. Emphasis has been on obtaining broadband acoustic data together with detailed information on the environmental parameters that affect acoustic propagation.

The propagation of sound in shallow water depends on many environmental parameters, such as the sea surface, the water medium, and the sea floor. Due to the multiple-reflected propagation paths, acoustic propagation characteristics are very sensitive to the boundary conditions, and the sound-speed structure of the water column controls the influence of these boundaries on propagation.

To obtain data under different environmental conditions, techniques have been developed and experiments have been conducted in different geographical locations and in various seasons over different types of bottom material and bathymetric features.

In order to be able to demonstrate the effects of the sea floor on propagation in shallow water, we have chosen here some characteristic bottom features over which the propagation measurements were made and compared them with the transmission-loss levels and optimum frequencies for propagation.

1   EXPERIMENTAL AND ANALYSIS TECHNIQUES

The experimental technique to measure transmission losses is shown in Fig. 1. The receiving ship would anchor and launch a vertical hydrophone string made of omnidirectional hydrophones spaced in such a way as to cover most of the water column. Hydrophones were usually suspended from a spar buoy and drifted away from the ship to minimize ship-radiated noise. A source ship moving away or towards the receiving ship at a fixed, pre-determined course would launch explosive charges (180 g TNT) at regular intervals and take XBTs (expendable bathythermographs) along the track. At the moment of the explosion a radio pulse was transmitted from the source ship to calculate travel times. The signals were pre-amplified near the hydrophones, then low-pass filtered on board and digitized with a 24 kHz or a 12 kHz sampling frequency and recorded on magnetic tapes. At the same time the signals were also transferred for on-line inspection and processing of a few channesl to a HP 2116 B, or recently with more channels to a HP 21 MX computer. Figure 2 shows the block diagram of the receiving chain.

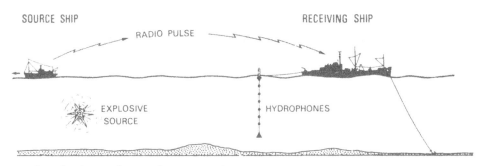

Fig. 1 Experimental set-up for transmission loss measurements

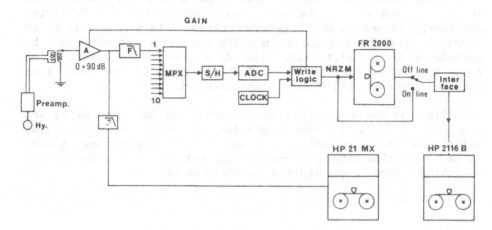

Fig. 2   Block diagram of the receiving chain

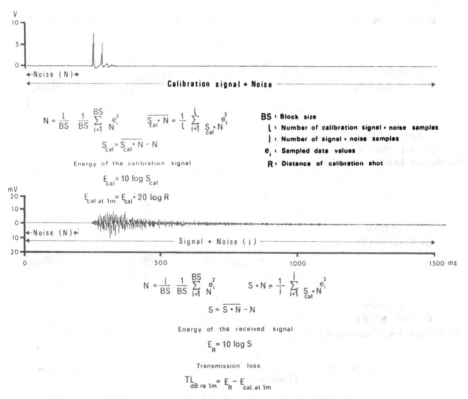

$$N = \frac{L}{BS} \cdot \frac{1}{BS} \sum_{i=1}^{BS} \frac{e_i^2}{N_i} \qquad \overline{S_{cal} + N} = \frac{1}{L} \sum_{i=1}^{L} \frac{e_i^2}{S_{cal} + N_i}$$

$$S_{cal} = \overline{S_{cal} + N} - N$$

Energy of the calibration signal

$$E_{cal} = 10 \log S_{cal}$$

BS : Block size
L : Number of calibration signal + noise samples
j : Number of signal + noise samples
e_i : Sampled data values
R : Distance of calibration shot

$$E_{cal\ at\ 1m} = E_{cal} + 20 \log R$$

$$N = \frac{j}{BS} \cdot \frac{1}{BS} \sum_{i=1}^{BS} \frac{e_i^2}{N_i} \qquad S + N = \frac{1}{j} \sum_{i=1}^{j} \frac{e_i^2}{S_{cal} + N_i}$$

$$S = \overline{S + N} - N$$

Energy of the received signal

$$E_R = 10 \log S$$

Transmission loss

$$TL_{dB\ re\ 1m} = E_R - E_{cal.\ at\ 1m}$$

Fig. 3   Transmission loss calculation

Transmission-loss calculations were then performed as shown in Figs. 3 and 4. Transmission losses reported in the following are energy losses in decibels with reference to a source level at 1 metre. For those signals in which the noise content could be considered to be stationary, the data have been corrected for noise by subtracting the measured noise energy of the same time-duration observed prior to the signal arrival.

Since transmission loss is a function of both frenquency and range, it was felt that the simple display that contains the major information was the isoloss contour in a frequency/range plane. In order to eliminate small-scale fluctuations, a two-dimensional smoothing procedure was applied, as shown in Fig. 5.

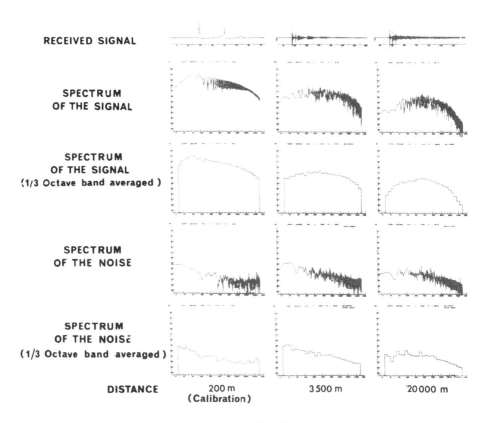

Fig. 4 Example for transmission loss analysis

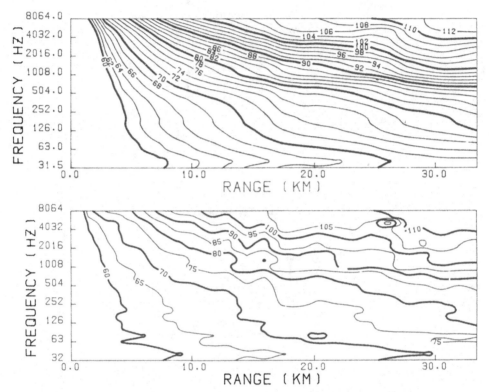

Fig. 5   Isoloss contours in frequency/range plane
before and after smoothing

To support the interpretation of acoustic data, environmental data that could effect acoustic propagation have been collected, as summarized in Fig. 6. Apart from XBTs taken by the source ship, the receiving ship usually takes TDS (temperature/depth/salinity) casts, monitors meteorologic and oceanographic conditions, and studies sea-floor characteristics of the area by means of cores, bottom samples, narrow-beam high-resolution sounder, side-scan sonar, seismic profilers, and stereo photographs.

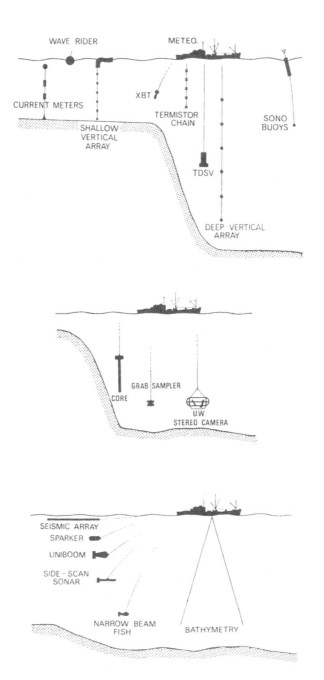

Fig. 6 *Summary of environmental data collected*
during acoustic measurements

2   PHYSICAL CONSIDERATIONS

Figure 7 shows schematically a simplified shallow-water environment. For the time being, we take the water column to be isovelocity. The water column essentially forms a wave guide bounded above by a virtually perfectly-reflecting surface (air) and below by the ocean-bottom sediment. Under these simplified conditions, acoustic propagation is dominated by the repeated interaction of the acoustic waves with the ocean bottom; each interaction can be viewed as a reflection process.

A typical bottom-reflection curve for the simplified environment under study is shown diagramatically in Fig. 8. The solid curve (a) represents reflection from a bottom without attenuation ($\alpha = 0$). Since the speed of sound is greater in the bottom, there exists a critical angle, θ_c; hence, rays with smaller grazing angles will be perfectly reflected and no acoustic energy will be transmitted into the ocean bottom. The dashed line (b) is the result of making the bottom lossy. Though there really is not a true critical angle there is still an approximate one, at the same angle as in the non-lossy case. Rays with angles larger than θ_c will be partially transmitted and therefore suffer reflection loss.

Now consider a point-source radiating in such an environment, as shown in Fig. 7. For the non-lossy bottom case, there will exist a cone of energy that couples into the water column and undergoes only cylindrical-spreading loss (minimum loss possible under these conditions). Paths associated with larger grazing angles will be severely attenuated, since each reflection involves a loss caused by part of the sound being transmitted into the bottom. When the bottom is lossy, the above is still generally true except that we will have more loss than cylindrical spreading in the propagation region below critical angle; however, our discussion for the higher grazing angles is not essentially altered. The region of small grazing angle is usually referred to as the "discrete" normal-mode region, "discrete" referring to the fact that the only paths that propagate are those that construct-

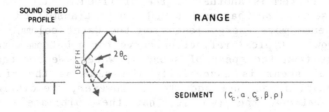

Fig. 7 Simplified shallow-water environment

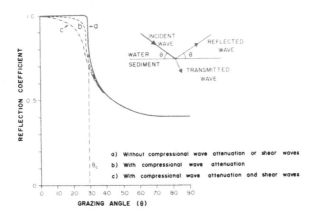

Fig. 8   Three examples of bottom reflectivity curves

ively interfere. The larger grazing-angle region is usually referred to as the continuous region, in that all angle paths exist although they die out rather rapidly. Normally, we are therefore concerned with the discrete region. In this case the sound speed in the bottom determines the size of the cone of energy $(2\theta_c)$ that can efficiently couple into the water channel and then the bottom loss determines to what extent the overall propagation loss for that portion of the energy will differ from the ideal case of cylindrical energy.

It is customary to think that bottoms with a high compressional speed afford better propagation conditions, since the cone is larger. But now the picture becomes more complicated. In reality, the ocean bottom has a certain degree of rigidity and hence can support shear and the propagation of shear waves (with velocity c_s).

For this discussion, let us say that the existence of shear waves in the bottom is another degree of freedom that the bottom sediment possesses, so that more sound can be transmitted into the bottom and hence more energy lost from the water column. Case c of Fig. 8 shows a typical reflection curve for a bottom with shear velocity less than the speed of sound in water. We see then that the effect of shear is essentially the same as the effect of bottom attenuation and just adds to it. However, the distinguishing factor between the two is that they ultimately have a different frequency dependence. Bottoms with high compressional speed tend to have high shear speeds; hence losses in the water column due to coupling to shear waves in the bottom become more important in the high compressional sound-speed region[1] [2] [3] where, if shear was neglected, very good propagation conditions might be expected.

3   EXPERIMENTAL EVIDENCE

3.1 Effects of Seasonal Changes

The sound-speed structure of the water column plays a major role on the effects of the sea floor on propagation. In summer, because of downward-refracting conditions, the propagation is mainly influenced by the sea-floor characteristics, whereas in winter the influence of the sea floor is reduced due to the upward-refracting conditions. Figure 9 shows this effect on two acoustic runs made over the same track under summer and winter conditions. It is clearly seen that in summer conditions losses are much higher than in the winter.

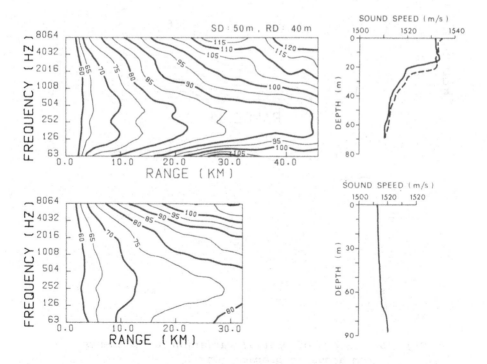

Fig. 9   Effects of seasonal changes

3.2 Effects of bottom material

In contrast to the deep-sea floor, the continental shelves mostly consist of a large variety of sediments with considerable spatial variation.

Figure 10 is an example of the bottom material. Within a distance of 60 km, material on the sea-floor changes from SAND-SILT-CLAY to SILTY SAND and then to SAND. The acoustic field also demonstrates this by an increased optimum propagation frequency (OPF) over the sandy bottom and by higher losses at low frequencies.

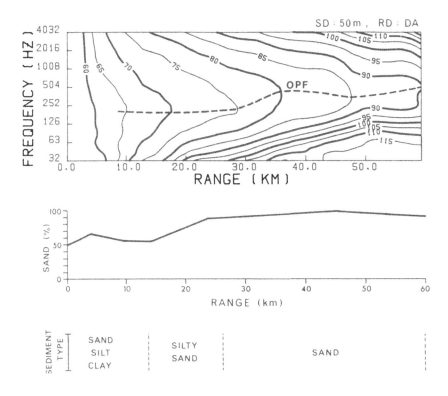

Fig. 10 *Effects of spatial variation of sediments on acoustic propagation*

Figure 11 displays the effects of two extremely different bottom materials on propagation. The upper figure presents the transmission losses measured over an area with high porosity, relatively low sound-speed material (clay). The lower figure shows how the optimum frequency increases with decreasing porosity (more rigid bottom, higher shear speeds).

3.3 Effects of Bathymetric Features

Steep slopes, different water depths, rough bottoms, and drastic changes in bathymetry such as shallow banks and narrow straits, are among the most important features that form the lower

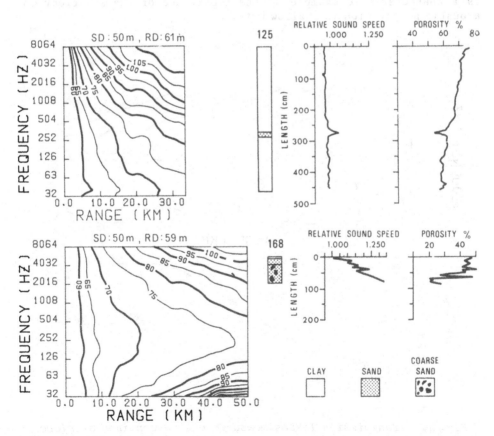

Fig. 11 Effects of two different bottom material on propagation

limit of the propagation medium and affect the propagation
conditions. Now let us see separately the effects of these
features on propagation.

3.3.1 Down-slope

     Figure 12 shows the transmission losses measured over a
continental slope. The receiver is situated in the deep-water
part and the source (50 m) moves towards the continental shelf. Up
to the range of 50 km, the propagation is characteristically
deep-water propagation: the lower the frequency the better the
propagation, with a convergence zone having a distance of
approximately 25 km. When the source moves into the continental
shelf at the range of 60 km, the propagation characteristics
change drastically. Low frequencies (below 60 Hz) interact more
with the shear-supporting bottom and hence are more severely
attenuated and create an optimum frequency around 100 Hz. This
is a characteristic example of the importance of the sea floor on
acoustic propagation in shallow water.

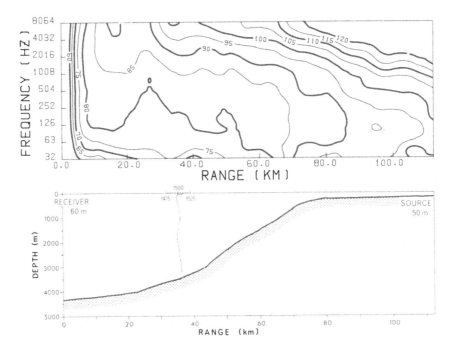

*Fig. 12   Transmission losses measured over a continental slope
           (shallow to deep)*

3.3.2 Up-slope

Figure 13 displays the acoustic field and the bathymetric profile. This run is from deep to shallow. We see from the figure that there are two basic regions: first a flat, shallow-water part where a characteristic optimum frequency around 200 Hz develops under very good propagation conditions (2 dB/10 km); then with a steep slope the losses increase drastically and the optimum frequency moves towards lower frequencies due to the change in water depth.

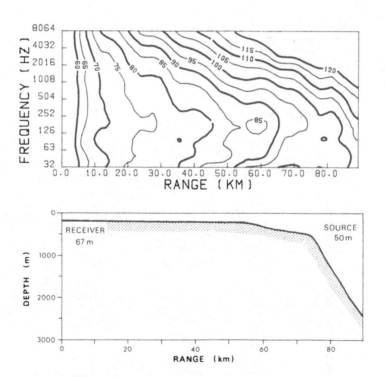

Fig. 13 Transmission losses measured over a continental slope (deep to shallow)

3.3.3 Water depth

Figures 14 and 15 compare the effects of water depth on propagation in two different areas. As can be seen, the greater the water depth the less the losses and the lower the optimum frequencies.

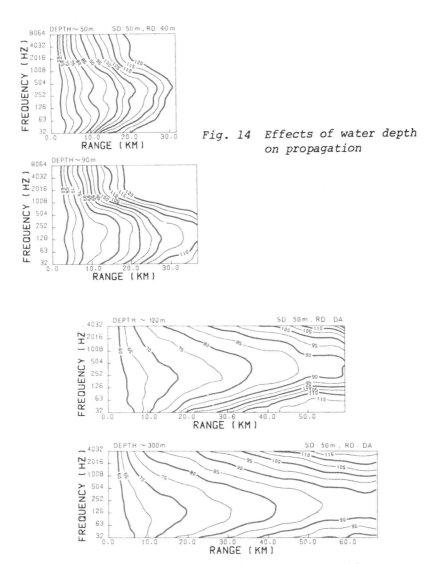

Fig. 14   Effects of water depth on propagation

Fig. 15   Effects of water depth on propagation

3.3.4 Shallow banks

Shallow banks are very often found over the continental shelves. To demonstrate the shadowing effect of these features, measurements were made by keeping the range constant at 16 km (Fig. 16) but varying the bearing of the source in such a way that the propagation path was partly obscured by a shallow bank of 16 m depth[4]. Figure 17 shows the transmission losses as functions of frequency and bearing angle. As expected, the low frequencies are most affected by the shallow bank. Optimum frequency increases approximately two octaves.

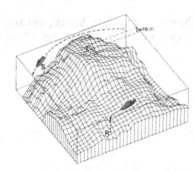

Fig. 16 *A shallow bank and geometry of acoustic measurements*

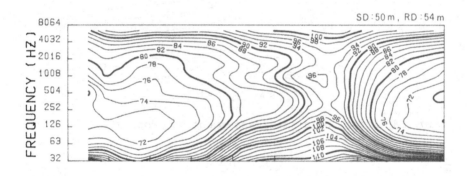

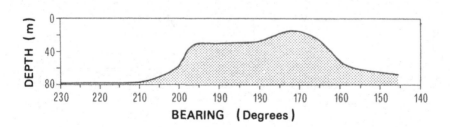

Fig. 17 *Transmission losses measured in the frequency/bearing angle plane and the silhouette of the shallow bank as seen from the receiver*

3.3.5 Sea-mounts

Figure 18 shows a run over a small sea-mount; this figure also shows the roughness spectrum measured over the track where acoustic measurements have been conducted. On the acoustic results the losses are much higher close to the sea-mount due to shadowing effect[5][6] and after a certain distance propagation starts to improve.

3.3.6 Rough bottom

Scattering because of bottom roughness can also be important the higher the frequency the more the scattering loss.   The effect

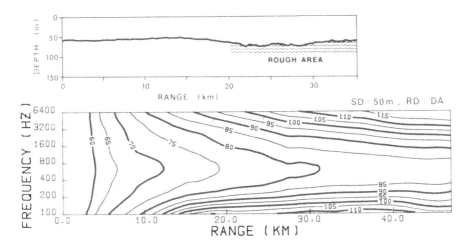

Fig. 18 Effect of a small sea-mount on propagation

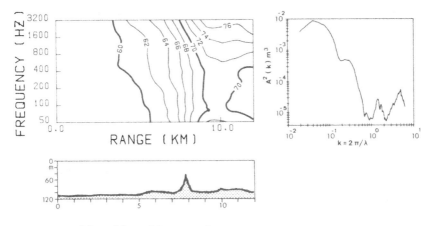

Fig. 19 Effects of bottom roughness

of roughness is one of the most difficult parameters to isolate from other factors that also contribute to high-frequency attenuation.

Figure 19 shows a run where during the first part the bottom is flatter. Note the difference in the loss contours over the rough area.

CONCLUSION

In order to describe the fundamentals of the shallow-water loss mechanism, we have presented the effects of some characteristic features of the sea floor on acoustic propagation. It is clear that the loss mechanism in shallow water is very complex and is generally the combination of more than one of the features described above.

To isolate the effect of a certain feature we have chosen simple cases where only one particular parameter was the dominant environmental factor, and have used SACLANTCEN's acoustic modelling facilities[7][8] to check them. Despite the wide extent of the cases covered, as in any duct propagation there exists an optimum frequency at which the total loss is a minimum.

The dependence of optimum propagation frequency on sea-floor characteristics is summarized in Fig. 20, which shows the optimum

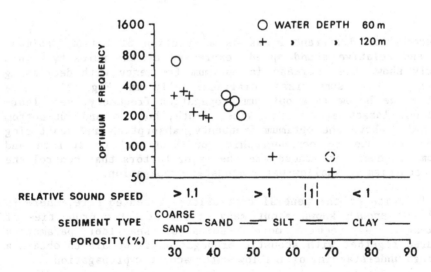

Fig. 20 The dependence of optimum propagation frequency
         on sea floor characteristics

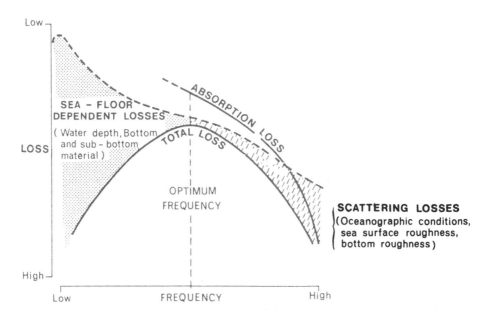

Fig. 21 *Diagrammatic summary of the major factors that*
         control the loss mechanism in shallow-water
         acoustic propagation

frequency for different areas as a function of bottom sediment
type and relative sound speed, expressed by the porosity. This
clearly shows the increase in optimum frequency with decreasing
porosity. As summarized diagramatically in Fig. 21, at a
fixed range below this optimum propagation frequency, sea-floor-
dependent losses are due to water depth, bottom, and sub-bottom
material; above the optimum frequency, absorption and scattering
losses are due to oceanographic condition and to surface and
bottom roughness and these are the major factors that control the
loss mechanism in shallow-water acoustic propagation.

In spite of the general conclusions presented here, there is
still not enough known about many of the acoustic properties of
sediments. We require more studies of sea-floor parameters
measured together with acoustic propagation if we are to obtain a
thorough understanding of shallow-water acoustic propagation.

REFERENCES

1. O. F. Hastrup, Some bottom-reflection loss anomalies near grazing and their effect on acoustic shallow-water propagation. (This volume).

2. F. Ingenito and S. N. Wolf, Acoustic propagation in shallow water overlying a consolidated bottom, J. Acoustical Society America 60:611-617 (1976).

3. F. Ingenito, R. H. Ferris, W. A. Kuperman, and S. N.Wolf, Shallow-water acoustics: summary report, NRL Rpt 8179, U.S.Naval Research Laboratory, Washington, D.C. (1978).

4. O. V. Olesen, Brüel and Kjaer Company, Copenhagen, Denmark. (Personal Communication).

5. H. Medwin and R. Spaulding, The seamount as a diffracting body. (This volume).

6. F. B. Jensen and W. A. Kuperman, Range-dependent bottom-limited propagation modelling with the parabolic equation. (This volume).

7. F. B. Jensen and W. A. Kuperman, Environmental acoustic modelling at SACLANTCEN, SACLANTCEN SR-34, SACLANT ASW Research Centre, La Spezia, Italy (1979). [AD A 081 853]

8. M. C. Ferla, G. Dreini, F. B. Jensen and W. A. Kuperman, Broadband model/data comparisons for acoustic propagation in coastal waters. (This volume).

BROADBAND MODEL/DATA COMPARISONS FOR ACOUSTIC PROPAGATION

IN COASTAL WATERS

M.C. Ferla, G. Dreini, F.B. Jensen, and W.A. Kuperman

SACLANT ASW Research Centre

La Spezia, Italy

ABSTRACT

Broadband propagation data collected by SACLANTCEN in various shallow water areas in the Mediterranean Sea and in the North Atlantic have been modelled using a normal-mode propagation model. The data cover a frequency range of 50 Hz to 6.4 kHz, with experiments performed under a variety of different environmental conditions including seasonal changes in sound-speed structure, different sea states, varying water depth (90-300 m) and bottom composition with range, propagation through oceanic fronts, etc. Conclusions of the model/data comparisons are presented with emphasis on difficulties encountered (e.g. lack of environmental information), on detected model limitations, and on inferred conclusions concerning the actual environments. The importance of bottom rigidity (shear) as a low-frequency loss mechanism is evident in many model/data comparisons, particularly through its effect on the optimum propagation frequency, which is found to increase with increasing shear speed.

INTRODUCTION

In this paper we present some examples of model/data comparisons that have been used to formulate a methodology for constructing geo-acoustic models of coastal-water areas. The comparison of broadband data, taken in various areas of the North Atlantic and the Mediterranean[1], with theory indicates that we can explain the main features of sound propagation in shallow-water areas.

Data were obtained from experiments using two ships. The receiving ship was anchored with a suspended vertical array covering most of the water column while the source ship steamed out in range dropping 180 g explosive charges. The experiments were performed during different seasons in water depths varying from 90 to 300 m. The data were processed in one-third octave bands from 50 Hz to 6.4 kHz.

The SACLANTCEN normal-mode model SNAP[2] was used in this study, though, in certain frequency regimes, other models were used as a cross-check[3]. The SNAP model handles range-dependent propagation in the adiabatic approximation. Input parameters for SNAP are shown in Fig. 1. Sound-speed profiles inputted in the model have been measured at sea. If a variable sediment sound-speed profile is used, it was obtained from a core together with density. Seismic profiling provided the depths of any significant bottom layering. The other parameters, such as compressional attenuation and shear speed of the bottom, were estimated from the literature[4-8] though final estimates were extracted from the propagation experiment itself. In particular, though there are very few measurements of the in-situ shear properties of the bottom, we have found that shear can be an extremely important factor limiting low-frequency propagation in certain shallow-water areas.

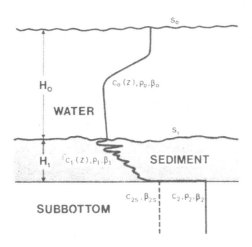

Fig. 1 Environmental input to the SNAP model

c_i : compressional speed β_i : compressional attenuation
c_{2s} : shear speed β_{2s} : shear attenuation
ρ_i : density s_i : rms roughness

In the first two sections we shall present some results comparing model calculations with experimental data from two different flat-bottom areas. Flat bottom refers to constant depth but may include some local random roughness. In Section III we present a similar study for a much more complicated environment: propagation over a sloping bottom where both bottom and water properties change with range. The conclusions presented at the end of the paper are based on our ability to explain the general features of sound propagation in these highly different environments.

I   PROPAGATION OVER A FLAT HOMOGENEOUS BOTTOM

In this section we present some results from an area of the Mediterranean where cores indicated that the bottom was nearly homogeneous. By homogenous we mean that small fluctuations in the bottom properties versus depth and range as measured by core analysis had no significant effect on computed propagation losses, and hence could be omitted from the model study. The data were taken in a water depth of 90 m in both summer and winter; the sound-speed profiles are shown in Fig. 2.

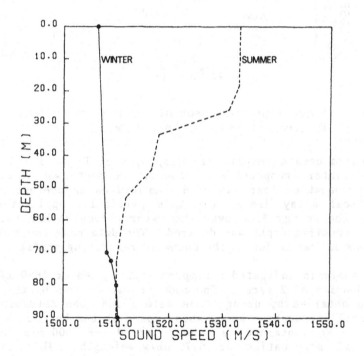

*Fig. 2   Sound-speed profiles from shallow-water area
          in the Mediterranean*

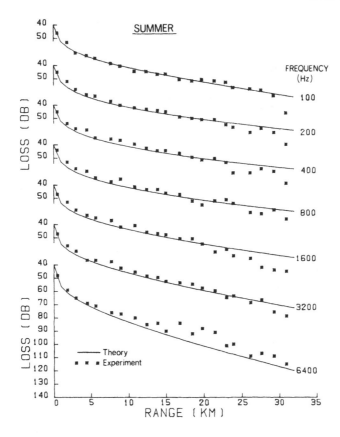

Fig. 3 *Model data comparison of depth-averaged losses
for summer. Source depth 50 m*

One-third octave results are displayed in Figs. 3 and 4 for summer and winter, respectively. The source depth was 50 m. The measured propagation loss has been computed as an average output of a vertical array since, for this particular application, a measure of the energy flux over the water column rather than the field at a specific depth was desired. The data show the expected greater loss in summer due to the downward refracting profile.

Core analysis indicated a compressional speed of 1650 m/s and a bottom density of 2 g/cm^3. The model results shown by the solid lines were obtained by using these values and then determining a compressional attenuation and a shear speed in the bottom to match the data. We obtained a shear speed of 600 m/s and a compressional attenuation of 0.75 dB/wavelength. This fitting assumes that the compressional attenuation is linear with frequency. The frequency dependence on propagation loss due to shear

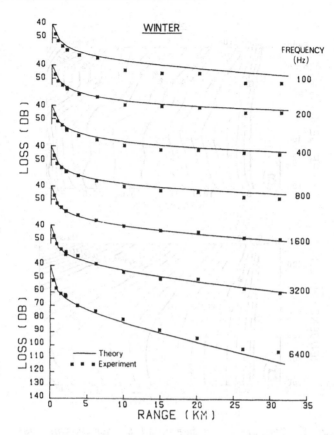

Fig. 4   Model/data comparison of depth-averaged losses
         for winter.   Source depth 50 m

is different from compressional attenuation; we found for this
particular data set that the inclusion of the shear properties was
necessary to explain the increasing loss at lower frequencies. We
also included a shear attenuation of 1.5 dB/wavelength. This
parameter, however, has only a minor effect on the computed
propagation losses.

The frequency dependence of propagation loss for these cases
is more clearly shown in Figs. 5 and 6 where loss is contoured
versus frequency and range. The propagation loss curves of Figs.
3 and 4 are obtained by taking horizontal cuts at the specific
frequencies of interest. Notice how the general shape of the
model calculation agrees with the experiment. We stress once more
that the inclusion of shear in the bottom was essential to obtain
agreement in the shapes of the curves at lower frequencies. In
particular, the optimum frequency of propagation[1] is highly
dependent on the shear speed in the bottom. The optimum frequency
is found to increase with increasing shear speed.

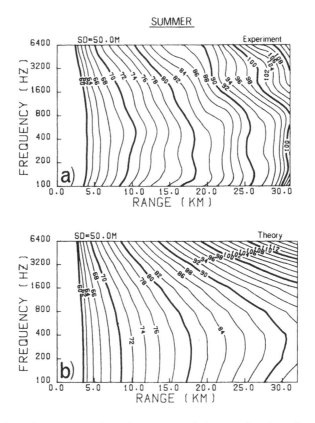

Fig. 5 Computed and measured transmission loss
        contours for summer

On the high-frequency end it was necessary to include
boundary roughness to obtain agreement between theory and
experiment. A surface rms roughness of 0.25 m for both summer and
winter was consistent with environmental conditions. A bottom rms
roughness of 0.1 m was included to provide better agreement with
the high-frequency summer data.

For the actual propagation levels, Figs. 5 and 6 indicate
better agreement between theory and experiment for winter than for
summer. We have found this to be the case also for other model/
data comparisons, presumably because the increased interaction
with the bottom in summer requires an extremely accurate knowledge
of the acoustic properties of the bottom. Nevertheless, the
agreement in shape of the curves for both summer and winter conditions indicates that normal-mode theory is indeed descriptive of
the general features of sound propagation in this area.

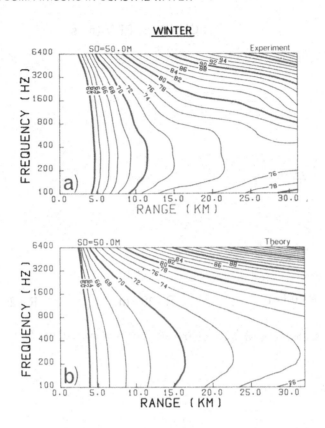

Fig. 6 *Computed and measured transmission loss*
contours for winter

II   PROPAGATION OVER A FLAT LAYERED BOTTOM

In the former section we demonstrated that we can obtain good
agreement between theory and experiment for depth-averaged
propagation loss as a function of range and frequency. Now we
would like to consider the variation of the acoustic field with
receiver depth. For this purpose we consider another data set
from the Mediterranean[9]. In this case the bottom was layered,
with the upper six metres essentially being mud with a lower speed
than that in the water column. The experiment was carried out in
110 m of almost isothermal water and the source depth was 50 m.

Sound-speed profiles obtained from cores of the upper few
metres of the bottom are shown in Fig. 7. An average geo-acoustic
model of the bottom extracted from cores and seismic profiling (for

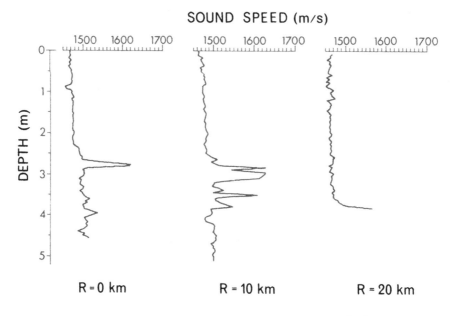

Fig. 7 *Sound-speed structure in bottom measured from cores*

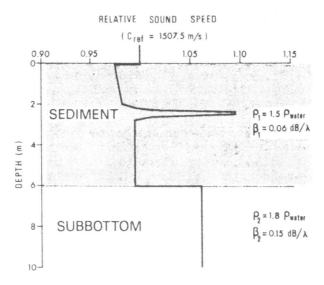

Fig. 8 *Geo-acoustic model derived from the core measurements*

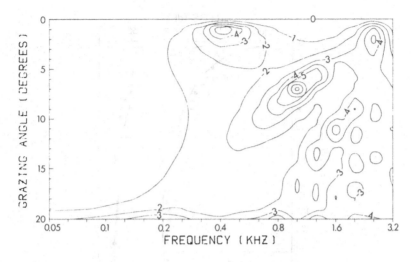

Fig. 9 Contoured reflection loss for bottom given in Fig. 8

the mean layer depth) is given in Fig. 8. Minor speed fluctuations have been averaged out considering their negligible effect on propagation as determined through a parametric study. Because of the lack of rigidity of mud, shear was not expected to be a significant parameter in this study and hence was neglected.

The complexity of the bottom is illustrated by the computed reflection loss versus grazing angle and frequency[9] shown in Fig. 9. The more familiar loss versus angle curves are obtained from this contour plot by making a vertical cut at the specific frequency of interest. Figure 9 shows the existence of a critical angle at about 18° to 20° at lower frequencies, whereas at higher frequencies we get a more complicated reflectivity pattern.

The model/data comparisons are shown in Figs. 10 and 11. The field near the surface and bottom boundaries is emphasized in both of these figures; these near-boundary regions are most severely influenced by complicated environmental conditions such as bottom layering. At 100 Hz we see excellent agreement between theory and data as a function of range and depth. Even at 1600 Hz the agreement is within a few decibels at a range of 20 km. No bottom roughness was used because of the muddy nature of the bottom. In addition, the sea was sufficiently calm to exclude any surface-roughness effects. Of course the two compressional attenuation values shown in Fig. 8 were derived from the propagation loss data, but it should be mentioned that these numbers are consistent with the literature[4-8].

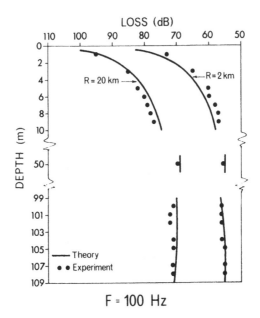

Fig. 10   Model/data comparison of transmission losses
near the sea surface and the sea floor.
Source depth 50 m, frequency 100 Hz

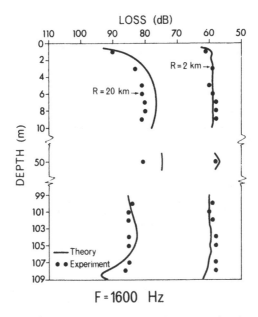

Fig. 11   Model/data comparison of transmission losses
near the sea surface and the sea floor.
Source depth 50 m, frequency 1600 Hz

To summarize: We have included bottom layering, which is an added complication over the environment studied in Section I; on the other hand, the muddy upper layer eliminated the complication of the existence of a shear wave in the upper layer of the bottom. With this change in environment, normal-mode theory still seems to provide an excellent description of propagation for this area.

III PROPAGATION IN A RANGE-DEPENDENT ENVIRONMENT

We now move to an extremely complicated environment in the North Atlantic where the depth, the sound-speed profile, and the bottom properties vary with range. The environment is schematically represented in Fig. 12. The water depth is seen to vary between 115 and 305 m over the length of the track. Two types of sound-speed profiles exist: an almost isovelocity profile out to the position of the ocean front (25 km) and then a profile characterized by a higher sound speed in the upper 30 m of the water column. The bottom can essentially be divided into two different types: a homogeneous hard sand bottom on the first 15 km followed by a softer layered bottom consisting of a silty top layer (2 m) overlying sand. Again the basic bottom parameters (see Table 1) were taken from cores, while shear speed and attenuation were estimated from the acoustic data.

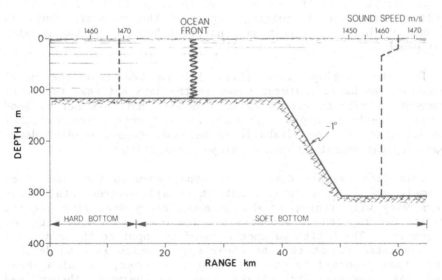

Fig. 12 Range-dependent environment from the North Atlantic

Table 1   Bottom parameters for range-dependent environment

		HARD BOTTOM	SOFT BOTTOM	
COMPRESSIONAL SPEED	(m/s)	-	1460	SEDIMENT (2 m)
DENSITY	(g/cm^3)	-	1.5	
COMPRESSIONAL ATT.	(dB/λ)	-	0.1	
COMPRESSIONAL SPEED	(m/s)	1700	1700	SUBBOTTOM
DENSITY	(g/cm^3)	2.0	2.0	
COMPRESSIONAL ATT.	(dB/λ)	1.0	1.0	
SHEAR SPEED	(m/s)	900	700	
SHEAR ATTENUATION	(dB/λ)	1.5	1.5	

We consider propagation both from left to right and vice versa for a source and receiver at 50 m depth. The data and model results shown in Figs. 13 and 14 are for down- and up-slope propagation, respectively. The range-dependent theoretical results were provided by the adiabatic normal-mode model SNAP.

For the down-slope case (Fig. 13) we have excellent agreement between theory and data. Before propagating over the slope we see that model and data show an optimum frequency of about 250 Hz; in addition the propagation losses, particularly above 100 Hz, are in agreement to within about 2 dB. In propagating over the slope we see an abrupt fall-off, which is actually geometrical spreading resulting from the increasing depth of the channel. This is followed by better propagation conditions because of the greater water depth.

For the up-slope case (Fig. 14) we see that the model reproduces the basic features shown in the data but that the level agreement is only to within about 5 dB and therefore not as good as the down-slope result. We have noticed this trend in other studies; that is, the adiabatic normal-mode theory handles down-slope propagation better than up-slope propagation[3].

This last example demonstrates that even in the case of an extremely complicated environment where all environmental parameters vary with range, adiabatic mode theory does give a quite accurate description of acoustic propagation over a wide range of frequencies. The bottom parameters used as input to the model all seem realistic except for the shear speed, which is considerably higher than reported in the literature. However, the high shear speed was necessary to obtain agreement between theory and experiment at low frequencies. Further comments on the importance of shear in marine sediments are given below.

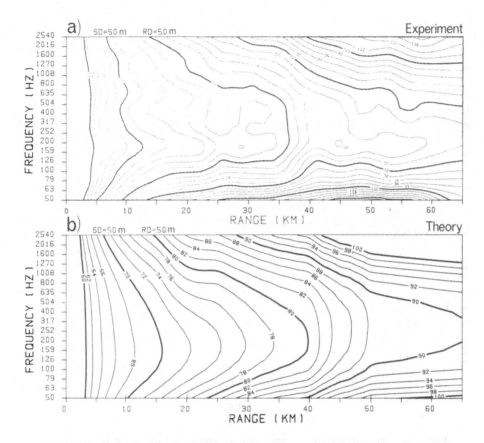

*Fig. 13 Measured and computed transmission loss contours
for down-slope propagation as referred to Fig. 12
(SD: Source Depth, RD: Receiver Depth)*

CONCLUSIONS

We have presented examples from three data sets showing
fairly good agreement between theory and experiment for increasing
environmental complexity. We may therefore conclude that the SNAP
normal-mode model works well and that it can give a quite accurate
description of acoustic propagation even in complicated ocean
environments. However, in general, good agreement with
experimental data can be obtained only by including such features
as bottom layering, bottom rigidity, scattering at rough boundaries, range-varying environments, etc.

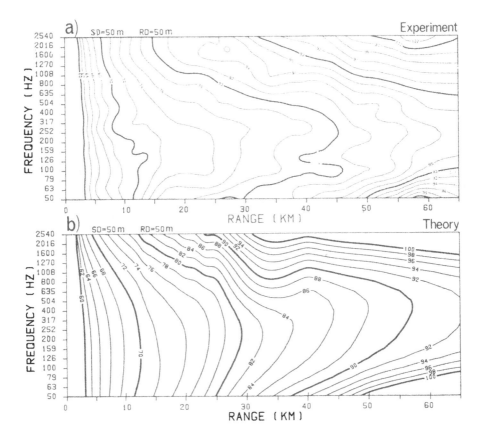

Fig. 14   Measured and computed transmission loss contours
for up-slope propagation as referred to Fig. 12

An important result of this study is the necessity to include
a low-frequency attenuation mechanism for the harder bottoms. The
inclusion of shear-wave propagation in these harder marine sediments seems to satisfy this requirement; however the shear-wave
velocities used in these theoretical studies appear to be greater
than what is reported in the literature[6-8]. Of course very few
in-situ measurements have been made of shear-wave velocities, and
complicating the problem is that we must know this velocity to
about 30 m depth in the bottom if we are interested in frequencies
as low as 50 Hz[10]. If, on the other hand, the speeds are not as
great as those mentioned in Sections I and III, then there must
exist an additional low-frequency attenuation mechanism not
included in the standard theory of elastic media whose plane-wave
attenuation coefficients vary approximately linearly with
frequency. This possible conclusion should provide an extra

motivation for the increasing research into more sophisticated sediment models[11-12], which predict an additional dilatational wave that has recently been observed in a laboratory experiment[13].

It is quite clear from this and other studies that the problem of predicting propagation loss in shallow water is equivalent to knowing the structure of the ocean bottom since there is sufficient evidence that we can handle water-column sound-speed profile effects. How then do we construct a geo-acoustic model of the ocean bottom? We feel that the only feasible way is a combination of a) sound-speed profiles and bathymetry, b) cores for determining the upper bottom layering, c) seismic profiling for the deeper layering, d) a broadband propagation experiment, and e) a sophisticated propagation model.

Some initial rough information on bottom composition, layering, etc. is needed, and this information can best be obtained from coring and seismic profiling. Then the data/model comparison is used to "fine-tune" bottom parameters until an acceptable agreement is obtained between theory and experiment. For broadband experimental data with a good depth and range coverage, the "solution" for the ocean bottom should be unique. However, with too many unknown parameters and a limited data set, one may find several combinations that seem to describe equally well the measured acoustic properties of the bottom. Therefore the more initial information that is available about the bottom, the better we can determine the actual bottom composition. Thus we need the acoustic experiment to construct a geo-acoustic model of a given area. Then what do we use the acoustic model for in terms of actual propagation predictions in shallow water? We feel that when bottom properties are known for a given area meaningful predictions can be made for various seasons and sea states and for different source/receiver combinations.

REFERENCES

1. T. Akal, Sea-floor effects on shallow-water acoustic propa-
    gation. (This volume).
2. F. B. Jensen and M. C. Ferla, SNAP: the SACLANTCEN normal-mode
    acoustic propagation model, SACLANTCEN SM-121, SACLANT ASW
    Research Centre, La Spezia, Italy (1979). [AD A 067 256]
3. F. B. Jensen and W. A. Kuperman, Environmental acoustical
    modelling at SACLANTCEN, SACLANTCEN SR-34, SACLANT ASW
    Research Centre, La Spezia, Italy (1979). [AD A 081 853]

4. T. Akal, Acoustical characteristics of the sea floor: experimental techniques and some examples from the Mediterranean Sea, in: "Physics of Sound in Marine Sediments", L. Hampton, ed., Plenum, New York, N.Y. (1974), pp. 447-480.

5. E. L. Hamilton, Compressional-wave attenuation in marine sediments, Geophysics 37:620-646 (1972).

6. E. L. Hamilton, Acoustic properties of the sea floor: a review, in: "Oceanic Acoustic Modelling, Proceedings of a Conference at La Spezia, Italy, 8-11 Sep 1975", Part 4: Sea Bottom, SACLANTCEN CP-17, W. Bachmann and R. B. Williams, eds., SACLANT ASW Research Centre, La Spezia, Italy (1975). pp.18-1 to 18-96. [AD A 020 936/1G1]

7. E. L. Hamilton, Attenuation of shear waves in marine sediments, J. Acoustical Society America 60:334-338 (1976).

8. E. L. Hamilton, V_p/V_s and Poisson's ratios in marine sediments and rocks, J. Acoustical Society America 66:1093-1101 (1979).

9. F. B. Jensen, Sound propagation in shallow water: a detailed description of the acoustic field close to surface and bottom, SACLANTCEN document in preparation, SACLANT ASW Research Centre, La Spezia, Italy.

10. F. B. Jensen, The effect of the ocean bottom on sound propagation in shallow water, in: "Sound Propagation and Underwater Systems, Proceedings of a Conference sponsored by the British Institute of Acoustics, Imperial College, London, 1978", R. H. Clarke, ed., Imperial College, London, U.K. (1978).

11. M. A. Biot, Theory of elastic waves in a fluid-saturated porous solid, J. Acoustical Society America 28:168-191 (1956).

12. J. M. Hovem, Attenuation of sound in marine sediments. (This volume).

13. T. J. Plona, Observation of a second bulk compressional wave in a porous medium at ultrasonic frequencies, Applied Physics Letters 36:259-261 (1980).

COMPUTER MODEL PREDICTIONS OF OCEAN BASIN REVERBERATION FOR LARGE UNDERWATER EXPLOSIONS

Jean A. Goertner

Naval Surface Weapons Center
White Oak
Silver Spring, Maryland   20910

ABSTRACT

Ocean basin reverberation results from interaction of the pressure wave from a large underwater explosion with the boundaries of the basin and of the seamounts and islands within it. Its net effect is to raise the ambient noise level at low frequencies for periods of from 30 minutes to 3 or 4 hours, depending on the acoustic source level and the size of the ocean basin.

A computer model has been developed to predict the reverberant field. This model has adequately matched experimental data for varying conditions in the North Atlantic, for explosive yields of up to ten tons. As an important test of its validity, we have applied the model for three high explosive shots, 0.5 to 1 kiloton in yield, one in the Atlantic and two in the North Pacific, a basin with much greater area and myriad seamounts and islands. The remarkably good agreement between model results and data indicates that the assumptions made in developing the reverberation model are valid not only in the Atlantic, but can be applied equally well to making predictions of the character of the reverberant field for much larger sources in the much more complex Pacific Ocean environment.

INTRODUCTION

At the Naval Surface Weapons Center in Silver Spring, Maryland, we have developed a computer model to predict the acoustic reverberation that will follow detonation of a large underwater explosion. This reverberation results from energy that is returned after interaction of the pressure wave with the continental shelf and the boundaries of the seamounts and islands that lie within the ocean basin. Its effect is to raise the ambient noise level in the basin for periods of from 30 minutes to several hours, a fact of considerable importance to those operating low frequency acoustic systems in the ocean environment.

Figure 1 shows broadband reverberation from two explosions detonated in the North Atlantic ocean ten years and fifty miles apart and both recorded on receivers located near Bermuda. You will note that, although the amplitudes are different for the two different yields, the character of the signals is very similar. The first 250 seconds of signal is continuous reverberation from the ocean surface and bottom. The large clumps of discrete arrivals that follow are what we call ocean basin reverberation.

The model was developed using 10-ton shot data from 10 burst locations in the North Atlantic Ocean, using two different receivers near Bermuda (Figure 2). In Figure 3 we see broadband recordings from four of these locations, each made on the same receiver. The arrivals marked "A" in each case are from the same region near the Bahamas. Here we see how geometry affects the character of the reverberation signal. Our general conclusions from this initial work were that reverberation is primarily dependent on geometric considerations; secondary in importance are acoustic source level and frequency.

The success of our comparisons with experimental data, as indicated in Figure 4 by a typical 10-ton case, has given us a great deal of confidence in the model's ability to predict the reverberant field in the North Atlantic. Although this is significant in itself, it is important that we test the applicability of the model to other ocean basins as well.

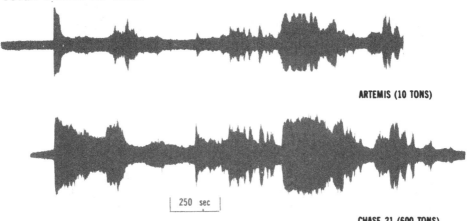

ARTEMIS (10 TONS)

| 250 sec |

CHASE 21 (600 TONS)

Fig. 1.   Artemis 1-2 (10 tons) and CHASE 21 (600 tons) detonated in the North Atlantic and recorded at Bermuda.

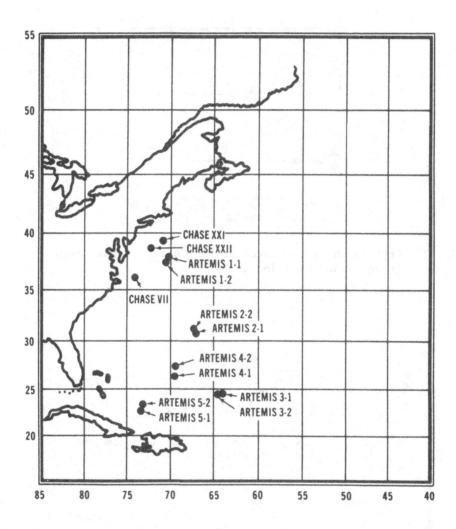

Fig. 2. Source locations for experimental data, North Atlantic Ocean

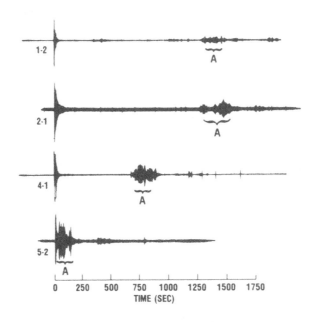

Fig. 3. Artemis shots detonated at four sites in the North
Atlantic and recorded at Bermuda

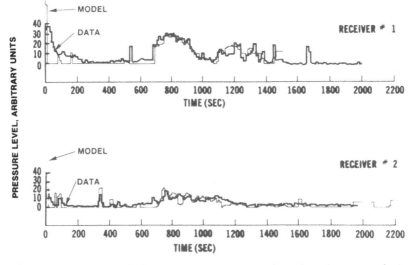

Fig. 4. Comparison of data and model results for Artemis 4-1
(10 tons)

Because of its much greater area (Figure 5) and the large
number of seamounts and islands it contains, as well as the fact
that there are experimental data available, we have selected the
North Pacific Ocean for this purpose.

We shall describe the application of the computer model to
prediction of reverberation for three shots in the CHASE (Cut Holes
and Sink 'Em) series: a 600-ton shot fired in the North Atlantic
and two shots (of 1/2 and one kiloton yield) fired in the North
Pacific. This set of data affords us an opportunity to study the
effect both of burst size and of ocean basin parameters in
determining the character of the reverberation signal.

COMMENTS ON THE MODEL

Basic input to the model for an ocean basin is in the form of
geographical coordinates, read at an appropriate depth, for points
along the continental shelf and along the boundaries of seamounts
and islands within the basin. For both the Atlantic and Pacific
we have used a depth of 1000 fathoms.

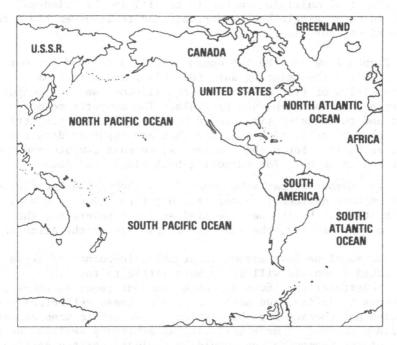

Fig. 5.   Map showing relative sizes of North Atlantic and North
          Pacific Oceans

At the basin perimeter we assume that energy is scattered from each shelf segment by a rectangular surface whose size is a function of the segment length, the contour depth and the segment slope. For seamount segments the reflecting surfaces are computed as trapezoids whose shape depends on the contour depth, the distance of the peak below the water surface and the slopes of the seamount sides.

Each single-bounce signal is computed as it travels along great-circle paths from the source, to a reflecting shelf or seamount segment, where we assume simple scattering, and then to the receiver. Each travel path is checked for obstruction by obstacles in the basin. Individual arrivals are modified for signal-spreading effects and absorption losses before all are summed at the receiver.

As explosion yield approaches a kiloton, there is sufficient energy available that significant reverberant signals can also be received via paths that have intersected two reflecting shelf segments. We call these double-bounce arrivals. In Figure 6 we see a model computation for 100 kilotons. Experience has shown that double-bounce signal levels (i.e., those that have interacted with three shelf reflectors) will be down 20 to 25 dB from single-bounce ones; triple-bounce signals will be down another 20 to 25 dB. Until the yield nears 100 kilotons, triple-bounce signals will contribute little to the total received reverberation level. The general effect of multiple bounces is to fill in the "windows" between blocks of single-bounce arrivals and to lengthen the total received signal duration.

In Figure 7 we see the two ocean basins as defined for our purposes. Since the existing data for this study were from explosive yields of from one-half to one kiloton, we have computed both single- and double-bounce arrivals. The computer memory required and the running time needed for the model are proportional to the number of reflecting segments that are input as data for the given ocean basin. For n shelf segments, we must compute and check n single-bounce paths. For computing both single and double bounces, the number of paths becomes n^2. Although the Pacific basin as we have defined it covers roughly three times the area of our North Atlantic basin, we have limited input points for the Pacific boundary to 199, the same number as used in the Atlantic.

In the model we have assumed that multiple-bounce arrivals from isolated seamounts will contribute little to the final received reverberation. Even so, each seamount requires three to ten segments to define, and paths to each of these reflectors must be computed and checked for obstruction. In order to keep computer costs within reason, we made initially an arbitrary decision as to the size of the seamounts that would be included for the Pacific. Initially, other than those near enough to the continental margins

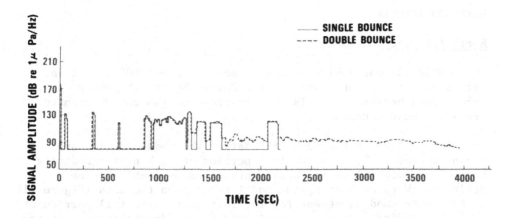

Fig. 6. Model computation for 100 kilotons: single and double
         bounces

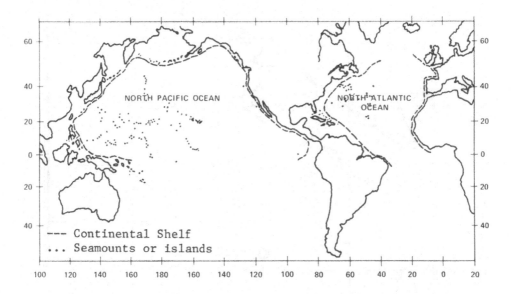

Fig. 7. Basin definition for Atlantic and Pacific

to be included as shelf reflectors, only 76 seamounts and islands
were used as basin input for the Pacific. In the Atlantic we used
50, which included all seamounts that rose to within 1000 fathoms
of the water surface. We shall consider this decision shortly.

DATA AND RESULTS

North Atlantic

CHASE XXI was a high explosive shot of about 600-ton yield, detonated off the east coast of the United States (Figure 8) in 1970. Reverberation signals were recorded at two bottom-mounted receivers near Bermuda.

Figure 9 shows the CHASE XXI data. The model results are shown in Figure 10, with the tail portion of each plot containing most of the significant returns contributed by double-bounce arrivals. When we overlay the model results on the data (Figure 11), we see quite good agreement for the most part. The tail portion of the plots indicates that we are indeed seeing double-bounce returns

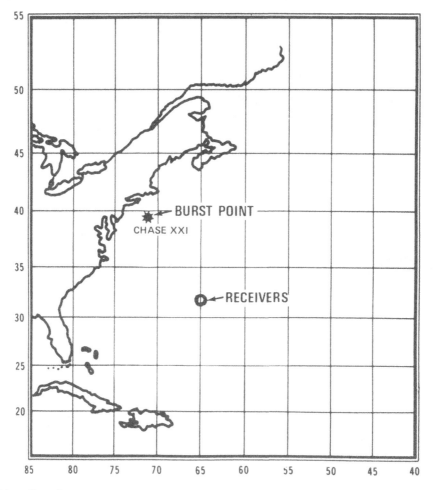

Fig. 8. Approximate shot and receiver locations for CHASE XXI

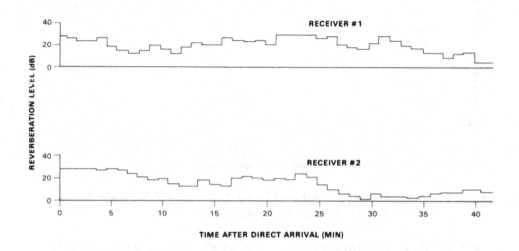

Fig. 9.   CHASE XXI data: one-minute averages, frequency = 20 Hz

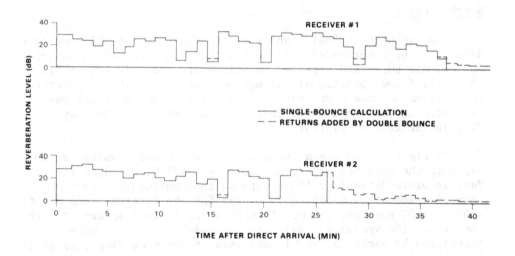

Fig. 10.   Model computation for CHASE XXI: one-minute averages,
           frequency = 20 Hz

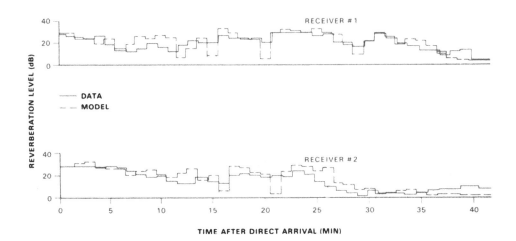

Fig. 11.   Comparison of data and model results for CHASE XXI:
           one-minute averages, frequency = 20 Hz

in the data. The late returns seen in the experimental records
arrive too early to be single-bounce returns from distant reflectors
such as the Mid-Atlantic Ridge.

North Pacific

The two high-explosive shots in the Pacific, for which we have
data, were both detonated off the west coast of the United States.
Approximate burst and receiver locations are shown in Figure 12.
CHASE XIX (about 500-ton yield) was detonated in 1970 and recorded
on a bottom-mounted hydrophone, while CHASE V (one kiloton) was
detonated in 1966 on a receiver suspended from FLIP, the Navy's
floating instrument platform.[1]

In Figure 13 we see a comparison of the model results (at 10
Hz) with the data for a 3-20 Hz band for CHASE V, computed for a
Pacific basin defined by 199 points on the continental shelf and
only 76 seamounts and islands. Except for the interval from about
4200 to 4600 seconds, the model results are in good agreement with
the data. The arrivals missing in the model prediction were
identified by Northrop, in his analysis of the recordings, as coming
from the Emperor Seamount Chain.

Figure 14 shows the location of the Emperor Seamounts relative
to the CHASE V burst. Twenty-three additional seamounts that were
borderline in size on the first pass were entered as basin input;
these included some of the Emperor Seamounts as well as some of the
Mid-Pacific Mountains, which are also shown here, in addition to
seamounts in other locations.

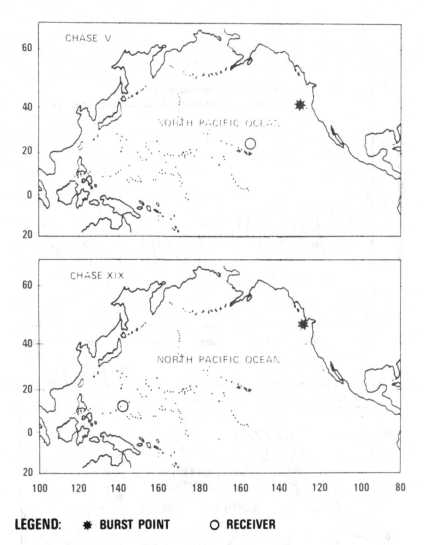

LEGEND: ✳ BURST POINT ○ RECEIVER

Fig. 12. Approximate source and receiver locations for Pacific shots

The insert in Figure 15 shows the portion of the model curve that was changed by addition of these 23 seamounts. We now see arrivals from the model that match the missing arrivals in the data curve. These have been contributed by the Emperor Seamounts and the Mid-Pacific Mountain chains. The other added seamounts added no significant arrivals that either improved or worsened the agreement.

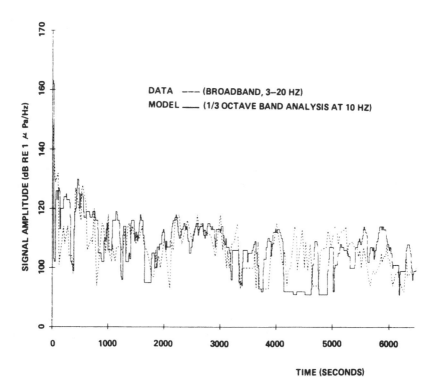

Fig. 13.   Comparison of data and model results for CHASE V at 10 Hz
           (199 shelf points, 76 seamounts and islands)

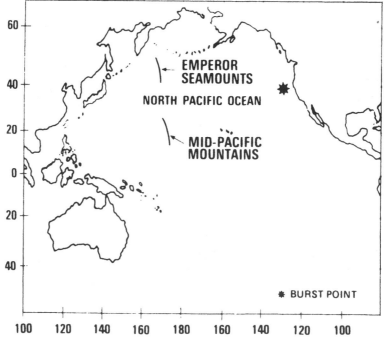

Fig. 14.   Location of Emperor Seamounts and Mid-Pacific Mountains
           relative to CHASE V burst location

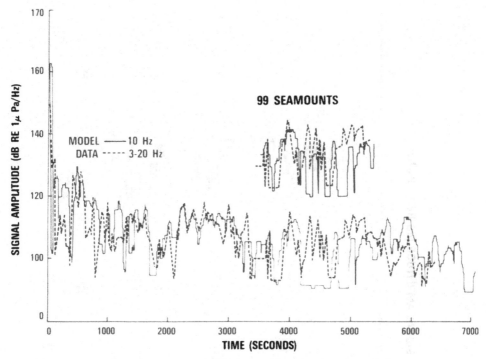

Fig. 15. Comparison of data and model results for CHASE V for
increased number of seamounts and islands

Although further basin refinement would probably improve the
model prediction even further, this input basin, with 99 seamounts,
has been the basis of all our Pacific predictions.

Next we see (Figure 16), in addition to the 10 Hz results at
the top, a comparison of model results for 100 Hz with CHASE V data
for the 20-200 Hz band. The lower source level and increased
absorption losses for the higher frequency combine to produce
reverberation of much lower level and decreased duration than we
see for 10 Hz.

The remarkably good agreement between model and data for
CHASE V are particularly significant for two reasons. First, this
shot was recorded, not on a bottom-mounted hydrophone, as were all
our other data, but on a receiver suspended relatively near the
surface. Second, the model exhibits the same frequency dependence
as the data. That is, the differences in reverberation at the two
different frequencies are fully accounted for by the accompanying
differences in source level and absorption.

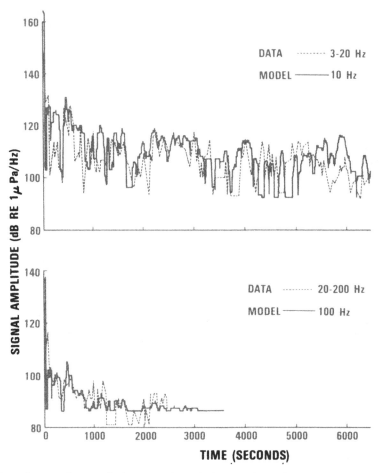

Fig. 16. Comparison of data and model results for CHASE V for two
          different frequencies (199 shelf points, 99 seamounts)

The only other high-explosive shot for which we have data is
CHASE XIX, for which the source-receiver separation was more than
twice that for CHASE V (see Figure 10). Figure 17 shows a comparison
of model and data for an analysis at 50 Hz. The differences in source
level and absorption losses for this shot, coupled with somewhat
greater travel distances for many signals, account completely for
the reduced number of arrivals and the generally lower signal level,
as compared with the 10 Hz result for CHASE V. The fact that the
experimental data and the model are correspondingly nondescript in
character further indicates that the model is successfully predicting
the reverberant field.

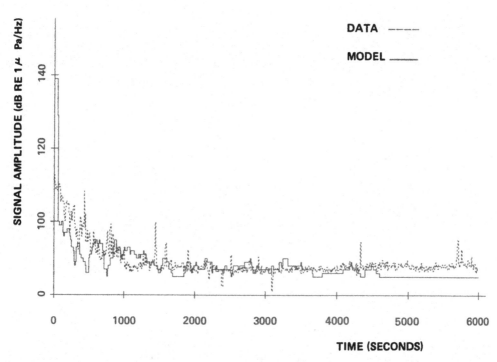

Fig. 17. Comparison of data and model results for CHASE XIX at 50 Hz

SUMMARY AND CONCLUSIONS

        Since nothing was changed in the way the model was applied in
these two very different ocean basins, the agreement with data that
was obtained overall is striking. These results support our previous
conclusion that for a given source level, which is dependent on the
yield and frequency of interest, geometric considerations are of
primary importance in determining the level and duration of the
signals as well as when they arrive at the receiver. The agreement
we have seen convinces us that we can make good first-order predictions concerning the reverberant character of the North Pacific as
well as of the North Atlantic Ocean.

        In other ocean basins, it is probably safe to assume that the
model in its present form would correctly predict which signals will
reach the receiver and when they will arrive. Signal levels,
however, might be grossly in error if propagation losses vary
significantly from the average conditions that seem to obtain in
the two basins we have considered here.

REFERENCES

1. J. Northrop, "Submarine Topographic Echoes from CHASE V,"
    Journal of Geophysical Research, Vol. 73, No. 12, 15 June 1968 .

FLUCTUATION STATISTICS OF SEA-BED

ACOUSTIC BACKSCATTER

P.A. Crowther

Marconi Space and Defence Systems Limited

Camberley, Surrey, U.K.

ABSTRACT

A theoretical model is developed that applies to statistics of
sea-bed backscatter, including non Rayleigh envelope distribution,
finite reverberation envelope auto-correlation at lag >1 pulse
width, and pulse-pulse envelope correlation. The model is based on
a two-phase patch surface assumption, generated by a Markov process,
and defined through parameters ρ = high scatter area fraction, γ =
high/low backscatter strength ratio, and λ = effective resolution
cell 1 patch size ratio. Experiments at 1.8, 4.1 and 8.1 kHz with
beamwidth 3.6^o to 17° at 5 sites and 3 range bands to minimum
grazing angle 2.5^o in 70 to 160 m water depth are analysed, indicating non-Rayleigh envelope distributions, power envelope auto-
correlations of 0.1 to 0.4 at >1 pulse lag, as expected. Close
experimental-model fitting requires $\rho = 10^{-4}$ to 2.10^{-2}, γ = 3.5 to 80,
and a patch dimension of 100 to 400 m. Factorial analysis on model
fitting parameters, within experimental sensitivity, indicated beamwidth dependence of statistics as predicted; site, grazing angle and
site-frequency dependence were all significant at <1% level; simple
frequency dependence was not significant.

1. INTRODUCTION

A frequent problem in active sonar is that of operation against
a background of natural backscatter from the sea boundaries. In
assessing this 'reverberation' it is customary to introduce the
concept of backscattering strength for different sources, and to

attempt to measure and model the same. Sea-bed backscattering
strength, defined in terms of the mean energy scattered per unit
area, is valuable in modelling general levels of reverberation, but
cannot adequately deal with more detailed questions, for example
estimation of realistic receiver operating characteristics and
assessment and comparison of processing methods. The reason is that
reverberation in general does not conform to simple universal probability distributions, as does thermal noise. This paper is a study
of some ways in which more realistic behaviour can be modelled. Some
of the impetus for this work stems from the analogous radar problem,
which has attracted considerable attention[1-8]. Strangely, acoustic
work appears to have been comparatively sporadic.

A model for the simplest fluctuation statistics is developed in
Section 2, based on what the author believes is a minimal number of
parameters - three - combining some degree of physical realism with a
maximal model simplicity. Experiments to test this model are described in Section 3, and the comparison is examined in Section 4.

2.   THEORETICAL MODEL

2.1 Definitions and General Nomenclature

We consider a sonar transmitting a short pulse such that at
transit time t a backscattered pressure envelope $E(t)$ is observed.
For simplicity and definiteness, the sonar is supposed to have a
resolution cell defined by 'top-hat' functions in range and bearing,
of width X and $Y = R.\Phi_R$ respectively, where $R = ct/2$ = range, c =
speed of sound, Φ = beamwidth. $H(\underline{r})$, with $<H> = 1$, is the normalized
local scattering strength at point $\underline{r}_0$.

$$H_A(r) = \Phi^{-1} \int_{-\Phi/2}^{\Phi/2} H(r,\psi) \, d\psi \qquad (1)$$

is the beam average of H, taking the beam centre at bearing $\psi = 0$
for convenience.

$$y(t) = E^2(t)/<E^2(t)> \qquad (2)$$

is the normalized power envelope, $<--->$ denoting an ensemble average
over all realizations of the sea-bed statistical process. Clearly
$<y> = 1$. $F(y)$ is the cumulative distribution function (d.f.) of y
(probability of sample lying below level y). $G(y) = 1-F$, the comple-
mentary d.f. (c.d.f.). We also introduce the cell scattering
strength s, defined as

$$s = \int_{-X/2}^{X/2} H_A(R + x) \, dx \, / X \qquad (3)$$

The theory uses the coefficient of variation of several variables, definable as

$$V_u \equiv V(u) = (<u^2>/<u>^2 - 1)^{\frac{1}{2}} \tag{4}$$

for variable u.

2.2 The Markov Model and Reverberation Distributions

We aim to estimate distribution and correlation functions of y, via the model which we now introduce. This hinges on 3 basic assumptions, as follows. Firstly, we assume that the local scattering strength is just 2 - valued - $H(\underline{r}) = H_0$ or H_1, where $H_0 = [1+\rho(\gamma-1)]^{-1}$, $H_1 = \gamma.H_0$, the distribution of areas on the sea-bed between the 2 states H_0 of probability $1-\rho$, and H_1 of probability ρ being by a patch process. Secondly, scatters within a given patch structure are assumed small and numerous, so that the Rayleigh distribution holds conditionally, given the cell scattering strength, s; this we may term the composition hypothesis:

$$G(y|s) = \exp(-y|s), \tag{5}$$

$$G(y) = \int_0^\infty \exp(-y/s) \, f(s)ds, \tag{6}$$

f(s) being the probability density function (p.d.f.) of s. The third assumption required is some method for generating the patch statistics and hence f(s). This we do via a simple Markov chain assumption; it can be introduced first in 1 dimension by considering a straight line drawn in the plane of the sea-bed. Along this line we may consider the 2-value function describing the scattering state of the bed at that point, 0 or 1. We assume that the probability of transition from state j to state i after traversing an infinitessimal length $\delta\chi$ is

$$t_{ij} = \begin{bmatrix} t_{00} & t_{01} \\ t_{10} & t_{11} \end{bmatrix} = \begin{bmatrix} 1 - \rho\delta\chi/L, & (1-\rho) \, \delta\chi/L \\ \rho\delta\chi/L, & 1-(1-\rho) \, \delta\chi/L \end{bmatrix}, \tag{7}$$

L being a characteristic patch length scale.

A first step to finding the distribution for s is to find that for β, the fraction of a line of length X which is in state 1, say. By considering starting in either state at one end of the line, and allowing for all possible transitions between states in passing along the line, and adding their contributions, it can be shown that the probability density function for β is

$$f(\beta) = \{(1-\rho)\ \delta(\beta) + \rho\delta(\beta-1) + 2\lambda^2\rho(\beta\rho+1-\beta)I_1(\mu)/\mu$$

$$+\ 2\lambda\ \rho I_0(\mu)\} \exp\ \{-\left[\beta + (1-\beta)\rho\right]\lambda\} \cdot (0 \leqslant \beta \leqslant 1) \qquad (8)$$

where δ denotes the Dirac δ function, $\lambda = X(1-\rho)/L$, the ratio of insonified length to effective patch length, $\mu = 2\sqrt{\rho\beta(1-\beta)/(1-\rho)}\cdot\lambda$, and where I_0, I_1 are hyperbolic Bessel functions of the first kind, defined by $I_0(x) = J_0(ix)$, $I_1(x) = -iJ_1(ix)$. In the simplified case of a 'top-hat' beam function, with very high resolution in one direction, say range, the insonified area at a given instance resembles a line of length $X = \Phi.R$ and we have:

$$s = \left[1 + (\gamma-1)\beta\right]/\left[1 + (\gamma-1)\rho\right], \qquad (9)$$

preserving $<s> = 1$, and therefore s would have the p.d.f.

$$f(s) = \frac{d\beta}{ds}\ f(\beta) = \left[\rho + (\gamma-1)^{-1}\right]\ f(\beta). \qquad (10)$$

We defer the case of a more truly 2-dimensional insonification area while we next consider coefficients of variation.

The 2-phase assumption can easily be shewn to imply

$$V_H^2 = \rho\ (1-\rho)(\gamma-1)^2/\left[1 + (\gamma-1)\rho\right]^2, \qquad (11)$$

and clearly

$$V_s^2 = <s^2> - 1 = X^{-2} \int_o^x \int_o^x <H(x)H(x')>\ dx\ dx'-1, \qquad (12)$$

but for the Markov process, we find

$$<H(x)\ H(x')> = 1 + <H^2-1>\exp\ (-\ |x-x'|/L), \qquad (13)$$

so that

$$V_s^2 = \phi(\nu)\ V_H^2, \qquad (14)$$

where $\nu = X/L = (1-\rho)\lambda$, $\qquad (15)$

and

$$\phi(\nu) = 2\left[\nu^{-1} -(1-e^{-\nu})/\nu^2\right] \doteq \left[1 + \nu/2\right]^{-1}, \qquad (16)$$

the final approximation holding for all ν to within 12%.

Eq (14) may be interpreted as stating that $\phi(\nu)$ measures the reduction in variance of s from that in H by integrating over a finite number of patch dimensions, and the approximation in (16) states that we obtain 1 effective degree of freedom for each 2L of length traversed. This now suggests a reasonable transition to 2-dimensions, we retain (8-12, 16), and generalise (14) to read

$$V_s^2/V_H^2 = (1+\nu/2)^{-1} = (1+X/2L)^{-1} (1+Y/2L)^{-1}, \tag{17}$$

where X = azimuthal cell width, Y = range cell width. λ is now defined in terms of ν, as

$$\lambda = \nu/(1-\rho). \tag{18}$$

In concluding this section, it is worth noting that the composition hypothesis links the moments of y and s as:

$$<y^n> = n ! <s^n>; \tag{19}$$

in the special case n = 2, this implies

$$V_y^2 = 2V_s^2 + 1 \tag{20}$$

2.3 Power Envelope Transit Time Lagged Correlation

We complete this sketch of the model by considering fluctuation time correlation. Using the azimuth integrated H function at range R, and defining

$$K_{HA}(r) = <H_A(R) H_A(R + r)>, \tag{21}$$

where we assume $|r| \ll R$, then by considering the analysis of instantaneous signal into real and quadrature (e.g. Hilbert transformed) components, p_1 and p_2 say, the sum of whose squares defines y(t), we can by appeal to the basic assumptions estimate $<p_i(t_1)p_j(t_2)>$ and $<p_i^2(t_1)p_j^2(t_2)>$, and hence estimate the time fluctuation statistics of y. The result is as follows; if

$$K(t_d) = <y(t_1) y(t_2)>, \tag{22}$$

where $t_d = |t_1 - t_2|$, and $t_d \ll |t_1|$, putting $R_d = \frac{1}{2}c |t_1 - t_2|$, we have

$$K(t_d) = Y^{-2} \int_{R_d-\frac{1}{2}Y}^{\frac{1}{2}Y} \int K_{HA} (r' - r'') \, dr' \, dr''$$

$$+ Y^{-2} \int_{-\frac{1}{2}Y}^{\frac{1}{2}Y} \int K_{HA} (R_d + r' - r'') \, dr' \, dr'' \quad (|R_d| < Y), \tag{23}$$

else

$$K(t_d) = Y^{-2} \int\limits_{-\frac{1}{2}Y}^{\frac{1}{2}Y} \int K_{HA} (R_d + r' - r'') \, dr' \, dr'' \; . \; (|R_d| > Y)$$

(24)

Further, if we make the reasonable approximation

$$K_{HA}(r) \doteq 1 + V_{HA}^2 \exp (- |r|/L),$$

(25)

where

$$V_{HA}^2 = V_H^2/(1 + X/2L),$$

(26)

then if $K(t_d)$ is standardised to the normalised correlation function,

$$C_y(t_d) = (<y(t_1) \, y(t_2)> - 1)/(<y^2> - 1)$$

(27)

we deduce

$$C_y(t_d) = \left[(1-\alpha)^2 \left[1 + K\phi(\overline{1-\alpha\tau}) \right] + K\phi(\tau) \exp (-\alpha\tau) \right] / \left[1 + 2K\phi(\tau) \right],$$

$$(\alpha < 1)$$

(28)

$$= K\phi(\tau) \exp (-\alpha\tau) / \left[1 + 2K\phi(\tau) \right] \; (\alpha > 1),$$

where $\alpha = R_d/Y$,   (lag/range resolution)

(29)

        $\tau = Y/L$,   (range resolution/patch length)

(30)

$\phi(\tau)$ is as defined in (16), and $K = V_{HA}^2$. Note that for a 'non patchy' scattering process, $\gamma \to 1$ or $\rho \to 0$ and therefore $K \to 0$; the fluctuation autocorrelation is then triangular and finite only for lag less than 1 pulse length, as for example shewn by Ol'Shevskii[9]. Eq (28) therefore produced a generalisation, shewing how auto-correlation of envelope, especially at lag > pulse length, can be used to help to estimate the model parameters, in particular the scale, L.

3.   EXPERIMENTAL

Experiments were conducted at 5 sites in the waters around the British Isles, between latitudes $50^\circ N$ and $58^\circ N$ in the autumn of 1975. The equipment consisted of a projector/receiver array towed at $\sim$4 km/hr, working in a 'side-scan' mode at four frequencies simultaneously, returns at 1.8, 4.1 and 8.1 kHz being used in analysis. The pulse length was 10 msec at each frequency, in a pulse repetition

TABLE 1: ARRAY BEAMWIDTHS

Frequency kHz	1.8	4.1	8.1
Beamwidth, Degrees	11.4 - 17.1	5.0 - 7.0	3.6 - 5.6

TABLE 2: SITES

Site no.	1	2	3	4	5
Sea Bed	Mud/Sand	Sand/Rock	Mixed	Mixed	Sand
Water Depth : m.	141	162	86	89	71
Area Traced : km × km.	22 × 19	34 × 4	25 × 14	22 × 19	19 × 10
Duration : hrs.	25	19	25	24	21

TABLE 3: POWER ENVELOPE CORRELATION PARAMETERS

Site no.	1	2	3	4	5
Correlation at lag $\gtrsim 1$ pulse length	0.32	0.33	0.17	0.20	0.12
L from subtangent, m.	120	100	100	100	80

period of 4 secs. By provision of different receive hydrophones, it was possible to obtain a range of beamwidths, listed in Table 1.

The 5 runs selected for analysis were in the depths and sea-bed types, classified by grab samples summarized in Table 2, the measurements being made over the stated area and durations.

Recorded data were processed by a realisation of (2) substituting a long term average for the ensemble average $<E^2(t)>$ - the exponential memory type of averaging being used with an averaging time of 400 pulses, equivalent to about 2,000m of path. y(t) was thus computed for each pulse at intervals of 10 msec over transit times t = 0.8 to 3.8 secs. In order to estimate range dependence of statistics, probability histograms for y were accumulated over all

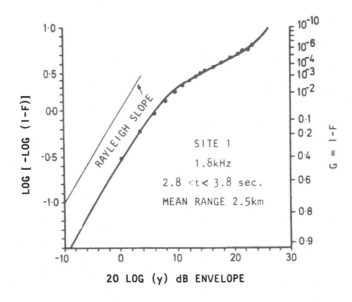

Fig. 1   Typical Cumulative Distribution on dB Rayleigh Scale ● Expt
———— Theoretical Formula

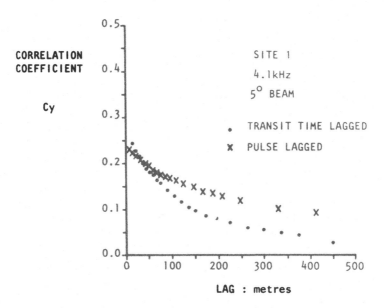

Fig. 2   Typical Within Pulse and Pulse Lagged Power Envelope Auto-
Correlations

samples, in 3 range groups, namely

   t = 0.8-1.8, 1.8-2.8, and 2.8-3.8 secs. respectively

A digital record of y(t) was also taken, for subsequent evaluation
of $C_y(t)$.

   A specimen result for F(y) is shewn as the points in Fig. 1,
plotted on a decibel Rayleigh paper. The Rayleigh slope is also
shewn for comparison. The continuous line shewn in Fig. 1 is a
best fit to the distribution, using the model. This distribution
is typical of all those found, in that (a) it appears to be approximately Rayleigh at low levels; (b) at a certain critical level, a
'knee' developes, the distribution above the knee being markedly
flatter than Rayleigh. The position of this knee varies from
$G = 1-F \sim 3.10^{-2}$ to 3.10^{-5}, usually being $\sim 10^{-3}$. It is more or less
sharp, according to conditions.

   A specimen power envelope auto-correlation is shewn in Fig. 2,
where the $C_y(t)$ function obtained has been plotted against range lag.
All the points on this plot are at greater than 1 pulse length separation, so that for 'non patchy' reverberation, the correlation would
be zero. Also shewn is the pulse-pulse correlation - correlation of
returns at equal transit times, from <u>different</u> pulses; for comparison
this is shewn on the same scale, the conversion from pulse lag to
spatial separation being effected using the elapsed time between
pulses, and the speed of tow. The correlation obtained using the two
methods is usually fairly similar, the pulse lagged values tending to
fall off somewhat more slowly with increase in lag than do the transit
lagged correlations, the reason probably being due to smudging by the
beamwidth. Observed correlations at just >1 pulse length separation
and fitted correlation length, defined as the subtangent to the
transit lagged correlation function at lag $\gtrsim 1$ pulse lengths are
listed in Table 3, taking an average over the 3 frequencies.

4.   FITTING TO THEORETICAL MODEL

4.1 Parameters

   The F(y) and C_y results were fitted to the model by first setting
the length scale L to give best agreement with the within pulse auto-
correlation. Secondly, an optimisation program was used to obtain a
best fit to the distribution F(y), defined as least squares error on
the decibel Rayleigh plot of Fig. 1 type, the optimisation being
essentially over γ and ρ, λ having been fixed by L and the known
beamwidths and pulse length. The theoretical distribution was
evaluated from (10, 6) by numerical integration, for each set of γ,
ρ values tested. The joint distribution of values of γ and ρ found
by this procedure is displayed in the scatter diagrams in Fig. 3.
The most important single valued combination of model parameters is

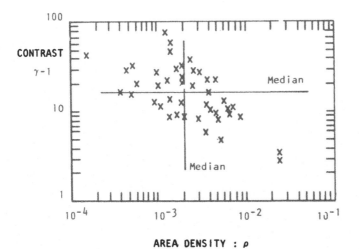

Fig. 3   Scatter Diagram of γ, ρ at Closest Fit to Experiment (All
Ranges and Frequencies)

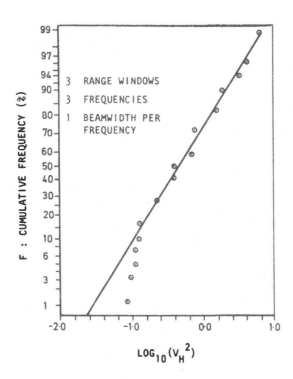

Fig. 4   Cumulative Distribution of V_H^2 From γ, ρ Parameters of Best
Fit

that giving V_H^2, the distribution of which over all the fitted runs being approximately log normal, as shewn in Fig. 4.

4.2 Factorial Dependence of Parameters

Finally, an attempt is made to study how the fluctuation parameters varied with site, range, frequency and beamwidth. The method employed is a slightly modified form Fisher's 'factorial analysis of variance'.[10] It was first found using this method that whilst beamwidth dependence of V_S^2 was significant, on computing V_H^2 from the model parameters of best fit, there ceased to be any significant dependence of the latter on beamwidth. This indicates that beamwidth dependence of fluctuation statistics, at least over the restricted range of beams available, was correctly modelled by the theory in that by definition V_H^2 should depend only on the bed, not the beamwidth.

Factorial analyses was performed using log (V_H^2) as the test variable describing the general acoustic nature of the scatterers, lower decile outliers being modified by inverse normal scores transformation, to render the test variable more Gaussian. The tests were done in 2 ways, firstly over the 4 factors Site (S), Range (R), Frequency (F), and beamwidth (B), at 5, 3, 2, 2 levels respectively, using samples of about 1/5 of each run, and secondly over the 3 factors S.R.F. at 5, 3, 3 levels respectively, using all data from each run. The conclusions were, for the shorter data groups that factors S, R, SR were all significant at <0.5% significance, and SF was significant at <1%, all other factors being found not significant at 5% level. For the full data groups, the factors S, R, SR were found significant at <0.5% level, the SF factor is now only significant at 10% level, and other factors are not significant at 10% level. The 2 analyses thus gave nearly the same conclusion, the only difference being in the reduced significance of the site-frequency interaction for larger data samples - i.e. effectively for larger sites. This appears to indicate local frequency dependence of the fluctuation statistics, becoming less pronounced over a larger sampled area on the sea-bed.

The range dependence established for V_H^2 is probably not inherent, there being no likely physical explanation of such an effect, but it can be interpreted as a sensitivity to grazing angle. Accordingly, in Fig. 5, a plot is given of the mean log (V_H^2) for all runs, and frequencies against the log of mean grazing angle for each of the 3 range groupings. The plot indicates an increase in V_H^2 as the angle approaches grazing incidence and is consistent with a law $V_H^2 \propto \theta^{-1}$ or θ^{-2}, θ being grazing angle. Possible explanations of this are greater acoustic contrast between different patch compositions, and icnreased highlight and/or shadow activity at the approach to grazing incidence.

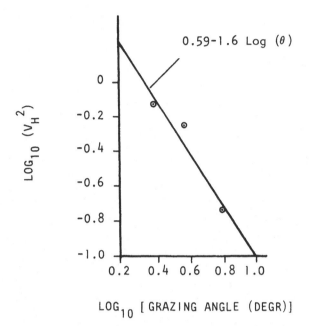

Fig. 5   Dependence of Mean Scatter Index Log (V_H^2) on Mean Grazing
Angle Over all Experimental Results

5.   SUMMARY, COMMENTARY AND CONCLUSIONS

An attempt has been made to develop a model capable of describing acoustic backscattering effects of higher order than can be dealt
with via a simple scattering strength. The model proposed is thought
to be as simple as possible subject to retaining physically realistic
features. It involves the idea of a patchy 2 state surface, which,
although a simplification, has some intuitive justification in dealing
with sea beds, where there are for example rock outcrops or stone or
shell patches on a sand or mud background. Some details of derivation
have had to be omitted from this paper through limitations of space
available.

Experimental work appears to confirm the essential nature of
model predictions regarding the probability distribution function of
sea bed returns, and the dependence of this on beamwidth, and of the
partial correlation of the power envelope of backscattering at greater
than 1 pulse length lag. Fitting the theoretical model to observation
has allowed estimates to be made of the appropriate values for the
model parameters and patch scales characterizing the experimental
areas. One general prediction of the theory is of a characteristic
'knee' in the cumulative distribution function. This is invariably

observed experimentally, although its shape is usually somewhat less
sharp than is predicted from the model. This is probably a consequence of the assumption of 'top-hat' beam functions and clear discrete
2 state scattering patches on the bed. Further work could be done
with real beam functions, still retaining the simplicity of the basic
model, to see whether noticeably smoother transitions in the distribution function occur.

Further work could also be done on the exact nature of the power
envelope auto-correlation function. The Markov assumption leads to
prediction of an exponential function, which is often not a bad
approximation to experiment. However, there are signs that in some
cases there is a longer scale 'tail' to the correlation function,
indicative of some longer scale structure, in addition to the ca 100m.
structure recorded above. Also, it remains to be seen how the
present structure and patch parameters determined for relatively
shallow water would alter for returns in ocean depths.

Another interesting unanswered question is whether effects
analoguous to those modelled for backscatter can be found to characterize propagation in areas where the latter is strongly influenced
by the sea bed, and whether similar scales and parameters could be
found.

ACKNOWLEDGEMENTS

Computer implementation of the statistical analysis on
experimental data was designed and executed by Mr M. Burns. This
work has been carried out with the support of the Procurement
Executive, Ministry of Defence.

REFERENCES

1.   G.V. Trunk, 'Radar Properties of Non-Rayleigh Sea Clutter'
         IEEE Trans Aerosp and Elect. Systs. AES-8, 196(1972),
         also ibid AES-7, 553(1971).
2.   R.R. Boothe, 'The Weibull Distribution Applied to Ground
         Clutter Backscatter Coefficient' U.S. Army Missile
         Command Report RE-TR-69-15 (June 1969).
3.   J.C. Dayley, W.T. Davies and N.R. Mills, 'Radar Return in High
         Sea States' U.S. Naval Research Lab. Rept. NRL-7142
         (Sept. 1970).
4.   N.W. Guinard et al 'NRL Terrain Clutter Study Phase I' U.S.
         Naval Research Lab. Rept. NRL 6487 (May 1967).
5.   A.M. Findlay, 'Sea Clutter Measurements by Radar Return
         Sampling', U.S. Naval Research Lab. Rept. NRL 6661
         (Feb. 1968).
6.   J.H. Blythe, D.E. Rice and W.L. Attwood, 'The CFE Clutter

Model with Application to Automatic Detection' Marconi
Review 32, p 185 (1969).

7.   G.V. Trunk and S.F. George, 'Detection of Targets in Non-
Gaussian Sea Clutter' IEEE Trans. Aerosp. and Elect.
Systems AES-6, p 620 (1970).

8.   G.B. Goldstein, 'False Alarm Regulation in Log-Normal and
Weibull Clutter' IEEE Trans Aerosp. and Elect. Systems
AES-9, p 84 (1973).

9.   V.V. Ol'shevskii, 'Characteristics of Sea Reverberation',
Eng. Trans. publ. Consultant Bureau, New York, 1967.

10.   R.A. Fisher 'Statistical Methods for Research Workers' Oliver
and Boyd, Edinburgh-London, 1941, and H. Scheffe, 'The
Analysis of Variance', Wiley, New York (1950).

MEASUREMENTS OF SPATIAL COHERENCE OF BOTTOM-INTERACTING

SOUND IN THE TAGUS ABYSSAL PLAIN

J.M. Berkson, R.L. Dicus*, R.Field, G.B. Morris,
and R.S. Anderson**

Naval Ocean Research & Development Activity
NSTL Station, MS 39529

*Present Address: Naval Research Laboratory
**Present Address: U.S. Naval Oceanographic Office

ABSTRACT

Coherence of bottom-interacting sound in the Tagus Abyssal
Plain was determined as a function of bearing and frequency (20 to
2000 Hz) for grazing angles between 11° and 13°. An acoustic experiment was performed in which SUS charges were dropped in a circular pattern of 28 km radius around a deep hydrophone. The
bottom-interacting arrival was isolated and processed to remove the
decorrelating effects of varying bubble-pulse periods. A spatial
coherence function was calculated between shot pairs corresponding
to 5, 9, and 13 km separation. For frequency bands of 20 to 500 Hz
and 1200 to 2000 Hz, the spatial coherence of the bottom-interacting
arrival is high. For the frequency band of 500 to 1200 Hz, the
spatial coherence is lower and more variable. The high coherence
values are consistent with the Eckart theory for scattering from
an interface having rms roughness less than 0.1 m. The sharply
tuned nature of the low coherence values in a discrete frequency
region suggests that the effect is due to interference from sediment
multi-paths rather than scattering from bottom roughness. Band-
pass filtered impulse responses show that the energy of the sediment-refracted arrival predominates at low frequencies and the
reflected arrival predominates at high frequencies. The interference effects occur in the middle frequency region where the two
types of arrivals have nearly equal amounts of energy, and may be
expected to depend on sediment properties, bottom topography, measurement geometry, and oceanographic environment.

INTRODUCTION

The interaction of acoustic energy with the ocean bottom has been studied extensively. Much of this work has focused on energy partitioning of the signal, such as measurement and prediction of reflection coefficients, rather than properties which describe the nature of the scattered signal, such as coherence. There have been few measurements of forward scattering from the ocean bottom,[1] until recently. Most of the coherence measurements were made at high grazing angles. These are summarized in reference 2. In general, coherence measurements are highly variable, ranging from near 0 to 1.0, and with the exception of the effects of small-scale roughness,[3] they are poorly understood and not easily predicted.

In this study, we have determined spatial coherence values for low grazing angle sound interacting at the ocean floor in the Tagus Abyssal Plain (Fig. 1). The measurements were conducted in a well-studied region[4] of the ocean bottom in an attempt to understand the physical processes of the bottom interaction which cause loss of signal coherence. Such fundamental understanding is essential to predict and to understand the utility and performance of acoustic sensor systems which use or perhaps even reject acoustic bottom interacting energy.

The Tagus Abyssal Plain ocean bottom is composed of two major turbidite sequences, each having thickness of 1.0 sec (two-way travel time) or greater.[4] In this experiment, the acoustic interaction with the bottom involves only the upper turbidite sequence. On seismic records, these sediments appear as flat stratified abyssal plain layering which begins to thin near the ridge at the basin margin. The abyssal plain boundary has been defined (Fig.1) by either the point of thinning or by the point of change of sediment type.[4] In this experiment, the area where sound interacts with the sea floor (Fig.1) falls within both definitions of the Abyssal Plain province. At the site, reflecting horizons occur at 0.09, 0.35, 1.09 (lower sequence boundary) and 2.0 sec (basement boundary). From analysis of bottom reflectivity data, Dicus[5] estimated the sediment sound velocity gradient in the upper turbidite to be 0.8 sec^{-1} and the attenuation to be 0.03 dB/m/kHz. From estimates of coherence of normal incidence echosounding and seismic data, Clay and Leong[6] estimated the rms roughness of the Tagus Abyssal Plain to be 0.1 to 1.0 meters.

In this paper, we present coherence measurements of bottom-interacting sound in the Tagus Abyssal Plain. In general, the high coherence values are consistent with a theory for scattering from an ocean bottom having an rms roughness less than 0.1 m. However, notches of low coherence in the 500 to 1200 Hz frequency band

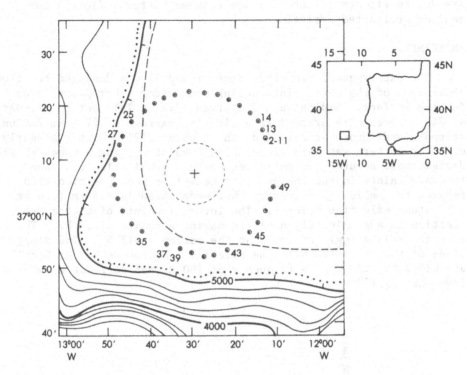

Fig.1. Acoustic experiment location, Tagus Abyssal Plain. Shot
locations (circles). Bottom reflection points (dashed
circle) occur within both definitions of abyssal plain
boundary: Change in sediment type (dotted line), or point
of thinning of sediment layering (dashed line). Bathymetry
and geology from Reference 4. Contour interval 200 meters.

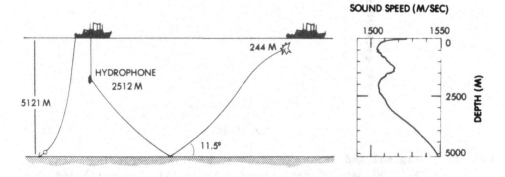

Fig.2. Experimental Geometry. Paths in sediment not shown.

are due to effects of interference between bottom-reflected and
sediment-refracted arrivals.

EXPERIMENT

The experimental geometry (Figs. 1 and 2) was designed to allow
separation of the bottom-interacting sound from other paths involving the surface. Explosive SUS charges (0.8 kg TNT) set to detonate
at 245 m depth were dropped in a circular pattern of 27.9 km radius
around a hydrophone at 2512 m depth in water 5121 m deep. A nearly
circular pattern was maintained using Raydist ranging between ships.
Navigational accuracy is estimated to be 5 m which is less than
the uncertainty in the changes in receiver or shot position with
respect to a ship. The average shot separation is 4.6 km (1.6 km
at bottom reflection points). The locus of points of bottom reflection is approximately a circle having a radius of 10 km. The
average grazing angle of sound at the bottom is 11.5°. The dimensions of the first fresnel zone calculated by the method of Kerr[7]
are 620 m by 2280 m at a frequency of 100 Hz and decrease with
frequency as $f^{-\frac{1}{2}}$.

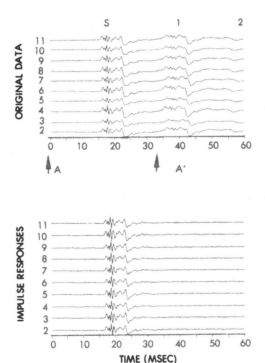

Fig. 3.   Original and deconvolved data, bottom-interacting path.
           First 60 msec of data shown. Shock (S), First (1) and
           Second (2) bubble pulses. Windowing for second processing
           method from A to A'. Shot numbers shown at left.

Signals were recorded analog on magnetic tape over the band from 0 to 2000 Hz. During processing, the bottom-interacting sound was isolated from other paths by travel time. The data were processed to remove the decorrelating effects of varying bubble pulse periods. To determine if the results were related to the type of processing, two methods were used. In the first method, the bottom interacting signals were deconvolved with a replica source signal by the method of Dicus[5,8] to produce bottom impulses (Figs. 3 and 4). In the second method, the primary signals were separated from the bubble pulses by time-windowing before the second bubble pulse. Such a method is successful for experimental geometries where the energy in these arrivals are clearly separable in time. Fig.3 shows such a time separation can be achieved for these Tagus data.

Once the primary bottom interacting signal is isolated by either method, a spatial coherence function γ can be calculated between two shots, a and b, where

$$\gamma^2_{ab}(\omega) = \frac{\left|\hat{S}_{ab}(j\omega)\right|^2}{\hat{S}_{aa}(\omega)\ \hat{S}_{bb}(\omega)} \qquad (1)$$

where

$S_{ab}(j\omega)$ = cross spectrum

$S_{aa}(\omega)$, $S_{bb}(\omega)$ = autospectra

$j = \sqrt{-1}$

ω = radial frequency

The caret indicates a smooth estimate. Smoothing may be accomplished by averaging estimates of either data segments or adjacent frequency bands.[9] Errors in estimating the coherence function may be minimized by proper smoothing[9-11] and by alignment of the two signal series in the time domain.[11] In this study, the estimates were obtained as an average of 16 adjacent frequency bands, a processing band of 78 Hz. For this particular averaging, incoherent time series will have an average coherence of about 0.2.[9,12]

Coherence spectra between adjacent shots of average separation 4.6 km around the circle (Fig. 1) are shown in Figs. 5 and 6. Each curve of the stacked display represents the coherence vs frequency between indicated shots. In general, the data exhibit very high coherence, nearly 1.0, except for notches of low coherence which appear predominantly in the middle frequency band of 500 to 1200 Hz. Very few notches occur above 1600 Hz or below 400 Hz. Similar

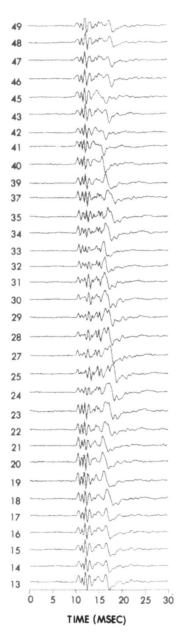

Fig.4.   Bottom Impulse Responses of Tagus Circle.   Gaussian low-
          pass input, -3 dB at 1.1 kHz.   First 30 msec shown.   Shot
          numbers shown at left.

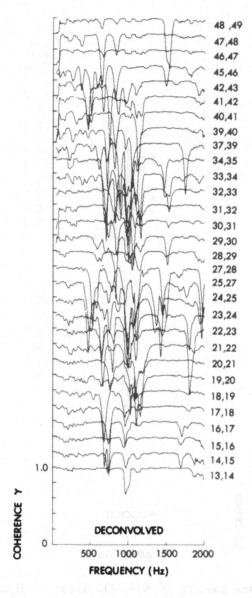

Fig. 5. Coherence Spectra (Deconvolved), Tagus Circle. Primary
         bottom event isolated by deconvolution. Adjacent shots of
         average separation 4.6 km shown at right. Stacked display,
         successive curves displaced 0.2.

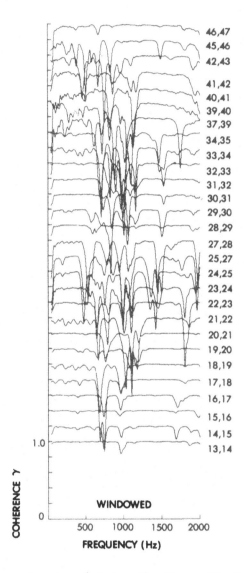

Fig. 6.   Coherence Spectra (windowed), Tagus Circle.   Primary bottom-
          event isolated by windowing before first bubble pulse. Ad-
          jacent shots of average separation 4.6 km shown at right
          Stacked display, successive curves displaced 0.2

results are obtained for both types of processing (Figs. 5 and 6).
Figure 7 shows a distinct but very small decrease in averaged coherence with increasing shot separation greater than 4.6 km.

Several factors other than bottom interaction may cause loss of
coherence. The relationship between coherence and signal-to noise
power ratio, S_n, for two receivers having common signal and uncorrelated noise is given by.[13,14]

$$\gamma_{ab} = \frac{S_n(\omega)}{S_n(\omega) + 1} \tag{2}$$

Thus, low signal-to noise ratios (e.g.at frequencies of explosive
source spectral nulls) may cause low coherence. For example, a
power ratio of less than 4 (6 dB) would result in coherence less
than 0.8, even if the signals alone had perfect coherence. However,
the high signal-to-noise ratios for the Tagus circle date (Figs.8 & 9)
indicate that the low coherence notches are not the result of this
cause, except for some shots at about 1000 Hz and near the 2000 Hz
roll off.

Other factors may cause loss of coherence. These include distortion of the wavefront by the medium and processing artifacts. A
test was performed to determine the net effect of these factors.
Ten shots (numbered 2 through 11 in Fig. 1) were dropped at one
location at an average range of 27.6 km from the receiving hydrophone. The maximum difference between shot depths estimated from
the bubble pulse period was 16 m. Coherences were determined between bottom-interacting signals processed by both methods. (A
similar test could not be performed for a waterborne path because
a suitable direct arrival does not exist at this range due to re-

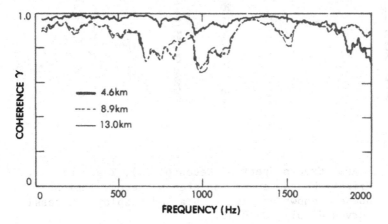

Fig. 7. Coherence averaged over the Tagus Circle for three shot
separations.

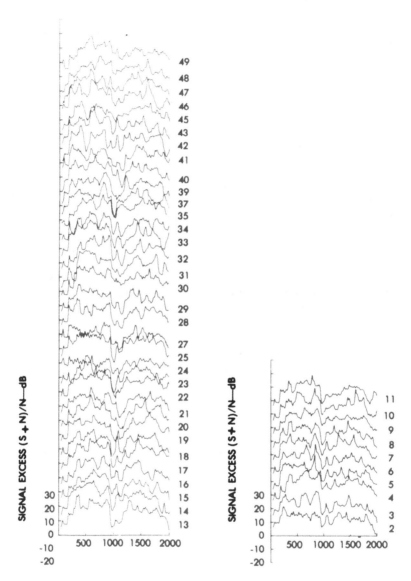

Fig. 8.   Signal Excess Spectra (Deconvolved), Tagus Circle.
Primary bottom event isolated by deconvolution.   Shot
numbers shown at right.   Stacked display, successive
curves displaced 10 dB.

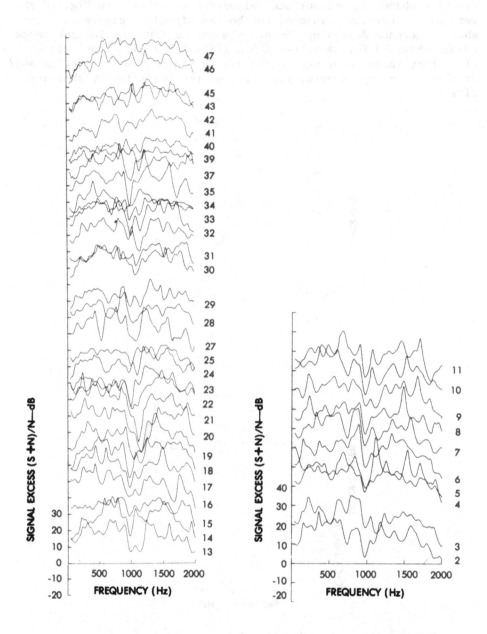

Fig. 9.   Signal Excess Spectra (windowed), Tagus Circle. Primary
          bottom event isolated by windowing.   Shot numbers
          shown at right.   Stacked display, successive curves
          displaced 10 dB.

fractive shadowing and surface reflection effects). In Fig. 10 the
very high coherence values of the bottom signals indicate that the
above coherence degrading factors are not causing the low coherence
values observed for separated shots (Figs. 5 and 6). Thus, we conclude that these coherence losses are due to bottom interaction and/
or changes in experimental geometry for the two adjacent shot arrivals.

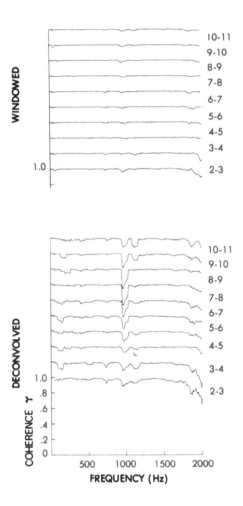

Fig. 10   Coherence Spectra Shots #2 - 11 (Same Location). Shot
           numbers shown at right. Stacked display, successive
           curves displaced 0.2. Primary bottom event isolated by
           deconvolution (below) or windowing before first bubble
           pulse (above).

DISCUSSION

Roughness at a reflecting interface causes loss in coherence. The coherent reflection coefficient, R_c, for reflection from a randomly rough interface having surface displacement distributed by a Gaussian probability density function is given by Eckart[15] as

$$R_c^2 = e^{-g}, \qquad\qquad (3)$$

where $\quad g = (2k_z\sigma)^2$,

$\qquad\qquad k_z = k \sin\theta$,

$\qquad\qquad k = $ acoustic wave number,

$\qquad\qquad \theta = $ grazing angle,

and $\qquad \sigma = $ rms roughness of interface within reflection zone.

Clay[3, 16] showed that average coherence between two sensors is an estimate of R_c^2, if the coherence loss is due only to rough surface scattering described by the Eckart theory and if the sensor spacing is enough for the two scattering areas on the bottom to be independent.

The spatial coherence value vs frequency for one of the Tagus shot pairs is shown in Fig. 11. In general, the high coherence values are consistent with the theory for scattering from an inter-

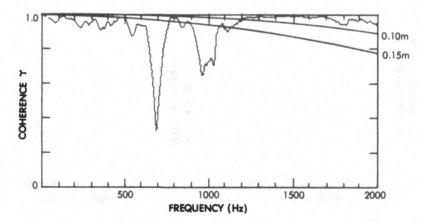

Fig. 11   Coherence Spectra, measured vs theoretical. Shots 18 and 19. Theoretical curves represent e^{-g} for roughness of 0.10 and 0.15 meters.

face having rms roughness less than 0.1 m. This is consistent with
the estimates of roughness by Clay and Leong.[6] However, the sharply
tuned nature of the low coherence values in a discrete frequency
region suggests a multipath interference cause rather than scattering from bottom roughness. Since major multipaths have been removed by windowing, a sediment multipath is likely.

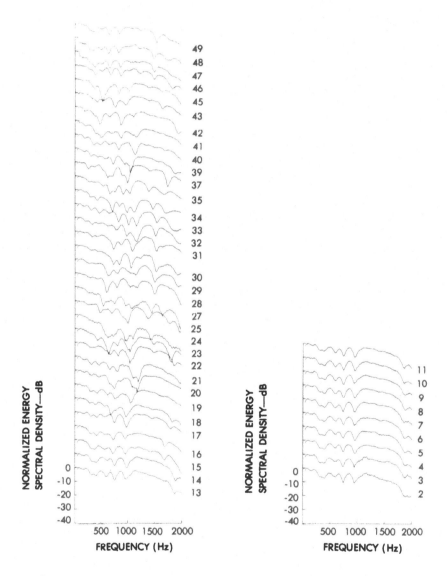

Fig. 12. Impulse Response Spectra, Tagus Circle. Primary bottom
          event isolated by deconvolution. Shot numbers shown at
          right. Each shot separately normalized. Stacked display,
          successive curves displaced 10 dB.

Interference between the bottom-reflected arrival and the bottom-refracted arrival cause oscillations in the bottom arrival spectra. (A similar effect has been observed for non-penetrating reflections from different bottom points[17]). Fig. 12 shows such oscillations with periodicities of approximately 200 Hz frequency in the mid-frequency region. (The 50 Hz periodicities due to the explosive source bubble pulse have been removed by deconvolution processing). At high frequencies, the reflected arrival (B) shown in Fig. 13 is dominant, and at low frequencies, the refracted arrival (D) is dominant. When the two arrivals have nearly equal energy, interference oscillations may be expected with periodicities of 160 to 290 Hz consistent with their observed travel time separations of 3.5 to 6.1 msec.

Figs. 5 and 6 show that interference notches are also observed in the coherence spectra at the middle frequencies. Such interference notches in coherence must result from differences in the interference patterns between two shots, since identical bottom spectra would result in nearly perfect coherence, as in shots numbers 2 - 11, Figs. 10 and 12.

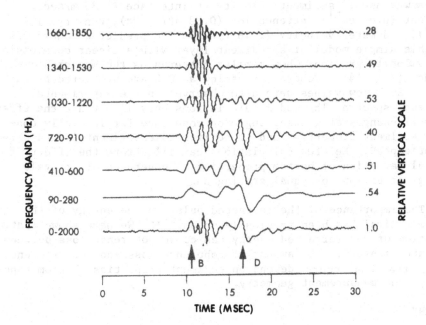

Fig. 13. Impulse Responses, Band Passed. Shot 2. Arrows identify the bottom-reflected arrival (B) and the sediment refracted arrival (D), which has undergone a 90° phase shift. Gaussian band-pass filter - 3 dB points labeled at left. 2 kHz impulse response, full response.

The coherence spectrum is determined by a combination of bottom properties and geometry. For example, average sediment properties and geometry determine the frequency band in which both arrivals are nearly equal in amplitude. Within this band, the exact character of the coherence spectra would be sensitive to either changes in bottom properties or changes in geometry. The following model calculation shows how small changes in grazing angle may affect coherence for a fixed bottom. A plane wave reflection model that calculates complex reflection coefficients from a viscoelastic multi-layer ocean bottom was modified to calculate coherence between two reflections at different grazing angles at the bottom. The model is a delta-matrix extension[18] of the Thompson-Haskell method. It assumes plane wave incidence, and approximates density and velocity gradients by thin homogeneous layers. (The plane wave approximation is best for zero velocity offset at the water-sediment interface, where the grazing angle of the refracted and reflected arrivals are nearly the same.) The coherence was calculated by Equation (1) for grazing angles corresponding to the Tagus data and averaging 16 adjacent frequency bands, a processing band of 80 Hz. The average difference in angle is 0.24° and the maximum difference is 0.8°. The following simple geoacoustic model, based on Ref. 5 and measured values was used: sediment velocity at interface (1533 m/sec), gradient (0.8 sec^{-1}), attenuation (0.03 dB/m/kHz), density (1.6 g/cm^3), and water velocity (1548 m/sec). The predicted coherences for this simple model of a sediment layer with a linear compressional wave velocity gradient show similar patterns as the measured coherence (Fig. 14). Coherences as low as 0.2 are predicted by the model. Such low values were also observed in the experimental coherence spectra. The model calculations only illustrate the effect of small changes of grazing angle on coherence for laterally constant sediment properties and close fits to experimental data were not attempted. Related calculations can illustrate the effect of lateral variation of bottom properties on coherence of arrivals having reflections of equal grazing angle.

The importance of the refracted pulse on the energy of the bottom arrival has been previously shown.[19,20] We have shown that reception of the refracted energy can cause coherence loss between separated sensors. The amount of coherence loss and the frequency at which the loss occurs depend on sediment properties, bottom topography, and measurement geometry.

SUMMARY

An acoustic experiment was performed in which explosive SUS charges were dropped in a circular pattern around a deep hydrophone. The bottom-interacting arrival was isolated and processed to remove the decorrelating effects of varying bubble-pulse periods. A spatial-coherence function was calculated between shot pairs corresponding

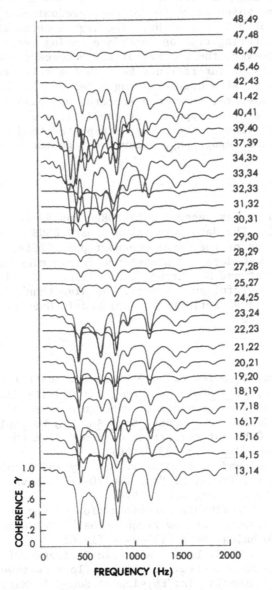

Fig. 14. Coherence Spectra (Theoretical), Tagus Circle. Predicted
coherences between bottom corresponding Tagus grazing
angles for model case of sediment layer with linear compressional wave velocity gradient of 0.8 sec^{-1}. Stacked
display, successive curves displaced 0.2.

to 5,9 and 13 km separations and having grazing angles 11 to 13°. Coherences were high for frequencies 20 to 2000 Hz except for low coherence values which occur in notches, predominantly in the middle frequency band of 500 to 1200 Hz. The high coherences are consistent with simple scattering from an interface having rms roughness less than 0.1 m. However, the notches of low coherence appear to be caused by effects of interference between the reflected and re-fracted arrivals. This origin is suggested by the character of the bottom-arrival spectra, coherence spectra, and band-limited impulse responses. It is also supported by predictions of a plane wave model. Coherence loss effects may be expected to depend on sediment properties, bottom topography, measurement geometry, and oceanographic environment.

ACKNOWLEDGEMENTS

This research was supported by NORDA and Naval Electronic Systems Command. The data were taken aboard USNS KANE and WILKES under the auspices of the Naval Oceanographic Office. We thank R. Winokur and W. Geddes for assistance in planning and carrying out the experiment and S. Marshall, K. Gilbert, O. Diachok, and C. Spofford for helpful suggestions. We also thank H. Bucker for providing the n-layer reflection model and I. Spencer for compu-tational assistance.

REFERENCES

1. C. W. Horton, A review of reverberation, scattering, and echo structure, J. Acoust. Soc. Am., 51:1049 (1972).
2. J. M. Berkson, Measurements of coherence of sound reflected from sediments, J. Acoust. Soc. Am., To be published.
3. C. S. Clay, Coherent reflection of sound from the ocean bottom, J. Geophys. Res.,71:2037 (1966).
4. W. F. Ruddiman and L. K. Glover,"Geology and Geophysics of The Tagus Basin,"Technical Note TN6120-1-76, Naval Oceanographic Office, Washington, D.C. (1976).
5. R. L. Dicus, "Preliminary Investigation of the Ocean Bottom Impulse Response at Low Frequencies," U.S. Naval Oceanographic Office Technical Note 6130-4-76 (1976).
6. C. S. Clay and W. K. Leong, Acoustic Estimates of the topography and roughness spectrum at the sea floor southwest of the Iberian Peninsula, in:"Physics of Sound in Marine Sediments," p 373, L. Hampton, ed., Plenum, N.Y., (1974).
7. D. E. Kerr,"Propagation of Short Radio Waves, MIT Radiation Laboratory Series, Vol. 13',"McGraw-Hill, N.Y. (1951).
8. R. L. Dicus, "Synthetic deconvolution of explosive source acoustic signals in colored noise,"U.S. Naval Oceanographic Office Technical Note 6130-3-76 (1976).

9. V. A. Benignus, Estimation of the coherence spectra and its
    confidence interval using Fast Fourier Transform, IEEE
    Trans. Audio Electro-acoustics,AU-17:145 (1969).

10. G. M. Jenkins and D. G. Watts, "Spectral Analysis and its
    Applications," Holden-Day, San Francisco (1968).

11. A. F. Seybert and J. F. Hamilton, Time delay bias errors in
    estimating frequency response functions, J. Sound and
    Vibration, 60:1 (1978).

12. G. C. Carter, C. H. Knapp, and A. H. Nuttall, Statistics of the
    estimate of the magnitude-coherence function, IEEE Trans.
    Audio Electro-acoustics, AV-21:388(1973).

13. G. C. Carter, C. H. Knapp, and A. H. Nuttall, Estimation of the
    Magnitude-Squared Coherence Function via Overlapped Fast
    Fourier Transform Processing, IEEE Trans. Audio Electro-
    acoustics,AV-21:337(1973).

14. M. R. Foster and N. J. Guinzy, The coefficient of coherence:
    Its estimation and use in geophysical data processing,
    Geophysics, 32:602 (1967).

15. C. Eckart, The scattering of sound from the sea surface, J.
    Acoust. Soc. Am., 25:566 (1953).

16. C. S. Clay, Personal communication. An exponent of 2 is
    incorrectly ommitted in the corresponding expressions in
    reference 3 (1979)

17. R. R. Goodman and A. Z. Robinson, Measurements of reflectivity
    by explosive signals, in: Physics of Sound in Marine Sediments,"
    p.537, L. Hampton, ed., Plenum, N.Y.(1974)

18. G. J. Fryer, Reflectivity of the ocean bottom at low frequency,
    J. Acoust. Soc. Am., 63:35 (1978).

19. J. S. Hanna, Short-range transmission loss and the evidence for
    bottom-refracted energy, J. Acoust. Soc. Am., 53:1686 (1973).

20. R. E. Christensen, J. A. Frank, and W. H. Geddes, Low frequency
    propagation via shallow refracted paths through deep ocean
    unconsolidated sediments, J. Acoust. Soc. Am., 57:1421 (1975).

EFFECTS OF MEDIUM FLUCTUATIONS ON UNDERWATER ACOUSTIC

TRANSMISSIONS IN A SHALLOW WATER AREA

Erik Sevaldsen

Norwegian Defence Research Establishment
Div. U
Horten, Norway

ABSTRACT

The paper describes a series of measurements in a shallow
water area south of Elba, Italy, in 1977 - 78. The purpose of the
experiment was to study the variability of one-way transmitted signals in shallow waters with

- time (duration of individual experiments and season of the
  year)

- space (source location, source and receiver depth)

- center frequency of transmission (1 to 6 kHz)

Fluctuations are observed both in amplitude, frequency and
phase (delay) of a signal. As a combined measure of fluctuations
we used spreading functions and their widths in frequency and time.
We also measured transmission loss and transmission loss time
variations over the duration of individual experiments (30 minutes).
The results showed small spreading in time. Three different classes
of frequency spreading ranging from less than 50 mHz to more than
1 Hz were observed. Sea surface roughness has been identified as
one source of fluctuations. Mean transmission loss is found to be
rather high and seems to be more dependent on bottom properties
than on range.

INTRODUCTION

A common experience in underwater acoustic measurements is the variability of received signals. Signals fluctuate in time and change with spatial variables and frequency.

During the last decade considerable efforts have been made to explain observed fluctuations in underwater acoustic signals and relate them to physical processes in the ocean [1,2,3,4]. It now seems clear that the dominant source of phase fluctuations in the deep ocean far from boundaries is internal gravity waves. Amplitude fluctuations cannot be accounted for completely this way, but oceanic finestructure may provide the additional ingredient necessary to obtain this [5].

The scope of our investigations has been to look into fluctuations in shallow waters. Conditions in shallow waters differ from deep waters mainly because of sound field interaction with the boundaries. The moving surface is an important source of fluctuations. Surface waves scatter incident acoustic waves spreading the energy and creating sidebands around a carrier [6,7,8]. Energy and bandwidth of the sidebands depend on sea state conditions.

Internal wave conditions in shallow waters will be different from deep waters. Wunsch and Webb [9] report deviations from the canonical Garrett-Munk frequency spectrum of internal wave energy density near topographic features and in canyons. Energy levels are found to be higher and there is pronounced anisotropy . The shape of the spectrum does not seem to change much. One would expect that vertical symmetry is lost in shallow waters.

Other sources of fluctuations than surface waves and internal waves are currents, oceanic finestructure and microstructure and platform motion. Radial platform motion has been compensated for.

We wanted to investigate experimentally the effects of the fluctuating shallow water medium on underwater acoustic transmissions. Specifically it was desired to measure the spreading in frequency and time of one-way transmitted signals. Also transmission loss and transmission loss time variations of the same signals have been measured.

By the medium we mean the water itself and its boundaries, surface and bottom and also bottom sediments and rocks.

The combined effect of the sound field interacting both with the boundaries, and with the shallow water internal wave field will cause results to differ from those found in deep waters. It is clear that the spreading in frequency often will be larger in shallow waters.

BASIC SIGNAL: TRANSMITTED SIGNAL:

COHERENT PULSE TRAIN   SUM OF ALL 6 BASIC SIGNALS
OF LINEAR FM PULSES   OR 4 OF THEM
PULSE LENGTH   .125 sec. OR 2 OF THEM
PULSE BANDWIDTH   500 Hz
PULSE REPETITION PERIOD 1.0 sec.
CENTER FREQUENCIES 1,2,3,4,5,6 kHz

RECEIVED SIGNALS

MATRIX FOR SPREADING FUNCTION CALCULATIONS
128 CONSECUTIVE PULSES (MATCHED FILTER OUTPUTS)

Fig. 1. Signal characteristics

EXPERIMENTS

We chose to treat the shallow water acoustic channel as a
linear, time-varying filter[10,11,12]. The impulse response of such
a filter is $h(\tau,t)$. We Fourier-transform this function with respect
to the time variable to find the channel spreading function
$\tilde{s}(\tau,\phi) = F\{h(\tau,t)\}$. The functions $h(\tau,t)$ and $s(\tau,\phi)$ both will have
random parts. They then must be described by their autocorrelation
functions,

$$R_S = R_S(\tau,\tau',\phi,\phi') = \overline{s(\tau,\phi) \cdot s^*(\tau',\phi')}$$

for the spreading function. To make this function tractable we introduce the WSSUS assumptions on the underlying random processes.
(WSSUS = wide sense stationarity in time, uncorrelated scattering in
frequency). These assumptions, which may be locally valid, result in

$$R_S = R_S(\tau,\phi) \cdot \delta(\tau'-\tau) \cdot \delta(\phi'-\phi), \ R_S(\tau,\phi) = \overline{|s(\tau,\phi)|^2}$$

This function is called the medium scattering function. If the
signal ambiguity function of an echo locating system is narrow compared to the scattering function and the target is a point target
the resolution in time and doppler of the system is given by the
scattering function. Under the WSSUS assumption an estimate of the
scattering function is obtained by averaging spreading functions
magnitude squared. To compute the spreading functions we need an
estimate of the channel impulse response. As estimates we used the
received signals correlated with the transmitted signal (a matched
filter). Characteristics of the transmitted signals are shown in
Fig. 1.

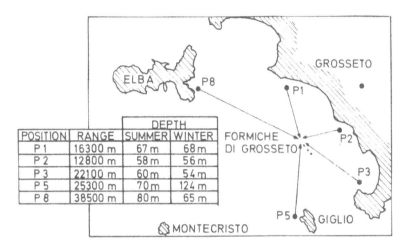

POSITION	RANGE	DEPTH	
		SUMMER	WINTER
P 1	16300 m	67 m	68 m
P 2	12800 m	58 m	56 m
P 3	22100 m	60 m	54 m
P 5	25300 m	70 m	124 m
P 8	38500 m	80 m	65 m

Fig. 2.   Map of the area with transmitter positions, ranges and
          depths.

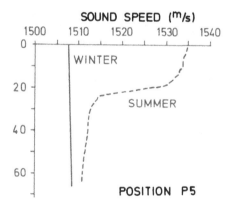

Fig. 3.   Typical sound speed profiles, measured in the area south
          of Elba.

It is also stated that 128 received pulse signals are used to form
the matrix over which we do the Fourier-transform on the time
variable.

Fig. 2 shows a map of the area with transmitting positions.
Two series of experiments were done, one in summer 1977 and one in
winter 1978. The winter trials were merely repetitions of those in
summer under different sound propagation conditions (see Fig. 3).
The transmitter directivity pattern was onmidirectional. The
receiving end was a moored vertical array. Spatial variability has
been achieved by transmitting from 5 different positions and from
3 different depths, usually bottom, 45 m and 15 m. Also hydrophone
depth has been variable, usually 40 m and 54 m. One individual run
lasted 30 minutes.

The experiments and the results are presented in more detail
by Sevaldsen [13].

RESULTS

The processing has been based on spreading functions. For the
scattering functions to exist the WSSUS assumption must be valid.
We tested our impulse response estimates for independence and
stationarity using a one sample runs test for indipendence and a
Kolmogorov-Smirnov two sample test for stationarity. The tests
indicated that conditions in most cases must be considered un-
stationary.

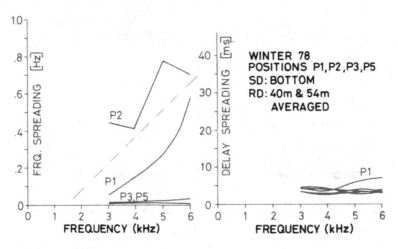

Fig. 4.   Spreading in frequency and delay. Sources on the bottom,
          Winter 78.

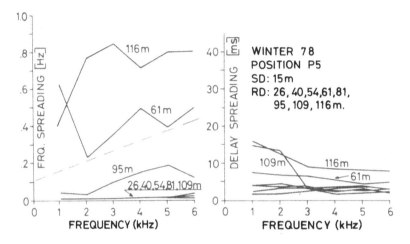

Fig. 5.   Spreading in frequency and delay.   Variable hydrophone
          depths, Winter 78.

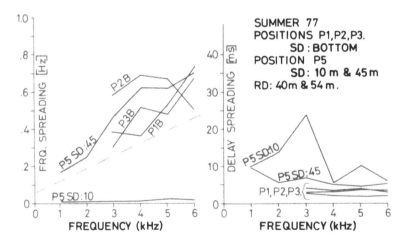

Fig. 6.   Spreading in frequency and delay, Summer 77

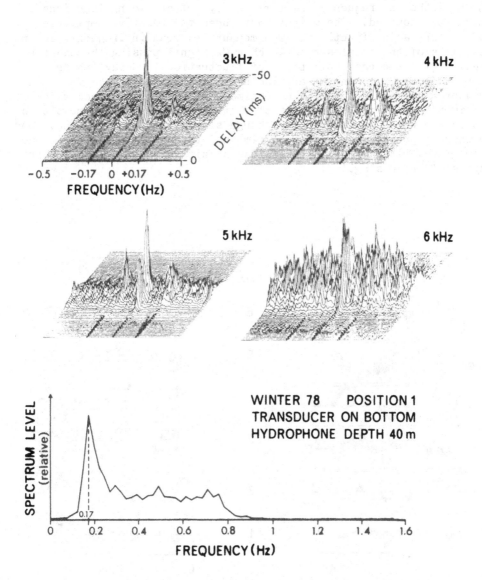

Fig. 7. Scattering functions with sidebands and wave spectrum.

Because of this the scattering function concept is strictly not
valid. It should be noted, however, that results based on the
(non-existing) scattering function usually are not inconsistent with
spreading function results. Below we present weighted averages of
3 dB widths in frequency and time (delay) of spreading functions
magnitude squared. The weights have been the relative amplitudes
of the spreading functions. The measured spread in frequency is an
estimate of the total spread which a CW signal passing through the
medium will undergo. All types of fluctuations - delay (phase),
amplitude and frequency - are included in this estimate. The
measured spread in delay is an estimate of the total spread which
an impulse will suffer because of multipath. The average multipath
structure which we observe may change slowly from ping to ping. Our
method of processing gives both time and frequency spreading at the
same time. It is thought to be less sensitive to noise than other
types of processing.

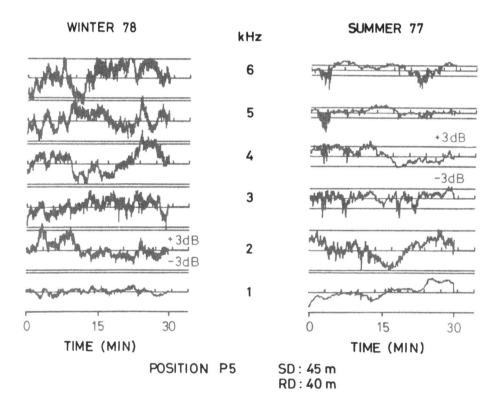

Fig. 8. Transmission loss variations.

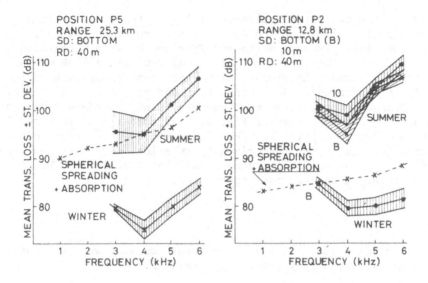

Fig. 9. Transmission loss mean value $\pm$ standard deviation.

Some spreading results are shown in Figs. 4 , 5 and 6. The results generally show small delay spreading, near 5 ms which correspond to one dominant arrival only. One case with 2 clearly separated equally strong arrivals are shown in Fig. 6 (P5 SD:10).

The frequency spreading results can be classified in 3 groups:

- very small spreading, $\Delta f < 0.05 - 0.1$ Hz
- relatively large spreading, $\Delta f \geqslant 1.0$ Hz
  partly unresolved because of undersampling
- intermediate cases

In Figs. 4 , 5 and 6 the curves above the dashed lines represent cases with large spreading. Because of undersampling they are missplaced (too low). The Pl curve (Fig. 4) represents an intermediate case where spreading in frequency is increasing strongly with frequency. Looking at the spreading functions we find that the peak of the wave spectrum shows up as sidebands around the main lobe. Fig. 7 shows the corresponding scattering functions with sidebands. Similar observations have been made by DeFerrari and Nghiem-Phu[7].

We also measured transmission loss on the signals used for spreading measurements.

In Fig. 8 are displayed typical examples of transmission loss
variations with time over the duration of a run (30 minutes). As
can be seen there is considerable difference between summer and
winter results. The variations are usually within ± 3 dB of the
mean both in summer and winter. There are some exceptions most
notably in summer where deviations from the mean can be 12 dB or
more. Transmission loss mean values plus-minus standard deviation
from a few cases are shown in Fig. 9. For comparison spherical
spreading plus absorption has been indicated. We observe that the
measured loss is higher than given by spherical spreading plus absorption in some cases. It seems clear that location and bottom
properties are more important than range.

DISCUSSION

    Among the results we will concentrate on what we have found
most interesting, namely frequency spreading.

    Our method of processing is sensitive to all types of time
fluctuations and provides an estimate of the total spreading effect
of the medium.

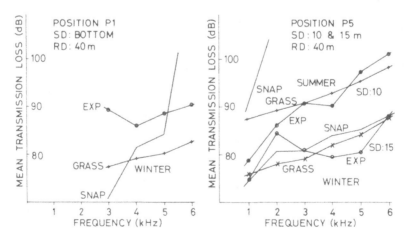

Fig. 10. Transmission loss comparisons: models and experiment.

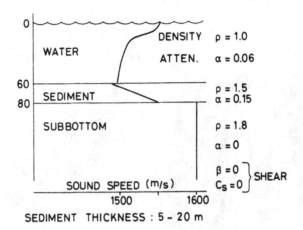

Fig. 11. Bottom model used in the SNAP calculations.

The results have been classified in 3 different groups:

- small spreading, Δf < 0.05 - 0.1 Hz
- large spreading, Δf ≥ 1.0 Hz (undersampled)
- intermediate cases

Most cases belong to the first two groups where changes with
frequency of transmission are small. The third group includes some
cases with spreading increasing strongly with frequency of transmission. We will first look at the third group represented here by
Fig. 7 and position P1, Fig. 4. From Fig. 7 is seen that the
spreading in this particular case is caused by surface wave modulation of the acoustic field. In this case more energy goes into
the sidebands at higher frequencies which explains the increase in
spreading with frequency. Clearly visible sidebands seem to occur
only with a fixed source and a peaked surface wave spectrum under
up- or down-wind conditions. Transversal transducer motion and wave
spectra without a dominant peak will cause smearing of sidebands.
No significant motion-induced spectral widening seems to have occured
in our case. Cross wind conditions cause less energy to be scattered
into sidebands.

In shallow waters with the sound field interacting both with
surface and bottom surface roughness is a very likely source of
frequency spreading. Our large spreading results are not inconsistent with such a hypothesis. But surface waves cannot explain all
cases of observed frequency spreading.

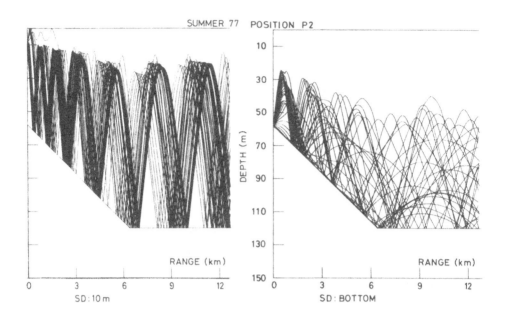

Fig. 12. Ray tracings from GRASS.

In summer with calm sea and a strongly downward refracting profile
we observe large spreading in most cases. Only with a shallow
transducer (at 10 m) do we find small spreading. And in winter with
rougher sea and stronger surface interaction we measure mostly small
spreading.

During some experiments we have observed that the spreading may
change from one group to another by changing hydrophone- or source
depth. The reason for this could be that the sound field is interacting with different parts of the medium belonging to different
sound transmission regimes. These regimes would be characterized by
different types of scattering, saturated or unsaturated, or different
degree of boundary interaction or both. The small spreading results
could be explained as deep water type amplitude and phase fluctuations caused by internal waves and oceanic finestructure. Large
spreading not caused by surface interaction could be due to the
acoustic field interacting strongly with part of the medium
characterized by strong (saturated) scattering. If these assumptions
hold it should be visible in ray plots or mode representations. A
narrow group of modes would correspond to low spreading. For comparison we have run two propagation computer models on our data:
GRASS, a range dependent ray trace model, and SNAP, a normal mode
model with range dependence in the adiabatic approximation.

Fig. 10 shows some examples of transmission loss comparisons with experiment. In general models and experiment compare well in winter, not so well in summer when bottom interaction is stronger. The transmission loss measurements have been included to provide a check on the environment and the parameters used in the model calculations. Fig. 11 describes the environment used by SNAP, the presumably best of the two models. A ray tracing example from GRASS is shown in Fig. 12. The left part represents the low spread case. One observes that there is more energy at the receiver depths 40 m and 54 m than in the other case. As a whole the ray tracings give few clues to understanding the observed spreading. Finally we show two SNAP model examples of energy versus arrival angle or mode number (Fig. 13). This time the case with small spreading is represented by the right hand figure.

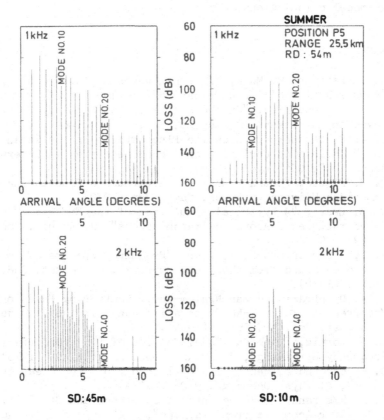

Fig. 13. Mode energy versus arrival angle or mode number (SNAP).

It is seen that small spreading is associated with a more concentrated or grouped mode structure than large spreading, as was
postulated. Lack of environmental measurements and knowledge of
bottom conditions prevent us from getting further. No model incorporating time variations in the environmental conditions has been
available to us.

CONCLUSION

We have observed time fluctuations in one-way transmitted
signals. Surface roughness has been identified as a source of
fluctuations in some cases but not all. To explain the rest of our
observations we assume that the acoustic field interacts with
different parts of the medium belonging to different sound transmission regimes (weak or strong scattering). The source of
scattering could be internal waves combined with advection of finestructure layers. Too sparse environmental sampling and lack of
suitable models prevent us from checking the assumptions
through model calculations.

REFERENCES

1. C. Garrett and W. Munk, "Internal Waves in the Ocean", for
    "Annual Review of Fluid Mechanics", (1979).
2. Y. J. F. Desaubies, "On the Scattering of Sound by Internal
    Waves in the Ocean", J. Acoust. Soc. Am. 64:1460 (1978).
3. Y. J. F. Desaubies, "Acousitc Fluctuations in the Ocean", for
    the "Congress on Caviation and Inhomogeneities in Underwater
    Acoustics", Göttingen (1979).
4. S. M. Flatté, "Sound Transmission through a Fluctuating Ocean",
    Cambridge University Press, Cambridge (1979).
5. T. E. Ewart, "A Numerical Simulation of the Effects of Oceanic
    Finestructure on Acoustic Transmissions", J. Acoust. Soc. Am.
    67:496 (1980).
6. W. I. Roderick and B. F. Cron, "Frequency Spectra of Forward-
    Scattered Sound from the Ocean Surface", J. Acoust. Soc. Am.
    48:759 (1970).
7. H. A. DeFerrari and Lan Nghiem-Phu, "Scattering Function
    Measurements for a 7-NM Propagation Range in the Florida Straits",
    J. Acoust. Soc. Am. 56:47 (1974).
8. J. G. Zornig, "Physical Model Studies of Forward Surface Scatter
    Frequency Spreading", J. Acoust. Soc. Am. 64:1492 (1978).
9. C. Wunsch and S. Webb, "The Climatology of Deep Ocean Internal
    Waves", J. Phys. Oceanogr. 9:235 (1979).
10. K. A. Søstrand, "Measurements of Coherence and Stability of
    Underwater Acoustic Transmissions", in "Proc. of Nato Advanced
    Study Institute on Signal Processing", Enschede (1968).

11. R. Laval, "Sound Propagation Effects on Signal Processing", in
     "Proc. of Nato Advanced Study Institute on Signal Processing",
     Loughborough (1972).
12. D. Costa and E. Hug, "An Estimate of the Scattering Function
     of an Undersea Channel", Alta Frequenza 45:245 (1976).
13. E. Sevaldsen, "Variability of Acoustic Transmissions in a
     Shallow Water Area", (Saclant Report in press, March 1980).

ANALYSIS OF TIME-VARYING SPATIALLY-VARYING

SOUND-PROPAGATION SYSTEMS

Lewis Meier

Systems Control, Inc.
Palo Alto, California, U.S.A.

ABSTRACT

It has proven very useful in the analysis of signal processing
for sonar systems to represent signals by cross-ambiguity functions
and system - such as a target or the medium - that modify the signal
by spreading or scattering functions. These functions measure respectively how signals are spread and how systems spread signals in time
and frequency. In such systems, sound-propagation paths play a significant part. This paper presents an extension of the temporal
theory to consider spatial variations: as before, signals are represented by cross-ambiguity functions and systems by spreading or
scattering functions; however, in the extension, spreads in space
and wave number are considered in addition. A major cause of such
spreads is often bottom interactions. Two versions of the spatial
theory are presented: a free-space version for use when ray theory
is applicable and a layered-media version for use when it is not.
In the layered-media version a normal-mode model of propagation is
assumed and spreads occur in mode number as well as wave number.
If the transmitted signal is assumed to be a plane wave modulated
by a narrow-band time signal, the ambiguity function for the transmitted signal is the product of the temporal ambiguity function of
the time signal and the beam pattern for the receiving array.

INTRODUCTION

The body of this paper is rather heavy with equations; therefore, in this introduction a summary of the results contained in
the body is given proceeded by the background of these results and
followed by some possible applications and extensions of the results.

Background

The use of linear time-varying system models has proven very useful in signal processing problems.[1] For example, the various parts -- medium and target -- in the signal path between transmitter and receiver in an active sonar system can be conveniently represented as time-varying systems.[2,3] The basic approach taken in this analysis is to characterize linear time-varying systems by a spreading function, which tells how the system spreads a signal in time and frequency. In the case of stochastic systems in which the spreading in time and frequency is uncorrelated in time and frequency, the spreading function is replaced by the scattering function, which in essence is its covariance.

In this approach, signals are characterized by the cross-ambiguity function, which measures how the signal is spread in time and frequency. The use of the cross-ambiguity function is quite reasonable since the output of a bank of matched filters is a cross-ambiguity function. The typical front-end of a radar or active sonar system is a bank of matched filters; furthermore, Fourier analysis typically used in passive sonar systems may be viewed as a bank of filters matched to frequency shifted versions of a CW signal; therefore, the first step in temporal signal processing is typically determination of a cross-ambiguity function.

The usefulness of the spreading, scattering and cross-ambiguity functions lies in the convenient relationships among them. In particular, the concatenation relationship for combining the spreading functions of two systems in series into the spreading function of the combination and the input/output relationship giving the cross-ambiguity function of the output of a system in terms of its spreading function and the cross-ambiguity function of its input are both a modified double convolution in time and frequency. Furthermore, for uncorrelated stochastic systems, the corresponding concatenation and input/output relationships are an ordinary double convolution in time and frequency. These relationships are illustrated in Figure 1.

Summary of Results

In sonar systems, major portions of the signal path are through sound propagating media. For such media, spatial variations are just as important as temporal variations. The purpose of this paper is to extend the ideas just presented from the analysis of time-varying linear systems to time-varying spatially-varying systems with particular emphasis on the effects of bottom interaction in coastal waters. Two versions of the theory will be presented: the free-space version and the layered-media version. The free-space version is applicable in problems, such as typical active sonar problems, in which ray theory is applicable. On the other hand, the layered-media version is applicable in problems, such as typical passive sonar problems, in which normal mode theory is required.

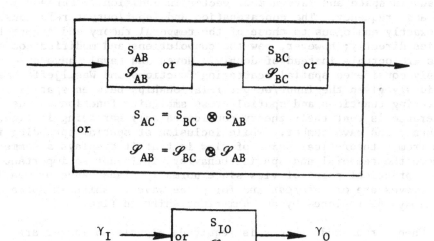

$$S_{AC} = S_{BC} \otimes S_{AB}$$

$$\mathscr{S}_{AB} = \mathscr{S}_{BC} \otimes \mathscr{S}_{AB}$$

$$\gamma_I \longrightarrow \boxed{\begin{array}{c} S_{IO} \\ \text{or} \\ \mathscr{S}_{IO} \end{array}} \longrightarrow \gamma_O$$

$$\gamma_O = S_{IO} \; \tilde{\otimes} \; \gamma_I$$

$$\text{or} \quad |\gamma_O|^2 = \mathscr{S}_{IO} \otimes |\gamma_I|^2$$

S = Spreading Function $\otimes$ = convolution

$\mathscr{S}$ = Scattering Function $\tilde{\otimes}$ = modified convolution

γ = Cross Ambiguity Function

Figure 1. Concatenation and Input-Output Relationships

The free-space version of the theory presented herein is a
straightforward extension of the temporal theory. Again systems
are characterized by spreading or scattering functions and signals
by cross-ambiguity functions; these will be called spatial spreading
functions, et. herein to distinguish them. They are concerned with
spreads in space and wave number vector in addition to spreads in
time and frequency. The concatenation and input/output relationships
are exactly analogous to those of the temporal theory and Figure 1
applies directly; however, now the convolutions and modified convolutions are octuple instead of double. Hoven and Laval[4] have previously considered spatial scattering functions and Wasiljeff[5] has
previously given the input/output relationships between spatial
scattering functions and spatial cross-ambiguity functions. One
difference is that their theory considered only spreading in time,
frequency and wave number. While inclusion of spatial spreading is
nice from a theoretical point of view in that it displays a symmetry
between the temporal and spatial behavior, it is not so important
from a practical point of view since normally plane waves or nearly
plane waves are of interest and for plane waves a shift in space
may always be replaced by an equivalent shift in time.

When normal mode theory is required, signals no longer are
appropriately viewed as sums of plane waves as in the free-space
version. Rather, they are appropriately viewed as sums of hybrid
waves that are plane waves as far as horizontal behavior is concerned,
but are standing waves as far as vertical behavior is concerned;
therefore, it is appropriate to replace the spreading in the vertical
direction and in the vertical wave number by spreading in more suitable variables. Since the standing waves are characterized by solutions to the Helmholtz equation, the vertical posiion at which the
Helmholtz equation is initialized and mode number are the suitable
variables. Again, systems are characterized by spreading or scattering functions and signals by cross-ambiguity functions and Figure
1 applies; however, the octuple convolutions and modified convolutions differ from those of the free-space version. Herein the
spreading functions, etc. of the layered-media version are referred
to as layered-media spreading functions, etc. to distinguish them.

Applications

Consider briefly the application of the theory just described
to sonar systems operating in coastal waters, where both the surface
and bottom have an important influence on sound propagation. Consider
first the passive sonar system illustrated in Figure 2. A convenient
fiction is to take the input to be a time modulated plane wave broadside to the receiving array. The source spreads this input in time,
frequency and direction so that without the medium it would appear to
be a signal extended in time, frequency and perhaps bearing coming
from a direction in general other than broadside. The medium will

further spread the signal in time, frequency and direction. If ray theory is applicable, the effects of the bottom and surface can be separated from that of the medium and be considered to spread the signal in time, frequency and direction for the paths interacting with them. If normal mode theory is required, the effects of the bottom and surface cannot be separated from that of the medium and both, in fact, become intimate parts of the spreading by the medium. Finally, the array may be moving and its elements may not be in their nominal positions relative to each other; hence, the array may also be viewed as spreading the signal in time, frequency and direction. For an active sonar the situation is the same except a target is now present, which again spreads the signal in time, frequency and direction.

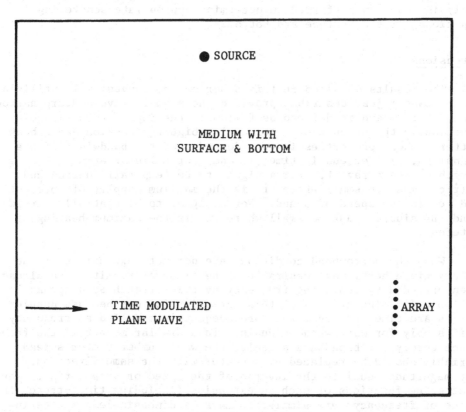

Figure 2. A Passive Sonar System

Clearly, all the parts of a sonar system can be represented by spreading or scattering functions. In the body of this paper, it is shown that the input cross-ambiguity function factors into the product of a temporal ambiguity function and a beam pattern. Using the relationships of Figure 1, the output cross-ambiguity function may be obtained from the input ambiguity function and the spreading or scattering functions. In a passive sonar, the spreading or scattering function of the source embodies the relevant properties it is desired to measure while in active sonar, the target spreading or scattering function serves that purpose. The spreading or scattering functions of other parts of the system determine how that information is modified. Temporal scattering functions have proven to be useful models of submarines for use in analysis of active sonar systems.[2,3] Extension of these models to the spatial domain should allow other submarine properties to be measured. Experimental determinations have been made of temporal scattering functions for the medium.[6] Problems have been encountered in these determinations because of motion of hydrophones and transducers, which might be alleviated by measuring this motion and determining either the free-space or layered-medium scattering functions. Finally, a great amount of analysis has been performed to determine appropriate scattering functions for the surface and bottom.[5]

Extensions

The results obtained in this paper may be conveniently utilized in the manner just described provided the signals involved are narrowband. Narrowband is defined as follows: the input cross-ambiguity function is the product of a temporal ambiguity function and a beam pattern, whose properties in turn depend upon the bandwidth of the signal B, its extend in time T and, for a linear array, the length of the array L. For a signal to be temporally narrowband $BT(\dot{r}/c)$ must be small where $\dot{r}$ is the maximum doppler of interest and c is the speed of sound. For a signal to be spatially narrowband $BL\sin\theta/c$ must be small where θ is the maximum bearing of interest.

When the narrowband conditions are not met, extensions of the theory given herein are desirable. The temporal results have already been extended by replacing frequency by time stretch as a spread variable.[2,3] The utility of this extension follows because doppler shifts are time stretches that are adequately modelled as frequency shifts only for narrowband signals. In a similar manner in the free-space theory for broadband signals, the wave number vector spread variable should be replaced by a vector with the same direction, but magnitude equal to the inverse of the speed of propagation. The scattering functions of such an extension (including time stretch in place of frequency) are similar to Laval's dimensionless scattering functions.[4] For the layered-medium version, the required extension

is not clear at this point. Certainly the horizontal wave number
vector should be replaced by a vector with the same direction and
magnitude equal to the inverse of the group velocity. It may also
be desirable to include phase terms involving the phase velocity.

REPRESENTATION OF SIGNALS

Ordinary temporal signals can be represented in either the time
domain or the frequency domain; so too, spatial signals can be represented in either the time-space domain or the frequency-wave number
space. The relationship between the two representations is that of
the four-dimensional Fourier transform

$$\hat{s}(\omega,\underline{k}) = \iint e^{-j(\omega t - \underline{k}\cdot\underline{x})} s(t,\underline{x}) \, dt \, d\underline{x}, \tag{1}$$

where $s(t,\underline{x})$ is the space-time representation of a signal, $\hat{s}(\omega,\underline{k})$
is the frequency-wave number representation of the same signal, t
is time, $\underline{x}$ is the vector (x,y,z) of space variables, ω is the
frequency and $\underline{k}$ is the vector (k_x, k_y, k_z) of wave numbers. It will
prove convenient, especially in the layered media version, to use
the frequency-wave number representation of signals.

FREE SPACE VERSION

In this subsection the spatial spreading function, the spatial
scattering function and the spatial ambiguity functions are defined
and the fundamental relationships between them given.

The Spatial Spreading Function

A time and space-varying free-space propagation system can be
thought of as spreading the input in time and frequency and in their
spatial analogs space and wave number; therefore, such a system can
be represented in terms of the spatial spreading function
$S^S(\Delta t, \Delta\underline{x}, \Delta\omega, \Delta\underline{k})$:

$$\hat{s}_0(\omega,\underline{k}) = \iiiint S^S(\Delta t, \Delta\underline{x}, \Delta\omega, \Delta\underline{k}) e^{-j(\omega\Delta t - \underline{k}\cdot\Delta\underline{x})}$$
$$\hat{s}_I(\omega-\Delta\omega, \underline{k}-\Delta\underline{k}) \, d\Delta t \, d\Delta\underline{x} \, d\Delta\omega \, d\Delta\underline{k}, \tag{2}$$

where $\hat{s}_I(\omega,\underline{k})$ and $\hat{s}_0(\omega,\underline{k})$ represent the input and output signals
respectively. Note that $e^{-j(\omega\Delta t - \underline{k}\cdot\Delta\underline{x})}\hat{s}_I(\omega-\Delta\omega,\underline{k}-\Delta\underline{k})$ represents the
input shifted first in frequency and wave number and then shifted
in time and space. Note also that shift in wave number physically
corresponds to a shift in speed and direction.

In the application of temporal signal processing theory to active sonar and radars, the reference signal for the bank of matched filters is the input signal whereas for passive sonars it is a CW pulse. Such reference signals can be written as the product $e^{j\omega_0 t} s_M(t)$ of a carrier and a modulation. An important application of the spatial theory is in the analysis and synthesis of arrays. For such situations, the x-axis can be selected as a convenient reference axis -- for example, perpendicular to the face of a planar array. Furthermore, in such situations, it is convenient to assume that the input is a plane wave along the x-axis modulated by

$$s_M(t) \sim s_M(t-c_0 x) e^{j(\omega_0 t - \underline{k}_0 \cdot \underline{x})}, \quad \text{where} \quad \underline{k}_0 = (\frac{\omega_0}{c_0}, 0, 0) \quad \text{and} \quad c_0 \quad \text{is}$$

the nominal speed of propagation -- and to assume the various parts of the propagation system -- such as the source, medium and target -- spread this input in time, space, frequency and wave number. For narrowband signals, this input may be approximated by $s_M(t) e^{j(\omega_0 t - \underline{k}_0 \cdot \underline{x})}$.

The Spatial Scattering Function

Now consider stochastic systems and assume that S^S is uncorrelated in all its arguments. In this case

$$E\{S^{S*}(\Delta t', \Delta \underline{x}', \Delta \omega', \Delta \underline{k}') \; S^S(\Delta t, \Delta \underline{x}, \Delta \omega, \Delta \underline{k})\}$$

$$= \mathscr{S}^S(\Delta t, \Delta \underline{x}, \Delta \omega, \Delta \underline{k}) \; \delta(\Delta t' - \Delta t) \; \delta(\Delta x' - \Delta x) \; \delta(\Delta \omega' - \Delta \omega) \; \delta(\Delta \underline{k}' - \Delta \underline{k}) \quad (3)$$

where $\mathscr{S}^S(\Delta t, \Delta \underline{x}, \Delta \omega, \Delta \underline{k})$ is the spatial scattering function.

The Spatial Cross-Ambiguity Function

A signal in free-space is appropriately characterized in terms of the output that would be obtained if it were passed through a band of space-time filters matched to frequency and wave-number shifted versions of a reference signal:

$$\gamma^S(\Delta t, \Delta \underline{x}, \Delta \omega, \Delta \underline{k}) = \frac{1}{(2\pi)^4} \iint e^{j(\omega \Delta t - \underline{k} \cdot \Delta \underline{x})}$$

$$\hat{s}_R^*(\omega - \Delta \omega, \underline{k} - \Delta \underline{k}) \; s(\omega, \underline{k}) \; d\omega \; d\underline{k} \quad (4)$$

where $s(\omega, \underline{k})$ represents the signal to be characterized and $\hat{s}_R^*(\omega, \underline{k})$ the reference signal. The function $\gamma^S(\Delta t, \Delta \underline{x}, \Delta \omega, \Delta \underline{k})$, which is a cross-correlation in time, space, frequency and wave number between the signal and the reference signal, is called the spatial cross-ambiguity function.

Now consider the typical signal processing at the front end of a sonar or radar system: a beam former and a bank of temporal matched filters that for narrowband signals takes the form

$$\gamma'(\Delta t, \Delta\omega, \Delta\underline{k}) = \int \sum e^{-j(\omega_0+\Delta\omega)(t-\Delta t)}$$
$$\cdot s_M^*(t-\Delta t) \, e^{j(\underline{k}_0+\Delta\underline{k})\cdot\underline{x}_i} \, s(t,\underline{x}_i) \, dt \qquad (5)$$

where $e^{+j\omega_0 t} s_M(t)$ represents the reference signal for the temporal-matched-filter bank, $\underline{x}_i$ $i=1,2,\ldots,$ are the positions of the ele-

ments of the receiving array, and $s(t,\underline{x}_i)$ represents the received

signal at the ith element of the receiving array. (Note that the beam forming and matched filtering can be performed in either order and further that in its most general form beam forming involves speed in addition to direction.)

On substitution of the relationship

$$s(t,\underline{x}) = \frac{1}{(2\pi)^4} \iint e^{j(\omega t-\underline{k}\cdot\underline{x})} \, s(\omega,\underline{k}) \, d\omega \, d\underline{k}, \qquad (6)$$

found by inverting (1) into (5), it becomes

$$\gamma'(\Delta t, \Delta\omega, \Delta\underline{k}) = \frac{1}{(2\pi)^4} \iint e^{j\omega\Delta t} \, \hat{s}_M^*(\omega-\omega_0-\Delta\omega)$$
$$\hat{s}_S^*(\underline{k}-\underline{k}_0-\Delta\underline{k}) \, \hat{s}(\omega,\underline{k}) \, d\omega \, d\underline{k}, \qquad (7)$$

where

$$\hat{s}_M(\omega) = \int e^{-j\omega t} \, s_M(t)dt, \qquad \hat{s}_S^*(\underline{k}) = \sum e^{j\underline{k}\cdot\underline{x}_i} \, . \qquad (8)$$

Comparison of (7) and (8) shows that

$$\gamma'(\Delta t, \Delta\omega, \Delta\underline{k}) = \gamma^S(\Delta t, \underline{0}, \Delta\omega, \Delta\underline{k}) \qquad (9)$$

or

$$s_R(\omega,\underline{k}) = \hat{s}_M(\omega-\omega_0) \, \hat{s}_S^*(\underline{k}-\underline{k}_0). \qquad (10)$$

Thus, the typical beamformer, temporal-matched-filter combination for narrowband siganls computes a spatial cross-ambiguity function. The fact that Δx is restricted to $\underline{0}$ is unimportant since for narrowband signals a shift in $\Delta\underline{x}$ can be closely approximated by an appropriate shift in Δt.

For the assumed input signal $s_M(t)\ e^{j(\omega_0 t - \underline{k}_0 \cdot \underline{x})}$

$$\hat{s}(\omega,\underline{k}) = \hat{s}_M(\omega-\omega_0)(2\pi)^3\ \delta(\underline{k}-\underline{k}_0) \tag{11}$$

and

$$\gamma^S(\Delta t,\underline{0},\Delta\omega,\Delta\underline{k}) = \left[\frac{1}{2\pi}\int e^{j\omega\Delta t}\ \hat{s}_M^*(\omega-\omega_0+\Delta\omega)\ \hat{s}_M(\omega-\omega_0)\ d\omega\right]\hat{s}_S(\Delta k) \tag{12}$$

The first factor on the right side of (12) is just the temporal ambiguity function for $s_M(t)\ e^{j\omega_0 t}$ and the second factor is the beam pattern for the array.

The Fundamental Relationships

Consider the modified octuple convolution $\otimes'_S$ defined by

$$a \otimes'_S b(\Delta t,\Delta\underline{x},\Delta\omega,\Delta\underline{k}) =$$

$$\iiiint e^{j[\Delta\omega'(\Delta t-\Delta t')-\underline{k}'\cdot(\Delta\underline{x}-\Delta\underline{x}')]}\, b(\Delta t-\Delta t',\Delta x-\Delta x',\Delta\omega-\Delta\omega',\Delta\underline{t}-\Delta\underline{k}')\cdot$$

$$a(\Delta t',\Delta\underline{x}',\Delta\omega',\Delta\underline{k};)\ d\Delta t'\ d\Delta\underline{x}'\ d\Delta\omega'\ d\Delta\underline{k}. \tag{13}$$

It is easy to show by direct substitution and algebraic manipulation that $\otimes'$ is associative but not in general commutative; furthermore, it is possible to rewrite (2) and (4) in terms of $\otimes'_S$. When this is done and use is made of the associative property of $\otimes'_S$ the following fundamental relationships may be derived:

$$\gamma_0^S(\Delta t,\Delta\underline{x},\Delta\omega,\Delta\underline{k}) = S_{IO}^S\ \otimes'_S\ \gamma_I^S(\Delta t,\Delta\underline{x},\Delta\omega,\Delta\underline{k}) \tag{14}$$

$$S_{AC}^S(\Delta t,\Delta\underline{x},\Delta\omega,\Delta\underline{k}) = S_{BC}^S\ \otimes'_S\ S_{AC}^S(\Delta t,\Delta\underline{x},\Delta\omega,\Delta\underline{k}) \tag{15}$$

where γ_0^S and γ_I^S are cross-ambiguity functions for a signal passed through the system characterized by S_{IO}^S and where S_{AC} characterizes the system formed by concatenating the system characterized by S_{BC}^S after the system characterized by S_{AB}^S. Note that because $\otimes'_S$ is not commutative the order of concatenation is important as is to be expected with time-and-space varying systems.

Use of (3) with (13), (14) and (15) yields the following fundamental relationships for uncorrelated stochastic systems:

$$E\{|\gamma_O^S(\Delta t,\Delta \underline{x},\Delta \omega,\Delta \underline{k})|^2\} = \mathscr{S}_{IO}^S \quad \circledast_S \quad E\{|\gamma_I^S(\Delta t,\Delta \underline{x},\Delta \omega,\Delta \underline{k})|\} \tag{16}$$

$$\mathscr{S}_{AC}^S(\Delta t,\Delta x,\Delta \omega,\Delta \underline{k}) = \mathscr{S}_{BC}^S \circledast_S \mathscr{S}_{AB}^S(\Delta t,\Delta \underline{x},\Delta \omega,\Delta \underline{k}) \tag{17}$$

where $\mathscr{S}_{IO}^S$ etc. correspond to s_{IO}^S etc. and $\circledast_S$ is the unmodified octuple convolution defined by

$$a \circledast_S b(\Delta t,\Delta \underline{x},\Delta \omega,\Delta \underline{k}) = \iint b(\Delta t-\Delta t',\Delta \underline{x}-\Delta \underline{x}',\Delta \omega-\Delta \omega',\Delta \underline{k}-\Delta \underline{k}')$$

$$a(\Delta t',\Delta \underline{x},\Delta \omega',\Delta \underline{k}') \ d\Delta t' \ d\Delta \underline{x}' \ d\Delta \omega' \ d\Delta \underline{k}' \tag{18}$$

It is easy to show that $\circledast_S$ is commulative as well as associative; hence the order of concatenation of $\mathscr{S}_{BC}$ and $\mathscr{S}_{AC}$ is unimportant -- an unsurprising result in view of the stationarity implied by the existence of scattering functions.

THE LAYERED-MEDIA VERSION

For ease of exposition only, a single layer will be considered, bounded above by a pressure release boundary and below by a semi-infinite propagating medium. Nominally, the upper boundary of the layer is the x,y plane, the lower boundary of the layer in a plane parallel to the x,y plane (and thus a constant depth below the nominal surface) and the layer is uniform in x and y.

It is assumed that signals in such a layer may be represented as a sum

$$s(t,\underline{x}) = \frac{1}{(2\pi)^3} \iiiint e^{j(\omega t - \underline{k} \cdot \underline{x})} \ \varphi(z;z_o,\omega,k) \ \tilde{s}(z_o,\omega,\underline{k},k)$$

$$dz_o \ d\omega \ d\underline{k} \ dk \tag{19}$$

of waves that take the form of a plane wave in the x-y plane and a standing wave in the z-dimension; where $\underline{x}$ is the vector (x,y), $\underline{k}$ is the vector (k_x,k_y), k is the mode number and $\varphi(z;z_o,\omega,k)$ is the normal mode, which satisfies the Helmholtz equation

$$\frac{d^2\varphi}{dz^2} + \left[(\frac{\omega}{c(z)})^2 - k^2\right]\varphi = 0 \tag{20}$$

with the boundary condition $(z_o;\omega,z_o,k)=0$,   where $c(z)$ is the nominal speed of sound profile for the layer.

For the nominal conditions $z_o=0$,   the constraint $\underset{\sim}{k}\cdot\underset{\sim}{k}=k^2$ applies and the boundary conditions at the bottom of the layer will be met for a discrete set $\{k_i\}$ of values of   k; however, when conditions vary from nominal there is spreading away from these nominal values of z_o, $\underset{\sim}{k}$ and   k.   In particular variations of the surface from nominal will cause spreading over z_o,   variations in the medium from nominal will cause spreading of $\underset{\sim}{k}$ away from values that obey $\underset{\sim}{k}\cdot\underset{\sim}{k}=k^2$ and variations in the bottom from nominal will cause spreading of   k outside the set $\{k_i\}$.

It is more convenient to work in the frequency, wave-number domain. Fourier transformation of (19) yields

$$\hat{s}(\omega,\underline{k}) = \iint \hat{\varphi}(k_z;z_o,\omega,k)\ \tilde{s}(z_o,\omega,\underset{\sim}{k},k)\ dz_o\ dk \tag{21}$$

where $\hat{\varphi}(k_z;z_o,\omega,k)$ is the Fourier transform of $\varphi(z;z_o,\omega,k)$

$$\hat{\varphi}(k_z;z_o,\omega,k) = \int e^{jk_z z}\ \varphi(z;z_o,\omega,k)\ dz \tag{22}$$

In general $\tilde{s}(z_o,\omega,k,k)$ will not be unique since in (21) two variables, z_o and   k,   are replaced by one variable k_z.   Nevertheless, by assumption there will exist at least one $\psi(k_z;z_o\omega,k)$ such that

$$\tilde{s}(z_o,\omega,\underset{\sim}{k},k) = \int \hat{\psi}^*(k_z;z_o,\omega,k)\ \hat{s}(\omega,\underline{k})\ dk_z \tag{23}$$

satisfies (21).

In the layered-media version, we will be interested in the shifts Δt and $\Delta\omega$ in time and frequency as before; however, in place of the shifts $\Delta\underline{x}$ and $\Delta\underline{k}$ in three dimensional space and wave number will be the shifts $\Delta\underset{\sim}{x}$ and $\Delta\underset{\sim}{k}$ in two dimension space and wave number and shifts Δz_o and Δk in upper boundary and mode number. The shifts $\Delta z_o,\Delta\omega,\Delta\underset{\sim}{k}$ and Δk are given by the operator

$\mathcal{O}[\cdot;\Delta z_o\Delta\omega,\Delta\underset{\sim}{k},\Delta k]$ defined by

$$\mathcal{O}[\hat{s}(\omega,\underline{k});\Delta z_o,\Delta\omega,\Delta k,\Delta k] = \iint \varphi(k_z;z_o,\omega,k)$$

$$\tilde{s}(z_o-\Delta z_o,\omega-\Delta\omega,k-\Delta k,k-\Delta k) \, dz_o \, dk$$

$$= \int\left[\iiint \varphi(k_z;z_o,\omega,k)\right.$$

$$\left. \hat{\psi}^*(k_z';z_o-\Delta z_o,k-\Delta k,k-\Delta k) \, dz_o \, dk\right]$$

$$\hat{s}(\omega,k,k_z') \, dk_z' \tag{24}$$

Note that this operator is additive, i.e.,

$$\mathcal{O}\{\ \mathcal{O}[\hat{s}(\omega,\underline{k});\Delta z_o,\Delta\omega,\Delta k,\Delta k]; \ \Delta z_o',\Delta\omega',\Delta k',\Delta k'\}$$

$$= \ \mathcal{O}[\hat{s}(\omega,\underline{k}); \ \Delta z_o+\Delta z_o',\Delta\omega+\Delta\omega',\Delta k+ k',\Delta k+\Delta k'] \tag{25}$$

The Layered-Media Spreading Function

In analogy with (2) a time-and-space-varying layered-media propagation system can be represented in terms of the layered-media spreading function $S^L(\Delta t,\Delta x,\Delta\omega,\Delta k,\Delta z_o,\Delta k)$:

$$\tilde{s}_o(z_o,\omega,\underline{k},k) = \iiiint\!\!\!\iint S^L(\Delta t,\Delta x,\Delta z_o,\Delta\omega,\Delta k,\Delta k) e^{-j(\omega\Delta t-\underline{k}\cdot\Delta x)}$$

$$\tilde{s}_I(z_o-\Delta z_o,\omega-\Delta\omega,k-\Delta k,k-\Delta k) \, d\Delta t \, d\Delta x \, d\Delta z_o \, d\Delta\omega \, d\Delta k \, d\Delta k \tag{26}$$

or from (21) and (24)

$$\hat{s}_o(\omega,\underline{k}) = \iiiint\!\!\!\iint S^L(\Delta t,\Delta x,\Delta z_o,\Delta\omega,\Delta k,\Delta k) \, e^{-j(\omega\Delta t-\underline{k}\cdot\Delta x)}$$

$$\mathcal{O}[\hat{s}_I(\omega,\underline{k}),\Delta z_o,\Delta\omega,\Delta k,\Delta k] \quad d\Delta t \, d\Delta x \, d\Delta z_o \, d\Delta\omega \, d\Delta k \, d\Delta k \tag{27}$$

where $\hat{s}_I(\omega,k)$ and $\hat{s}_o(\omega,\underline{k})$ represent the input and output signals respectively. Note that (27) represents the output as a sum of versions of the input shifted in first upper boundary, frequency, wave number and mode number followed by shifts in time and space.

In analogy with the free-space version it is convenient in radar and sonar problems in a layered medium to assume that the signal is a hybrid plane-standing wave along the x-axis modulated by

$$s_M(t) -- s_M(t-U_1 x)e^{j(\omega_0 t - k_{\sim 1} \cdot x)} \quad \varphi(z;0,\omega_0,k_1), \quad \text{where}$$

$k_{\sim 1} = (k_1,0), k_1 = V_1/\omega_0$ and U_1 and V_1 are the group and phase

velocities of the lowest normal mode for the nominal media -- and to assume that the various parts of the propagation system such as the source, medium and target, spread the input in time, space, upper boundary, frequency, wave number and mode number. Note that the spreading function for the source would tell how the signal is spread in direction and over the various discrete modes of the nominal layered medium and that the spreading function for the medium would tell how the variations in the actual medium from the nominal medium would spread this signal. For narrow band signals this input may be approximated by

$$s(t)e^{j(\omega_1 t - k_{\sim 1} \cdot x)} \quad \varphi(z;0,\omega_1,k_1).$$

The Layered-Media Scattering Function

Now consider stochastic systems and assume that s^L is uncorrelated in all its arguments. In this case

$$E\{s^{L*}(\Delta t',\Delta x',\Delta z_0',\Delta \omega',\Delta k',\Delta k') \quad s^L(\Delta t,\Delta x,\Delta z_0,\Delta \omega,\Delta k,\Delta k)\}$$

$$= \mathscr{S}^L(\Delta t,\Delta x,\Delta z_0,\Delta \omega,\Delta k,\Delta k) \, \delta(\Delta t'-\Delta t) \, \delta(\Delta x'-\Delta x)$$

$$\delta(\Delta z_0'-\Delta z_0) \, \delta(\Delta \omega'-\Delta \omega) \, \delta(\Delta k'-\Delta k) \, \delta(\Delta k'-\Delta k) \qquad (28)$$

where $\mathscr{S}^L(\Delta t,\Delta x,\Delta z_0,\Delta \omega,\Delta k,\Delta k)$ is the layered media scattering function.

The Layered-Media Cross-Ambiguity Function

In analogy with (4), it is convenient to characterize signals in terms of the output that would be obtained if it were passed through a bank of space-time filters matched to upper boundary, frequency, wave number and mode number shifted versions of a reference signal:

$$\gamma^L(\Delta t,\Delta x,\Delta z_0,\Delta \omega,\Delta k,\Delta k) = \frac{1}{(2\pi)^4} \iiiint e^{j(\Delta \omega \Delta t - \Delta k \cdot \Delta x}$$

$$\tilde{s}_R^*(z_0-\Delta z_0,\omega-\Delta \omega,k-\Delta k,k-\Delta k) \cdot \tilde{s}(z_0,\omega,k,k) \, dz_0 \, d\omega \, dk \, dk \qquad (29)$$

or

$$\gamma^L(\Delta t, \Delta \underset{\sim}{x}, \Delta z_o, \Delta \omega, \Delta \underset{\sim}{k}, \Delta k) = \frac{1}{(2\pi)^4} \iint e^{j(\omega \Delta t - \underset{\sim}{k} \cdot \Delta \underset{\sim}{x})}$$

$$\mathcal{O}'[\hat{s}_R(\omega, \underline{k}), \Delta z_o, \Delta \omega, \Delta \underset{\sim}{k}, \Delta k]^* \; \hat{s}(\omega, \underline{k}) \; d\omega \; d\underline{k} \qquad (30)$$

where

$$\mathcal{O}'[\hat{s}_R(\omega, \underline{k}), \Delta z_o, \Delta \omega, \Delta \underset{\sim}{k}, \Delta k] = \iint \hat{\psi}(k_z; z_o, \omega, k)$$

$$\tilde{s}_R(z_o - \Delta z_o, \omega_o - \Delta \omega, \underset{\sim}{k} - \Delta \underset{\sim}{k}, k - \Delta k) \; dz_o \; dk \quad (31)$$

and $s(\omega, \underline{k})$ and $s_R(\omega, \underline{k})$ represent the signal to be characterized and the reference signal respectively. The function $\gamma^L(\Delta t, \Delta \underset{\sim}{x}, \Delta z_o, \Delta \omega, \Delta \underset{\sim}{k}, \Delta k)$ is called the layered-media cross-ambiguity function.

Now consider a typical signal processing front end for a sonar or radar system: a beam former and a bank of temporal matched filters that for narrowband signals in a layered media take the form

$$\gamma''(\Delta t, \Delta \omega, \Delta \underset{\sim}{k}; z_o, k) = \int \sum e^{-j(\omega_o - \Delta \omega)(t - \Delta t)}$$

$$s_M^*(t - \Delta t) e^{j(\underset{\sim}{k}_1 - \Delta \underset{\sim}{k}) \cdot \underset{\sim}{x}_i} \psi^*(z_i; z_o, \omega_o; k) \; s(t, \underline{x}_i) \; dt \quad (32)$$

where

$$\psi(z; z_o, \omega, k) = \frac{1}{2\pi} \int e^{-jk_z z} \; \hat{\psi}(k_z; z_o, \omega, k) \; dk_z , \qquad (33)$$

$e^{j\omega_o t} s_M(t)$ represents the reference signal for the temporal-matched filter bank, $\underline{x}_i$ $i=1,2\ldots$ are the positions of the elements of the receiving array (note $\underline{x}_i$ is the vector (x_i, y_i, z_i) and $\underset{\sim}{x}_i$ is the vector (x_i, y_i)) and $s(t, x_i)$ represents the received signal at the ith element of the receiving array.

On substitution of (6) into (32), it becomes

$$\gamma''(\Delta t, \Delta \omega, \Delta \underset{\sim}{k}; z_o, k) = \frac{1}{(2\pi)^4} \iint e^{j\omega \Delta t} \; \hat{s}_M^*(\omega - \omega_o - \Delta \omega)$$

$$\hat{s}_L^*(\underset{\sim}{k} - \underset{\sim}{k}_1 - \Delta \underset{\sim}{k}, k_z; z_o, k) \; d \quad d\underline{k} \qquad (34)$$

where

$$\hat{s}_L(\underset{\sim}{k},k_z;z_o,k) = \sum e^{j(\underset{\sim}{k}\cdot\underset{\sim}{x}_i + k_z z_i)} \; \psi(z_i;z_o,\omega_o,k). \tag{35}$$

Comparison of (34) and (30) shows that

$$\gamma''(\Delta t,\Delta\omega,\Delta\underset{\sim}{k};z_o,k) = \gamma^L(\Delta t,\underset{\sim}{0},0,\Delta\omega,\Delta\underset{\sim}{k},0) \tag{36}$$

for

$$\hat{s}_R(\omega,\underline{k}) = \hat{s}_M(\omega-\omega_o)\; \hat{s}_L(\underset{\sim}{k}\cdot\underset{\sim}{k}_1,k_z;z_o,k) \tag{37}$$

Thus, a typical beam former, temporal-matched-filter combination for narrowband signals computes a layered-media cross-ambiguity function.

Again, the fact that $\Delta\underset{\sim}{x}$ is restricted to $\underset{\sim}{0}$ is unimportant since for narrowband signals a shift in $\Delta\underset{\sim}{x}$ can be closely approximated by a shift in Δt. Of more significance is the fact that Δz_o and Δk are restricted to zero. In general

$$\mathscr{O}[s_M(\omega-\omega_o)\; s_L(\underset{\sim}{k}-\underset{\sim}{k}_1,k_z,0,k_1);\; \Delta z_o,\Delta\omega,\Delta\underset{\sim}{k},\Delta k]$$

$$\neq \hat{s}_M(\omega-\omega_o-\Delta\omega)\; \hat{s}_L(\underset{\sim}{k}-\underset{\sim}{k}_1-\Delta\underset{\sim}{k};\Delta z,k_1-\Delta k); \tag{38}$$

however, for an array that is repetitive in z_i with the subarrays closely enough spaced that the sums over z_i in (32) and (35) can be approximated by an integral equality holds in (38). In this case

$$\gamma''(\Delta t,\Delta\omega,\Delta\underset{\sim}{k};\Delta z_o,k_1+k) = \gamma^L(\Delta t,\underset{\sim}{0},\Delta z_o,\Delta\omega,\Delta\underset{\sim}{k},\Delta k), \tag{39}$$

for

$$\hat{s}_R(\omega,\underline{k}) = \hat{s}_M(\omega-\omega_o)\; \sum_i e^{j\underset{\sim}{k}_1\cdot\underset{\sim}{x}_i}\; \hat{\psi}(k_z;0,\omega_o,k_1). \tag{40}$$

For the signal $s_M(t)e^{j(\omega_o t-\underset{\sim}{k}_1\cdot\underset{\sim}{x})}\; \varphi(z;z_o',\omega_o,k')$

$$\hat{s}(\omega,\underline{k}) = \hat{s}_M(\omega-\omega_o)(2\pi)^2\; \delta(\underset{\sim}{k}-\underset{\sim}{k}_1)\; \varphi(k_z;z_o';\omega_o,k') \tag{41}$$

and

$$\gamma^L(\Delta t,\underset{\sim}{0},0,\Delta\omega,\Delta\underset{\sim}{k},0) + \left[\frac{1}{2\pi}\int e^{j\omega\Delta t}\; \hat{s}_M^*(\omega_o-\omega_o+\Delta\omega)\; \hat{s}_M(\omega-\omega_o)\; d\omega\right]$$

$$\cdot \sum e^{j\Delta\underset{\sim}{k}\cdot\underset{\sim}{x}_i}\; \psi^*(z_i;z_o,\omega_o,k)\; \varphi(z_i,z_o',\omega_o,k'), \tag{42}$$

which again is the product of a temporal ambiguity function and a beam pattern.

The Fundamental Relationships

Consider the modified convolution $\otimes_L'$ defined by

$$a \otimes_L' b(\Delta t, \Delta x, \Delta z_o, \Delta \omega, \Delta k, \Delta k) =$$

$$\iiiint e^{j[\Delta\omega'(\Delta t-\Delta t')-\Delta k'\cdot(\Delta x-\Delta x')]}$$

$$b(\Delta t-\Delta t', \Delta x-\Delta x', \Delta z_o-\Delta z_o', \Delta\omega-\Delta\omega', \Delta k-\Delta k', \Delta k-\Delta k')$$

$$a(\Delta t', \Delta x', \Delta z_o', \Delta\omega, \Delta k, \Delta k) \; d\Delta t' \; d\Delta x, \; d\Delta z_o', \; d\Delta\omega \; d\Delta k' \; d\Delta k'$$

$$\text{(43)}$$

It is easy to show as with (13) that $\otimes_S'$ is associative but not in general commutative; furthermore, it is possible to rewrite (26) and (29) in terms of $\otimes_L'$. When this is done and use is made of the associative property of $\otimes_S'$, the following fundamental relationships may be derived:

$$\gamma_o^L(\Delta t, \Delta x, \Delta z_o, \Delta\omega, \Delta k, \Delta k) = S_{IO}^L \otimes_L' \gamma_I^L(\Delta t, \Delta x, \Delta z_o, \Delta\omega, \Delta k, \Delta k) \qquad \text{(44)}$$

$$S_{AC}^L(\Delta t, \Delta x, \Delta z_o, \Delta\omega, \Delta k, \Delta k) = S_{BC}^L \otimes' S_{AC}^L(\Delta t, \Delta x, \Delta z_o, \Delta\omega, \Delta k, \Delta k) \qquad \text{(45)}$$

where γ_I^L and γ_o^L are input and output cross-ambiguity functions for a signal passed through the system characterized by S_{IO}^L and and where S_{AC}^L characterizes the system formed by concatenating the systems characterized by S_{BC}^L and S_{AB}^L.

Use of (28) with (43), (44) and (45) yields for uncorrelated stochastic systems:

$$E\{|\gamma_o^L(\Delta t, \Delta x, \Delta z_o, \Delta\omega, \Delta k, \Delta k)|^2\} =$$

$$= \mathscr{S}_{IO}^L \otimes_L E\{|\gamma_I^L(\Delta t, \Delta x, \Delta z_o, \Delta\omega, \Delta k, \Delta k)|^2\} \qquad \text{(46)}$$

$$\mathscr{S}_{AC}^L(\Delta t, \Delta x, \Delta z_o, \Delta\omega, \Delta k, \Delta k) =$$

$$= \mathscr{S}_{BC}^L \otimes_L \mathscr{S}_{AB}^L (\Delta t, \Delta x, \Delta z_o, \Delta\omega, \Delta k, \Delta k) \qquad \text{(47)}$$

$$a \underset{L}{\otimes} b(\Delta t, \Delta \underset{\sim}{x}, \Delta z_0, \Delta \omega, \Delta \underset{\sim}{k}, \Delta k) =$$

$$\iiiint\!\!\int b(\Delta t - \Delta t', \Delta \underset{\sim}{x} - \Delta \underset{\sim}{x}', \Delta z_0 - \Delta z_0, \Delta \omega - \Delta \omega', \Delta \underset{\sim}{k} - \Delta \underset{\sim}{k}', \Delta k - \Delta k')$$

$$a(\Delta t', \Delta \underset{\sim}{x}', \Delta z_0', \Delta \omega', \Delta \underset{\sim}{k}', \Delta k')$$

$$d\Delta t'\ d\Delta \underset{\sim}{x}'\ d\Delta z_0'\ d\Delta \omega'\ d\Delta \underset{\sim}{k}'\ d\Delta k' \tag{48}$$

It is easy to show that $\otimes$ is commutative as well as associative; hence, the order of concatenation of $\mathscr{S}_{BC}$ and $\mathscr{S}_{AC}$ is unimportant.

REFERENCES

1. K.A. Sostrand, Mathematics of the Time Varying Channel, NATO ASI on Signal Processing, Enschede (1968).
2. L. Meier, "A Brief Resume of Deterministic Time-Varying Linear System Theory with Application to Active Sonar Signal Processing Problems," SACLANTCEN Report (In Publication).
3. L. Meier, "A Brief Resume of Stochastic Time-Varying Linear System Theory with Application to Active Sonar Signal Processing Problems," SACLANTCEN Report (In Publication).
4. R. Laval, Time-Frequency Space Generalized Coherence and Scattering Functions, in: "Aspects of Signal Processing," G. Tacconi, ed., Reidel, Dordrecht, Holland (1977).
5. A. Wasiljeff, Influence of Time-and-Space-Variant Random Filters on Signal Processing, in: "Aspects of Signal Processing," G. Tacconi, ed., Reidel, Dordrecht, Holland (1977).
6. D. Costa and E. Hug, An Estimation of the Scattering Function of an Undersea Acoustic Channel, Alta Frequenza XLV:4 (1976).

PHASE-COHERENCE IN A SHALLOW WATER WAVEGUIDE

Ulrich E. Rupe

SACLANT ASW Research Centre

La Spezia, Italy

ABSTRACT

A fundamental property in determining the response of an array is the spatial coherence across the array. Modelling the shallow-water waveguide in terms of normal modes, the phase-coherence factor among modes is known to control the model interference term. This factor is defined using a generalized van Cittert-Zernicke theorem. The resulting expression for the phase coherence is calculated from a spatially-extended single-frequency source, each source element generating m modes. Since single source elements are assumed to radiate incoherently, the complex degree of coherence from an extended source is calculated without the explicit use of an averaging process. Results for a top-to-bottom array are presented for different bottom types, source locations, and sea-surface irregularities.

INTRODUCTION

A shallow-water waveguide includes the boundary conditions of the bottom interaction for any type of signals transmitted into such a waveguide. When signals from a source area S are mapped onto a receiving area R by way of a dispersive channel, the received waveform is a distorted replica of the original, which is expressed in terms of the partial coherence in the receiving area. As in fact the partial coherence is a function of receiver separation and receiver location in the waveguide, the coherence function bears information about source location. In knowing the spatial coherence from measurements, the goal of array processing is to extract information about sources.

In this paper the partial coherence for a vertical array is calculated for a normal-mode solution. In order to reconcile various definitions for the coherence, the basic relationship between source and receiving area is re-examined. A simple expression for the coherence function is obtained by using the generalization of Zernicke[1, 2, 5]. The attractive feature of this coherence function is that it operates with quantities determined from experiments.

The SNAP model[3], in connection with Clay's extension[4] to the normal-mode solution for the influence of an irregular sea-surface, is used to present properties of a partially coherent wavefield. A number of examples of practical interest are given. In particular, the spatial vertical coherence as a function of source location and the properties of the transmitting medium is demonstrated.

I   PHASE COHERENCE

Suppose at two spatially separated points P_n, P_m in an acoustic wavefield, the complex amplitudes c_n, c_m are produced due to the source element dS, Fig. 1. The powers E_n, E_m at these points are assumed to be known by measurement. Considering now disturbances from P_n, P_m reaching to point P, we may ask in what

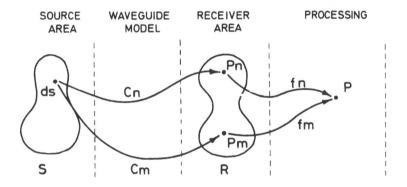

Fig. 1 *General source-receiver geometry*
Signals from source elements are mapped
onto the receiver area and combined by
the processing

way the two disturbances are combined at P. For this, consider f_n and f_m, the complex amplitudes produced at P by the disturbances P_n, P_m; then f_n, f_m express the phase properties of paths P_nP and P_mP. Clearly the superposition at point P of those disturbances will depend on the phase correlation and the power in the receiving area. To calculate this power at P, an additional factor has to be known, which is defined as the phase coherence[5]. Since the complex amplitudes c_n, c_m at the points P_n, P_m are originated in the source element dS — implying that c_n and c_m are coherent, but not necessarily in phase (because they might be originated at a different time) — the amplitude at the point P due to the source element dS is $c_nf_n + c_mf_m$, permitting the calculation of the power E

$$dE = (c_nf_n + c_mf_m)(c_n^*f_n^* + c_m^*f_m^*)dS$$
$$= |c_n|^2|f_n|^2\ dS + |c_m|^2|f_m|^2\ dS + 2Re[c_nc_m^*f_nf_m^*]\ dS, \tag{1}$$

where the star denotes complex conjugates. The different elements of the source are assumed to be mutually incoherent statistically independent with zero mean value; thus integrating over the source area yields

$$E = E_n|f_n|^2 + E_m|f_m|^2 + 2\sqrt{E_mE_n}\ Re\{\gamma_{nm}f_nf_m^*\}, \tag{2}$$

where
$$E_n = \int_S |c_n|^2\ dS \quad\text{and}\quad E_m = \int_S |c_m|^2\ dS$$

are the total powers at P_n, P_m due to the whole source area S. The phase coherence between P_n and P_m is now defined[5] as:

$$\gamma_{nm} = (E_nE_m)^{-\frac{1}{2}} \int_S c_nc_m^*\ dS. \tag{3}$$

For the sake of explanation the complex quantities f_n f_m and γ_{nm} are rewritten in the form:

$$f_n = W_n\ exp\{ik\phi_n\} \tag{4.1}$$

$$f_m = W_n\ exp\{ik\phi_m\} \tag{4.2}$$

$$\gamma_{nm} = C_{nm}\ exp\{ik\psi_{nm}\}, \tag{4.3}$$

where W_n, W_m are amplitude weights, $k\phi_n$ and $k\phi_m$ are the phase introduced by the ways P_nP and P_mP. In writing the absolute amplitudes

$$A_n = W_n\sqrt{E_n} \qquad\qquad (5.1)$$

$$A_m = W_m\sqrt{E_m}, \qquad\qquad (5.2)$$

Eq. 2 becomes

$$E = A_n^2 + A_m^2 + 2 A_n A_m C_{nm} \cos\{k[\psi_{nm}+(\phi_n-\phi_m)]\} \qquad (6)$$

The factor C_{nm} of Eq. 4.3 is called the modulus of the coherence and ψ_{nm} is called the phase of the coherence, indicating the differential phase of the field points considered. It is worth-while noting that Eq. 6 is general in the way that it combines the disturbances by knowing only the coherence γ_{nm} and the total powers E_n, E_m. In fact, since the phase coherence in terms of modulus C_{nm} and phase ψ_{nm} is calculated from the complex ampli-tudes produced at the points P_n, P_m by means of the paths SP_n and SP_m, the phase-coherence expresses the properties of these paths. Considering the point P as part of the receiving system, the amplitudes and the phase differences $(\phi_n-\phi_m)$ are needed to super-impose the disturbances arriving from P_n and P_m at P. When P is displaced, making the phase difference zero, this is equivalent to forming a broadside beam, and only the phase coherence of the medium is observed. If $C_{nm} = 0$, the product term Eq. 6 is zero, and the intensity P is now the incoherent sum of the two intensities at P_nP_m. If $C_{nm} = 1$ perfect coherence is implied and the phase term will amount to a displacement, indicating the relative advance of the phase at point P_n relative to P_m.

II   THE NORMAL-MODE SOLUTION

The calculation of the phase coherence is carried out according to Eq. 3, which in this expression is independent of any model used for producing the complex amplitudes c_n and c_m. To calculate the phase-coherence γ_{nm} for a pair of points of the acoustical field in a waveguide the complex pressure is expressed in terms of normal modes[3,4], permitting the numerical solution. This complex pressure for points (r,z) is expressed as the sum of the depth and range-dependent part

$$P(r,z) = K \sum_m D^m(z, k^m) \, R^m(r, k^m), \qquad (7)$$

where
$$K = \omega\rho(z_0) / H\sqrt{8\pi}$$

and k^m is the horizontal wavenumber for the mth mode.
The depth function is expressed by

$$D^m(z, k^m) = u^m(z)u^m(z_0) / \sqrt{k^m} \qquad (8)$$

and the range-dependent part is

$$R^m(r, k^m) = r^{-\frac{1}{2}} \exp\{i[k^m r - \pi/4]\} . \qquad (9)$$

Since the depth function is range independent it can be replaced by constants yielding

$$P(r,z) = K \sum_m \frac{A^m(z)}{\sqrt{r}} \exp\{i[k^m r - \pi/4]\}. \qquad (10)$$

This represents the complex pressure at a range r and depth z. The concrete values for a given waveguide are normally evaluated numerically.

In calculating the phase-coherence between different points with given coordinates (r_i, z_i), (r_j, z_j) in the waveguide, we return to Eq. 3. Inserting the complex amplitude given by Eq. 10 yields

$$\gamma_{ij} = C \int_S P_n P_m^* \, dS \qquad (11)$$

$$= C \int_S \left\{ \sum_n \frac{A^n(z_i)}{\sqrt{r_i}} \exp[i(k^n r_i - \tfrac{\pi}{4})] \right\} \left\{ \sum_m \frac{A^m(z_j)}{\sqrt{r_j}} \exp[-i(k^m r_j - \tfrac{\pi}{4})] \right\} dS$$

$$= C \int_S \sum_n \sum_m \frac{A^n(z_i)A^m(z_j)}{r_i r_j} \exp[i(k^n r_i - k^m r_j)] dS$$

where $C = 1/[\ P_n^{\ 2}\ P_m^{\ 2}]^{\frac{1}{2}}$ is a normalizing factor. For the sake of simplicity the double sum and the product of amplitudes are given different indices without changing the order of summation and the integration. Including the mode-dependent phase ϕ^{nm} for a slightly irregular sea surface[4], Eq. 11 is written

$$\gamma_{nm} = C \int_S \sum_{n,m} A_{ij}^{nm} \exp[i\phi^{nm}] \exp[i\theta_{ij}^{nm}]dS, \qquad (12)$$

where
$$A^{nm} = A^n(z_i)A^m(z_j) / r_i r_j$$

and
$$\phi^{nm} = k^n r_i - k^m r_j$$

are short notations for the amplitude and phase at points (r_i, z_i), (r_j, z_j) with horizontal wave numbers k^n, k^m.

The expression γ_{nm} is called the phase coherence between two points taken in a waveguide containing all the cross terms of the modes in the waveguide.

In deriving Eq. 12 use has been made of a generalized van Cittert-Zernicke theorem[3] known as the Hopkins formula[5]. It is derived for the situation where the medium is inhomogeneous or consists of adjacent homogenous regions. The complex amplitude P_n, Eq. 11, is the disturbance that would occur at the point (r_i, z_i) from a single frequency point-source having zero phase at the origin. Thus Eq. 12 is an expression for the phase coherence γ_{ij} due to an extended single-frequency source, each source element generating m modes. The usefulness of Eq. 12 is the same as the van Cittert-Zernicke theorem, that is, it permits the calculation of the complex degree of coherence from an incoherent source area.

III RESULTS

The geometry and the parameters of the waveguide model furnished for SNAP to calculate the complex amplitudes across the vertical array are shown in Fig. 2. In all of the following examples, the array extends from top to bottom. The bottom parameters are taken from[6][7]. The modulus and phase of the vertical coherence for three different bottom types are shown in Fig. 3. The results are presented as contoured levels of the modulus and phase of the coherence for all the spatial combinations of the receivers. The range for the levels is from 0 to 1 for the modulus and from -2π to $+2\pi$ in radians for the phase. The complex coherence levels are symmetric about the diagonal for a

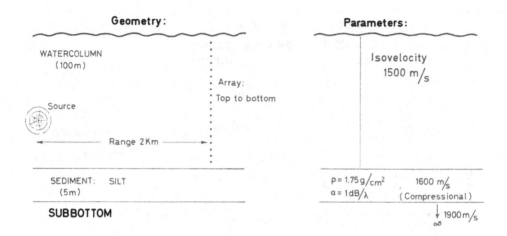

Fig. 2 The waveguide model used to furnish the SNAP-model calculation

vertical array and are symmetric about both diagonals for a horizontal array. The change of modulus and phase parallel to the diagonal line (corresponding to a fixed receiver distance exploring the water column) shows clearly the inhomogeneity in the vertical plane. For all examples except that in Fig. 5, the sea-surface irregularities are modelled with an rms wave height of 1 m. In Fig. 4 a silt-bottom[6], with the parameters as indicated, is successively mismatched for density, attenuation, and compressional speed, showing that bottom parameters have to be known as accurately as possible when matching a vertical array to the eigenfunctions of the waveguide. The spatial variations of the phase are much more sensitive to mismatching than is the modulus of the coherence.

The dependence of the vertical coherence on the waveheight is essentially a decrease of coherence among the modes contributing to the acoustic field. Clay[4] has demonstrated that the attenuation in the slightly irregular waveguide depends on waveheight, water column, and wave-vector components. The influence of rms waveheight is shown in Fig. 5 for waveheights of 2.5 m and 5 m. An interesting example is shown in Fig. 6, where two sources are made to depart from the middle of the water column towards the top and bottom. The vertical variation of coherence is now due to the fact that the mode excitation is different for different source locations, hence the interference pattern across the vertical plane changes accordingly. In fact, acoustic propagation in terms of modes shows range-dependent interference with relative minima and maxima. The vertical coherence is demonstrated in Fig. 7 for different ranges associated with two relative minima and one

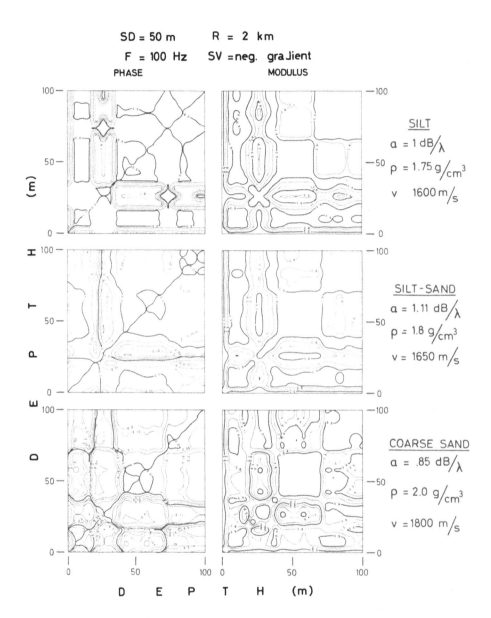

Fig. 3 Dependence on bottom parameter of vertical coherence. Dependence of vertical coherence on bottom parameters[6,7], as indicated on right. The negative gradient of the sound-speed profile affects larger coherent areas below the middle of the 100 m water column. Source depth 50 m

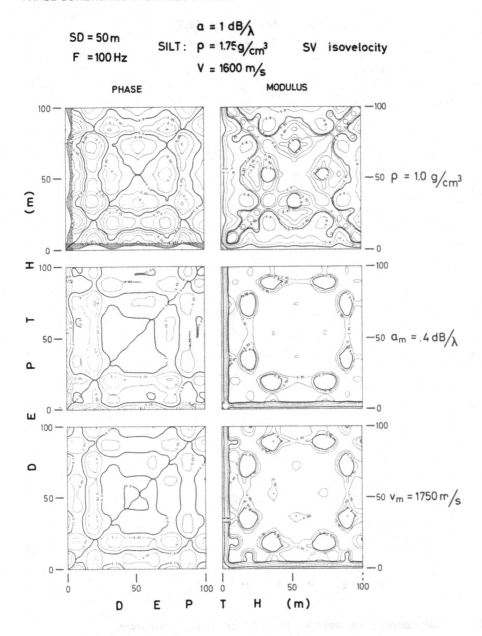

Fig. 4 *Dependence on mismatch of bottom parameters. Bottom
parameters successively mismatched for density,
attenuation, and compressional speed, showing that
bottom parameters have to be known as accurately as
possible when matching a vertical array to the waveguide.*

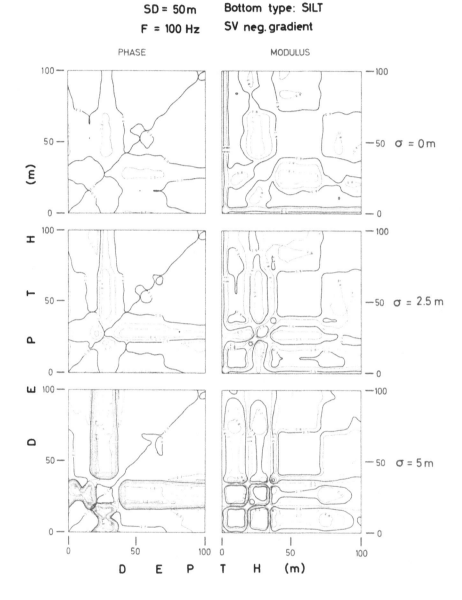

SD = 50 m   Bottom type: SILT

F = 100 Hz   SV neg. gradient

Fig. 5 Dependence on waveheight of vertical coherence.
Attenuation in a slightly irregular waveguide depends on
waveheight, water column, and wavevector components[4] .
The rms waveheight dependence is shown for σ = 2.5 m and
σ = 5 m.   The area below the middle of the water column
is relatively unaffected by surface irregularities.

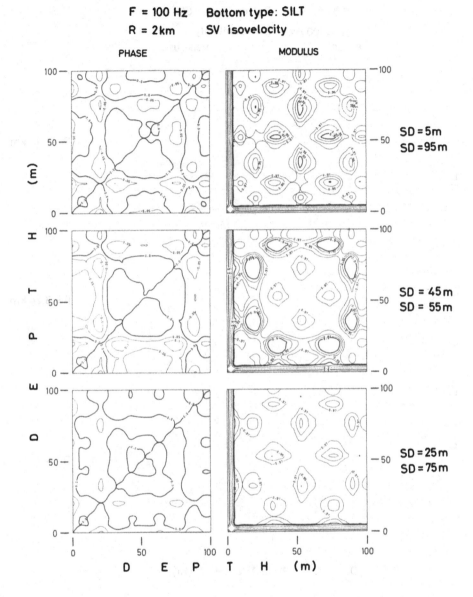

F = 100 Hz   Bottom type: SILT

R = 2km   SV isovelocity

Fig. 6 *Dependence on source locations of vertical coherence.
Two sources are made to depart from the middle of the
water column towards the top and bottom.   Vertical
variation of coherence is now due to the fact that
the mode excitation is different for different source
locations, hence the interference pattern across the
vertical plane changes accordingly.*

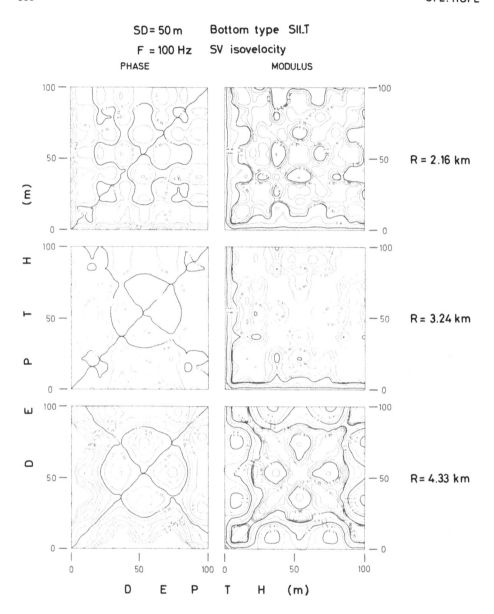

SD = 50 m   Bottom type SILT

F = 100 Hz   SV isovelocity

PHASE   MODULUS

R = 2.16 km

R = 3.24 km

R = 4.33 km

DEPTH (m)

D E P T H (m)

Fig. 7 Dependence on range of vertical coherence. Vertical
coherence for indicated ranges is associated with two
relative minima (R = 2.16 km and R = 4.33 km) and one
relative maximum (R = 3.24 km) of the 1st and 2nd mode.
The modulus and phase variations of vertical coherence
are evidently less for the interference maximum than
for the minimum.

relative maximum of the 1st and 2nd mode. The modulus and phase variations of vertical coherence are evidently less for the interference maximum caused by constructive interference than for the minimum, which is effected by destructive interference.

CONCLUSIONS

Detailed examination of the partial coherence and its application to the normal-mode solution of a shallow-water wave guide, as demonstrated for various examples, shows that the phase-coherence function is sensitive to mismatch of waveguide parameters. A guide to the selection of receiver depth is obtained in the sense that areas with small variations of coherence correspond to small wavefront distortions.

REFERENCES

1. F. Zernicke, The concept of degree of coherence and its application to optical problems, Physica 5:785-795 (1938).
2. M. Born and E. Wolfe, "Principle of Optics", 3rd ed., Pergamon Press New York, N.Y. (1965): pp.508-513.
3. F. B. Jensen and M. C. Ferla, SNAP: the SACLANTCEN normal-mode acoustic propagation model, SACLANTCEN SM-121, SACLANT ASW Research Centre, Italy (1979). [AD A 067 256]
4. C. S. Clay, Effect of a slightly irregular boundary on the coherence of a waveguide propagation, J. Acoustical Society America 36:883-837 (1964).
5. H. H. Hopkins, The concept of partial coherence in optics, Proc. Royal Society of London A208: 263 (1951).
6. E. L. Hamilton, Compressional wave attenuation in marine sediments, Geophysics 37:620-646 (1972).
7. W. Bachmann and R. B. Williams, eds., "Ocean Acoustic Modelling, Proceedings of a Conference at La Spezia, Italy, 8-11 Sep 1975", Part 4: Sea bottom, SACLANTCEN CP-17, SACLANT ASW Research Centre La Spezia, Italy (1975). [AD A 020 936/1G1]

THE INFLUENCE OF UNKNOWN BOTTOM PARAMETERS

ON BEARING ESTIMATION IN SHALLOW WATER

Richard Klemm

SACLANT ASW Research Centre

La Spezia, Italy

ABSTRACT

The spatial dispersion of the sound energy in the shallow-water channel causes bearing estimates obtained by conventional beamformers or similar estimators to be biased towards broadside direction. The bias depends on such environmental parameters as the sound-speed profile and bottom parameters. Error-free bearing estimates can be obtained if the sound field is entirely known a priori. In practice, however, the bottom parameters are usually unknown. This paper discusses the problem of mismatch between the actual bottom parameters and assumptions on the processing side. Comparisons are made with an alternative method using a test source for estimating the actual channel response.

INTRODUCTION

The use of horizontal line arrays as sonar receiver has received considerable attention in the recent past. In particular, when implemented as a towed hydrophone array a couple of advantages are obvious, such as high gain, low towing-ship noise level, and the possibility of spatial towing-ship noise suppression. However, bearing estimation by a linear array may involve some complications when performed in shallow water. In shallow water sound propagation can be described in terms of normal modes. Each of the modes is interpreted by the horizontal line array as a plane wave arriving from a certain vertical incident angle given by the modal wave-number. Therefore, for any bearing angle other than broadside, conventional bearing estimators based on a beamformer will obtain

bearing estimates biased towards broadside. An unbiased bearing
estimate can be obtained by using so-called generalized power
estimators if the statistics of the waveguide are known.

It is the purpose of this paper to explore how far bearing
estimation can be improved beyond the use of just a conventional
beamformer. In particular, the question is discussed of how
sensitive the bearing estimators are to imprecise a priori
knowledge about the channel parameters. The bottom parameters are
of special interest, because they are usually not as well known as
the water column (water depth, sound-speed profile). The use of a
priori knowledge about the sound field for processor design is
problematic anyway because the processor may become range and depth
dependent, which is not desirable in general. We assume for
simplicity in the following that the field and the processing are
range independent and that the source depth is known.

I   SIGNAL MODEL

In the following investigation high-resolution power estimators
are used for bearing estimation. The sound field and the processor
will be simulated by a normal-mode sound-propagation model.
Therefore we have to derive the covariance matrix due to a point
source from the outputs of the given computer model. The SNAP-
model[1] used in this investigation computes a set of M modal
horizontal wave number k_n and modal amplitudes A_n for a set of
environmental input parameters, such as receiver/source geometry,
frequency, and channel parameters. The sound pressure of a
monochromatic source received at distance r can be written as

$$p(r,z,z_o,t) = e^{-(j\omega t + \frac{\pi}{4})} \sum_{N=1}^{M} A_n e^{jk_n r}. \qquad (1)$$

The modal amplitudes are

$$A_n = a\frac{\omega\rho^2}{H} \sqrt{\frac{1}{8\pi r}} \frac{u_n(z_o)u_n(z)}{\sqrt{k_n}} e^{-\alpha_n r}, \qquad (2)$$

where a is the source strength, $\omega = 2\pi f$, ρ = water density,
H = water depth, r = range, z = receiver depth, z_o = source
depth, α_n = modal attenuation coefficient, $u_n(.)$ = normal mode
function[2]. Each of the wave components in Eq.1 is associated
with a modal vertical incident angle γ_n defined by

$$\cos\gamma_n = k_n/k_o, \quad k_o = \omega/c(z) ,$$

c(z) being the sound speed at receiver depth.

To introduce the geometry of a linear horizontal array with uniform spacing d we have to replace the range r in Eqs. 1 & 2 by

$$r = r_o + d_i, \quad d_i = d \cdot i \cdot \cos\beta, \quad i = 0 \ldots N-1, \qquad (3)$$

N being the number of sensors and β the angle between the array axis and the horizontal direction of wave propagation. The output signal at the i-th sensor is consequently

$$x_i = C_i \, e^{-(j\omega t + \frac{\pi}{4})} \sum_{n=1}^{M} A_n(i) e^{j k_n (r_o + d_i)} \qquad (4)$$

where the C_i are the complex gain factors of individual channels. We assume in the following that $C_i = 1$, $\forall i$.

The covariance matrix $\underline{S}$ of received signals contains the elements

$$\rho_{i\ell} = E\{x_i x^*\} = \sum_{n=1}^{M} A_n(i) e^{j k_n (r_o + d_i)} \sum_{m=1}^{M} A_m(\ell) e^{-j k_m (r_o + d_i)}$$

$$= \sum_{n=1}^{M} A_n(i) A_n(\ell) e^{j k_n (d_i - d_\ell)}$$

$$+ \sum_{n=1}^{M} \sum_{m=1}^{M} A_n(i) A_M(\ell) e^{j[(k_n - k_m) r_o + k_n d_i - k_m d_\ell]} \qquad (5)$$
$$n \neq m$$

The first term of Eq. 5 is the incoherent summation of the radiation in all modes; the second expression is the mode interference term.

So far the model is entirely deterministic. Let us now introduce some randomness due to channel fluctuations. It has been shown by several authors[2][3] that scattering by randomly varying boundaries causes the individual modal components of the received signals to be random in amplitude and phase. This results in a slight increase of modal attenuation coefficients α_n in Eq. 2 and a decrease of the mode interference term in Eq. 5. The increase of the α_n is taken into account by the SNAP model[1]. The mode-interference term we suppose to be zero in the following. Neglecting furthermore the range dependence of the modal amplitudes (small aperture), Eq. 5 simplifies to

$$\rho_{i\ell} = \sum_{n=1}^{M} A_n^2 \, e^{j k_n (d_i - d_\ell)}$$

For broadside ($\beta = 90°$) we get $d_i = 0$ $\forall i$. Inserting this in Eqs. 6 and 2 the elements of $\underline{S}$ become independent of i and l:

$$\rho_{il} = \sum_{n=1}^{M} A_n^2(r_o) = M \cdot A_n^2(r_o)$$

That means that the signal covariance matrix $\underline{S} = (\rho_{il})$ has rank 1. In other words, wavefronts due to sources at broadside appear to be coherent because of the rotational symmetry of line arrays.

In addition, spatial white noise independent of the signal is assumed in order to model roughly non-directive kinds of noise (ambient, surface, flow, receiver, reverberations). Now the covariance matrix of signal and noise is $\underline{R} = \underline{S} + P_w \underline{I}$ where P_w is the white-noise power. $\underline{R}$ is now positive definite even for broadside.

II   HIGH-RESOLUTION PARAMETER ESTIMATION

Estimation of an unknown parameter of a vector process, such as the random vector of the output signals of the array channels, is frequently carried out by applying a steering vector $\underline{h}(\theta)$ to some representation of the data vector and by variation of θ until the output power becomes maximum. This method is convenient particularly if the data are a non-linear function of the parameter. High resolution (sensitivity to mismatch between steering vector and signal component in the data) is obtained if the data are represented by a vector or a matrix orthogonal or approximately orthogonal to the steering vector when matched to the signal[4]. In the following we will use the following expression

$$P(\beta) = |\underline{r}^* \; \underline{h}(\beta)|^{-2}, \tag{7}$$

where β means bearing, $P(\beta)$ is the output power and $\underline{r}^*$ is the first row of the inverse covariance matrix. Notice that for stationary data Eq. 7 becomes the well-known maximum entropy method (MEM). A generalized version of Eq. 7 can be obtained by replacing the beamformer vector $\underline{h}(\theta)$ by a beamformer matrix $\underline{H}(\beta)$

$$P(\beta) = (\underline{r}^* \; \underline{H}(\beta)\underline{r})^{-1} \tag{8}$$

Equation (8) can be used to identify a random wavefront as given by Eq. 6. For $\underline{H}(\beta)$ a complete set of all possible signal matrices $\underline{S}(\beta)$ has to be used.

Figure 1 may illustrate the difference between the methods (7) and (8) when applied to a signal of the form of Eq. (6). The true

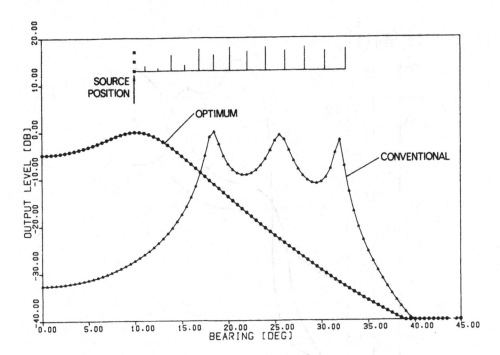

Fig. 1 Conventional and optimum bearing estimation

source position is indicated by three asterisks ($\overset{*}{\underset{*}{*}}$), the vertical lines in the subplot indicate the modal powers A_n^2 (in dB), i.e. the spatial response of the channel as seen by the line array. The conventional method (7) looks simply for the energy maximum and, hence, obtains a considerable bearing error (apart from yielding more than 1 maximum), whereas method (8) obtains only one power maximum at the true position of the target.

III MISMATCH OF BOTTOM PARAMETERS

The use of a generalized power estimator such as (8) requires a priori knowledge about the sound propagation channel in order to use a modelling program for the design of the steering matrices $\underline{H}(\beta)$. The important parameters of the water column such as SVP and water depth are easily obtained, however, the parameters of the bottom are usually not precisely known. This may cause mismatch between the actual bottom parameters and those assumed in the steering matrices of the array processor (8). A couple of examples as given in Figs. 3 to 5 may give some idea of the sensitivity of (8) to mismatch in bottom parameters.

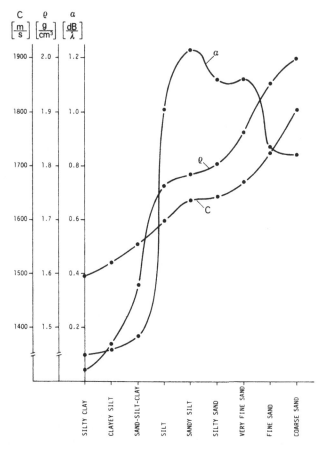

Fig. 2 Bottom parameters

Figure 2 shows the parameters ρ (density), α (compressional attenuation and c (sound velocity) of some bottom types. Figures 3 to 5 show (apart from the conventional MEM) results obtained by more or less mismatched generalized power estimators. The upper subfigure shows the target position and the spatial channel response. One curve shows the power response of a perfectly matched generalized MEM-processor. The other two figures on the right indicate the channel responses as assumed on the processor side.

In Fig. 3 the actual channel has a hard bottom (coarse sound); consequently a rather high number of modes are excited. If the processor is perfectly matched to the bottom type an error-free bearing estimate is obtained. Assuming a sandy silt type bottom a certain bearing offset and broadening of the power response is observed. The assumption of a very soft bottom (clayey silt) leads

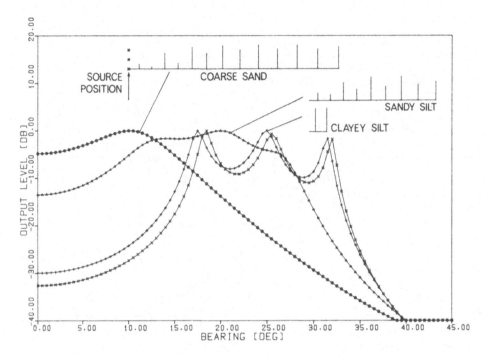

Fig. 3 Mismatch of bottom parameters

to a similar result like the conventional MEM (7). The reason is obvious: as only two modes are excited the steering matrix $\underline{H}(\beta)$ approaches a dyadic $\underline{h}(\beta)\underline{h}^*(\beta)$ which reduces (8) to (7). The three bottom types in Fig. 4 show rather similar channel responses which results in rather good bearing estimates. Figure 5 shows just the opposite of Fig. 3, namely a channel with soft bottom, but the assumption of harder bottom types for the processor. In this example the conventional MEM is the best approximation, which we expected since the conventional beamformer is the closest approach to the response of a channel with clayey silt bottom.

IV   COMPARISON WITH THE TEST SOURCE METHOD

The test source method is a procedure to calibrate the array processor adaptively on the instantaneous (short time average) channel response, i.e. the measured test covariance matrix is approximated by a sin ax/ax-Toeplitz matrix where a∿cosβ. The method consists of the following steps:

1.   Have a test source at known bearing.

2.   Estimate the first row of the covariance matrix due to the test source signal.

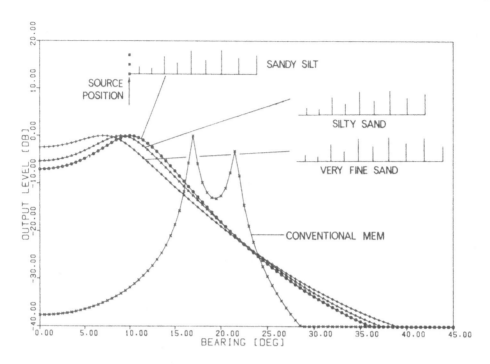

Fig. 4 Mismatch of bottom parameters

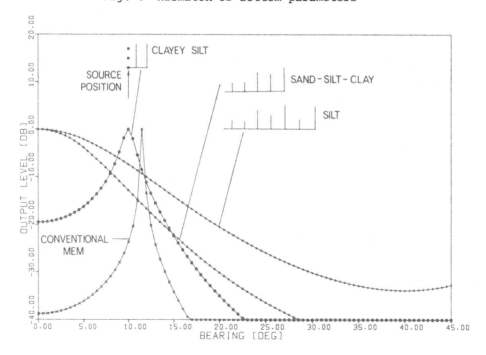

Fig. 5 Mismatch of bottom parameters

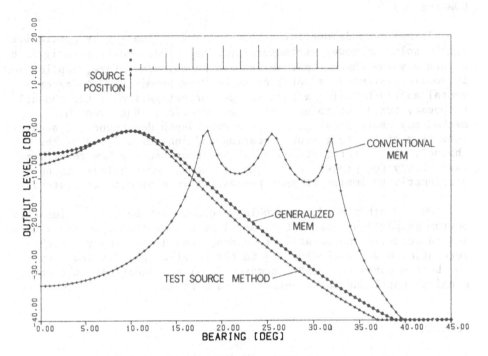

Fig. 6   The test source method

3.   Approximate numerically the estimated covariance by
     a sin ax/ax function by variation of a (estimation
     of signal spread).

4.   Insert the sin ax-beamformer matrix in a generalized
     power estimator such as (8) and steer it over the test
     source.

5.   Observe the bearing error and correct for it in the
     beamformer matrix.

For more details see[6]. This steering matrix describes a beamformer with bearing dependent beamwidth which copes approximately
with the bearing dependent width of the channel response.

     An example is shown in Fig. 6. Rather good agreement between
the test source method and the optimum generalized MEM can be
observed.

CONCLUSIONS

It has been demonstrated that bearing estimation by generalized
high resolution power estimators such as (8) depends seriously on how
accurately the channel parameters are known. Therefore, application
in mobile systems does not seem to be very promising. An experimental approach (using a test source for estimation of the channel
response) turns out to be the better solution. But even this
method may cause problems due to source depth dependence, inaccurate
estimates of the test source bearing and inhomogeneities of the
channel. In summary, it can be shown theoretically how to utilize
wave theory for processor design; it seems, nevertheless, to be
problematic to implement such processors in mobile sonar systems.

On the other hand, the whole procedure may be used to identify
bottom properties: Use a horizontal or even better a vertical array,
and an acoustic source at certain range and depth, apply a high
resolution power estimator (8) to the received signals and vary
the bottom parameters (SVP, compressional and shear attenuation,
density) until the power output shows a peak.

REFERENCES

1. F.B. Jensen and M.C. Ferla, "SNAP, the SACLANTCEN Normal
       Mode Acoustic Propagation Model," SACLANTCEN SM-121,
       La Spezia, Italy, SACLANT ASW Research Centre, 1979.
2. W.A. Kuperman and F. Ingenito, "Attenuation of the coherent
       component of sound propagating in shallow water with rough
       boundaries." Journal Acoustical Society America, 61, 5,
       1977: 1178-1187.
3. C.S. Clay, "Effect of a Slightly Irregular Boundary on
       the coherence of waveguide propagation." Journal
       Acoustical Society America, 36, 1964: 833-837.
4. R. Klemm, "High resolution analysis of non-stationary data
       ensembles." Proc. of the EUSIPCO-80, to be held in
       September 1980, Lausanne, Switzerland.
5. R. Klemm, "Use of generalised resolution methods to locate
       sources in random dispersive media," IEE Proc., 127,
       Pt F, 1, 1980: 34-40.
6. R. Klemm, "Low error bearing estimation by horizontal line
       arrays in shallow water," to be published in IEEE Trans. AES.

LIST OF CONTRIBUTORS

T. Akal
SACLANT ASW Research Centre
Viale San Bartolomeo 400
19026 La Spezia
ITALY

J.E. Allen
Naval Oceanographic Office
NSTL Station
Bay St. Louis, MS 39522
U.S.A.

R.S. Anderson
Naval Oceanographic Office
NSTL Station
Bay St. Louis, MS 39522
U.S.A.

J.F. Andrews
Dept. of Oceanography
University of Hawaii at Manoa
2525 Correa Road
Honolulu, HI 96822
U.S.A.

A.B. Baggeroer
Dept. of Ocean Engineering
Massachusetts Institute
 of Technology
Cambridge, MA 02139
U.S.A.

J.H. Beebe
Applied Research Laboratory
P.O. Box 30
State College, PA 16801
U.S.A.

J.M. Berkson
Naval Ocean Research and
 Development Activity
NSTL Station
Bay St. Louis, MS 39529
U.S.A.

H.O. Berktay
University of Bath
School of Physics
Claverton Down
Bath BA2 7AY
U.K.

D.M.F. Chapman
Defence Research Establishment
 Atlantic
P.O. Box 1012
Dartmouth, N.S., B2Y 3Z7
CANADA

N.R. Chapman
Defence Research Establishment
 Pacific
Forces Mail Office
Victoria, B.C., VOS 1B0
CANADA

A.B. Coppens
Dept. of Physics and Chemistry
Naval Postgraduate School
Monterey, CA 93940
U.S.A.

L.R. Cox
Applied Research Laboratories
The University of Texas
  at Austin
Austin, TX 78712
U.S.A.

P.A. Crowther
Marconi Space and Defence
  Systems Ltd.
Chobham Road, Frimley
Camberley, Surrey GU16 5PE
U.K.

R.L. Dicus
Naval Research Laboratory
Washington, DC 20375
U.S.A.

F.R. DiNapoli
Naval Underwater Systems Center
New London, CT 06320
U.S.A.

D.J. Dodds
HUNTEC ('70) Ltd.
P.O. Box 1006
Dartmouth, N.S., B2Y 4A2
CANADA

G. Dreini
SACLANT ASW Research Centre
Viale San Bartolomeo 400
19026 La Spezia
ITALY

D.D. Ellis
Defence Research Establishment
  Atlantic
P.O. Box 1012
Dartmouth, N.S., B2Y 3Z7
CANADA

H.H. Essen
Universität Hamburg
Institut für Geophysik
Bundestrasse 55
2 Hamburg 13
WEST GERMANY

M.C. Ferla
SACLANT ASW Research Centre
Viale San Bartolomeo 400
19026 La Spezia
ITALY

R. Field
Naval Ocean Research and
  Development Activity
NSTL Station
Bay St. Louis, MS 39529
U.S.A.

H.G. Frey
Applied Research Laboratories
The University of Texas
  at Austin
Austin, TX 78712
U.S.A.

G.V. Frisk
Woods Hole Oceanographic
  Institution
Woods Hole, MA 02543
U.S.A.

J.A. Goertner
Naval Surface Weapons Center
White Oak
Silver Spring, MD 20910
U.S.A.

D.F. Gordon
Naval Ocean Systems Center
San Diego, CA 92152
U.S.A.

J.J. Hanrahan
Naval Underwater Systems Center
New London, CT 06320
U.S.A.

O.F. Hastrup
SACLANT ASW Research Centre
Viale San Bartolomeo 400
19026 La Spezia
ITALY

P. Herstein
Naval Underwater Systems Center
New London, CT 06320
U.S.A.

W.R. Hoover
David W. Taylor Naval Ship
   Research and Development
   Center
Bethesda, MD 20084
U.S.A.

R.E. Houtz
Lamont-Doherty Geological
   Observatory of Columbia
   University
Palisades, NY 10964
U.S.A.

J.M. Hovem
Electronics Research
   Laboratory
The University of Trondheim
O.S. Bragstads Plass 6
N-7034 Trondheim-NTH
NORWAY

P. Humphrey
Dept. of Oceanography
University of Hawaii at Manoa
2525 Correa Road
Honolulu, HI 96822
U.S.A.

F.B. Jensen
SACLANT ASW Research Centre
Viale San Bartolomeo 400
19026 La Spezia
ITALY

R. Klemm
Forschungsinstitut für Funk
   und Mathematik
Abt. Elektronik
D 5307 Wachtberg-Werthhoven
WEST GERMANY

F.C. Kögler
Geologisch-Paläontologisches
   Institut der Universität
Olshausenstrasse 40/60
2300 Kiel
WEST GERMANY

W.A. Kuperman
SACLANT ASW Research Centre
Viale San Bartolomeo 400
19026 La Spezia
ITALY

Y. Labasque
Société d'Etudes et Conseil AERO
5 Avenue de l'Opera
75001 Paris
FRANCE

R. Laval
Société d'Etudes et Conseil AERO
5 Avenue de l'Opera
75001 Paris
FRANCE

M.H. Manghnani
Hawaii Institute of Geophysics
University of Hawaii at Manoa
2525 Correa Road
Honolulu, HI 96822
U.S.A.

S.T. McDaniel
Applied Research Laboratory
P.O. Box 30
State College, PA 16801
U.S.A.

H. Medwin
Dept. of Physics and Chemistry
Naval Postgraduate School
Monterey, CA 93940
U.S.A.

L. Meier
Systems Control Inc.
1801 Page Mill Road
Palo Alto, CA 94304
U.S.A.

P.D. Milholland
Hawaii Institute of Geophysics
University of Hawaii at Manoa
2525 Correa Road
Honolulu, HI 96822
U.S.A.

G.B. Morris
Naval Ocean Research
  and Development Activity
NSTL Station
Bay St. Louis, MS 39529
U.S.A.

A.H.A. Moustafa
University of Bath
School of Physics
Claverton Down
Bath BA2 7AY
U.K.

T.G. Muir
Applied Research Laboratories
The University of Texas
  at Austin
Austin, TX 78712
U.S.A.

A. Nagl
Physics Department
Catholic University
Washington, DC 20064
U.S.A.

E.A. Okal
Dept. of Geology and
  Geophysics
Yale University
P.O. Box 6666
New Haven, CT 06511
U.S.A.

J.S. Papadakis
Mathematics Department
University of Rhode Island
Kingston, RI 02881
U.S.A.

D. Potter
Naval Underwater Systems Center
New London, CT 06320
U.S.A.

D. Rauch
SACLANT ASW Research Centre
Viale San Bartolomeo 400
19026 La Spezia
ITALY

U.E. Rupe
SACLANT ASW Research Centre
Viale San Bartolomeo 400
19026 La Spezia
ITALY

J.V. Sanders
Dept. of Physics and Chemistry
Naval Postgraduate School
Monterey, CA 93940
U.S.A.

F. Schirmer
Universität Hamburg
Institut für Geophysik
Bundesstrasse 55
2 Hamburg 13
WEST GERMANY

S.O. Schlanger
Hawaii Institute of Geophysics
University of Hawaii at Manoa
2525 Correa Road
Honolulu, HI 96822
U.S.A.

B. Schmalfeldt
SACLANT ASW Research Centre
Viale San Bartolomeo 400
19026 La Spezia
ITALY

E. Sevaldsen
Norwegian Defence Research
  Establishment
P.O. Box 115
N-3191 Horten
NORWAY

R.P. Spaulding, Jr.
Dept. of Physics and Chemistry
Naval Postgraduate School
Monterey, CA 93940
U.S.A.

C.W. Spofford
Science Application Inc.
8400 Westpark Drive
McLean, VA 22102
U.S.A.

A. Stefanon
Istituto di Biologia del Mare
Riva 7 Martiri 1364/A
Venezia
ITALY

J. Talandier
Laboratoire de Géophysique
Commissariat à l'Energie Atomique
Papeete, Tahiti
FRENCH POLYNESIA

L.A. Thompson
Applied Research Laboratories
The University of Texas
  at Austin
Austin, TX 78712
U.S.A.

J.N. Tjötta
Matematisk Institutt
Allegt. 53-55
5000 Bergen
NORWAY

S. Tjötta
Matematisk Institutt
Allegt. 53-55
5000 Bergen
NORWAY

H. Überall
Physics Department
Catholic University
Washington, DC 20064
U.S.A.

H.F. Weichart
PRAKLA-SEISMOS GMBH
P.O. Box 4767
Haarstrasse 5
D-3000 Hannover
WEST GERMANY

F. Werner
Geologisch-Paläontologisches
  Institut der Universität
Olshausenstrasse 40/60
2300 Kiel
WEST GERMANY

D.E. Weston
Admiralty Underwater Weapons
  Establishment
Portland, Dorset DT5 2JS
U.K.

D. White
Naval Ocean Research and
  Development Activity
NSTL Station
Bay St. Louis, MS 39529
U.S.A.

K. Winn
Geologisch-Paläontologisches
  Institut der Universität
Olshausenstrasse 40/60
2300 Kiel
WEST GERMANY

D.H. Wood
Naval Underwater Systems Center
New London, CT 06320
U.S.A.

G. Ziehm
Forschungsanstalt der
  Bundeswehr für Wasserschall
  und Geophysik
Klausdorfer Weg 2-24
23 Kiel 14
WEST GERMANY

SUBJECT INDEX